Geochemistry of Marine Sediments

Geochemistry of Marine Sediments

DAVID J. BURDIGE

PRINCETON UNIVERSITY PRESS

PRINCETON AND OXFORD

Copyright © 2006 by Princeton University Press

Published by Princeton University Press, 41 William Street, Princeton,
New Jersey 08540

In the United Kingdom: Princeton University Press, 3 Market Place, Woodstock,
Oxfordshire OX20 1SY

All Rights Reserved

ISBN-13: 978-0-691-09506-X
ISBN-10: 0-691-09506-X
Library of Congress Control Number: 2006925778

British Library Cataloging-in-Publication Data is available

This book has been composed in Utopia

Printed on acid-free paper, ∞

pup.princeton.edu

Printed in the United States of America

1 3 5 7 9 10 8 6 4 2

FOR JULI, BENJAMIN, AND EMILY

≈ Contents ≈

Preface	xv
Common Abbreviations and Symbols	xvii

CHAPTER ONE
Introduction — 1

CHAPTER TWO
The Components of Marine Sediments — 5
  2.1 DETRITAL COMPONENTS — 5
  2.2 BIOGENIC COMPONENTS — 8
    2.2.1 Biogenic Carbonates — 9
    2.2.2 Biogenic Silica — 10
    2.2.3 Distribution of Biogenic Components in Marine Sediments — 10
  2.3 AUTHIGENIC MINERALS — 12
    2.3.1 Nonbiogenic Carbonates — 13
    2.3.2 Mn Crusts, Layers, and Nodules — 13
    2.3.3 Phosphorites — 14
    2.3.4 Sulfides — 15
  2.4 CLAYS AND CLAY MINERALS — 15
    2.4.1 Distribution of Clay Minerals in Surface Marine Sediments — 18
    2.4.2 Ion Exchange/Adsorption — 20
  2.5 THE CLASSIFICATION OF MARINE SEDIMENTS AND SEDIMENTARY REGIMES — 24

CHAPTER THREE
Isotope Geochemistry — 27
  3.1 INTRODUCTION — 27
  3.2 PRINCIPLES OF ISOTOPE FRACTIONATION — 28
    3.2.1 Terminology — 30
    3.2.2 Equilibrium Isotope Exchange Reactions — 31
  3.3 ISOTOPE FRACTIONATION IN INORGANIC MATERIALS IN NATURE — 32
    3.3.1 Isotope Fractionation in the Hydrosphere and in Ice Cores — 32

3.3.2 Isotope Fractionation during Clay Mineral Formation   34
3.3.3 Oxygen and Carbon Isotopes in Calcite   35
3.4 CARBON ISOTOPES IN ORGANIC MATTER   36
3.4.1 Photosynthesis   37
3.4.2 Respiration (Early Diagenesis in Sediments)   38
3.5 OXYGEN AND CARBON ISOTOPES IN SEDIMENT PORE-WATERS   38
3.5.1 Carbon Isotopes   38
3.5.2 Oxygen Isotopes   39
3.6 NITROGEN ISOTOPES   39
3.7 SULFUR ISOTOPES   40
3.8 RADIOACTIVE ISOTOPES   40
3.8.1 Basic Principles   40
3.8.2 Radiocarbon   43

CHAPTER FOUR
Physical Properties of Sediments   46
4.1 GRAIN SIZE   46
4.2 POROSITY AND SEDIMENT DENSITY   47
4.3 PERMEABILITY   55

CHAPTER FIVE
An Introduction to Transport Processes in Sediments   59
5.1 DIFFUSION   59
5.2 SEDIMENT ACCUMULATION, STEADY STATE, AND THE FRAME OF REFERENCE FOR PROCESSES IN MARINE SEDIMENTS   61
5.3 AN INTRODUCTION TO BIOTURBATION AND BIOIRRIGATION   65
5.4 TIME AND SPACE SCALES OF SEDIMENT PROCESSES   67
5.5 THE CLASSIFICATION OF MARINE SEDIMENTS ON THE BASIS OF THEIR FUNCTIONAL DIAGENETIC CHARACTERISTICS   70

CHAPTER SIX
Models of Sediment Diagenesis   72
6.1 THE GENERAL DIAGENETIC EQUATION   72
6.1.1 Diffusion   74
6.1.2 Advection, Sediment Compaction, and Bioturbation   78
6.1.3 Adsorption   83
6.2 SOLUTIONS TO THE DIAGENETIC EQUATION   84
6.2.1 Boundary Conditions   86
6.3 SOLUTIONS TO SPECIFIC DIAGENETIC EQUATIONS   87
6.3.1 Organic Matter Remineralization without Bioturbation   88

6.3.2 Organic Matter Remineralization with Bioturbation	89
6.3.3 Organic Matter Remineralization Coupled to Sulfate Reduction	91
6.3.4 Ammonium Production in Anoxic Sediments	92
6.3.5 Determination of Sediment Accumulation Rates	95

CHAPTER SEVEN
Biogeochemical Processes in Sediments — 97

- 7.1 BACTERIAL METABOLISM: GENERAL CONSIDERATIONS — 98
- 7.2 BACTERIAL RESPIRATION AND BIOGEOCHEMICAL ZONATION IN SEDIMENTS — 99
- 7.3 BACTERIAL RESPIRATION: SPECIFIC PROCESSES — 105
  - *7.3.1 Aerobic Respiration* — 105
  - *7.3.2 Denitrification* — 105
  - *7.3.3 Manganese and Iron Reduction* — 107
  - *7.3.4 Sulfate Reduction* — 110
  - *7.3.5 Methanogenesis* — 111
- 7.4 CHEMOLITHOTROPHIC REACTIONS — 114
  - *7.4.1 Aerobic Processes* — 114
  - *7.4.2 Anaerobic Processes* — 116
  - *7.4.3 Linkages between Chemolithotrophic and Organic Matter Remineralization Processes* — 116
- 7.5 THE DISTRIBUTION OF ORGANIC MATTER REMINERALIZATION PROCESSES IN MARINE SEDIMENTS — 120
  - *7.5.1 Depth Scales of Biogeochemical Zonation* — 120
  - *7.5.2 General Trends with Water Column Depth or Sediment Type* — 124
- 7.6 DYNAMICS OF ORGANIC MATTER DECOMPOSITION IN SEDIMENTS — 134
  - *7.6.1 General Considerations* — 134
  - *7.6.2 Anaerobic "Foodchains"* — 135
  - *7.6.3 Dynamics of Organic Matter Decomposition under Mixed Redox Conditions* — 139

CHAPTER EIGHT
Quantifying Carbon and Nutrient Remineralization in Sediments — 142

- 8.1 MODELS OF ORGANIC MATTER DECOMPOSITION IN SEDIMENTS — 142
- 8.2 SEDIMENT BUDGETS FOR REACTIVE COMPONENTS — 150
  - *8.2.1 Theoretical Considerations* — 151

CONTENTS

8.2.2 Sediment Nutrient Budgets Using Cape Lookout Bight as an Example ... 153
8.3 CARBON BURIAL IN SEDIMENTS ... 161
8.4 LAYERED AND COUPLED MODELS OF SEDIMENT DIAGENESIS ... 162

CHAPTER NINE
An Introduction to the Organic Geochemistry of Marine Sediments ... 171
9.1 GENERAL CONSIDERATIONS ... 172
9.2 CONCENTRATIONS AND SOURCES OF ORGANIC MATTER IN MARINE SEDIMENTS ... 174
9.3 THE BULK CHEMICAL COMPOSITION OF MARINE SEDIMENT ORGANIC MATTER ... 175
9.4 AMINO ACIDS ... 179
9.5 CARBOHYDRATES ... 189
9.6 LIGNINS ... 193
9.7 LIPIDS ... 194
9.8 HUMIC SUBSTANCES AND MOLECULARLY UNCHARACTERIZED ORGANIC MATTER ... 204
9.8.1 Black Carbon ... 206
9.8.2 Molecularly Uncharacterized Organic Matter (MU-OM): General Considerations ... 207
9.8.3 Geopolymerization: The Formation of Humic Substances ... 209
9.8.4 Selective Preservation of Refractory Biomacromolecules ... 212
9.8.5 Physical Protection ... 213
9.9 ORGANIC NITROGEN DIAGENESIS IN SEDIMENTS ... 215

CHAPTER TEN
Dissolved Organic Matter in Marine Sediments ... 218
10.1 GENERAL OBSERVATIONS ... 218
10.2 DIAGENETIC MODELS OF PORE-WATER DOM CYCLING IN SEDIMENTS ... 227
10.3 PORE-WATER DOM COMPOSITIONAL DATA ... 228
10.3.1 Short-Chain Organic Acids ... 230
10.3.2 Carbohydrates ... 231
10.3.3 Amino Acids ... 231
10.4 FLUXES OF DOM FROM MARINE SEDIMENTS ... 232
10.5 DOM ADSORPTION AND SEDIMENT–ORGANIC MATTER INTERACTIONS ... 234

Chapter Eleven
Linking Sediment Organic Geochemistry and Sediment Diagenesis — 237
- 11.1 THE SOURCES OF ORGANIC MATTER TO MARINE SEDIMENTS — 237
  - 11.1.1 Carbon and Nitrogen Isotopic Tracers of Organic Matter Sources — 238
  - 11.1.2 Elemental Ratios as Tracers of Organic Matter Sources — 241
  - 11.1.3 Spatial Trends in the Sources of Organic Matter to Marine Sediments: Marine versus Terrestrial — 244
  - 11.1.4 Other Sources of Organic Matter to Marine Sediments: Black Carbon and Recycled Kerogen — 249
  - 11.1.5 Production of Bacterial Biomass in Sediments — 250
- 11.2 THE COMPOSITION OF ORGANIC MATTER UNDERGOING REMINERALIZATION IN MARINE SEDIMENTS — 253
  - 11.2.1 Pore-Water Stoichiometric Models for Nutrient Regeneration/Organic Mater Remineralization — 254
  - 11.2.2 Benthic Flux and Sediment POM Stoichiometric Models for Nutrient Regeneration — 260
  - 11.2.3 The Composition of Organic Matter Undergoing Remineralization: Elemental Ratios and Stable Isotopic Composition — 261
  - 11.2.4 The Composition of Organic Matter Undergoing Remineralization: Organic Geochemical Composition — 265

Chapter Twelve
Processes at the Sediment-Water Interface — 271
- 12.1 THE DETERMINATION OF BENTHIC FLUXES — 272
- 12.2 DIFFUSIVE TRANSPORT AND THE BENTHIC BOUNDARY LAYER — 274
- 12.3 SEDIMENT-WATER EXCHANGE PROCESSES IN PERMEABLE SEDIMENTS — 283
- 12.4 BIOTURBATION — 286
  - 12.4.1 General Considerations — 286
  - 12.4.2 Models of Bioturbation — 289
  - 12.4.3 Nonlocal Sediment Mixing — 299
- 12.5 BIOIRRIGATION — 302
  - 12.5.1 The Diffusive Openness of Bioirrigated Sediments — 313
  - 12.5.2 Methods for Quantifying Bioirrigation in Sediments — 316
  - 12.5.3 Rates of Bioirrigation in Marine Sediments — 319
- 12.6 OTHER SEDIMENT-WATER INTERFACE PROCESSES: METHANE GAS EBULLITION — 326

Chapter Thirteen
Biogeochemical Processes in Pelagic (Deep-Sea) Sediments ... 328
- 13.1 Organic Matter Remineralization ... 328
- 13.2 Trace Metal Diagenesis ... 332
- 13.3 Manganese Nodules and Crusts ... 344
- 13.4 Diagenesis of Opaline Silica ... 352
- 13.5 Diagenesis of Calcium Carbonate ... 359

Chapter Fourteen
Nonsteady-State Processes in Marine Sediments ... 373
- 14.1 General Considerations ... 373
- 14.2 Periodic Input Processes ... 374
- 14.3 Seasonality in Sediment Processes ... 378
- 14.4 Diagenetic Processes in Deep-Sea Turbidites ... 382
  - *14.4.1 Organic Geochemical Studies of Turbidite Diagenesis* ... 391
- 14.5 Multiple Mn Peaks in Sediments: Nonsteady-State Diagenetic Processes Associated with Paleoceanographic Changes ... 395
  - *14.5.1 Multiple Mn Peaks and the Glacial-Holocene Transition* ... 400
  - *14.5.2 Multiple Mn Peaks and Pleistocene Climate Cycles* ... 402
  - *14.5.3 Multiple Mn Peaks in Holocene Sediments* ... 404

Chapter Fifteen
The Controls on Organic Carbon Preservation in Marine Sediments ... 408
- 15.1 Organic Matter–Mineral Interactions ... 412
- 15.2 The Role of Oxygen in Sediment Carbon Remineralization and Preservation ... 417
- 15.3 The Role of Benthic Macrofaunal Processes in Sediment Carbon Remineralization and Preservation ... 419
- 15.4 Oxygen Exposure Time as a Determinant of Organic Carbon Preservation in Sediments ... 421
  - *15.4.1 What Exactly Does Sediment Oxygen Exposure "Mean"?* ... 425
  - *15.4.2 Organic Carbon Burial and Controls on Atmospheric O_2* ... 428
- 15.5 The Composition of Organic Matter Preserved in Marine Sediments and the Fate of Terrestrial Organic Matter in Marine Sediments ... 432

15.6 THE RELATIONSHIP BETWEEN PHYSICAL PROTECTION, OXYGEN EXPOSURE, AND POSSIBLE ABIOTIC CONDENSATION REACTIONS IN SEDIMENT CARBON PRESERVATION ... 439

CHAPTER SIXTEEN
Biogeochemical Processes in Continental Margin Sediments.
I. The CO_2 System and Nitrogen and Phosphorus Cycling ... 442
   16.1 PORE-WATER pH AND CARBONATE CHEMISTRY UNDER SUBOXIC AND ANOXIC CONDITIONS ... 442
   16.2 SEDIMENT NITROGEN CYCLING ... 452
      16.2.1 Benthic DON Fluxes ... 463
   16.3 SEDIMENT PHOSPHORUS CYCLING ... 464
      16.3.1 Formation of Authigenic CFA and Phosphorus Burial in Sediments ... 474

CHAPTER SEVENTEEN
Biogeochemical Processes in Continental Margin Sediments.
II. Sulfur, Methane, and Trace Metal Cycling ... 478
   17.1 SEDIMENT SULFUR CYCLING ... 478
      17.1.1 Sulfur Burial Efficiency ... 486
      17.1.2 Long-Term Changes in the Sedimentary Sulfur Cycle ... 489
   17.2 METHANOGENESIS AND ANAEROBIC METHANE OXIDATION ... 490
      17.2.1 Shallow (Coastal) Sediments ... 490
      17.2.2 Continental Margin Sediments ... 493
   17.3 TRACE METAL CYCLING ... 500

CHAPTER EIGHTEEN
Linking Sediment Processes to Global Elemental Cycles:
Authigenic Clay Mineral Formation and Reverse Weathering ... 509
   18.1 SEDIMENT SILICA BUDGETS ... 514
   18.2 FINAL THOUGHTS ... 515

Appendix
Some of the Field Sites Discussed in the Text ... 517

References ... 521

Index ... 593

≈ *Preface* ≈

For more than a decade I have taught a graduate-level course in marine sediment geochemistry that covers many of the topics discussed in this book. The goal now is to present this material to a broader group of students and other interested individuals. In this book I present the fundamentals of marine sediment geochemistry and discuss the ways we can quantify geochemical processes occurring in recent marine sediments.

In my mind, Bob Berner's (1980) *Early Diagenesis, A Theoretical Approach* was the first book to present a clear and concise discussion of how geochemical processes in recent marine sediments can be quantitatively studied. However, tremendous advances have been made in this field since this book was published in 1980. Thus I feel there is the need for a book like this one that picks up (in some senses) where *Early Diagenesis* left off. Other books published since 1980 examine a number of the topics presented here, and do so in an excellent manner (Boudreau, 1997; Boudreau and Jørgensen, 2001; Schulz and Zabel, 2000). However, overall, they do not provide the reader with as broad a view of marine sediment geochemistry as I hope I have presented here.

There are many people I need to thank for their direct and indirect assistance in writing this book. First and foremost, I would like to thank Joris Gieskes, Ken Nealson, Paul Kepkay, and Chris Martens for their advice and guidance over the years. I would never have gotten to the point of writing this book if I hadn't been fortunate to have worked with these fine scientists during the early stages of my scientific career. Over the years many other colleagues and friends provided me with stimulating conversations, good (and some bad) ideas, and a good laugh or two when needed. At the risk of leaving anyone out I won't list these individuals here, but you all know who you are. Thanks for everything.

A part of this book was written while on sabbatical from Old Dominion University, and I would like to thank Larry Atkinson and the Center for Coastal Physical Oceanography at ODU for providing me with an office in which to hide and write. Much of the book was also written while trying to juggle the standard teaching, research, and service activities that come with my faculty position. Thanks go to the three people who served as my department chair during this period (Jim Sanders, Tom Royer, and Dick Zimmerman) for their patience

PREFACE

and understanding. I would also like to thank Princeton University Press for their patience during the entire process of writing this book.

In the course of this project, many people graciously provided me with unpublished manuscripts, answered my email questions, and hunted through old computers and data notebooks to uncover previously published data that is replotted in the book. With the rapid advances in computer technology and software over the past few years, I quickly discovered that data "archaeology" is often not a trivial task! In any event, thanks here go to Bob Aller, Marc Alperin, Dave Archer, Will Berelson, Neal Blair, Bernie Boudreau, Liz Canuel, Jeff Cornwell, Steve Emerson, Yves Gélinas, Marty Goldhaber, Mark Green, Per Hall, Markus Huettel, Christelle Hyacinthe, Rick Jahnke, Karen Johannesson, Bo Barker Jørgensen, Mandy Joye, Pete Jumars, Val Klump, Carla Koretsky, Joel Kostka, Karima Khalil, George Luther, Bill Martin, Carrie Masiello, Larry Mayer, Jim McManus, Jack Middelburg, John Morse, Filip Meysman, Christophe Rabouille, Kathleen Ruttenberg, Dan Schrag, Howie Spero, Bjorn Sundby, Brad Tebo, Phillipe Van Cappellen, and Claar van der Zee.

Kim Krecek assisted me with the preparation of some of the figures in the book, and entered a large number of references into EndNote. Debbie Miller of Academic Technology Services at ODU also did a superb job of drafting many of the figures in this book, and put up with my many requests for "just one more change." Don Emminger (also of Academic Technology Services) was a real life-saver in helping me get the figures into their final format.

Several people read either all or parts of the first draft of the book and I would like to thank them for their useful comments: Xinping Hu, Scott Kline, Will Berelson, Anitra Ingalls, Rick Murray, Mike Krom, Clare Reimers, and the students in OCEN 613 *Geochemistry of Marine Sediments* (spring 2005): Joy Davis, Hussain Abdulla, Krista Stevens, and Pete Morton. A very special thanks also goes to Bernie Boudreau for his extremely thorough review of the first draft of the book. This is a much better book because of his time and effort, although in the end all of the mistakes, errors, and omissions are still my responsibility.

Finally, a special thanks goes to my parents and sisters for their support over the years. And last but not least, my wife, Juli, and my children, Ben and Emily, showed extraordinary patience and understanding during the entire project. I can't express my appreciation enough.

Commonly Used Abbreviations and Symbols

Throughout the text (and the literature in general) these symbols are generally used to define the following parameters/quantities.

BE	Burial efficiency (eqn. 8.19)
BFE	Benthic flux enrichment factor (eqns. 12.27 and 12.32)
D	Diffusion coefficient
D_b	Bioturbation (or biodiffusion) coefficient
D_s	Bulk sediment diffusion coefficient (D values corrected for sediment tortuosity)
DBL	Diffusive boundary layer
DOM	Dissolved organic matter
DON	Dissolved organic nitrogen
DOP	Degree of pyritization
F	Formation factor (used in the determination of sediment tortuosity; see eqn. 6.11)
$\mathscr{F}$	A factor used to convert solid phase sediment concentration units to pore water concentration units (see eqn. 6.41)
gdw (or g_{dw})	Grams sediment dry weight
HMW	High molecular weight
J	Flux across the sediment-water interface
$\mathscr{K}$	Hydraulic conductivity (eqn. 4.7)
L	Sediment mixed layer depth
$\mathscr{L}$	The stoichiometric ratio of the moles of sulfate reduced : carbon oxidized during bacterial sulfate reduction
LMW	Low molecular weight
MU-OM	Molecularly uncharacterized organic matter
ox	The average carbon oxidation state in particulate organic matter (eqns. 11.6 and 11.12)
OET	Oxygen exposure time
POC	Particulate organic carbon
POM	Particulate organic matter
R_{cox}	The depth-integrated rate of sediment organic carbon oxidation. Note that in some sediments this is the

COMMONLY USED ABBREVIATIONS AND SYMBOLS

	portion of the benthic ΣCO_2 flux due to respiration (also see eqn. 11.13). In some works, this term is referred to as C_{ox}.
r_n	The C/N ratio of organic matter undergoing remineralization
TCO_2 (also DIC or ΣCO_2)	Total dissolved inorganic carbon ($= [CO_2] + [HCO_3^-] + [CO_3^{2-}]$)
TOC	Total organic carbon (note that the letter G is often used as a symbol for TOC in diagenetic equations)
TN	Total nitrogen
TOM	Terrestrial organic matter
α or $\alpha(z)$	The depth-dependent non-local bioirrigation coefficient
φ	Sediment porosity (cm^3_{pw}/cm^3_{ts}). Note that in many texts ø is often used as the symbol for sediment porosity. However, here and in other recent works this other form of phi (φ) is used for porosity, to unambiguously define these different parameters.
φ_s	Sediment solid fraction (cm^3_{ds}/cm^3_{ts}; $= 1-\varphi$)
ø	Sediment grain size
θ^2	Sediment tortuosity factor (see eqn. 6.12)
ω	Sedimentation rate
Ψ	A geometric parameter that incorporates reaction geometry considerations into α (i.e., solute transport by bioirrigation; see eqn. 12.31)
subscript p or pw	Pore water
subscript ds or s	Sediment (solid phase)
subscript dw	Sediment dry weight
subscript ts	Total sediment (pore water plus solid phase)

Geochemistry of Marine Sediments

≈ **CHAPTER ONE** ≈

Introduction

THE PROCESSES OCCURRING in the upper several meters of marine sediments[1] have a profound effect on the local and global cycling of many elements. For example, the balance between carbon preservation and remineralization represents the key link between carbon cycling in active surface reservoirs in the oceans, in the atmosphere, and on land, and carbon that cycles on much longer, geological time scales—in sedimentary rock, and in coal and petroleum deposits (Berner, 1989; Hedges, 1992). Denitrification in marine sediments, i.e., the reduction of nitrate to gaseous N_2, is an important component of the global nitrogen cycle, and on glacial-interglacial time scales may play a role in regulating the oceanic inventory of reactive nitrogen (Ganeshram et al., 1995; Codispoti et al., 2001). On more local scales, nitrogen and phosphorus remineralization in coastal and estuarine sediments can provide a significant fraction of the nutrients required by primary producers in the water column (Klump and Martens, 1983; Kemp and Boynton, 1984). In deep-sea sediments, trace metal remineralization may play a role in the growth and genesis of manganese nodules (Glasby, 2000). Similarly, in coastal and estuarine sediments subjected to elevated anthropogenic inputs of certain toxic metals, sediment processes affect the extent to which these sediments represent "permanent" versus "temporary" sinks for these metals (e.g., Huerta-Diaz and Morse, 1992; Riedel et al., 1997).

Understanding processes occurring in surficial marine sediment is also important in the accurate interpretation of paleoceanographic sediment records, since sediment processes can sometimes significantly alter the primary "depositional" signal recorded in the sediments (e.g., Martin and Sayles, 2003). At the same time, temporal changes in ocean conditions can lead to the occurrence of nonsteady-state conditions in sediments (Wilson et al., 1985; Finney et al., 1988). The ability to recognize and accurately quantify nonsteady-state processes in sediments may therefore provide important paleoceanographic

---

[1] Throughout the book, this portion of the sediments is referred to as surface or surficial marine sediments.

information that is complementary to that obtained using more traditional tracer approaches such as carbon or oxygen isotopes.

The geochemistry of marine sediments is controlled by both the composition of the material initially deposited in the sediments and the chemical, biological, or physical processes that affect this material after its deposition. These processes fall within the general category of what is commonly referred to as early diagenesis (*sensu* Berner, 1980). Since these processes occur in the upper portions of the sediments, temperatures are generally not elevated above bottom water values. Sediment pore spaces are also still water saturated,[2] although in some sediments gas bubbles may also occur (e.g., see section 12.6).

More importantly, though, a key fact that has emerged in the past 20–30 years of research in marine sediment geochemistry is that the oxidation, or remineralization, of organic matter deposited in sediments is either the direct or the indirect causative agent for many early diagenetic changes. Thus in many ways, we are actually examining the biogeochemistry of these sediments. Much of this organic matter remineralization is mediated by bacteria, since marine sediments often become anoxic (i.e., devoid of oxygen) close to the sediment-water interface (generally <1 cm in coastal sediments to several centimeters or more in some deep-sea sediments). At the same time, surficial marine sediments are often colonized by benthic macrofauna such as burrowing clams and shrimp and tube-dwelling polychaetes. The presence of these benthic macrofauna and their resulting activities can also have a profound effect on sediment geochemistry (e.g., Aller, 1982b).

Given the key role that organic matter remineralization plays in many early diagenetic processes, significant efforts have gone into understanding and quantifying these processes. Such studies have taken both organic and inorganic approaches, with the latter often carried out through studies of the pore-water chemistry of remineralization products or reactants. Studies of pore-water geochemistry are particularly useful in this effort because they are very sensitive indicators of diagenetic changes occurring in the sediments. As an example of this, Berner (1980) notes that a 20% increase of dissolved calcium in the pore waters from the dissolution of calcium carbonate is

---

[2] As will be discussed in chapter 3, the water found in these pore spaces is referred to as *pore waters* or *interstitial waters*.

roughly equivalent to a decrease of only 0.02% $CaCO_3$ by weight. While the former is easily measurable, the latter is not. Thus, a great deal of effort has gone into the study of pore-water geochemistry and the development of diagenetic models of the processes affecting pore water solutes.

Historically, there has been more of a tendency to use inorganic geochemical studies to quantify rates of sediment carbon remineralization processes. However, an increasing number of workers have also begun to use organic geochemical measurements to examine the rates of these processes. Such efforts have built important links between inorganic and organic geochemical approaches to the study of sediment biogeochemistry. They have also played a major role in advancing not only what we know about sediment geochemical processes, but also how we approach their study.

The remainder of this book is divided up as follows. Chapters 2–6 contain a basic introduction to the study of marine sediment geochemistry. These chapters also begin to discus the ways we can quantify processes occurring in sediments using mathematical models of early diagenesis. Chapters 7–12 further examine sediment organic matter remineralization and early diagenetic processes from the standpoint of: the potential reactions that may occur; the relationships between these reactions, e.g., thermodynamic vs. kinetic controls; the composition and reactivity of sediment organic matter; and the role that external factors play in controlling these reactions, e.g., carbon rain rate to the sediments or bioturbation.

Chapters 13–17 build on these previous chapters in more specific discussions examining processes occurring in pelagic and continental margin sediments. The division of the material presented here is perhaps somewhat arbitrary since changes in sediment geochemical processes are clearly a continuum as one moves from deep-sea to nearshore settings (e.g., see discussions in section 7.5.2). Nevertheless, I believe that this approach is as good as any other to present this material.

Chapter 13 describes processes occurring in pelagic sediments; this discussion then leads to a discussion in chapter 14 of nonsteady-state, or time-dependent, diagenetic processes occurring in sediments. By presenting a discussion of nonsteady-state processes in a separate chapter the intent is not to suggest that the occurrence of nonsteady-state conditions is "unusual," or the exception, as compared to steady-state conditions. In fact, evidence increasingly suggests that the opposite is the case, and that true steady-state conditions may be far less common in marine sediments than has been previously assumed.

Chapter 15 builds on much of what has been discussed in earlier chapters by examining the controls on organic carbon preservation in marine sediments, a process that occurs largely in continental margin sediments. Chapters 16 and 17 further examine processes occurring in continental margin, coastal, and estuarine sediments from the standpoint of the sediment cycling of trace metals and nutrients. The book concludes by examining sediment biogeochemical processes in the context of the global cycles of the major elements.

An appendix at the end of the book briefly describes many of the field sites discussed in the text.

In writing this book I have assumed that the reader has some basic knowledge of geology, chemistry, and biology. Readers who come across unfamiliar terms or concepts may want to consult introductory texts in these fields. In contrast, many readers may not be as familiar with some of the concepts of chemical oceanography that are brought into the discussions here. Several good texts have been published in this area that readers may find useful (Broecker and Peng, 1982; Libes, 1992; Millero, 1996; Pilson, 1998; Chester, 2000; Gianguzza et al., 2002). Where appropriate I cite these works in the text.

≈ **CHAPTER TWO** ≈

The Components of Marine Sediments

MARINE SEDIMENTS comprise three basic types of inorganic material: detrital material, biogenic material, and authigenic material. Detrital (terrigenous) components originate on land while biogenic components (mainly calcium carbonate and opaline silica) are generally produced by organisms in the water column. Authigenic components form *in situ* in the water column, on the sediment surface, or in the sediments.

Detrital and biogenic material generally predominate in most marine sediments. Authigenic components are quantitatively less important in terms of the total sediment input, although they can play important roles in the geochemical cycling of elements such as Fe, Mn, P, and S. They can also be important indicators of biogeochemical processes occurring in the sediments. In a similar vein, the relatively small amount of organic matter found in most marine sediments, generally less than a few percent, is not an indication of its importance in many early diagenetic processes in sediments.

In this chapter I briefly introduce these major components of marine sediments and discuss clay mineralogy and geochemistry in slightly more detail.

2.1 DETRITAL COMPONENTS

The vast majority of the detrital material found in marine sediments is formed on land by chemical weathering and erosion (physical weathering). A much smaller component of this detrital material is of volcanic or cosmogenic (extraterrestrial) origin. Physical weathering involves the mechanical break-up of continental material, while chemical weathering involves several different reactions that continental rocks undergo. These include simple dissolution of salt deposits, (e.g., NaCl) or limestone ($CaCO_3$), and oxidation-reduction reactions, such as pyrite oxidation (FeS_2). However, perhaps the most important type of chemical weathering is the incongruent dissolution of silicate rocks by carbonic acid. In its most general sense, this reaction can be written as,

$$\text{cation-rich Al-silicates} + H^+ \to \text{cation-poor clays} + SiO_2 + \text{cations} \qquad (2.1)$$

where "cation-rich Al-silicates" is a shorthand abbreviation for any number of minerals found in continental rocks (e.g., felspar, mica, or pyroxenes). The protons in this reaction generally come from carbonic acid (H_2CO_3), primarily produced by the hydration of atmospheric CO_2 in rainwater. However, vascular plants and other soil organisms also produce CO_2 and organic acids that can be involved in mediating weathering reactions (Berner et al., 2003).

Clay minerals produced by weathering on land are transported to the oceans primarily by rivers or winds (eolian transport). In addition to the occurrence of detrital clays in marine sediments, new (authigenic) clay minerals may also form in the oceans and in sediments. Authigenic clays form by processes generally referred to as reverse weathering. The significance of these processes will be discussed in chapter 18.

Maps such as that in fig. 2.1 illustrate the distribution of different types of surface sediments in the world's oceans. Generally, a sediment is defined by the component that is more than 30–50% of the total sediment mass (see section 2.5 for further details). Because these sediment types are defined in a relative sense, the degree of dilution of one component versus another is an important consideration. Thus, even though primary production (and therefore production of biogenic material) is high in coastal environments, there are even higher inputs of detrital material from the nearby continents. These sediments are therefore predominantly terrigenous. In contrast, biogenic sediments tend to dominate further offshore, in part because of their distance from these detrital sources; here detrital clays are diluted out by the more rapid sedimentation of biogenic material. However, there are some open ocean settings where clays again predominate, and this occurs because of the absence of significant biogenic sediment deposition (see section 2.2.3).

In maps such as that in fig. 2.1, detrital components are often further separated into three subgroups: ice-rafted debris, terrigenous material, and red clay (sometimes also referred to as deep-sea clay). The distinction made here is based on both the composition of the material as well as its mode of transport to the oceans.

DETRITAL COMPONENTS

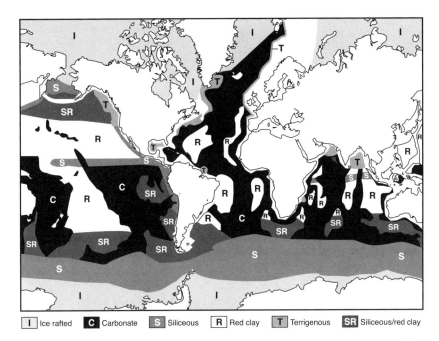

| I | Ice rafted | C | Carbonate | S | Siliceous | R | Red clay | T | Terrigenous | SR | Siliceous/red clay |

Figure 2.1 The distribution of modern (surficial) marine sediments. Modified after Barron and Whitman (1981).

Ice-rafted debris is transported to the oceans by glacial erosion, and consists largely of material formed by the mechanical weathering of source rocks in high-latitude regions (Barron and Whitman, 1981). Lithic fragments and quartz are common in these sediments, as is the clay mineral chlorite. The lack of soil-forming processes in these regions precludes the occurrence of chemical weathering processes that would destroy chlorite found in the metamorphic and igneous source rocks (Barron and Whitman, 1981; Srodon, 1999).

Terrigenous sediments are primarily transported to the oceans by rivers and are found along continental margins and near the mouths of rivers. This material consists of sands, silts, and clays that result from both mechanical and chemical weathering of continental rocks. Further details on the classification of these sediments is given in section 3.4.1.

Deep-sea red clays are fine-grained clays that are actually more chocolate brown, rather than red, in color (the term red clay appears to be largely historical, apparently dating back to the original description of these sediments during the HMS *Challenger* expedition of

7

1873–76; Kennett, 1982). The coloration of these clays is due largely to amorphous iron oxyhydroxide coatings on the clays. Most of these clays appear to be transported to the deep sea by eolian transport (Barron and Whitman, 1981; Kennett, 1982), and consistent with this mode of transport is the fact that relative clay content (<2 μm particle size; see section 4.1) is positively correlated with water depth (Premuzic et al., 1982). Red clays predominate in many open ocean (pelagic) settings because there is little dilution of the clays by other (generally biogenic) components (see discussions below for further details). However, given their great distance from continental sources, these sediments have extremely low sediment accumulation rates.

2.2 Biogenic Components

Biogenic components are defined here as inorganic phases produced by living organisms. The term generally used to describe this process is biomineralization (Lowenstam and Weiner, 1989). Biologically induced formation of minerals results in a wide variety of minerals formed by a diverse group of organisms. In marine sediments, carbonates (e.g., calcite and aragonite) and opal (amorphous silica) are the predominant biominerals. However, phosphate minerals, Fe(III) oxyhydroxides, magnetite (Fe_3O_4), and manganese oxides may, under some circumstances, also be of significance.

In general, there are two types of biomineralization: organic matrix–mediated precipitation and biologically induced precipitation. In the former case, an organism produces an organic structure or framework that aids in the precipitation process. This mechanism appears to be responsible for the biological precipitation of opal or calcium carbonate.

This organic matrix likely serves two purposes. First, it acts as a structural framework or template that aids in inducing precipitation by chelating the appropriate cations and anions. Second, the matrix may also protect the precipitate from dissolution. This would be particularly important, for example, in siliceous tests because the oceans are everywhere undersaturated with respect to opaline silica.

These organic matrices are composed predominantly of proteins and carbohydrates, with proteins playing the more significant role in mediating precipitation and carbohydrates being more important as structural components. Siliceous and calcareous organisms produce proteins of slightly different amino acid composition, in part because of the different chemistries of the minerals they produce. Matrices associated with carbonate mineral precipitation tend to be rich in acidic amino acids, such as aspartic acid and glutamic acid (Lowenstam and Weiner, 1989), since the carboxyl groups in these amino acids can bind Ca^{2+} and enhance carbonate precipitation. In contrast, siliceous organisms produce a matrix that is rich in glycine and hydroxyl-containing amino acids, such as serine and threonine (Hecky et al., 1973). However, silica precipitation appears to be mediated by long-chain polyamines attached to cationic polypeptides associated with the diatom cell wall (Kröger et al., 2001).

In contrast, biologically induced precipitation involves intra- or extracellular formation of minerals without the aid of an organic matrix, although mineral production is clearly related to some biological process. Examples of this include the bacterial production of manganese oxides as a result of changes in the local pH around a cell or even the production of metal sulfides as a result of bacterial sulfate reduction (sulfide production).

2.2.1 Biogenic Carbonates

The predominant forms of calcium carbonate found in marine sediments are the minerals calcite and aragonite. Both have the same chemical formula ($CaCO_3$), but differ in their crystal structure, with aragonite being 1.5 times more soluble than calcite (based on 25°C K_{sp} values; Morse and Mackenzie, 1990).

Calcium carbonate found in open ocean sediments is dominated by calcite tests produced by organisms such as cocolithophores (planktonic algae that produce calcite-platelets called cocoliths). Given their small size, cocoliths are often referred to as nannofossils. Biogenic calcite is also produced by benthic and pelagic foraminifera (animals). Numerically, cocolithophores are much more abundant than foraminifera (Morse and Mackenzie, 1990). The majority of the aragonite in open ocean settings is produced by pteropods (planktonic gastropods) found in tropical and subtropical waters. However, because of its high solubility, aragonite is generally not well preserved in most deep-sea sediments (see section 13.5).

In shallow waters, carbonates tend to be dominant only in tropical and subtropical sediments (Morse and Mackenzie, 1990), although minor amounts of carbonate can be found in nearshore temperate terrigenous sediments (e.g., see Green and Aller, 1998, and references therein). In contrast to open ocean sediments, shallow-water carbonate-rich sediments are composed mainly of aragonite and Mg-calcites, which contain up to 30 mole% Mg in place of Ca. In contrast, open ocean biogenic calcite generally contains <1 mole% Mg. Shallow water carbonates are composed primarily of disintegrated skeletal remains of benthic organisms such as corals, echinoids, molluscs, bryozoans, and calcareous red and green algae. Also found in shallow water carbonate sediments are oöids, small spherical to oval carbonate grains that often have an internal concentric or radial (laminated) structure. The origin of oöids (biological vs. chemical) remains controversial (see discussions in Morse and Mackenzie, 1990).

2.2.2 Biogenic Silica

The predominant form of biogenic silica in marine sediments is amorphous hydrated silica ($SiO_2 \cdot nH_2O$) or opal. Diatoms (unicellular algae) and radiolaria (surface and deep-water zooplankton) produce the majority of the biogenic silica found in marine sediments. Diatoms tend to dominate in high-latitude regions, while radiolaria are more important in low-latitude regions (Barron and Whitman, 1981).

In contrast to calcium carbonate, the surface oceans are everywhere undersaturated with respect to biogenic silica. As a result, the vast majority of the opal produced in surface waters dissolves in the water column during particle sinking (Nelson et al., 1995). Additional information on the factors controlling silica preservation in sediments is discussed below and in section 13.4.

On "long" time scales ($\sim 10^6$–10^7 yrs) biogenic silica in marine sediments undergoes recrystallization to quartz (the more stable phase of SiO_2). The discussion of this process is beyond the scope of this book (see Kastner, 1981, and references therein for further details).

2.2.3 Distribution of Biogenic Components in Marine Sediments

Biogenic silica and calcite are produced predominantly by organisms that live in the surface ocean. However, the factors controlling their distribution in surficial marine sediments differ. These distributions are briefly discussed here, with additional details presented in chapter 13.

Silica is a limiting nutrient in the oceans and is generally found at extremely low concentrations in surface ocean waters. Therefore, to a first order, patterns of silica distribution in surface sediments appear to be controlled by the input of silica to the surface ocean, which is dominated by the upwelling of silica-rich deeper waters (e.g., Libes, 1992; Nelson et al., 1995). As a result sediments underlying upwelling regions, such as the equatorial Pacific, the Antarctic, and the west coast of South and Central America, tend to be silica rich (fig. 2.1). In these regions the production of biogenic silica in the surface ocean is apparently sufficiently large that its export flux exceeds water column dissolution, allowing silica to accumulate in the sediments.

Because the oceans are everywhere undersaturated with respect to opal, greater than ~50% of the opal produced in surface waters dissolves in the upper 100 m of the water column during particle sinking (Nelson et al., 1995). This dissolution continues through the deeper water column and in the sediments, and globally, the ratio of net sediment burial to surface water production is ~3%. This ratio also appears to be strongly bimodal, with values of 15–25% for siliceous oozes and values of near zero for other sediments (also see Tréguer et al., 1995). Many of the details of how this more efficient preservation occurs in silica-rich sediments are poorly understood (Nelson et al., 1995; also see section 13.4).

In contrast, the production of biogenic calcite is more evenly distributed in the surface ocean than that of silica, although regions of high carbonate production and high silica production are often similar (Barron and Whitman, 1981). At the same time, though, the distribution of calcium carbonate in surface marine sediments is less closely linked with carbonate production in the surface waters.

Although surface ocean waters are supersaturated with respect to calcite and aragonite, the solubility of both minerals increases with decreasing temperature and increasing pressure (i.e., with depth in the water column; e.g., Broecker and Peng, 1982; Millero, 1996). Bottom-water circulation and water column remineralization (production of aqueous CO_2) also both contribute to a decrease in calcium carbonate solubility with increasing water column depth (see section 13.5). As a result of all of these factors, deep waters of the oceans often become undersaturated with respect to calcium carbonate. The depth where the deep waters become undersaturated (the calcite saturation horizon) versus the depth of the entire water column then controls, to a first order, the fraction of biogenic calcite produced in the surface waters that escapes dissolution as this material sinks (e.g., Broecker and Peng, 1982). Furthermore, at some depth below the saturation horizon (termed the calcite compensation depth or CCD), the waters become sufficiently undersaturated that all biogenic carbonate sinking though the water column dissolves before reaching the sediments. Sediments below the CCD are therefore devoid of any biogenic calcite, in spite of the occurrence of carbonate production in the surface waters (again see section 13.5 for further details).

The depths of both the saturation horizon and the CCD decrease as one moves from the Atlantic to the Pacific because of the interplay between surface water carbonate production, water column remineralization processes, ocean circulation, and deep-water dissolution (e.g., Broecker and Peng, 1982; Millero, 1996). Calcareous sediments are consequently more prevalent in the Atlantic than in the Pacific. They also tend to follow the topographic "highs" of the mid-ocean spreading centers.

Although much of the calcite and virtually all of the aragonite produced in surface waters dissolves before it reaches deep-sea sediments, calcareous oozes are common in many places (as noted above). Furthermore, some fraction of the calcite that escapes dissolution in the water column and is deposited in marine sediments eventually undergoes dissolution in the sediments on early diagenetic time scales. Understanding sediment calcite dissolution and preservation is of interest in part because carbonate in marine sediments represents the largest buffer for neutralizing anthropogenic CO_2 produced by fossil fuel combustion (Walker and Kasting, 1992; Archer et al., 1998). Furthermore, patterns of calcite preservation in sediments may provide important information on the past history of the oceans (e.g., Archer, 1991a).

2.3 Authigenic Minerals

Authigenic minerals form either in the water column, on the sediment surface, or in the sediments either by direct precipitation from seawater or through chemical reactions with existing sediment materials.

Historically these processes were thought to be entirely inorganic (abiotic) in nature, although recent studies have shown direct or indirect biological involvement in the formation of some of these minerals. Quantitatively, these minerals are generally not important in terms of the total sediment input. However, they can be very important in terms of the geochemical cycling of elements such as Fe, Mn, P, and S. They can also be important indicators of biogeochemical processes occurring in the sediments and can provide certain types of paleoceanographic information (see discussions in Kastner, 1999, and chapters 14, 16, and 17).

2.3.1 Nonbiogenic Carbonates

An important authigenic carbonate mineral phase found in recent marine sediments is dolomite, or $CaMg(CO_3)_2$. Thermodynamically, dolomite is the stable carbonate mineral in seawater, given its Ca^{2+} and Mg^{2+} concentrations. However, dolomite is found only in a limited number of environments (i.e., certain anoxic sediments, hypersaline environments, algal mats) where it appears to occur as both a primary precipitate and a replacement phase of precursor biogenic carbonates (Kastner, 1999). In contrast, dolomite is a major component of many sedimentary rocks (Morse and Mackenzie, 1990). This "dolomite problem" is of great interest and importance in sedimentary geology and geochemistry, although a discussion of this problem is beyond the scope of this book (see Morse and Mackenzie, 1990, for further details).

In some circumstances Mn^{2+} can substitute in the calcite or dolomite structure producing mixed phases that may be thought of solid solutions between calcite and rhodochrosite ($MnCO_3$) (Suess, 1979; Aller, 1980b; Middelburg et al., 1987; Burke and Kemp, 2002). The significance of these authigenic phases in terms of manganese geochemistry is discussed in section 13.2.

2.3.2 Mn Crusts, Layers, and Nodules

Authigenic manganese and iron oxides occur in marine sediments as crusts, layers, and nodules. Their occurrence results from the fact that the solubilities of Mn and Fe vary under differing redox conditions, with reduced forms being more soluble than oxidized forms.

Manganese nodules are quite common on the surface of deep-sea sediments, with aerial coverages that can exceed 50% (Glasby, 2000). They can also be found buried in deep-sea sediments and are common in many high-latitude freshwater lakes in the northern United States, Canada, and Russia. These nodules are composed of Fe and Mn oxides that often occur as alternating layers with some sort of nucleating agent (e.g., a shark's tooth) at the nodule's center. Manganese nodules have attracted scientific and economic interest for quite some time, in part because they contain high concentrations of trace metals such as Ni, Cu, and Co.

Other forms of authigenic Mn, and to a lesser extent Fe, oxides often occur as layers just above the "redox boundary" of marine sediments, as a result of diagenetic concentration during sediment Mn redox cycling (Burdige, 1993; also see sections 7.3.3 and 13.2). Extremely high concentrations of these oxides (both Fe and Mn) are also found in metalliferous sediments that occur in close proximity to mid-ocean ridges (sea-floor spreading centers; Dymond et al., 1973; Chester, 2000). Hydrothermal vent fluids contain elevated levels of both Mn^{2+} and Fe^{2+} as compared to ambient seawater (Elderfield and Schultz, 1996) and these oxides may form directly when vent fluids contact oxygenated seawater. The oxidation of massive sulfide deposits that form near hydrothermal vents can also contribute to the metal oxides found in metalliferous sediments (Edmond et al., 1979b; Hannington et al., 1995).

2.3.3 Phosphorites

Phosphorites are sediments containing significant amounts of authigenic phosphate minerals, primarily in the form of carbonate fluoroapatite (CFA, also known as the mineral phase francolite; Föllmi, 1996; Ruttenberg, 2003). Phosphorites can form as nodules at the sediment-water interface and within the sediments. They tend to be associated with sediments underlying coastal upwelling areas, e.g., the Peru margin and the Namibian shelf, and can also be an important component of some ancient sediments and rocks. Authigenic CFA formation is an important sink for reactive phosphorus in the oceans (Ruttenberg, 2003), and the mechanisms by which these minerals form will be discussed in section 16.3.

2.3.4 Sulfides

In highly reducing (anoxic) marine sediments hydrogen sulfide (H_2S) forms as a result of bacterial sulfate reduction, described approximately as

$$2CH_2O + SO_4^{2-} \rightarrow H_2S + 2HCO_3^- \qquad (2.2)$$

(see section 7.3.4 for further details). This sulfide often reacts with either dissolved Fe^{2+} or solid Fe oxides to produce sulfide minerals. These sulfide minerals include iron monosulfides such as FeS (mackinawite), greigite (Fe_3S_4), or pyrite (FeS_2), the thermodynamically stable sulfide phase in the absence of oxygen (Berner, 1984)). In sediments iron monosulfides or greigite may also be intermediates in the formation of pyrite (see section 17.1 for further details). Sulfide minerals play an important role in iron and sulfur cycling, as well as in the cycling of trace metals incorporated into these sulfide mineral phases.

2.4 Clays and Clay Minerals

In the context of this discussion the term *clay* has meaning in terms of both particle size and mineral composition (Grim, 1968; Hathaway, 1979). Clays represent a class of minerals known as phyllosilicates (layered silicates), and clays are also defined as particles with a diameter <2 μm (see section 3.4.1). At the same time, though, there are clay minerals that are >2 μm in diameter, and materials other than clays in the <2 μm size fraction. However, phyllosilicates dominate this fine particle size fraction in marine sediments and strongly influence its physical and chemical properties.

The basic building blocks of clay minerals are sheets of corner-sharing SiO_4 tetrahedra and sheets of gibbsite-like ($Al(OH)_3$) or brucite-like ($Mg(OH)_2$) octahedral layers (fig. 2.2). In these octahedral layers other cations, Fe in particular, can substitute for Al and Mg. Trioctahedral sheets contain divalent cations (Mg^{2+}, Fe^{2+}), and require that all octahedra be filled to provide a charge balance for the sheet. Dioctahedral sheets contain trivalent cations (e.g., Al^{3+}, Fe^{3+}), and require that only two out of every three octahedra be filled to balance charge for the whole sheet. Traditional clay minerals incorporate both octa-

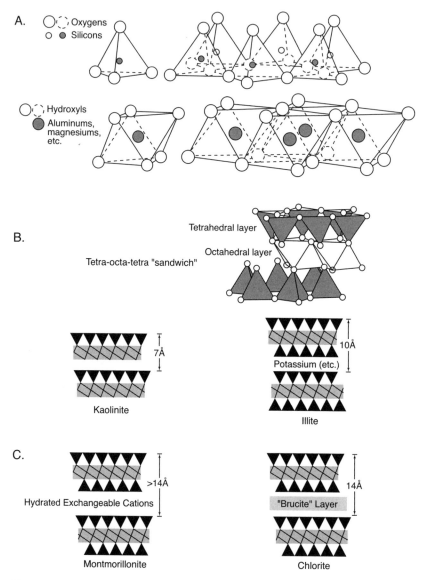

Figure 2.2 A. (Top) The 3-dimensional structure of a single SiO_4 tetrahedral unit (*left*) and that of a sheet structure of corner sharing tetrahedral units (*right*). (Bottom) The 3-dimensional structure of an individual octahedral unit (*left*) and that of a sheet structure of linked octahedral units (*right*). B. The 3-dimensional structure of the 2:1-joined octahedral and tetrahedral sheets found in the illite, chlorite, and montmorillonite (smectite) clay

hedral and tetrahedral sheets. Gibbsite and brucite are "end-member" minerals containing only tri- or dioctahedral sheets, and strictly speaking are not clay minerals, in spite of the fact that they are considered as such in many discussions. In contrast to the fairly common cation substitution in octahedral sites, Al^{3+} (and rarely other cations) only occasionally substitutes for Si^{4+} in tetrahedral sites.

Because the basic building blocks of clay minerals are these sheet structures, these minerals tend to exist as small platelike particles (platelets). In addition, the relatively high surface area of these platelets plays an important role in controlling the initial porosity of many marine sediments (see section 4.2 for details).

The simplest clay mineral is the 1:1 layer group, which contains layers of joined silica tetrahedra and Al-octahedra (fig. 2.2). In the marine environment, kaolinite is the dominant mineral in this group. The 2:1 layer group consists of sheets in which two layers of Si-tetrahedra have one octahedral layer (both di- and tri-octahderal forms) sandwiched in between. In these minerals Al^{3+} can substitute for Si^{4+} in some fraction of the tetrahedra, causing a charge imbalance that is overcome by the inclusion of cations in the interlayer spaces. Illites (also known as the mica group) are clay minerals with this basic structure. Another class of clay minerals in the 2:1 group are smectites (or montmorillonites), in which there is cation substitution in both tetrahedra and octahedra. In these minerals both cations and water molecules can be found in the interlayer region, and organic compounds can also be adsorbed here. The 2:1:1 layer (or chlorite) group consists of a 2:1 trioctahedral structure that contains an octahedral sheet in the interlayer position. The octahedral sheets in the interlayer positions may be either Fe- or Mg-containing. Finally, mixed

---

groups. C. Two-dimensional representations of the 1:1 kaolinite clay mineral group (consisting of sheets of 1:1 layers of octahedral and tetrahedral units), the 2:1 illite clay mineral group, the 2:1 montmorillonite (smectite) clay mineral group, and the 2:1:1 chlorite clay mineral group. Note that the "brucite" layer here refers to a single sheet of octahedral units such as that shown in A. Also shown to the right of each clay structure is the basal spacing between complete layers in the clay. These spacings are generally determined by X-ray diffraction techniques, and are often diagnostic of these clay mineral groups. Modified after Grim (1968) and Seibold and Berger (1996).

layer clay minerals exist that are interstratified at the unit cell level. Random interstratification is common in marine sediments (e.g., in clays of detrital origin), while regular interstratification is rare unless late-stage (burial) diagenesis has occurred (Hathaway, 1979). Smectite-illite combinations are most common, with mainly smectite layers and fewer illite layers.

2.4.1 Distribution of Clay Minerals in Surface Marine Sediments

The distribution of clay minerals in surface marine sediments is determined mainly by detrital processes, given that their primary source is the continents (see section 2.1 and discussions in Griffin et al., 1968; Hathaway, 1979; Chester, 2000). Furthermore, in most deep-sea settings isotopic evidence suggests that most of the clays found there were formed on the continents and have undergone little post-depositional alteration (i.e., reverse weathering-type recrystallization or precipitation; see section 3.3.2 for details).

The map in fig. 2.3 shows clay mineral distribution in the <2 μm size fraction of surface sediments for the four main clay groups: chlorite, illite, kaolinite, and montmorillonite (smectite). In looking at this map it is also important to remember that in many areas these clay minerals are diluted by biogenic materials, and only in certain areas will clays then be the dominant material in the surface sediments (e.g., see fig. 2.1). Examples of where this occurs include sediments underlying low productivity regions such as the Sargasso Sea in the north Atlantic and north Pacific Ocean sediments that are below the CCD (see sections 2.2.3 and 13.4).

The distribution of chlorite in surface sediments (fig. 2.3) shows that these clays are found primarily at high latitudes. This occurs because of the mechanical weathering of chlorite-bearing continental rocks in these regions, along with the lack of soil-forming processes that lead to minimal chemical alteration of these rocks. In contrast, kaolinite is found predominantly in low-latitude sediments. It is the product of intense chemical weathering in equatorial and subtropical regions that is then carried to marine sediments by rivers or eolian transport.

Illites and smectites (fig. 2.3) are the most abundant clay minerals in marine sediments. Illite concentrations are generally higher in Northern Hemisphere sediments than in Southern Hemisphere sediments, because of their continental source. Smectites are an alteration (weath-

CLAYS AND CLAY MINERALS

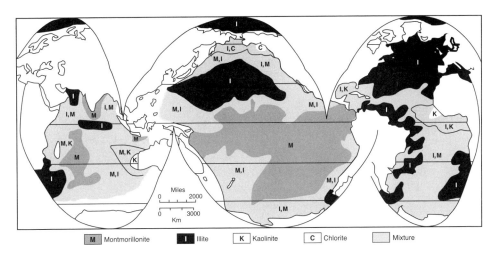

Figure 2.3 The distribution of clay minerals in the <2 μm size fraction of modern (surficial) marine sediments. In areas labeled *Mixture*, no one clay mineral group exceeds 50% of the total. Modified after Seibold and Berger (1996) based on data from Griffin et al., (1968).

ering) product of volcanic material found in the oceans. In contrast then to other clay minerals, most marine smectites appear to be authigenic in origin (Kastner, 1981). There are also continental (detrital) smectites found in marine sediments, although these smectites are generally mixed-layer smectites that contain 60–80% smectite plus illite, versus authigenic smectites that are almost pure smectites.

In surface sediments, authigenic Fe-rich montmorillonites form through the reaction of amorphous iron oxyhydroxides (which precipitate out of hydrothermal solutions) and biogenic silica. Their concentrations are highest in the Pacific and Indian Oceans, reflecting the greater amount of hydrothermal activity associated with mid-ocean spreading centers in these oceans.

Other authigenic smectites may form from basalt and volcanic glass at high temperatures at mid-ocean ridge spreading centers, and at low temperatures away from these spreading centers. These low-temperature reactions occur through interactions between basement basalt and sediment pore waters, and appear to have a significant effect on the chemical composition ($\delta^{18}O$, Mg^{2+}, K^+, and Ca^{2+}) of deep-sea sediment pore waters on long time and depth scales, i.e., hundreds of meters of sediment (Gieskes, 1981; McDuff, 1981; Kastner, 1999).

Low-temperature formation of authigenic clay minerals was first proposed by Sillén (1967) as the primary mechanism that balances the input of major ions and bicarbonate to the oceans from rivers (also see Mackenzie and Garrel, 1966). The occurrence of these processes and their impact on elemental cycles has remained controversial since their initial discussion, in part because, until recent years, strong evidence in support of the occurrence of these processes was lacking (see recent discussions in Michalopoulos and Aller, 2004). These processes will be discussed in further detail in chapter 18.

2.4.2 Ion Exchange/Adsorption

For some clay minerals, substitution within octahedral (e.g., Mg^{2+} or Fe^{2+} for Al^{3+}) and tetrahedral (e.g., Al^{3+} for Si^{4+}) sites leads to a net negative charge in the clay that is balanced by interlayer cations. These cations can undergo ion exchange reactions in response to changing environmental conditions. Similarly, surface charge can develop on clays by several mechanisms, allowing for additional ion adsorption on the clay surfaces. In one case, broken bonds at the edges of clay surfaces leads to this charge. In another case, reactions can occur with exposed hydroxyl groups of Al-octahedra at clay surfaces. Finally, similar reactions can occur with surface Si-tetrahedra, after their hydration to form silanol (-Si-$(OH)_2$) groups. If we define these surface groups as $\equiv$MOH (where M is either Al or Si, and for simplicity we have shown only one of the hydroxyl groups for each case) then these surface hydroxyl groups can undergo the following acid-base reactions:

$$\equiv MOH + H^+ \rightleftharpoons \equiv MOH_2^+ \tag{2.3}$$

$$\equiv MOH \rightleftharpoons \equiv MO^- + H^+ \tag{2.4}$$

This amphoteric behavior implies that the surface charge of these minerals will be a function of the solution pH (see Stumm and Morgan, 1996, or Li, 2000, for a further discussion of this surface complexation model). We can then define the pH of zero point of charge (pH_{zpc}) as the pH at which the concentration of positively and negatively charged surface complexes are equal, and the surface therefore has no net charge. In cases where the environmental pH is less than pH_{zpc} the surface will be positively charged. Similarly when the natural

TABLE 2.1
Values of the pH of Zero Point of Charge for Selected Materials
(Data from Stumm and Morgan, 1996)

Material	pH_{zpc}	Net Surface Charge at pH 7.5–8.2
SiO_2	2.0	–
Montmorillonite	2.5	–
δ-MnO_2	2.8	–
Kaolinite	4.6	–
α-$Al(OH)_3$	5.0	–
Fe_3O_4	6.5	–
γ-Fe_2O_3	6.7	–
β-MnO_2	7.2	–
Calcite*	8.2 (8–9.5)	+
$Fe(OH)_3$ (amorph.)	8.5	+
α-Al_2O_3	9.1	+

*As Stumm and Morgan (1996) discuss, the pH_{zpc} for calcite is a function of the pCO_2 of the solution. The first value was calculated for a solution in equilibrium with air ($pCO_2 = 10^{-3.5}$ atm). At $pCO_2 = 1$ atm the calculated pH_{zpc} drops to ~6.5. The range in parentheses was determined experimentally with calcite in equilibrium with air (Somasundaran and Agar, 1967).

pH is greater than pH_{zpc} the surface will be negatively charged. Values of pH_{zpc} for clays and other selected solids are shown in table 2.1. Given typical pH values in marine sediments (~7.5–8), this then implies that most common solids in marine sediments (clays, metal oxides, silica) will have a net negative surface charge and will therefore preferentially adsorb dissolved cations from solution.

In clay minerals, adsorption and ion exchange reactions are generally discussed in terms of the clay's cation exchange capacity (CEC). Values of CEC vary with the type of clay (see table 2.2). Clays that have a large degree of substitution in the mineral structure will have large CEC values, as a result of the large number of interlayer cations needed for charge balance. Conversely, a clay such as kaolinite that has no interlayer cations will have a low CEC value.

Ion exchange or adsorption reactions can be expressed with the following simple equation:

$$n A^{m+}_{(aq)} + m B^{n+}_{(s)} \rightleftharpoons m B^{n+}_{(aq)} + n A^{m+}_{(s)} \tag{2.4}$$

COMPONENTS OF MARINE SEDIMENTS

TABLE 2.2
Cation Exchange Capacities (CEC) of Selected Clay Minerals (Modified after Faure, 1998)

Clay Mineral	CEC (meq/100 g, at pH 7.0)
Kaolinite	3–15
Chlorite	10–40
Illite	10–40
Smectite	70–100
Vermiculite	100–150

where (aq) denotes ions in solution and (s) denotes ions adsorbed or in cation exchange sites. The equilibrium constant for this reaction can then be expressed as

$$K = \frac{a_{B,aq}^m \cdot a_{A,s}^n}{a_{B,s}^m \cdot a_{A,aq}^n} \tag{2.5}$$

where a is the activity of each ion in solution or on the solid. This equation can be rewritten as

$$\left(\frac{\bar{C}_B}{C_B}\right)^m = K^* \left(\frac{\bar{C}_A}{C_A}\right)^n \tag{2.6}$$

where C represents dissolved concentrations, $\bar{C}$ represents concentrations of the adsorbed/ion exchanged component and K^* incorporates the thermodynamic K value above and the activity coefficients of A and B in both phases.

An examination of eqn. (2.6) suggests three possible cases. The first is a situation in which the concentration of one ion is much greater than the other, regardless of the charge of the ions. This could be, for example, the case of dissolved Mn^{2+} (ion B) exchanging with dissolved Na^+ or Ca^{2+} (ion A) in an anoxic marine pore water. In this case, $C_A \gg C_B$ and therefore $\bar{C}_A \gg \bar{C}_B$. As a result there is then little change in the concentration of A (in either phase) when A and B exchange for one another and the ratio $(\bar{C}_A/C_A)^n$ is roughly constant. Eqn. (2.6) can be re-written as

$$\bar{C}_B = K' C_B \tag{2.7}$$

where K' is a function of K^*, $\overline{C}_A$ and C_A, m and n. This equation is identical to the simple linear adsorption isotherm that can be derived from other adsorption models. This linear adsorption isotherm is also commonly used to incorporate adsorption in sediment diagenesis models, as is discussed in section 6.1.3.

The second case is one in which the concentrations of A and B are of the same order of magnitude and the charges of A and B are the same (i.e., $m = n$). Here, eqn. (2.6) reduces to

$$\left(\frac{\overline{C}_B}{C_B}\right) = K^*\left(\frac{\overline{C}_A}{C_A}\right) \quad (2.8)$$

and the ratio of ions adsorbed to the solids is directly proportional to the concentration ratio in solution, regardless of the absolute solution concentrations.

The third case is one in which the concentrations of A and B are of the same order of magnitude and their charges are different. Looking at this for the simple case of Na$^+$ and Ca^{2+} exchange, eqn. (2.6) can be written as

$$\left(\frac{\overline{C}_{Na}}{C_{Na}}\right)^2 = K^*\left(\frac{\overline{C}_{Ca}}{C_{Ca}}\right) \quad (2.9)$$

When such a clay suspension is diluted, and assuming that K^* remains constant with this dilution, each of the $\overline{C}/C$ ratios increases, although because the Na ratio increases as its square, the Ca ratio must increase more rapidly to maintain the equality of eqn. (2.9). This then implies that $\overline{C}_{Ca}$ must increase more than $\overline{C}_{Na}$ and demonstrates the common observation that dilution leads to the selective uptake of higher valence cations relative to lower valence cations (e.g., see discussions in Sayles and Mangelsdorf, 1977). Similarly, when clays come in contact with a more concentrated solution, they selectively take up lower valence cations as compared to higher valence cations.

In the marine environment, this latter effect is most important when clays are transported from freshwaters (rivers) to the ocean. Using both standard clays and Amazon River suspended sediments, Sayles and Manglesdorf (1977, 1979) have shown that the primary ion exchange reactions that occur when these materials are introduced into seawater is the uptake of Na$^+$ and release of Ca^{2+}. Furthermore, mass balance calculations carried out by these authors suggest this Na$^+$ uptake represents up to ~20% of the sodium river flux.

2.5 THE CLASSIFICATION OF MARINE SEDIMENTS AND SEDIMENTARY REGIMES

The classification of marine sediments has been undertaken in several different ways. All of these approaches have their advantages and disadvantages in that they may emphasize one aspect of the characteristic of a sediment at the expense (or omission) of other characteristics. Some of the more "geochemical" classifications are based on the occurrence of different organic matter remineralization processes in sediments (Berner, 1981a; section 7.2), or are based on the functional diagenetic characteristics of a sediment (Aller, 1998; section 5.5).

From a sedimentological standpoint, the classification scheme of Dean et al. (1985) is useful in describing the composition of deep-sea sediments. However, this classification scheme is less informative in terms of understanding most early diagenetic processes in these sediments. This approach starts with the basic assumption that there are three end-member sediment components that make up the vast majority of all material found in deep-sea sediments: a biogenic calcareous component; a biogenic siliceous component; and a nonbiogenic component. The predominant sediment component (> 50%) then determines its overall designation. If this component is nonbiogenic, it is classified as *sand, silt,* or *clay* on the basis of grain-size classifications (see section 4.1). Biogenic sediments are referred to as oozes and are defined as either *nannofossil* or *foraminifera* oozes (calcareous) or *diatom* or *radiolarian* oozes (siliceous). Additional modifiers can also be used to further describe other minor sediment components (Dean et al., 1985). However, in spite of the development of this classification scheme, many geochemical papers still refer to deep-sea sediments simply as calcareous oozes, siliceous oozes, or red clays.

In principle, the Dean et al. (1985) approach could also be used to characterize nearshore or estuarine sediments. However, as noted above, in these sediments detrital terrigenous materials generally overwhelm biogenic components, given the close proximity of these sediments to terrigenous sources. Thus these sediments are often simply classified according to Folk's scheme discussed in section 4.1.

CLASSIFICATION OF MARINE SEDIMENTS

Finally, marine sediments can be characterized in terms of different sedimentary regimes (see general discussions in Kennett, 1982; Open University, 1989; Seibold and Berger, 1996; Chester, 2000). For the basis of this discussion, the sea floor is initially divided into three major topographical regions: the ocean basins, the continental margin, and mid-ocean ridges.

The continental margin represents the transition between oceanic crust and continental crust. It is divided further into the continental shelf, continental slope, and continental rise. The boundaries between these regions are generally defined by the sea-floor slope, and to a lesser extent by water depth or distance from the continents. However, the latter description of continental margins is strongly controlled by whether the margin is an active or passive margin. Thus passive continental margins (e.g., the Atlantic) are generally wider than active continental margins (e.g., the Pacific). Although the continental margins represents ~20% or so of the total sea floor, ~50–60% of the total sediment volume is contained on the margins. Analogous discussions in section 8.3 and chapter 15 also show that the majority of marine sediment organic carbon burial and preservation (as well as remineralization) occurs in continental margin sediments.

The continental shelf is the seaward extension of the continents and its outer limit is defined by the shelf edge, or shelf break, where there is a sharp increase in the sea-floor slope. The water depth at which this break occurs averages ~130 m, but can be as deep as 500 m. Below the continental shelf is the continental slope, with the continental rise found at the base of the continental slope. Again, changes in sea-floor slope define this boundary, which occurs at 1,500–3,500 m water depth (with 2,000 m being a rough average). In many parts of the ocean, the continental slope and rise are cut by steep, V-sided valleys or submarine canyons. These represent important conduits for the transport of sediment from continental shelves to deeper waters by turbidity flows (see section 5.2 for further details).

Below the continental rise (and the continental margin in general) lies the ocean basin floor, or the deep sea. The average water depth at which this transition occurs ranges from 3,000 to 5,000 m. The floor of the ocean basin is generally bisected by the mid-ocean ridges, where new oceanic crust is formed and the sea floor shallows to water depths of ~2,700 m.

In discussing the geochemistry of marine sediments several other terms are often used. One is the description of sediments (or marine environments) as being coastal or nearshore. These terms are less rigorously defined than those discussed above, although nearshore or coastal sediments can be thought of as shelf sediments deposited in close proximity to continental land masses, such that they are strongly influenced by these land masses. Such environments include estuaries, bays, lagoons, and (perhaps) river deltas.

Other terms that are often used are *pelagic* and *hemipelagic*. Pelagic sediments are generally synonymous with deep-sea sediments and, for the most part, contain material that settles out of (or sinks through) the water column. Hemipelagic sediments are found along the lower portions of the continental margin in shallower water depths and often contain higher amounts of materials delivered by bottom (horizontal) processes such as, for example, turbidity flows.

The overall thickness of marine sediments (i.e., from the sediment surface to the oceanic crust/basement basalt) varies by location. Its overall average in the world oceans is ~500 m, although its average is >1 km in the Atlantic and <1 km in the Pacific. At the same time, sediment accumulation rates vary among different sedimentary regimes. Deep-sea sediments generally have low sedimentation rates (less than ~5–10 cm/10^3 yr) and as one moves onshore, sedimentation rates increase to values as high as a few centimeters per year. Between these two regions, a majority (~70%) of the continental shelf represents nonaccumulating, well-sorted relict sands that were deposited during the low stand of sea level during the last ice age (Emery, 1968).

≈ CHAPTER THREE ≈

Isotope Geochemistry

IN RECENT YEARS isotope geochemistry has become an integral part of marine sediment geochemistry. This chapter introduces the basic concepts of isotope geochemistry and briefly discusses their application to the study of marine sediments. This discussion is not meant to be comprehensive (for this see, e.g., Faure, 1986; Hoefs, 1997), but will set the stage for later discussions in this book.

3.1 INTRODUCTION

Isotopes are atoms of the same element with the same atomic number but different atomic masses. They have the same number of protons in the nucleus (which defines their atomic number and chemical behavior) but have a different number of neutrons (hence the differences in their atomic mass). We can divide different isotopes into the categories of "stable" isotopes and "unstable" isotopes, based on whether or not they undergo radioactive decay.

The chemical characteristics of the different isotopes of an element are very similar but not identical. This occurs because the slight mass differences of the isotopes affect the way they respond in either chemical reactions or physical processes that are ultimately mass dependent at the molecular or atomic level. These differences lead to isotope "fractionation" during such processes, and the study of stable isotope biogeochemistry attempts to use this fractionation to extract information about biogeochemical processes in nature.

In contrast, radioactive isotopes are isotopes whose nuclei are unstable and undergo radioactive decay over time. This process involves the loss of nuclear particles that change the number of protons and/or neutrons in the nucleus, such that the nucleus reaches a stable configuration. Radioactive decay of an element has a characteristic decay constant or half-life ($t_{1/2}$, the amount of time it takes for half of the available material to decay away). As such, these isotopes can be used to "tell time" and quantify the rates of geologic processes.

TABLE 3.1
The Abundances and Masses of the Stable Isotopes of Elements Discussed in this Chapter (Modified after Faure, 1998)

Element	Stable Isotopes	Mass (amu)	Average Abundance (%)
Hydrogen	$^{1}_{1}H$	1.0078	99.985
	$^{2}_{1}H$ (D)	2.0140	0.015
Carbon	$^{12}_{6}C$	12.0000	98.90
	$^{13}_{6}C$	13.0034	1.10
Nitrogen	$^{14}_{7}N$	14.0031	99.63
	$^{15}_{7}N$	15.0001	0.37
Oxygen	$^{16}_{8}N$	15.9949	99.76
	$^{17}_{8}O$	16.9991	0.04
	$^{18}_{8}O$	17.9992	0.20
Sulfur	$^{32}_{16}S$	31.9721	95.02
	$^{33}_{16}S$	32.9715	0.75
	$^{34}_{16}S$	33.9679	4.21
	$^{36}_{16}S$	35.9671	0.02

The abundance and masses of the stable isotopes of the elements that will be discussed in this chapter are listed in table 3.1. For oxygen and sulfur (which have multiple stable isotopes), one generally examines the $^{18}O/^{16}O$ and $^{34}S/^{32}S$ ratios. Also, while C, H, N, and S all have radioactive isotopes, geochemical studies have focused on tritium (^{3}T), the radioactive isotope of hydrogen ($t_{1/2} = 12.5$ yr) and ^{14}C ($t_{1/2} = 5,730$ yr). This occurs because the radioactive isotopes of these other elements have half-lives that are generally too short for their use in geochemical studies at natural abundance levels.

3.2 Principles of Isotope Fractionation

Isotope fractionation is defined as the partitioning of isotopes between two substances or phases that causes these materials to have different isotopic compositions. In general, three phenomena can cause such fractionation in nature:

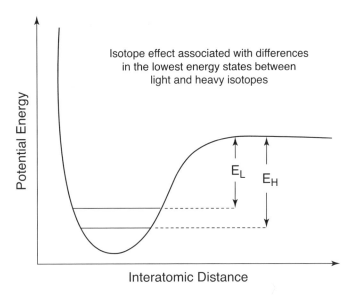

Figure 3.1 A schematic potential energy diagram illustrating the energy difference for a reaction in which a bond is broken between one atom and two different stable isotopes of a second atom (L and H = light and heavy). As one moves to the right along this potential energy curve, the reaction is complete (i.e., the bond between the two atoms is broken). Because E_L is greater than E_H (discussed in the text), it takes less energy to break the bond involving the lighter isotope. Modified after Bigeleisen (1965).

- equilibrium isotope exchange reactions (that simply involve the redistribution of isotopes of an element among different molecules that contain the element);
- irreversible kinetic processes (in which rates of reaction depend on the isotopic composition of the reactants and products);
- physical-chemical processes such as diffusion or evaporation/precipitation (in which the interplay between mass differences and temperature or concentration gradients leads to fractionation).

The fractionation of stable isotopes during a chemical reaction can be explained in part by differences in the molecular vibrational energy levels of bonds formed between one atom and different isotopes of a second atom (see fig. 3.1). Based on quantum mechanics the energy level of these bonds differs because the vibrational frequency of the bond is inversely proportional to the masses of the different atoms. In general, the lowest vibrational energy state is $1/2\, h\nu$ above the minimum value on the potential energy/interatomic distance curve, where h is Plank's constant and v is the vibrational frequency of the bond. Since v is inversely proportional to mass, the bond with the heavy isotope of the second atom will have a slightly lower zero-point energy than the one with the light isotope. Therefore it will take more energy to break the bond with the heavy isotope, and the bond with the light isotope will be more readily broken (again see fig. 3.1). As an example of this, during bacterial sulfate reduction (rxn. 2.2), $^{32}SO_4^{-2}$ is preferentially reduced relative to $^{34}SO_4^{-2}$, and pore-water sulfate becomes progressively heavier with depth as a result of this fractionation (also see sections 3.7 and 16.1). Similarly, during photosynthesis, the organic matter that is produced is enriched in ^{12}C relative to the starting bicarbonate or CO_2 (also see section 3.4.1).

During physical processes the fractionation of isotopes is also related to these slight mass differences, although in a slightly different way. Using an ideal gas as an example, we know based on statistical mechanics that at a given temperature all molecules of the gas have the same kinetic energy ($= 1/2\ mv^2$) where m is the mass of the molecule and v is its velocity (Faure, 1998). Thus for a molecule in this gas mixture containing a light (L) and a heavy (H) isotope,

$$v_L/v_H = \sqrt{m_H/m_L} \qquad (3.1)$$

implying that the velocity of the molecule with the light isotope will be faster than the one with the heavy isotope. Such considerations therefore lead to a concentration of the heavy isotope in the liquid phase during evaporation, because the light isotope evaporates more easily. Similarly, a light isotope has a slightly larger diffusion coefficient than a heavy isotope.

3.2.1 Terminology

Isotopic compositions are generally expressed as the isotope ratio in a sample (R_{smp}) relative to that in a common standard (R_{std}),

$$\delta_x = \frac{R_{smp} - R_{std}}{R_{std}} \cdot 1000 = \left(\frac{R_{smp}}{R_{std}} - 1\right) \cdot 1000 \qquad (3.2)$$

where R is the ratio of the heavy isotope to the light isotope, and δ_x is expressed as parts per thousand, or per mil (‰). Because isotope effects lead to relatively small changes in isotopic composition, they are expressed as this per mil difference compared to a known standard. This approach leads to more accurate isotopic determinations and also helps to reduce errors due to instrumental variability during isotope measurements. If δ_x is greater than zero, then the sample is enriched in the heavy isotope (relative to the standard), and conversely if δ_x is less than zero then the sample is depleted in the heavy isotope. A common set of internationally recognized standards is used in these measurements (Hoefs, 1997; Bickert, 2000) to minimize uncertainty (and confusion) in the comparisons of isotope measurements made in different labs.

3.2.2 Equilibrium Isotope Exchange Reactions

In these processes there is no net change in the chemical system (i.e., no chemical reactions occur), and the isotopes simply redistribute themselves until equilibrium is reached, e.g.,

$$H_2{}^{18}O + 1/3 CaC^{16}O_3 \rightleftharpoons H_2{}^{16}O + 1/3 CaC^{18}O_3 \tag{3.3}$$

The equilibrium expression for such processes can be described with either K or α, where K is a true equilibrium constant, and α, the isotope fractionation factor, is the ratio of the isotopic compositions of the two substances,

$$\alpha_{A-B} = R_A/R_B \tag{3.4}$$

For rxn. (3.3) this would then be written as

$$\alpha_{CaCO_3 - H_2O} = (^{18}O/^{16}O)_{CaCO_3} / (^{18}O/^{16}O)_{H_2O} \tag{3.5}$$

Both K and α depend on temperature, with less fractionation generally seen at higher temperatures. Furthermore, K and α are related by,

$$\alpha = K^{1/n} \tag{3.6}$$

where n is the number of atoms being exchanged. Since fractionation factors are generally close to 1, they too can be expressed on a per mil basis as,

$$\varepsilon_{A-B} = (\alpha_{A-B} - 1) \cdot 1000 \tag{3.7}$$

and for a given reaction isotope fractionation can also be expressed by Δ_{A-B} where

$$\Delta_{A-B} = \delta_A - \delta_B \qquad (3.8)$$

For α values less than ~1.010, Δ and α are related by

$$\Delta_{A-B} \approx 10^3 \ln \alpha \qquad (3.9)$$

For ideal gases, ln α can be shown both theoretically and experimentally to be a function of temperature, according to

$$10^3 \ln \alpha = A + \frac{B}{T} + \frac{C}{T^2} \qquad (3.10)$$

where A, B and C are constants and T is temperature (°K). Relationships similar to eqn. (3.10) also hold for many mineral-water pairs and in part form the basis for using stable isotopes in paleothermometry studies (see discussions in sections 3.4.2 and 3.4.3 and in Savin and Yeh, 1981; Faure, 1998).

3.3 Isotope Fractionation in Inorganic Materials in Nature

3.3.1 Isotope Fractionation in the Hydrosphere and in Ice Cores

During the global hydrologic cycle the ^{18}O and D (= ^2H) in water are fractionated as the water moves through various reservoirs (e.g., seawater, atmospheric water vapor). For reasons discussed above, evaporation of water from the ocean causes the water vapor that forms to be depleted in ^{18}O and D as compared to seawater. Similarly, precipitation (i.e., condensation of atmospheric water vapor as its temperature decreases) forms a condensate (precipitation) that is enriched in ^{18}O and D relative to the water vapor. This occurs primarily because evaporation and condensation can be considered reversible equilibrium processes.

When using these observations to examine the isotopic composition of meteoric waters (i.e., precipitation that has undergone this meteorologic cycle) one observes that water vapor that forms over the equatorial regions of the oceans and moves northward is initially

ISOTOPE FRACTIONATION IN INORGANIC MATERIALS IN NATURE

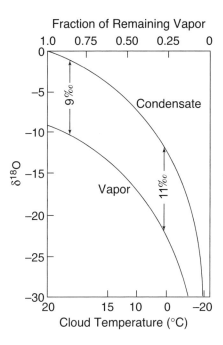

Figure 3.2 The $\delta^{18}O$ of precipitation (condensate) and the remaining water vapor in clouds as a function of the fraction of remaining water vapor, assuming that the process can be defined as a Rayleigh distillation process (modified after Dansgaard, 1964). A temperature axis is also shown on the lower x-axis, based on the assumption that this air mass forms over equatorial regions and moves northward (see discussions in Broecker, 1974, for further details). This decrease in temperature also leads to an increase in the isotope fractionation during condensation, as is shown here.

depleted in ^{18}O (and D) as a result of isotope fractionation during evaporation. As this water vapor begins to lose water by precipitation, the initial heavy condensate that forms has an isotopic composition almost exactly the same as the original seawater (fig. 3.2). However, because this condensation does not completely and instantaneously reverse evaporation, the remaining vapor becomes isotopically lighter. Therefore, rain that subsequently forms also becomes increasingly depleted in ^{18}O and D. At a first-order level, changes in the isotope composition of the water vapor in an air mass and in its condensate (both as a function of the fraction of water vapor remaining) can be viewed as a Rayleigh distillation process (Faure, 1998). In this model, the isotopic ratio of the remaining water vapor after condensation (R_v) is related to its original ratio (R_v^o) and the fraction of water vapor remaining according to,

$$R_v = R_v^o \, f^{\alpha_{l,v}-1} \tag{3.11}$$

where $\alpha_{l,v}$ (as defined by eqn. 3.4) is the isotopic ratio in the liquid (condensate) divided by that in the remaining vapor. Although other processes are also important (e.g., re-evaporation of meteoric water from the Earth's surface) the results in fig. 3.2 provide a reasonable explanation of the observed trends in meteoric water (see Broecker, 1974; Faure, 1998, for further details).

As a result of these processes meteoric waters have distinctly lower $\delta^{18}O$ and δD values than seawater. The isotopic composition of rainwater also shows spatial trends, becoming increasingly light with increasing latitude (and/or decreasing mean annual air temperature) for reasons discussed above.

These observations have implications in several important areas of geochemistry. The first involves the use of the isotopic composition of marine clays to determine their origin (see below). The second is that the $\delta^{18}O$ of seawater shows systematic variations with salinity that allows it to be used as a water mass tracer (see discussions in Broecker, 1974; Bickert, 2000). Finally, the temperature dependence of the fractionation of meteoric water during its formation allows the $\delta^{18}O$ (and δD) of water in ice to be used as a paleothermometer in Pleistocene ice core studies (see discussions in Alley, 2000).

3.3.2 Isotope Fractionation during Clay Mineral Formation

Studies of oxygen and hydrogen isotopes in clay minerals can be useful in distinguishing between detrital and authigenic clays found in marine sediments. The ability to carry out such an analysis is based on two considerations. The first is that most minerals forming at low temperatures generally do so in equilibrium with their surrounding water, and undergo subsequent isotopic exchange very slowly in a new environment (Savin and Yeh, 1981). Since meteoric waters are depleted in ^{18}O and D as compared to seawater, detrital and authigenic clays should show similar isotopic differences (e.g., detrital clays that form in equilibrium with meteoric waters should have less positive $\delta^{18}O$ values than authigenic clays that form in equilibrium with seawater). Secondly, clay mineral formation enriches clays in ^{18}O relative to the water of formation ($\Delta_{\text{clay-water}}$ or $10^3 \ln \alpha$ are both positive), and greater fractionation occurs with decreasing temperatures (i.e., $\Delta_{\text{clay-water}}$ in the oceans will be larger than $\Delta_{\text{clay-water}}$ on land). These observations further imply that clay minerals that form on land and are subsequently transported to the oceans (i.e., detrital clays) should be more depleted in ^{18}O (i.e., should have less positive $\delta^{18}O$ values) than authigenic clays that form *in situ* in the deep sea.

Consistent with these arguments, $\delta^{18}O$ values of deep-sea clays generally range from 11 to 20‰ and are similar to values observed for clays in soils and in Mississippi River sediments (see discussions in Savin and Yeh, 1981). In contrast, estimates of the $\delta^{18}O$ values of clays in equilibrium with bottom seawater (at 1°C) range from 23 to 31‰ (Savin and Yeh, 1981). Authigenic smectites that do form near mid-ocean ridges by the alteration of submarine volcanic material generally have $\delta^{18}O$ values that range from 26 to 34.5‰ (Savin and Yeh, 1981). However, in spite of these observations of authigenic smectites in deep-sea sediments, it still appears that most clay minerals in deep-sea sediments are of detrital origin, rather than formed by reverse weathering–type reactions. The significance of this observation in the context of recent studies re-examining the occurrence of reverse weathering in marine sediments will be discussed further in chapter 18.

3.3.3 Oxygen and Carbon Isotopes in Calcite

The earliest work in isotope geochemistry began with studies of oxygen isotopes in sedimentary carbonates (e.g., Epstein et al., 1951), attempting to use the $\delta^{18}O$ of the carbonate as a paleothermometer to estimate the temperature at which the carbonate was formed. Subsequent work by Emiliani (1955) used this technique to analyze the $\delta^{18}O$ of planktonic foraminifera in deep-sea sediment cores and to demonstrate the major features of glacial-interglacial climate change during the Quaternary. Later studies applied spectral analysis to similar data from two southern ocean cores and obtained convincing evidence for Milankovitch astronomical forcing of glacial cycles based on changes in the Earth's orbital parameters (see Imbrie and Imbrie, 1979, for an excellent overview of the history of this work).

This approach starts with the basic assumption that calcium carbonate precipitated by organisms is in isotopic equilibrium with the seawater in which the organisms grow. If one then knows (or can estimate) the isotopic composition of this water, then temperature relationships such as eqn. (3.10) can be used to estimate the temperature of formation of the carbonate material.

However, there are several potential problems with this approach. The first is the fact that only a few species of calcareous organisms precipitate their calcite or aragonite shell in isotopic equilibrium with seawater. Therefore, understanding these vital effects (i.e., isotope disequelibria associated with either the metabolism of the organism or other kinetic factors) has also been an important aspect of research in this field (see Hoefs, 1997, for further details).

At the same time, because the $\delta^{18}O$ of seawater can change over geologic time it has also been recognized for quite some time that the $\delta^{18}O$ of marine carbonates records both the temperature and the isotopic composition of water in which it forms (e.g., see Schrag et al., 2002). Since changes in the isotopic composition of seawater result from changes in global ice volume (because an increase in ice volume enriches seawater in ^{18}O as a result of fractionation during evaporation), the $\delta^{18}O$ of marine carbonates potentially records both temperature changes as well as changes in global ice volume (see discussions in Libes, 1992; Shackleton, 2000; Zachos et al., 2001). A variety of strategies have been devised to separate these two components of the $\delta^{18}O$ of marine carbonates from the overall signal, although all have certain limitations (Shackleton, 2000; Zachos et al., 2001; Schrag et al., 2002).

The carbon isotope records of calcareous shells in marine sediments can be similarly difficult to interpret. Work to date suggests that the $\delta^{13}C$ of foraminifera shells in sediments is a rough approximation of the bicarbonate concentration of the waters in which the organisms grew, suggesting that foraminifera isotope records should reflect changes in the oceanic and global carbon budgets (Savin and Yeh, 1981). Thus, studies examining the $\delta^{13}C$ of benthic foraminifera can be used to examine paleoceanographic changes in bottom water circulation (e.g., Curry et al., 1988) as well as changes in the distribution of carbon between the ocean and the terrestrial biosphere (e.g, Shackleton, 1977). Similarly, studies examining the difference between the $\delta^{13}C$ values of benthic and planktonic foraminifera appear to provide information on both the strength of the oceanic biological pump along with past values of atmospheric CO_2 (Shackleton et al., 1983, 1992; Mix, 1989).

3.4 Carbon Isotopes in Organic Matter

Fractionation of carbon isotopes in organic matter occurs as a result of organic carbon production (e.g., photosynthesis or biomass production by heterotrophs) and heterotrophic respiration. In general, fractionation is driven by kinetic isotope effects that lead to enrichments of ^{12}C in reaction end-products. Thus in the simplest sense photosynthesis produces organic matter that is enriched in ^{12}C; i.e., it has a negative $\delta^{13}C$, while respiration produces isotopically light CO_2.

3.4.1 Photosynthesis

The isotopic composition of new organic carbon produced by photosynthesis or by heterotrophic carbon assimilation depends on the isotopic composition of the carbon source, isotope effects associated with its assimilation and uptake, and isotope effects associated with cellular biosynthesis. These processes are discussed in detail elsewhere (Fogel and Cifuentes, 1993; Hayes, 1993; Hoefs, 1997; Popp et al., 1997), and will be only briefly summarized here.

Most plants utilize the enzyme RuBP carboxylase and the Calvin cycle to incorporate CO_2 into new biosynthetic products. These plants are referred as C_3 plants because of the production of 3-carbon intermediates during this process. The vast majority of terrestrial and marine plants (including algae and submerged seagrasses) use this photosynthetic pathway. Plants referred to as C_4 plants use an alternate enzyme (PEP carboxylase) in the first step of CO_2 fixation, and because this enzyme discriminates less strongly against ^{13}C (−2.2‰ versus −27‰; Fogel and Cifuentes, 1993), C_4 plants tend have less negative $\delta^{13}C$ values (see table 7.1). C_4 plants include some temperate and tropical grasses as well as saltmarsh grasses such as *Spartina alterniflora*. A final group of plants (known as CAM plants) use both C_3 and C_4 bioysnthetic pathways. These plants include succulents and other semiarid desert plants. Although most marine and terrestrial plants are C_3 plants, they generally have different carbon isotopic concentrations, as discussed in section 11.1.1. The ability to use these differences to elucidate sources of organic matter in marine sediments is also discussed in this section and in section 11.1.3.

In addition to examining the isotopic composition of bulk organic matter, the ability to separate organic compounds chromatographically and to examine the isotopic composition of the individual components of complex organic mixtures has become more common (e.g., Hayes et al., 1990). This approach provides important information on organic matter production, preservation, and diagenetic alteration that cannot be obtained from bulk organic matter isotopic analyses (e.g., Macko et al., 1993; also see discussions in section 11.1.1 for further details). Similar techniques can also be used to examine the ^{14}C content of these separated organic molecules, as is also discussed in section 3.8.2.

3.4.2 Respiration (Early Diagenesis in Sediments)

In general, there appears to be little carbon isotope fractionation during organic matter remineralization in sediments (Macko et al., 1993; Hedges et al., 1997; Popp et al., 1997; Bickert, 2000), although data from some sediments do suggest the possibility of fractionation occurring during organic matter diagenesis (Alperin et al., 1992b; Martens et al., 1992; Aller and Blair, 2004). This apparent lack of significant fractionation during diagenesis would seem to imply that all organic matter in sediments is homogeneous; i.e., that there are not different fractions of organic matter with different degrees of reactivity as well as different isotopic compositions (see, for example, discussions in Prahl et al., 1997, of this problem in terms of the degradation of marine and terrestrial organic matter in Madeira Abyssal Plain[1] turbidite sediments). Such considerations will also be re-examined in greater detail in later sections (sections 11.2.3 and 14.4.1).

In principle, the lack of a significant carbon isotope shift during diagenesis aids in the use of carbon isotopes in the determination of the provenance (sources) of sedimentary organic matter (see discussions in section 11.1.3). However, as is also discussed in this section, care must be taken in this interpretation.

3.5 Oxygen and Carbon Isotopes in Sediment Pore Waters

3.5.1 Carbon Isotopes

In marine sediments stable carbon isotopes in the pore water DIC (dissolved inorganic carbon) pool are affected by both organic matter oxidation and calcite dissolution. As discussed above, most organic matter in marine sediments is significantly depleted in ^{13}C as compared to both sedimentary carbonates (which have δ^{13}C values that are generally around 0‰ (±~2‰; Morse and Mackenzie, 1990) and seawater DIC (which has a δ^{13}C value of ~0.5–2‰; Libes, 1992). Therefore the analysis of the δ^{13}C of DIC in sediment pore waters is a powerful tool for establishing the relative contributions of calcite dissolution and organic matter remineralization to the pore-water DIC

---

[1] See the appendix for a brief description of these (and other) sediment sites discussed in the text.

pool (McCorkle et al., 1985; McNichol et al., 1991; Bauer et al., 1995; Martin et al., 2000). At the same time, because organic matter oxidation adds isotopically light carbon to the DIC in sediment pore waters, the $\delta^{13}C$ of the shells of benthic foraminifera may or may not accurately reflect bottom water (versus pore water) conditions, depending on whether the forams live on or in the sediments (see discussions in McCorkle et al., 1985; Mix, 1989).

In organic-rich marine sediments, pore-water carbon isotope concentrations are also affected by the processes of methanogenesis (methane production) and methylotrophy (methane consumption/oxidation). Methanogenesis via CO_2 reduction (see section 7.3.5) produces methane that is isotopically light (biogenic methane has $\delta^{13}C$ values that are $<\sim-50‰$) and DIC that is consequently enriched in ^{13}C (Alperin et al., 1992a; Boehme et al., 1996). Similarly methane oxidation leads to increasingly heavy methane as total methane concentrations decrease, and the production of isotopically light DIC from light biogenic methane (Alperin and Reeburgh, 1984).

3.5.2 Oxygen Isotopes

In deep-sea sediments, gradients in $\delta^{18}O$ in the pore waters (i.e., in the H_2O itself) over hundreds of meters (i.e., tens of millions of years) have been observed and interpreted as occurring as a result of alteration reactions involving basement basalt or volcanic material dispersed in the sediments (Gieskes and Lawrence, 1981). On much shorter time (and sediment depth) scales analyses of $\delta^{18}O$ and δD in sediment pore waters have been used to estimate the isotopic composition of seawater during the last glacial maximum (see most recently Schrag et al., 2002). Such information is important in helping better constrain oxygen isotope paleotemperatures from marine carbonates (see section 3.3.3 for further details).

3.6 NITROGEN ISOTOPES

Nitrogen has two stable isotopes (table 3.1). In the oceans the processes of nitrification (see section 7.5.1), denitrification (see section 7.3.2), and nitrogen fixation (the production of biologically available nitrogen [e.g., ammonium or amino acids] from N_2 gas) play major roles in controlling nitrogen isotope distributions. These processes all show kinetic isotope effects, although the α value for nitrogen fixation is small relative to that for either denitrification or nitrification (1.000–1.004 versus 1.02–1.04; see discussions in Montoya, 1994). However, it is important to note that this relatively large fractionation of nitrogen (nitrate) during denitrification appears to occur only in lab studies and in the open ocean (water column settings), but not in marine sediments (Brandes and Devol, 2002). This point will be discussed in further detail in section 15.2.

As was discussed above for carbon, there appears to be little fractionation of nitrogen isotopes during organic matter remineralization in sediments (see Macko et al., 1993; Altabet and Francois, 1994; Bickert, 2000). Nevertheless, because $\delta^{15}N$ values in marine versus terrestrial organic matter can show wide variations (i.e., the "end-member" values are not well constrained), care must be taken in using this approach to examine the provenance of sedimentary organic matter (see section 11.1.1 for further details).

3.7 Sulfur Isotopes

Sulfur has four stable isotopes (table 3.1) and in sulfur isotope studies one generally looks at the $^{34}S/^{32}S$ ratio (expressed as $\delta^{34}S$). The fractionation of sulfur isotopes occurs mainly from kinetic isotope effects associated with bacterial sulfate reduction and pyrite weathering (oxidation). Both processes discriminate against heavy sulfur, such that sulfate reduction produces isotopically light sulfide and leaves behind heavy sulfate. Similarly, pyrite weathering produces light sulfate and leaves behind heavy pyrite. These processes will be discussed in further detail in section 17.1. Sulfur isotopes in sediment organosulfur compounds have also been examined to further understand the source(s) of this material (see discussions in Goldhaber, 2003, for further details).

3.8 Radioactive Isotopes

3.8.1 Basic Principles

In contrast to stable isotopes, unstable, or radioactive, isotopes undergo spontaneous transformations that release both nuclear particles and radiant energy. This process is generally referred to as radioactivity. Radioactive decay occurs by several different processes, and often a radioactive element decays through a series of decay reactions until a stable atomic configuration is reached. The factors controlling the type of radioactive decay that a particular radionuclide undergoes generally depend on whether there is an "excess" of protons or neutrons in the unstable nucleus.

One type of radioactive decay, α decay, involves the emission of an α particle (or $^4_2$He nucleus) and some amount of energy (usually in the form of γ-rays). Since α-decay involves the loss of protons and neutrons, the daughter product is a different element than the parent; for example,

$$^{238}_{92}U \rightarrow {}^{234}_{90}Th + {}^4_2He + Q \tag{3.12}$$

where Q represents the release of energy. This type of decay is common among relatively heavy radionuclides.

Another type of radioactive decay, β decay, involves the loss a high-energy electron (β^- particle) plus γ-rays. As a result of this electron loss from the nucleus, a neutron is converted to a proton. Again, this type of decay results in a change in the element after radioactive decay, e.g.,

$$^{14}_{6}C \rightarrow {}^{14}_{7}N + \beta^- + Q \tag{3.13}$$

A third type of decay, positron emission, involves the transformation of a proton in the nucleus into a neutron and a high-energy positron (β^+). This positron is subsequently destroyed (and converted to energy) when it collides with an electron. This process decreases the atomic number of a nucleus without changing its atomic mass.

The last type of radioactive decay, electron capture, leads to a decrease in the atomic number of a nucleus, again with no change in its atomic mass. Here, the unstable nucleus captures one of its low-energy electrons (e.g., in the 1s shell), which then reacts with a nuclear proton, forming a neutron. When this low-energy electron is replaced with an electron from a higher energy outer shell, the element also emits high-energy, short-wavelength radiation, or x-rays.

The rate at which radioactive decay occurs is directly proportional to the number of radioactive nuclei present, or

$$\frac{dN}{dt} = -\lambda N \tag{3.14}$$

where N is the number of nuclei of this radioactive element that are present, and λ is the decay constant (units of inverse time). Integration of this equation yields the characteristic radioactive decay equation,

$$N(t) = N_o e^{-\lambda t} \tag{3.15}$$

where N_o is the radioactivity at $t = 0$. Radioactive decay therefore leads to an exponential decrease in the activity of a radioisotope over time. The time scales over which this decay occurs are often discussed in terms of the radionuclide's half-life ($t_{1/2}$), or the time it takes for half of the radioactive element to decay away. If we substitute $1/2N_o$ for $N(t)$ in eqn. (3.15), then

$$t_{1/2} = \frac{\ln 2}{\lambda} = \frac{0.693}{\lambda} \tag{3.16}$$

Thus elements with large decay constants have short half-lives. Although not commonly used, one can also define the mean life (τ) of a radionuclide as $1/\lambda$.

The types of radioisotopes found in nature can be divided into three broad categories. The first are naturally occurring radioisotopes with a primordial source (i.e., they were formed early in the Earth's history, as the planet formed). The most abundant of these are ^{45}K, ^{235}U, ^{238}U, and ^{232}Th, with half-lives that range from 0.71 to 14 billion years. These U and Th isotopes decay through a series of α- and β-decay reactions, ultimately producing stable isotopes of lead (Pb). The half-lives for these reactions range from $\sim 10^{-4}$ sec to 10^5 yrs (e.g., see Faure, 1986; Libes, 1992; or any general text in geochemistry or chemical oceanography).

The second of these groups of radioisotopes are cosmogenic radioisotopes. They are produced by spallation reactions associated with cosmic ray bombardment of the atmosphere (primarily from the sun). Examples of these isotopes include ^{14}C (naturally occurring; see next section), ^{10}Be, ^7Be, and ^{32}Si. Finally, a third group of radioisotopes found in nature are manmade, produced primarily by fission reaction or nuclear weapons. These include ^{90}Sr, ^{137}Cs, ^{239}Pu, as well as bomb radiocarbon (^{14}C) and tritium (^3H).

In the simplest sense, the use of radioisotopes to study geochemical processes depends on the half-life of the isotope versus the time scale of the process of interest. If the half-life is short in comparison to the time scale of the process, then all of the radionuclide will have decayed away before it can be of any use in the study of this process. In contrast, if the half-life is long compared to that of this process, then no change in activity will be detected during the process. Consequently the use of radioisotopes to tell "time" in many geochemical systems requires that these times scales "match" one another. In sediment systems, this approach then allows one to take advantage of the simple radioactive decay of an element predicted by eqn. (3.15) to quantify the rates of sediment accumulation (sedimentation; see section 6.4.5) or the rates of bioturbation (see section 12.4).

For multiple decay reactions (e.g., the U-Th decay series') many of the steps of these series involve large differences between the half-lives (decay constants) for sequential decay reactions. For example, in one portion of the decay series of ^{238}U to ^{206}Pb,

$$\ldots \rightarrow {}^{226}\text{Ra} \xrightarrow{t_{1/2} = 1620 \text{ yr}} {}^{222}\text{Rn} \xrightarrow{t_{1/2} = 3.82 \text{ d}} {}^{218}\text{Po} \ldots \quad (3.17)$$

the radioactive decay of Rn is much more rapid than that of Ra. As a result, the decay of the parent radioisotope becomes the rate-limiting step in this process. Such systems eventually reach a steady state in which the activity of ^{222}Rn (daughter decay) equals that of ^{226}Ra (parent decay or daughter production). This is termed secular equilibrium, and therefore implies the parent:daughter activity ratio = 1.

At the same time, because of differences in the chemistry of the daughter and parent elements involved in such secular equilibria, this parent:daughter ratio may not be equal to 1. Here, this departure from secular equilibrium in conjunction with radioactive decay can further be used to quantify rates of geochemical processes. As we will discuss in section 12.5.2, because Ra is particle reactive while Rn is a soluble inert gas, ^{222}Rn production in sediments, and its departure from secular equilibrium in sediment pore waters, can be used to quantify rates of sediment-water exchange in many sediments (Kipput and Martens, 1982; Berelson et al., 1987a; Martin and Banta, 1992).

3.8.2 Radiocarbon

In addition to its two stable isotopes, carbon has a radioactive isotope (^{14}C or radiocarbon) with a half-life of 5,730 yr. The primary source of

TABLE 3.2
Interconversion of Units Used to Report ^{14}C Concentrations*

F (Fraction Modern)	^{14}C age (yr)	Δ^{14}C
2	>modern	+2,000‰
1.07	>modern	+70‰
0.9	870 yr	−100‰
0.05	24,800 yr	−950‰
0	>45,000 yr	−1,000‰

*See Stuiver and Polach (1977) or Faure (1986) for additional details on these ^{14}C concentration units.

natural radiocarbon is cosmogenic production in the atmosphere. However, nuclear weapons testing in the 1950s and 1960s also led to a large influx of so-called bomb radiocarbon into the atmosphere (e.g., see discussions in Broecker and Peng, 1982; Bauer and Druffel, 1998; Pearson et al., 2001). For this reason radiocarbon measurements can be used to both date natural samples (as discussed above) and to discriminate between organic carbon that was produced either before or after about 1960.

Radiocarbon activity can be expressed in several different ways. One common way to express these values uses the Δ^{14}C terminology, defined as the per mil (‰) deviation of the ^{14}C activity of a sample as compared to that in nineteenth-century wood (Stuiver and Polach, 1977; Faure, 1998). Radiocarbon activities can also be expressed as the fraction (F) of modern (post-1950) carbon. Samples with Δ^{14}C values greater than 0‰ and F > 1 represent recent, "post-bomb" material, while samples with negative Δ^{14}C values and F < 1 are considered "pre-bomb" material that has not incorporated significant amounts of bomb radiocarbon. These values and conventional ^{14}C ages can be interconverted using equations described elsewhere (Stuiver and Polach, 1977; Faure, 1998; also see table 3.2). As can be seen in table 3.2, only samples in which F < 1 (or Δ^{14}C < 0) can be used for conventional radiocarbon dating as discussed above. It is also important to recognize that the conversion of conventional ^{14}C ages discussed here to absolute (calendar) ages requires corrections broadly related to what are referred to as "reservoir effects." These corrections take into account secular variation in the initial ^{14}C content of the atmosphere (and other relevant natural reservoirs) over time (see Libes, 1992; Faure, 1998, for further details).

In recent years ^{14}C analyses using accelerator mass spectrometry (AMS) have become an increasingly common tool in geochemistry, allowing for the analysis of microgram amounts of carbon-containing materials. As we will see in section 11.1.1, this technique can be used to examine the ^{14}C content of chromatographically separated organic molecules, providing important age constraints in studies using these compounds to understand oceanic and sedimentary carbon cycling (e.g., Eglinton et al., 1997; Pearson et al., 2001).

≈ **CHAPTER FOUR** ≈

Physical Properties of Sediments

IN DISCUSSING THE physical properties of sediments, we start with the simple observation that a marine sediment is composed of two basic elements: seawater and sediment particles. These particles are often loosely packed together (at least in surficial marine sediments) and the voids between the particles form the pore spaces (fig. 4.1). In water-saturated marine sediments, this water constitutes the sediment pore waters.[1] While bulk parameters of sediments can be defined by this simple model, additional information about the sediment microstructure is also needed to understand many physical and mechanical properties of sediments as well as transport processes affecting dissolved solutes in sediment pore waters (Bennett et al., 1991; Breitzke, 2000).

4.1 GRAIN SIZE

Grain size is quantified using the ϕ (phi) scale and is based on particle size classes nominally defined as clay, silt, sand, pebble, cobble, and boulder (fig. 4.2). However, some sediments are heterogeneous mixtures of particles, in spite of the fact that physical processes can often separate sediment deposits by grain size into fairly homogeneous mixtures (i.e., a pure sand or silt). Such heterogeneous sediments may be defined more broadly as a gravel, sand, or mud, based on the size distribution of particles in the sediment. Since gravels are rare in most marine sediments, we will focus here on sands and muds.

As fig. 4.2 shows, sands are defined as materials of a size between 62 μm and 2 mm. In addition to subdividing sands into categories that range from very fine to very coarse, one can also divide a sand into sorting classes based on the standard deviation of the frequency distribution of ϕ values in a given sample. Samples with a small standard deviation around a mean particle size are considered

---

[1] Note that the term interstitial water is often used interchangeably with the term pore water.

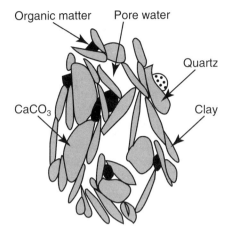

Figure 4.1 A view of a typical marine sediment at high magnification indicating interstitial (or pore) waters and various sediment particles and organic matter (modified after Boudreau, 1997a).

well sorted, while samples with a large standard deviation around this mean are defined as poorly sorted (Friedman and Sanders, 1978).

The term *mud* is generally used to define fine-grained sediments containing mixtures of clay- and silt-sized particles. Folk (1980) presents a general classification for clastic sediments that provides a more precise definition of mud (fig. 4.3).

4.2 Porosity and Sediment Density

Porosity (φ) is defined as the ratio of the volume of water-filled void spaces in a sediment to the total volume of sediment:

$$\varphi = \frac{\text{volume of interconnected water}}{\text{volume of total sediment}} \tag{4.1}$$

As Berner (1980) notes, this term is not the same as total porosity since it does not account for isolated fluid-filled pores. However, isolated pores are generally rare in surface marine sediments, and this definition is essentially the same as that for total sediment porosity. This definition also assumes that all pore spaces are filled with water and is therefore not valid for unsaturated soils.

Strictly speaking φ is a dimensionless quantity, although for many sediment calculations we will discuss here it is important to remember that porosity actually has units of $\text{cm}^3_{\text{pore water}}/\text{cm}^3_{\text{total sediment}}$.

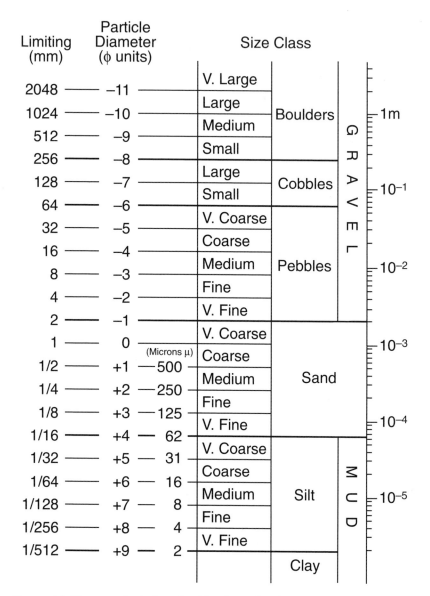

Figure 4.2 The standard size classification of sediments (ϕ scale; redrawn after Friedman and Sanders, 1978).

POROSITY AND SEDIMENT DENSITY

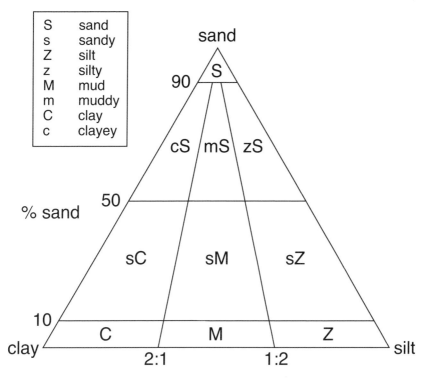

Figure 4.3 The classification of clastic sediments (modified after Folk, 1980).

Similarly, since we define a sediment as simply water and sediment particles we can define the parameter φ_s ($= 1 - \varphi$) as the volume fraction of sediment particles. This parameter then has units of $cm^3_{dry\ sedimet}/cm^3_{total\ sediment}$.

Sediment density is another important property of marine sediments. The bulk (or wet) sediment density is defined in the traditional sense as the mass of total (wet) sediment divided by the total sample volume. Porosity and bulk sediment density are closely related by the following equation,

$$\rho_b = \varphi \rho_w + (1 - \varphi)\rho_s \qquad (4.2)$$

where ρ values are the densities of the bulk sediment (ρ_b), the pore water (ρ_w) and the sediment particles (ρ_s). Bulk sediment densities will vary from values close to 1 (for high porosity, fine-grained sediments), to values >2 for low porosity sediments.

49

Because of the connection between porosity and sediment density, porosity can be determined by weight or volume measurements (Breitzke, 2000). In the former case, if a sediment aliquot is dried to constant weight, then the mass of water (m_w) and mass of dried sediment plus salt (m_d) can be used to estimate the sediment porosity:

$$\varphi = \frac{m_w/\rho_w}{m_w/\rho_w + (m_d - (S \cdot m_w))/\rho_s} \quad (4.3)$$

where the mass of the dried sediment has been corrected for the amount of dried salt in the sample using the pore-water salinity (S) and mass of water (m_w). Values of ρ_s can be either directly determined on dried sediments or estimated from the literature based on knowledge of the sediment composition (table 4.1).

The initial porosity of a fine-grained sediment is largely controlled by grain size and mineralogy (Berner, 1980). Fine-grained sediments have a relatively high specific surface area because of their small particle size and corresponding large surface area-to-volume ratio. As a result, surface chemical electrostatic interactions among the particles are important as these small particles are packed together. For clay minerals, which constitute the bulk of the fine-grained fraction, these interactions lead to sufficiently strong electrostatic repulsion among individual clay platelets that the clays initially aggregate face-to-edge (fig. 4.4A), likely as a result of differences in charge density (and sign) on different portions of the clay particle surfaces (Bennett et al., 1991). This then causes these sediments to have a relatively open structure with a high porosity (fig. 4.5; Meade, 1966; Engelhardt, 1977; Berner, 1980).

In contrast, the initial porosity of more coarse-grained sands is less a function of such surface interactions because of the lower surface area of sand particles. Here, geometric considerations play a much more important role. Based on such considerations, and depending on the nature of particle packing, a "sediment" of uniform spheres, regardless of size, will have a porosity that ranges from 0.46 (in an open packing arrangement) to 0.26 (in a closest packing arrangement; (Friedman and Sanders, 1978). For comparison, well-sorted sands have φ values between 0.35 and 0.46 (fig. 4.5; Engelhardt, 1977). Such natural sands apparently do not attain porosities of closest packing mixtures, primarily because the slightly angular nature of sand grains provides some resistance to grain re-orientation (Berner, 1980). At the

TABLE 4.1
Dry Sediment Densities of Selected Minerals and Marine Sediments

Material	ρ_s (g cm^{-3})
Pure Mineral Phases	
Quartz[a]	2.65
Calcite[a]	2.71
Aragonite[a]	2.95
Felspar[a]	2.54–2.76
Chlorite[a]	2.70
Kaolinite[a]	2.60
Heavy minerals (e.g., magnetite, pyrite, zircon, garnet, tourmaline)[b]	3.07–5.18
Mixed Sediments*	
Terrigenous (continental shelf and slope) sediments	
Fine sand (92% sand, 4% silt, 3.4% clay)[c]	2.71
Sandy mud (32% sand, 41% silt, 27% clay)[c]	2.65
Mud (5% sand, 41% silt, 54% clay)[c]	2.70
Deep-sea red clay (0.1–3.9% sand, 19–59% silt, 37–81% clay)[c]	2.74
Deep-sea diatomaceous ooze (3–8% sand, 37–76% silt, 17–60% clay)[c]	2.46
Deep-sea calcareous ooze (3–27% sand, 40–76% silt, 8–56% clay)[c]	2.66

* Percentages of sand, silt and clay are based on the size fraction definitions in fig. 4.2.
a. From Dietrich and Skinner (1979).
b. From Friedman and Sanders (1978).
c. From Hamilton (1976).

same time, poor sorting of sands also allows smaller grains to sit in the interstices between larger grains, thus potentially lowering the porosity.

With depth, sediment compaction (decreasing porosity) can occur as a result of the growing weight (overburden) of the overlying sediment. Almost all fine-grained sediments undergo significant compaction within a few centimeters of burial (fig. 4.6), as this overburden overcomes the repulsive charges between sediment particles, and clay platelets begin to rearrange to an overlapping face-to-face orientation (fig.4.4B; Bennett et al., 1991). Such changes in porosity with depth can have an important impact on sediment diagenesis, as is discussed in further detail in section 6.1.2. Sands, in contrast, generally undergo

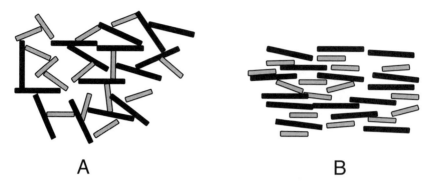

Figure 4.4 A schematic representation of clay mineral platelets before (A) and after (B) compaction (modified after Meade, 1966).

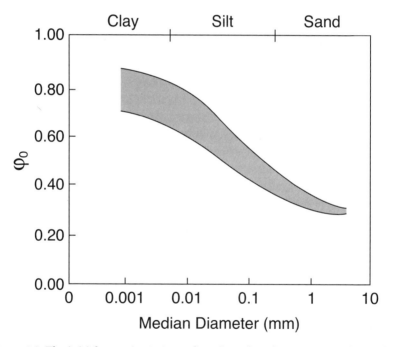

Figure 4.5 The initial porosity (φ_o) as a function of median grain size for surficial terrigenous marine sediments. The increase in porosity with decreasing grain size reflects the increasing amounts of clay minerals in these sediments (modified after Berner, 1980, based on data presented in Meade, 1966).

POROSITY AND SEDIMENT DENSITY

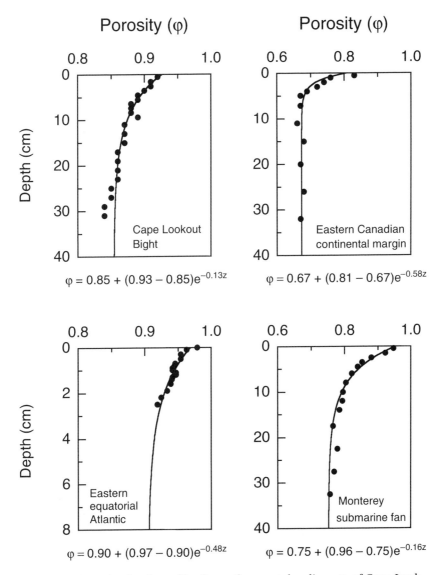

Figure 4.6 Porosity depth profiles from: the coastal sediments of Cape Lookout Bight (water depth = 8 m; data from Klump and Martens, 1989); Eastern Canadian continental margin sediments (water depth = 830 m; data from Mulsow et al., 1989); sediments on the Monterey (CA) submarine fan (water depth = 3,592 m; data from Reimers et al., 1992); Eastern equatorial Atlantic sediments (water depth = 5,075 m; data from Archer et al., 1989b). Note the different depth scales for the Eastern equatorial Atlantic sediment plot. The

minimal compaction and consequently minimal porosity changes with depth because of the nature of their initial packing (Berner, 1980).

In fine-grained sediments, there apparently is an increase in the number of stable contacts between sediment particles with increased compaction, and this results in a continual decrease in the compressibility of the sediment (Engelhardt, 1977). Therefore, the rate of compaction decreases with sediment depth, and porosity values often approach a "constant" value. In many cases, this decrease in porosity is modeled as an empirical function of the form,

$$\varphi = \varphi_\infty + (\varphi_o - \varphi_\infty)e^{-\alpha z} \qquad (4.4)$$

where the subscript o and ∞ indicate porosities at the sediment surface and at some great depth, and α is an empirical attenuation constant. In the upper meter of most marine sediments typical values of φ_o and φ_∞ are ~0.9 and ~0.7, respectively (both ± ~0.1; e.g., Berner, 1980; Ullman and Aller, 1982; Mulsow et al., 1998). Boudreau and Bennett (1999) examined the depth-dependence of porosity and sediment compaction using a rheological approach, and they suggested that a theoretically based alternative for the depth-dependence of porosity is

$$\varphi_s = \frac{(\varphi_s)_o (\varphi_s)_\infty}{(\varphi_s)_o + ((\varphi_s)_\infty - (\varphi_s)_o)e^{-\beta z}} \qquad (4.5)$$

where φ_s is the solid volume fraction in the sediments (equal to $1 - \varphi$ as defined above). The attenuation coefficient β is a function of several properties of the sediments, and in principle can be defined in-

---

best-fit lines for each dataset are shown on each plot and also listed below the plot (see the discussion of eqn. 4.4 for further details). The Cape Lookout Bight, Canadian margin, and Monterey fan results were obtained by sectioning sediment cores and drying sediment, as discussed in the text (e.g., see the discussion of eqn. 4.3). The equatorial Atlantic results were obtained with an *in situ* resistivity probe (Andrews and Bennett, 1981; Archer et al., 1989b). This probe actually determines the sediment formation factor F (see eqn. 6.11), and values of F are converted to porosity values with eqn. 6.14 (note that the value of n in this equation is determined experimentally for the sediments being studied).

dependently using data on the effective sediment stress or compressibility. In practice however it can also be used as an adjustable fitting parameter by fitting porosity data to eqn. (4.5)(B. Boudreau, pers. comm.).

Another factor possibly affecting the porosity of marine sediments is the presence of benthic macrofauna, and their resulting bioturbation or bioirrigation of the sediment. In the rheological porosity model described above, Boudreau and Bennett (1999) suggest that the parameters that define β in eqn. (4.5) will be affected by the presence or absence of bioturbation, although the details of how this might occur are not explicitly discussed. The possible relationship between bioturbation and porosity gradients will be discussed in further detail in section 6.1.2.

4.3 Permeability

Permeability is defined as the capacity of a porous medium to transmit fluid in response to a pressure gradient (Lerman, 1979). Permeability plays a role in marine sediment geochemistry in coarse-grained sands where high permeabilities lead to the occurrence of significant pore-water advection in response to pressure gradients in the sediments (e.g., see discussions in Huettel and Webster, 2001). In contrast, fine-grained sediments generally have much lower permeabilities, and here diffusion or bioturbation dominates pore-water transport processes.

In a one-dimensional sense, pressure-driven advective flow is governed by Darcy's Law,

$$q = \frac{k}{\varphi \eta} \frac{\partial p}{\partial z} \qquad (4.6)$$

where q is the flow rate, k is the permeability factor, η is the pore water viscosity, and $\partial p/\partial z$ is the pressure gradient. This pressure gradient may be generated by wave action or interactions between bottom currents and surface sediment structures (physical surface roughness or biogenic structures; see section 12.2 for further details).

The permeability k is also related to the hydraulic conductivity ($\mathcal{H}$) by

$$\mathcal{H} = k\rho g/\eta \qquad (4.7)$$

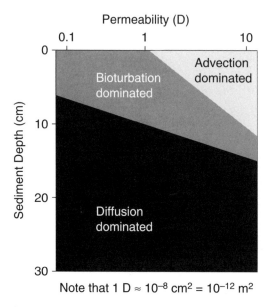

Figure 4.7 A schematic representation of the possible depth dependence of various pore water transport processes (diffusion, bioturbation, and physical advection) as a function of sediment permeability. Note that there is some overlap in these three regions, such that diffusion and advection can also occur in the region where bioturbation dominates (modified after Huettel and Gust, 1992a, and Huettel and Webster, 2001).

where g is the acceleration due to gravity and ρ is the fluid density (Lerman, 1979). The parameter k has units of length squared and is a complex function of sediment particle size, porosity, and particle packing; i.e., it depends on the texture or fabric of the porous media. Permeability is often measured in darcy units (1 D ≈ 10^{-8} cm^2).

Measured values of k vary over several orders of magnitude, with values in sands generally being greater than 1 D and values in fine-grained silts and clays being less than 10^{-4} D (e.g., Lerman, 1979; Huettel and Gust, 1992a). In examining the relative importance of different transport processes in sediments, Huettel and coworkers (e.g., Huettel and Gust, 1992a) have argued that pressure-driven pore-water advection (flow) is likely to be important in coarse-grained sands where k values exceed ~1 D (fig. 4.7). The significance of these processes has only recently been appreciated and will be discussed in further detail in section 12.3. At the same time, in fine-grained marine

TABLE 4.2
Directly Measured (det.) versus Calculated (calc.) Permeabilities in Sieved and Natural Sediments

Sediment Type*	φ	Mean Grain Size (μm)	k (calc., D)**	k (det., D)	k (calc.)/k (det.)
Sieved Sediment					
Coarse quartz sand[a]	0.317	730	202	168	1.2
Medium quartz sand[a]	0.324	370	56	56	1.0
Fine quartz sand[a]	0.346	220	32	15	2.1
Coarse silt (glass beads)[a]	0.402	50	2.5	1.9	1.3
Sand[b]	0.2–0.47	187–1500	2.8–445	7–455	0.4–3.0
Fine silt (diatom powder)[a]	0.81	10	8.2	3.4	2.4
Very fine silt (kaolinite)[a]	0.645	4	0.19	0.0026	73
Natural Sediment					
Medium quartz sand[a]	0.328	210	53	54	1.0
Well-sorted sand[c]	0.4	350	121	51	2.4
Intertidal sand[b]	0.20–0.21	187–2000	3–300	11–1030	0.2–0.3

*Descriptions of the sediments are taken from the original references.
**Calculated k values were determined using eqn. (4.8) assuming that the particle radius is half the mean grain size.
(a) Porosity, mean grain size, and directly determined k values from Huettel and Gust (1992b).
(b) Porosity, mean grain size, and directly determined k values from Huettel and Rusch (2000). For ease of presentation I have not listed the results from individual analyses of the sediments used in these experiments. For the sieved sands and intertidal sands the ranges listed here are for n = 10 of each sand type.
(c) Porosity, mean grain size, and directly determined k values from Huettel et al. (1998).

sediments with lower permeabilites, diffusion and/or bioturbation tend to dominate pore water transport (also see chapter 5).

Permeability can be either directly measured, e.g., with a constant head permeameter (Means and Parcher, 1964), or calculated using one of several equations, some more theoretically based than others (Lerman, 1979; Boudreau, 1997a). One of these, the Carman-Kozeny equation for spherical particles is

$$k = \frac{\varphi^3}{(1-\varphi)^2} \cdot \frac{r^2}{45} \tag{4.8}$$

where r is the particle radius (note that here we would consider porosity, φ, to be a dimensionless quantity, such that if r has units of cm, then k would have units of cm^2).

The use of eqn. (4.8) to estimate the permeability of a sediment is not necessarily unreasonable, as can be seen in table 4.2, where measured and calculated permeabilities agree to within a factor of ~2–5. Since none of these equations are well defined in high porosity (i.e., fine-grained or muddy) sediments (Boudreau, 1997a), the poor agreement between measured and calculated k values for the very fine-grained silt is not surprising. At the same time the low measured permeability of such sediments also minimizes the importance of Darcian flow relative to other transport processes such as diffusion or bioturbation. Thus, in low-porosity, high-permeability sediments where pressure-drive advective flow is likely to be important (fig. 4.6), the assumptions of equations such as eqn. (4.8) that can be used to calculate k values appear to be reasonably valid.

≈ CHAPTER FIVE ≈

An Introduction to Transport Processes in Sediments

QUANTIFYING TRANSPORT PROCESSES in marine sediments is crucial for understanding sediment geochemical processes. In this chapter I will introduce some of the transport processes that are most relevant in marine sediment biogeochemistry. This discussion will set the stage for more quantitative discussions of these processes in subsequent chapters.

5.1 Diffusion

Diffusion is a process by which matter is transported from one part of a system to another as a result of random particle (Brownian) motions (Crank, 1975). Diffusion occurs in response to concentration gradients (the change in concentration over some distance) and acts to eliminate these gradients. In many marine sediments, diffusion represents the predominant transport process by which ions and molecules move in pore waters.

The classic experiment illustrating diffusion involves setting up a tall glass cylinder in which the lower half is filled with an iodine solution and the upper half with clear water, in such a way as to inhibit any convective mixing of the two fluids (fig. 5.1). At the start of the experiment there is a sharp boundary between the two solutions, although over time the upper solution becomes darker, the lower solution lighter, and a color gradient (from clear to dark) develops across the central region of the cylinder. After sufficient time, the solution in the cylinder will be of uniform color from top to bottom. Overall, this experiment demonstrates that because of the I_2 concentration gradient in the cylinder, iodine diffuses from the bottom to the top of the cylinder, eventually producing a solution that is homogeneous in I_2 concentration.

During diffusion, the Brownian motion of each molecule typifies what is known as a "random walk." This means that while we *can* calculate the mean distance a molecule will travel in a given time period,

TRANSPORT PROCESSES IN SEDIMENTS

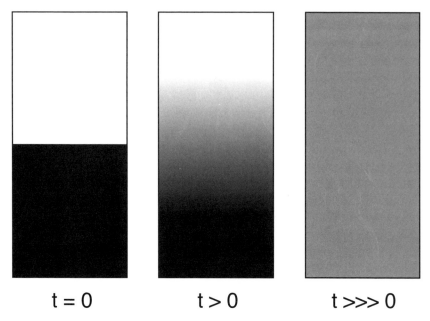

Figure 5.1 The diffusion of I_2 molecules from the lower, dark solution to the upper, clear solution of a hypothetical tall cylinder.

we *cannot* determine the exact direction that the molecule will move in this time period (Crank, 1975). Interestingly, however, although the motion of each molecule is random, there is still net unidirectional I_2 transfer in response to the concentration gradient until the solution is homogeneous from top to bottom.

In thinking about how this occurs, we consider a horizontal plane through the solution and two thin equal volumes of solution, one just below and one just above this plane. Although we cannot determine which way any *particular* I_2 molecule will move during a given time period, we do know that, on average, a definite fraction of molecules in the lower volume will cross the plane from below and an equal fraction of molecules from the upper volume will cross the plane from above. Since there are more I_2 molecules in the lower section than in the upper, there will then be a net flux of iodine from the bottom of the cylinder to the top as a result of this random motion. Thus, the random motion of the iodine molecules and their concentration gradient lead to a net transfer or flux (at the macroscopic level) of I_2 from the lower solution to the upper solution.

In general, quantifying diffusion in any system requires that we know something about the concentration gradient of the component of interest since this gradient defines the net direction of diffusion. This can be done using Fick's First Law of Diffusion, which states that the net transfer (or flux) of this component is proportional to its concentration gradient, or

$$J = -D\frac{dC}{dz} \qquad (5.1)$$

where J is the diffusive flux[1] (units of mass per unit area per unit time), C is the concentration (mass per unit volume), z is the direction of the maximum concentration gradient, and D is a proportionality constant known as the diffusion coefficient (area per unit time).

Diffusion coefficients vary with the charge and size of a molecule, with larger molecules generally having smaller diffusion coefficients (Burdige et al., 1992a; Boudreau, 1997a). They also vary as a function of temperature (increasing with increasing temperature) and the viscosity of the medium (decreasing with increasing viscosity; Li and Gregory, 1974; Boudreau, 1997a). Diffusion coefficients for many substances (inorganic cations or anions, organic compounds, and dissolved gasses) can be found in numerous references (Li and Gregory, 1974; Boudreau, 1997a; Schulz and Zabel, 2000). Additional details about diffusion coefficients and the role of diffusion in sediment processes can be found in the next chapter.

5.2 Sediment Accumulation, Steady State, and the Frame of Reference for Processes in Marine Sediments

Whether this occurs as a slow steady process, or episodically in pulses or bursts of activity, new particles constantly accumulate at the sediment-water interface of many sediments. In such depositional environments, we term this process *sedimentation* or *sediment accumulation*. Note that many sediment systems appear to be nondepositional in nature; these will be discussed in later chapters. The occurrence of sediment accumulation implies that particles at the sediment surface

---

[1] Regardless of the transport mechanism, a *flux* is defined as the transfer of some amount of mass per unit area per unit time (e.g., mol m^{-2} sec^{-1}).

are eventually buried below the sediment surface as new material continues to accumulate. Since the sediment-water interface is defined as the frame of reference for most geochemical studies, this then implies that there is a net downward flux (or burial) of sediment material as a result of sedimentation.

Sediment accumulation also leads to the addition of new sediment pore water at the sediment-water interface. This implies that there is also downward pore-water advection as a result of sediment accumulation. In the absence of sediment compaction (see below) or the occurrence of an externally impressed flow in the sediments, e.g., due to groundwater or hydrothermal advection, rates of pore water advection and rates of sediment burial are identical. However, in fine-grained sediments, where postdepositional compaction may be important (see below) the rates of pore water advection and solid phase sediment burial not only vary with sediment depth, but also can be uncoupled from each other. This point will be discussed in greater detail in the next chapter.

Also related to these processes is the concept of steady-state versus nonsteady-state conditions in a sediment. As shown in fig. 5.2, steady-state conditions exist when a sediment property (e.g., concentration, rate, etc.) at a given depth relative to the sediment-water interface does not vary with time. We also see that during this steady state if we track a fixed sediment parcel over time (i.e., with burial) this property does change with time. Taken together, these two observations imply that these changes occur in such a way that the depth profile of this property (again relative to the sediment-water interface) does not change. As Berner (1980) notes, steady-state with respect to one sediment property generally implies steady-state with respect to other properties, although this is not a necessary condition.

In contrast, nonsteady-state conditions exist when a sediment property at a given depth does vary with time (fig. 5.2). In the example shown here we see that the property of a given sediment parcel when tracked over time (with burial) does not change, although the overall profile (relative to the sediment-water interface) does change with time. The causes of nonsteady-state conditions in sediments are numerous and the example shown here is but one way that it may occur (e.g., also see sections 5.5 and 6.3, and chapter 14 for further details). However, the primary difference between steady-state and nonsteady-state sediment conditions lies in whether sediment depth profiles relative to the sediment-water interface do or do not change with time.

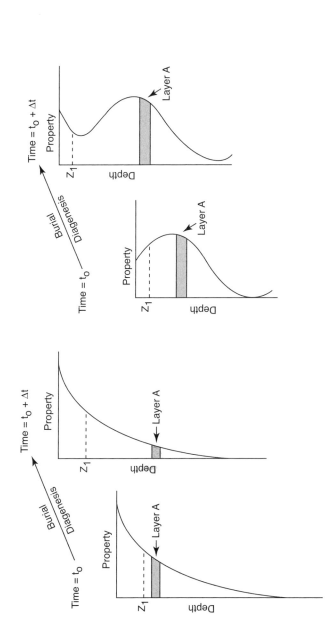

Figure 5.2 (Left profiles) A schematic illustration of steady-state diagenesis for a nonconservative (reactive) species. With burial, the value of this property does not change with time at either the sediment-water interface (e.g., z_1). However, this property does change with time if we follow a particular sediment layer (e.g., layer A). (Right profiles) A schematic illustration of nonsteady-state diagenesis for a stable, conservative (nonreactive) species. In contrast to the steady-state situation, with burial here the value of the property *does* change with time at both the sediment-water interface or at any other fixed depth relative to the sediment-water interface (e.g., z_1). At the same time, though, the property p does not change if one follows a particular sediment layer with time (e.g., layer A). Modified after Berner (1971, 1972).

We often think of sedimentation as occurring in a vertical sense, resulting from the sinking of particles through the water column to the sediment-water interface, regardless of whether these particles are ultimately derived from land or are produced in surface ocean waters. Under many circumstances this may indeed be the case, and the bulk of this sedimentation appears to occur via relatively "large" particles ($\sim 10^2$–10^4 μm in diameter or length) that are ultimately biologically derived, e.g., planktonic remains or fecal pellets. Inorganic (lithogenic) material is also incorporated into these aggregates because of "packaging" processes that occur during particle sinking (see Fowler and Knauer, 1986, for a review).

While the implicit assumption is that particle sinking leads to a continuous, or near continuous, rain of particles to the sediment surface, this is not necessarily always the case. In coastal sediments, the relatively shallow water column depths and seasonal cycles of primary productivity generally lead to a pronounced seasonality in the deposition of organic matter to the sediments (e.g., see observations in Kemp and Boynton, 1992; Roden and Tuttle, 1996; Gerino et al., 1998). Similarly, seasonal storm events in these shallow water systems may also be important (e.g., Canuel and Martens, 1993) and in some situations may also lead to seasonal variability in total sediment accumulation rates. In some cases one may be able to characterize these time-varying processes in sediment diagenesis models with a seasonally averaged sedimentation rate. At the same time, though, seasonality in sediment organic matter deposition may play a role in the observed seasonality of biogeochemical processes in these sediments (e.g., benthic fluxes of nutrients) and may need to be modeled as such (see discussions in chapter 14).

In open ocean (deep-sea) settings, seasonality, and perhaps interannual variability in surface water productivity may lead to analogous temporal variability in the deposition of organic matter to such sediments (see discussions in Sayles et al., 1994; Smith et al., 1996; Soetaert et al., 1996b; Gobeil et al., 1997; Rabouille et al., 2001). However, the extent to which such seasonality is imparted in sediment biogeochemical processes depends on the relationship between the time scales of organic matter remineralization and sediment transport processes (again, see chapter 14 for further details).

The discussion above assumes that most sediment accumulation occurs by vertical processes, that is, particulate material sinking through the water column. However, in some sedimentary environments lateral transport of both organic matter and lithogenic material may also be important (Walsh et al., 1988; Jahnke et al., 1990; Berelson et al., 1996; Hensen and Zabel, 2000; Keil et al., 2004). This can occur somewhat diffusely, and perhaps more regularly, as lateral transport of resuspended sediment, or may occur more coherently, and perhaps episodically, as downslope gravity-driven sediment flows (see Kennett, 1982, for a general discussion).

An important example of such gravity-driven flows are turbidity currents. These are powerful, short-lived events containing dilute mixtures of sediment and water that are denser than the surrounding seawater; the motion of these flows is generally maintained by internal turbulence (Kennett, 1982). *Turbidites* represent sediments transported from continental margins to deep ocean (abyssal) sediments by such gravity-driven turbidity flows. Turbidite emplacement is discontinuous, and generally leads to the "instantaneous" deposition of a geochemically homogeneous turbidite section above and below slowly accumulating pelagic sediments (deposited via particulate material sinking through the water column). The composition of turbidite sediments is, for the most part, very different than that of the surrounding pelagic sediment (e.g., turbidites generally contain elevated levels of total organic carbon, TOC), and diagenetic processes in turbidite sediments are generally nonsteady state (see section 14.4). As we will see in the subsequent discussion, the study of geochemical processes in turbidites has yielded important insights in many areas of marine sediment geochemistry. Similar types of nonsteady-state events may also occur in shallower sediments as a result of episodic depositional events caused by large storms or floods (Deflandre et al., 2002).

5.3 An Introduction to Bioturbation and Bioirrigation

In many marine sediments, the upper portion of the sediments is inhabited by bacteria as well as other higher organisms. The activities of these benthic macrofauna and meiofauna can not only transport material in surface sediments, but also affect redox conditions in the region of the sediments they inhabit. Their activity therefore has a profound impact on the sediment cycling of elements such as carbon, nitrogen, sulfur, and redox-sensitive trace metals (also see chapter 12 and table 15.1).

TRANSPORT PROCESSES IN SEDIMENTS

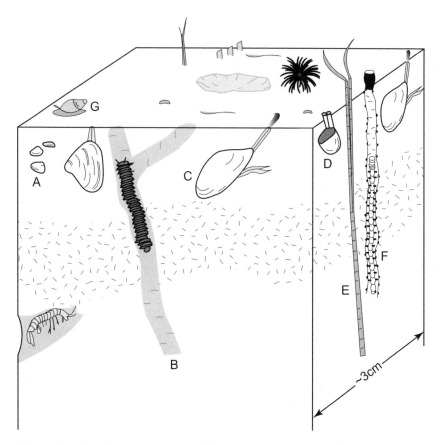

Figure 5.3 A schematic representation of the major benthic macrofauna found in a typical coastal marine sediment (based on Aller, 1980a). The organisms shown here include the deposit feeding bivalves *Nucula annulata* (A) and *Yoldia limulata* (C), the suspension feeding bivalve *Mulnia lateralis* (D), the mobile burrowing polychaete *Nephtys incisa* (B), the sedentary polychaetes *Spiochaetopterus oculatus* (E) and *Clymenella* sp. (F), and the deposit feeding gastropod *Nassarius trivitarus* (G).

There are a great variety of macrofaunal activities (fig. 5.3). Some organisms simply crawl through the sediments or along the sediment surface in search of food, mixing sediment and sediment pore waters as a result of their activities. In some cases these organisms also ingest and subsequently defecate sediment materials. Such processes are commonly referred to as *bioturbation* and lead to the creation of a mixed (or bioturbated) layer in the surface sediments (see section 12.4 for further details).[2]

Other deposit-feeding macrofauna live in fixed sediment burrows and move material from a fixed depth a few centimeters below the sediment surface to the surface, forming burrow mounds. This process is generally referred to as *nonlocal* or *conveyor-belt transport* (e.g., Boudreau, 1986b). Organisms referred to as *reverse conveyor-belt feeders* carry out a similar type of nonlocal transport in the opposite direction, taking surface sediment and moving it down to some depth in the sediments (see section 12.4.1 for further details).

Finally, organisms known as water column suspension feeders and tubeworms build sediment burrows and flush these burrows with overlying bottom waters. This process does not affect transport of the surrounding sediment particles, but does enhance the exchange of pore waters and bottom waters. This process is generally referred to as bioirrigation and will be discussed in section 12.5.

5.4 Time and Space Scales of Sediment Processes

In discussions of sediment processes it is often useful to have a general idea of the time and space scales over which certain processes operate. For diffusion, this is described by the Einstein-Smoluchowski equation,

$$L^2 = 2Dt \tag{5.2}$$

where strictly speaking, L is the mean of the squared displacement of a concentration impulse after some elapsed time t (Boudreau, 1997a). Thus for a typical bulk sediment diffusion coefficient of 5×10^{-6} cm^2 sec^{-1}, the distance this ion will diffuse in one day is ~1 cm. Similarly the characteristic time it will take this ion to diffuse 10 cm is ~116 days.

As we will see in chapters 6 and 12, in many cases we can quantify bioturbation (sediment mixing by organisms) by assuming that it is a random, diffusive-like process (e.g., Boudreau, 1986a). Under these circumstance we can then define a bioturbation coefficient (generally

---

[2] In some works the term *bioturbation* is used in a much broader sense to describe all of the activities of benthic macrofauna in surface sediments. However here (as in many other works) I will attempt to use this term in the more specific sense that is discussed here.

defined as D_B) that is analogous to a molecular diffusion coefficient for a dissolved species. In this case, eqn. (5.2) can also be used to examine the characteristic time and length scales of this diffusive-like, biological particle mixing.

Since sedimentation (sediment accumulation or burial) or pore-water advection has units of length over time, we can define the time and space scales over which these processes operate as,

$$L = vt \tag{5.3}$$

where v can be either the sedimentation rate or the rate of pore water advection.

Another parameter that examines the relative significance of different transport processes in sediments is the Peclet number,

$$Pe_H = vL/D \tag{5.4}$$

This term is a dimensionless parameter that compares the relative importance of advective and diffusive transport over the characteristic length scale L (Boudreau, 1997a). If Pe_H is less than 1 then diffusion (or bioturbation) dominates transport, and if it is greater than 1 then advection dominates.

Using the approaches outlined above we can examine the relative importance of these transport processes in different marine sediment settings. For example, the results in table 5.1 clearly indicate that in deep-sea sediments bioturbation is a much more important transport process than is sedimentation in the surface mixed layer, based on either the Peclet number for these processes or their characteristic time scales. In this analysis a shorter characteristic time for a process implies that an ion/particle responds more rapidly to that process, hence its dominance (for the application of this approach in studies of pelagic and hemipelagic sediments also see Emerson et al., 1985; Smith et al., 1993; Boudreau, 2000). An important outcome of calculations such as these is that they often allow one to omit the less important of two transport processes from quantitative diagenetic models of sediment processes with minimal impact on the accuracy of the calculation. This can then often allow one to obtain an exact analytical (versus numerical) solution to a diagenetic model (see chapter 6 and Boudreau,1997a, for further details).

TABLE 5.1
Scaling Calculations for Transport Processes in Marine Sediments

Transport Processes[a]	Characteristic Time Scale (yr)	
	t ($L = 1$ cm)	t ($L = 10$ cm)
Diffusion ($D_s = 5 \times 10^{-6}$ cm² sec⁻¹ [= 158 cm² yr⁻¹])[b]	0.003	0.32
Sediment accumulation or pore water advection[c]		
Deep-sea sediment ($\omega = 0.002$ cm yr⁻¹)	500	5,000
Coastal sediment ($\omega = 1$ cm yr⁻¹)	1	10
Deep-sea sediment bioturbation ($D_B = 0.4$ cm² yr⁻¹)[b]	1.25	125
Peclet Numbers[d]		
	Pe_H ($L = 1$ cm)	Pe_H ($L = 10$ cm)
Pore water advection relative to diffusion		
Deep-sea sediment	1.3×10^{-5}	1.3×10^{-6}
Coastal sediment	6.3×10^{-3}	6.3×10^{-4}
Sedimentation relative to bioturbation (deep-sea sediment)	0.0025	0.025

a. Values in parentheses are typical of these parameters in the appropriate sediments.
b. Calculated using eqn. (5.2).
c. Calculated using eqn. (5.3).
d. Calculated using eqn. (5.4) and values listed above.

The ability to similarly define the relative importance of transport processes and *in situ* reactions that occur in the sediments is also sometimes of use. For example, the Damkohler number (Da) allows one to estimate the relative importance of diffusion (or bioturbation) versus *in situ* reactions in sediments. The exact form of the equation for Da depends on the kinetics of the reaction of interest; for a first-order reaction, i.e., a reaction in which the rate of consumption or production equals $-kC$, Da is given by

$$Da = kL^2/D \tag{5.5}$$

where D can be either D_s or D_b (the bulk sediment diffusion coefficient [section 6.1.1] or the bioturbation coefficient [see section 12.4]). As in the formulation above, if $Da < 1$ then diffusion (or bioturbation) dominates, while if $Da > 1$ then the *in situ* reaction dominates (Boudreau, 1997a).

5.5 The Classification of Marine Sediments on the Basis of Their Functional Diagenetic Characterisitics

Biogeochemical processes in sediments are strongly affected by physical processes such as those described in this chapter. Based on these considerations, continental margin and shelf sediments can be described in terms of several functional types (Aller, 1998, 2004; also see discussions in McKee et al., 2004).

The first two of these sediment types are steadily accumulating sediments either with or without a bioturbated or bioirrigated zone. In the former case, the overlying waters may be oxic or anoxic (e.g., Black Sea sediments), while in the latter case, the overlying waters contain some amount of dissolved oxygen. Other sediment types include nonaccumulating sediments (in a net sense) referred to by Aller (1998) as bypass zones, and advectively permeable sediments that may (percolation zone) or may not (filtration units) be significantly impacted by submarine ground water discharge (see sections 4.3 and 12.2 for further details).

The last of these sediments are fluid muds[3] and other highly mobile sediment deposits, which appear to act as massive, suboxic fluidized bed reactors. This type of biogeochemical behavior occurs in Amazon continental shelf sediments as well as in other muddy continental margin deltas (e.g., the Gulf of Papua, Papua New Guinea), mobile tropical mud belts (e.g., the Amazon-Guianas mudbelt), and perhaps some estuarine sediments in general (Alongi, 1995; Aller, 1998, 2004; Zhu et al., 2002; Aller et al., 2004; McKee et al., 2004). An important aspect of many of these continental margin settings is that they generally occur in association with major rivers that have relatively

---

[3] Fluid muds are generally regarded as a high-concentration suspended sediment (>10 g l^{-1}) that forms under conditions in which sediment deposition is too fast to permit the formation of a consolidated mud bed (e.g., Kineke et al., 1996). Fluid muds are generally not considered part of the underlying consolidated sediment, and they can be rapidly remobilized and reworked on times scales as short as weeks.

high freshwater discharge rates. As a result, estuarine processes, which would normally take place in a physically confined estuary, take place offshore, on the adjacent continental shelf. Similarly, because these large rivers have high suspended sediment loads, these river-dominated ocean margins receive large inputs of both lithogenic and biogenic material (see McKee et al., 2004, for a recent discussion of such margin settings).

On the Amazon continental shelf, these sediments show very high rates of net sediment accumulation (up to ~1 cm yr^{-1}), but are also physically reworked by shelf currents and tides to sediment depths ranging from ~0.5 to 2 m. This reworking occurs on daily to seasonal time scales (see discussions in DeMaster and Aller, 2001; McKee et al., 2004).

From a biogeochemical perspective, physical reworking of such sediments has a profound impact on both sediment carbon cycling (i.e., preservation versus remineralization) as well as the oceanic cycles of elements such as U, F, and K by, for example, promoting the occurrence of authigenic mineral formation (or reverse weathering; Michalopoulos and Aller, 1995; also see discussions in Aller et al., 1996, and chapter 18). This reworking continually reoxidizes the sediments by periodically re-exposing them to oxygen-containing bottom waters, leading to nonsteady-state conditions in the sediments (Mackin et al., 1988), and the continual reoxidation of reduced Fe, Mn, or S produced in the sediments. Thus, in comparison to sediments in low-energy environments (which do not undergo this type of intense physical reworking), and which have comparable rapid rates of net sediment accumulation, Amazon margin sediments are, roughly speaking, "poised" at suboxic (iron reduction), rather than anoxic (sulfate reducing) conditions (see sections 7.5.2 and 7.6.3 and Aller et al., 1986).

Physical reworking of these sediments may also enhance the remineralization of sediment organic matter and lead to relatively low carbon burial efficiencies,[4] again as compared to comparable low-energy sediments (Aller, 1998). This may occur in a number of ways that will be discussed in further detail in sections 7.6 and 11.2.4 and chapter 15.

---

[4] See section 8.3 for a definition of carbon burial efficiency.

≈ CHAPTER SIX ≈

Models of Sediment Diagenesis

IN THIS CHAPTER we will begin to examine the quantitative description of early diagenetic processes that occur in sediments. This discussion will begin with steady-state models, being fairly general with regards to the kinetics of biogeochemical processes occurring in the sediments (see section 5.2 for a general discussion of the concept of steady state). In later chapters, as we continue to discuss biogeochemical processes that occur in sediments, the level of complexity of the models will increase accordingly.

6.1 THE GENERAL DIAGENETIC EQUATION

To quantitatively examine sediment diagenesis we start with Fick's Second Law of Diffusion (fig. 6.1), which states that the concentration change with time in an infinitesimally small box results from the change in the flux (J) through the box. In a purely diffusive system, and assuming a constant diffusion coefficient, this then implies that

$$\frac{\partial C}{\partial t} = -\frac{\partial J}{\partial z} = \frac{\partial}{\partial z}\left(D\frac{\partial C}{\partial z}\right) = D\frac{\partial^2 C}{\partial z^2} \tag{6.1}$$

where J is defined here using Fick's First Law of Diffusion (see section 4.1).

In sediments we modify this equation somewhat to what has been referred to as the general diagenetic equation (Berner, 1980),

$$\frac{\partial \hat{C}}{\partial t} = -\frac{\partial J}{\partial z} + \Sigma R_i \tag{6.2}$$

where $\hat{C}$ is the concentration of a solid or liquid (pore water) component (in units of mass per unit volume of total sediment), J is the flux of this component (mass per unit area of total sediment per unit time), and ΣR_i is the sum of the rates of all processes affecting this component (mass per unit volume of total sediment per unit time). In

Fick's Second Law

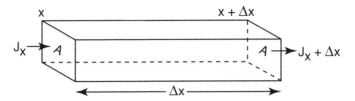

Figure 6.1 A derivation of Fick's second law of diffusion (modified after Berner, 1980).

If M = mass, then in this rectangular box:

$$\frac{\Delta M}{\Delta t} = -A(J_{x+\Delta x} - J_x)$$

For concentration C:

$$\Delta C = \Delta M / A \Delta x$$

Substituting with Fick's First Law:

$$\frac{\Delta C}{\Delta t} = \frac{D[(\partial C/\partial x)_{x+\Delta x} - (\partial C/\partial x)_x]}{\Delta x}$$

Passing to the limit $\Delta x \to 0$:

$$\frac{\partial C}{\partial t} = D\frac{\partial^2 C}{\partial x^2}$$

words, this equation states that with time, in an infinitesimally small sediment box, changes in the concentration of a component in the box result from any changes in its flux through the box, plus what is produced or consumed in the box. This equation is defined in one direction, since vertical (z direction) gradients in sediments are generally considered to be more important than horizontal (x and y directions) gradients (Berner, 1980; Boudreau, 2000).

We can therefore think of eqn. (6.2) as an extension of Fick's Second Law where in sediments J also includes nondiffusive fluxes. The specific fluxes we need to consider in sediments are both diffusive and advective fluxes as well as fluxes associated with bioturbation and/or bioirrigation. Thus we can define J as

$$J = -D\frac{\partial \hat{C}}{\partial z} + v\hat{C} + \mathcal{B} \qquad (6.3)$$

where v is the velocity of sediment or particles (relative to the sediment-water interface) and $\mathcal{B}$ is a general term (to be defined later) for fluxes associated with all types of bioturbation (J_b), and bioirrigation (J_i). For solids, advection results from sedimentation and compaction (i.e., depth changes in porosity), while for pore waters several possible processes may lead to their advection.

All concentrations $\hat{C}$ are expressed here as an amount per unit volume of total sediment (solid particles plus pore waters). However pore water concentrations (C) are generally determined as amounts per unit volume of pore water, and solid concentrations (S) as some amount per gram dry weight of sediment. This then leads to the following equation for converting between these different pore water concentration units:

$$\hat{C}_p = \varphi C \tag{6.4}$$

where φ is the sediment porosity. For solids, the relevant conversion equation is,

$$\hat{C}_s = (1 - \varphi)\rho_{ds}S = \varphi_s \rho_{ds} S \tag{6.5}$$

where ρ_{ds} is the dry sediment density and $\varphi_s \equiv 1 - \varphi$. As we will see below, similar corrections must also be made to ΣR_i. We will also see that incorporating these changes in units into the general diagenetic equation is not always as simple as just replacing the various $\hat{C}$ values with the terms in eqns. (6.4) and (6.5).

In the remaining derivations in this chapter we will neglect (for now) bioirrigation and nonlocal sediment mixing; bioturbation is, therefore, the only process quantified in $\mathfrak{B}$. For reasons that will be discussed in greater detail in chapter 12, we also will assume that bioturbation can be modeled as a random, diffusive-like process (Boudreau, 1986a) such that the fluxes associated with bioturbation can be expressed as,

$$J_b = -D_b \frac{\partial \hat{C}}{\partial z} \tag{6.6}$$

where D_b is the bioturbation coefficient (also see section 6.1.2).

6.1.1 Diffusion

The equations above include a bulk sediment molecular diffusion coefficient D_s for pore-water constituents. This coefficient is obtained from an infinite dilution diffusion coefficient that has been corrected for temperature, viscosity, and sediment tortuosity.

Most tabulations of diffusion coefficients (e.g., Li and Gregory, 1974; Schulz and Zabel, 2000) present infinite-dilution diffusion coefficients (D or $D°$) at a limited number of temperatures, although Boudreau (1997a) presents equations that allow one to calculate infinite-dilution diffusion coefficients at any temperature. Using a temperature-corrected $D°$ value to obtain a diffusion coefficient in seawater then involves a correction for the difference in the viscosity (μ) of seawater versus pure (distilled) water. At a fixed temperature, this can be done with the following equation (Li and Gregory, 1974),

$$\frac{D^{sw}}{D^o} \approx \frac{\mu^o}{\mu^{sw}} \qquad (6.7)$$

where sw indicates a value in seawater of a given salinity and o indicates infinite-dilution (pure water) values. These viscosity values can either be calculated (Boudreau, 1997a) or taken from standard tables available in many references.

Determining D^{sw} at some temperature involves first obtaining the infinite-dilution diffusion coefficient at the temperature of interest, and then correcting this value for pore water viscosity using eqn. (6.7). However, for seawater with a salinity of 35, the viscosity ratio μ^o/μ^{sw} ranges from 0.951 at 0°C to 0.932 at 25°C (Boudreau, 1997a). Therefore this viscosity correction has a minimal effect on calculated free solution D^{sw} values (see similar discussions in Li and Gregory, 1974). As such, an examination of the literature suggests that often this correction is omitted.

Diffusion in sediment pore waters may also be affected by ion-ion interactions in the multicomponent seawater/pore water mixture. For example, "cross-coupling" may occur such that the gradient of one ion can affect the diffusion of another ion. However, under most circumstances, the effects of these ion-ion interactions on diffusion in pore waters are generally small (Berner, 1980; Boudreau, 1997a).

More importantly, diffusion in sediment pore waters is strongly affected by the presence of particles in the sediment matrix. Therefore, quantifying diffusion in sediments requires that we take these particles into account, because the path an ion may travel in the sediment pore waters is hindered by their presence (fig. 6.2). This phenomenon is referred to as *tortuosity* and can have a significant impact on the effective diffusivity of a compound in sediment pore waters.

Diffusion coefficients in sediment pore waters are corrected for sediment tortuosity using the equation

$$D_s = D^{sw}/\theta^2 \qquad (6.8)$$

MODELS OF SEDIMENT DIAGENESIS

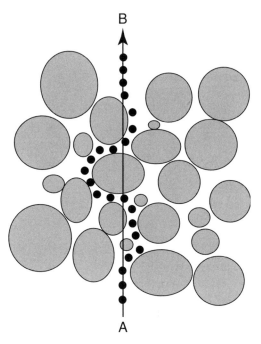

Figure 6.2 The concept of tortuosity in marine sediments. Because of the presence of sediment particles a solute diffusing from point A to point B must traverse a *tortuous* path (the dotted line) rather than the more direct path indicated by the solid line (from Boudreau, 1997a, and used with permission of the author).

where D_s is the bulk sediment diffusion coefficient and θ^2 is the square of the dimensionless tortuosity coefficient. Since by definition $\theta^2 \geq 1$ this leads to diffusion in sediment pore waters being "hindered" by sediment tortuosity. While it is hard (if not impossible) to theoretically calculate or directly measure tortuosity (Boudreau, 1997a), tortuosity can be determined indirectly in marine sediments using resistivity measurements (McDuff and Ellis, 1979; Andrews and Bennett, 1981; Archer et al., 1989b). This approach is based on the fact that conductivity (which is the inverse of resistivity) is the physical analog of diffusion. It therefore allows one to estimate a quantity known as the formation factor (F),

$$F = \frac{\text{bulk sediment resistivity}}{\text{pore water [alone] resistivity}} \qquad (6.9)$$

The formation factor and tortuosity are related by,

$$\theta^2 = \varphi F \qquad (6.10)$$

leading to

$$D_s = \frac{D^{sw}}{\varphi F} \qquad (6.11)$$

In addition to the development of techniques used to directly determine tortuosity (see references above), theoretical and empirical relationships that relate θ^2 and φ have also been proposed. In part this has been done because of the relative ease with which one can determine porosity versus tortuosity. Historically, many workers have used the general equation

$$F = \varphi^{-n} \qquad (6.12)$$

which for $n = 2$ is known as Archie's Law. This equation has been shown to work reasonably well for loosely consolidated sands and sandstones (Berner, 1980). In deeply buried, compacted marine sediments (Deep Sea Drilling Project [DSDP] cores collected over sediment depths to several hundred meters), a very similar value of n (= 1.8) is also observed (Manheim and Waterman, 1974). However, in surficial ($<\sim 1$ m), high porosity, fine-grained sediments it also appears that a higher value of n (2.5–3) is more appropriate (Andrews and Bennett, 1981; Ullman and Aller, 1982; Archer et al., 1989b). Based on all of these observations Ullman and Aller (1982) proposed a "segmented" form of eqn. (6.12), in which $n = 2$ for low porosity sediments (φ less than ~ 0.7), and $n = 2.5$–3 for sediments in which $\varphi > 0.7$. Boudreau (1997a) recently re-examined much of this data by fitting them to a wide range of empirical and theoretical equations for θ^2 versus φ. His conclusion was that the modified Weissberg relation

$$\theta^2 = 1 - ln(\varphi^2) \qquad (6.13)$$

provides the best fit to the available data, although differences in the goodness of this fit with a fit to an unsegmented form of eqn. (6.12) with $n \approx 2$ is relatively small. In summary, it appears that both eqn.

(6.13) and Ullman and Aller's segmented form of eqn. (6.12) are equally valid for most sediments.

6.1.2 Advection, Sediment Compaction, and Bioturbation

Advection of sediment solids and pore waters relative to the sediment-water interface can be caused by particle sedimentation, compaction upon burial, or bioturbation (see sections 5.2 and 5.3). An externally impressed flow, e.g., groundwater or hydrothermal flow, can also lead to pore water advection, although for now we will neglect this possibility.

At the same time, though, if we assume that porosity is constant with sediment depth then as discussed in section 4.2 this also implies that there is no compaction with burial. The assumption of some constant, mean sediment porosity is one that is often taken in sediment diagenesis models; in fact, much of the later discussion in this chapter will employ such an assumption. This approach is often taken because of its significant mathematical simplification of the diagenetic equations (see below). At the same time, in high porosity, organic-rich sediments the inclusion of depth-dependent sediment porosity in pore-water diagenesis models has been shown to have an insignificant effect on calculated rates (see discussions in Martens and Klump, 1984; Klump and Martens, 1989).

However, in other situations this approach may not be appropriate. One example of this occurs in the interpretation and modeling of solid phase sediment profiles (particularly near the sediment-water interface). Here we see that if φ decreases from 0.9 to 0.7 (see section 4.2) then φ decreases by only ~20% yet φ_s (= $1 - \varphi$) increases by a factor of 3 (= 300%; e.g., see discussions in Jahnke et al., 1986; Archer et al., 1989b; Mulsow et al., 1998). Similarly, in Cape Lookout Bight sediments Kipphut and Martens (1982) observed that modeling ^{222}Rn profiles to obtain parameters describing sediment-water exchange led to different results with and without depth-variable porosity in their model equations. This occurs because compaction with depth increases solid phase ^{226}Ra concentrations per unit volume of total sediment, and since ^{226}Ra is the parent radioisotope of ^{222}Rn, this then affects radon production rates in the sediments (see section 12.5.1 for further details).

The results of some modeling studies yield equivocal results in terms of the effects of compaction (and porosity gradients) on model

THE GENERAL DIAGENETIC EQUATION

results (see discussions in Rabouille and Gaillard 1991a; Dhakar and Burdige, 1996). Nevertheless, many recent coupled models of sediment diagenesis (see section 8.4) include depth-dependent porosity in the model equations, if only for completeness (e.g., Boudreau, 1996; Soetaert et al., 1996c).

If we choose not to ignore compaction with sediment depth, then we must consider the effects of depth-dependent porosity (e.g., see fig. 4.5) on advective and diffusive terms in the diagentic equations. This is generally done by assuming steady-state compaction (Berner, 1980), or

$$\frac{\partial \varphi}{\partial t} = \frac{\partial \varphi_s}{\partial t} \equiv 0 \tag{6.14}$$

When these equations are coupled with total mass and pore-water conservation equations, this then implies that

$$\frac{\partial (\varphi v)}{\partial z} = \frac{\partial (\varphi_s \omega)}{\partial z} = 0 \tag{6.15}$$

where v is the pore water burial velocity and ω is the sediment burial velocity,[1] and because of compaction, both are now functions of sediment depth (Berner, 1980; Meysman et al., 2005). Integration of these equations yields,

$$\varphi v = \varphi_\infty v_\infty \tag{6.16a}$$

and

$$\varphi_s \omega = \varphi_{s,\infty} \omega_\infty \tag{6.16b}$$

where φ_∞ and and $\varphi_{s,\infty}$ represent the asymptotic values of φ and φ_s obtained, for example, by fitting porosity depth profiles as in fig. 4.5.

---

[1] In the literature terms such as *sediment burial velocity* (or *sediment burial rate*), *sediment accumulation rate*, and *sedimentation rate* are often used interchangeably although, strictly speaking, they represent different quantities. Consistency in their use is more than just a question of semantics, and may have contributed to confusion in the literature regarding the treatment of porosity gradients in bioturbation models (see Meysman et al., 2005, for a discussion of this problem and its resolution; also see the discussion below where these results are discussed).

This then implies that while φ, φ_s, v, and ω all vary with sediment depth, the products φv and $\varphi_s \omega$ do not. Furthermore, since these two products can now be treated as constants, we will see that in some diagenetic equations (e.g., eqns. 6.22 and 6.23) they can be moved outside of the appropriate derivatives of advective flux terms.

An alternate way to view the observations in eqn. (6.16b) is to note that if the mass flux to the sediment surface (F^s_{sed}) is constant with time, then,

$$F^s_{sed} = \omega \, \varphi_s \rho_s \tag{6.17}$$

and therefore the terms in eqn. (6.16b) are both equal to F^s_{sed}/ρ_s (see Meysman et al., 2005, for further details). F^s_{sed} is also formally defined as the sedimentation rate, indicating that the sedimentation rate is a mass flux and not a mass velocity (Berner, 1980; Meysman et al., 2005). The conversion between sedimentation rates (in units of, e.g., g m^{-2} yr^{-1}) and sediment burial velocities (in units of, e.g., cm yr^{-1}) is carried out using eqn. (6.17) or its variants. Finally, the sediment accumulation rate (ω_{acc}) represents the rate of upward movement of the sediment-water interface relative to a fully consolidated base layer (i.e., the sediment depth at which φ and φ_s reach their asymptotic values φ_∞ and $\varphi_{s,\infty}$). Based on this definition it can be shown (Meysman et al., 2005) that the sediment accumulation rate equals the sediment burial velocity at great depth,

$$\omega_{acc} \equiv \omega_\infty = F^s_{sed}/(\rho_s \varphi_{s,\infty}) \tag{6.18}$$

The pore-water velocity v can also be described in a similar fashion. To do this, I note that at sediment depths where there is assumed to be no (or minimal) compaction that pore waters and sediments move downward at the same rate and therefore $v_\infty \approx \omega_\infty$. Applying this observation to eqns. (6.16a) and (6.17) it can be shown that

$$\varphi v = \frac{\varphi_\infty}{\varphi_{s,\infty}} \frac{F^s_{sed}}{\rho_s} \tag{6.19}$$

Returning briefly to the discussion at the beginning of this section, in the situation where compaction is absent (or assumed to be insignificant) φ and φ_s do not vary with depth. They are therefore treated as constants in the diagenetic equations. Based on the equations above it can also be shown that here the pore-water burial velocity (v), the sediment burial velocity (ω), and the sediment accumulation rate (ω_{acc}) are all constant with depth and equal to one another (and can similarly be treated as constants in diagenetic equations).

In sediments that are bioturbated it is necessary to also examine the relationship between bioturbation and compaction and changes with depth in sediment porosity. These parameters are strongly intertwined since compaction creates porosity gradients while under some circumstances bioturbation can destroy porosity gradients (based on the extent to which organisms mix solid particles alone versus solids plus pore waters). Past studies have also suggested several possible ways that macrofauna may increase porosity at depth, e.g., pelletization of sediments during feeding or water injection into sediments during burrowing (Rhoads, 1974; Aller, 1978; Mulsow et al., 1998; Ingalls et al., 2000). However, there does not appear to be a wealth of data on this subject, and the occurrence of this is likely to be a function of the type of benthic activity, e.g., mobile deposit feeding versus bioirrigating, sedentary deposit feeders (Rhoads, 1974).

In examining this topic, Boudreau (1986a) defined two end-member situations for the effects of bioturbation on sediment porosity. In one case, termed *interphase mixing*, bioturbation intermixes both pore waters and solids in a way that decreases porosity gradients. In the second case, bioturbation mixes solid particle and pore waters in a way that does not affect porosity, which is assumed to be controlled solely by physical compaction and sediment accumulation. This case is referred to as *intraphase mixing*. At the same time, these two types of mixing are not necessarily mutually exclusive and may occur at the same time. This could be a result of, for example, differences in the mixing or feeding behaviors of the different organisms inhabiting a sediment; here each type of mixing would provide separate contributions to the overall rates of sediment mixing and their impact on sediment porosity.

As might be expected, these different modes of sediment mixing are incorporated differently into equations for diffusive-like fluxes associated with bioturbation. In the case of interphase mixing, bioturbation affects gradients in both porosity and sediment solids, and J_b in eqn. (6.6) is expressed as

$$J_b = -D_b \frac{\partial \varphi_s \rho_{ds} S}{\partial z} \quad (6.20)$$

Similarly, in the case of intraphase mixing only gradients in solid sediment particles are affected by bioturbation (Boudreau, 1997a), and J_b in eqn. (6.6) is expressed as

$$J_b = -\varphi_s D_b \frac{\partial \rho_{ds} S}{\partial z} \qquad (6.21)$$

Although evidence to date is equivocal on the occurrence of intra- versus interphase mixing in sediments, past modeling studies have used both the intraphase and interphase formalisms for biological mixing, often with little justification or explanation. Determining which of these mixing models more accurately represents what occurs in marine sediments is of some importance for several reasons, including the fact that these different models can lead to significantly different sediment mixing rates (i.e., bioturbation coefficients) when they are used to model sediment depth profiles of naturally occurring and manmade radiotracers (e.g., Mulslow and Boudreau, 1998).

Recently Meysman et al. (2005) have rigorously re-examined these bioturbation models. They have shown that under the conditions of steady-state porosity (defined below) there exists only one correct form of diagenetic equation that includes bioturbation, regardless of the mode of biological mixing. In particular, they have observed under these conditions that φ_s should be placed outside of the differential in the modified form of Fick's First Law (i.e., that eqn. 6.21 is actually valid for both interphase and intraphase mixing).

In examining these results it is important to note that here steady-state porosity not only implies steady-state compaction, but also assumes that φ and φ_s are not dependent variables in the model. Rather, they are treated here as known functions of depth; i.e., porosity depth profiles are measured and the resulting data (or curve-fitting results obtained from the data) are used as model input parameters.

As Meysman et al. (2005) note, the reason these assumptions lead to the same flux equation regardless of the type of mixing is because with this approach the effects of biological mixing on sediment porosity (and porosity gradients) are now essentially embedded in the porosity profile. These observations also imply that by examining porosity profiles it should be possible to differentiate between the two types of mixing based on the shape of the profile. This approach therefore has the potential to more critically examine the relative importance of these two types of biological mixing in bioturbated sediments.

THE GENERAL DIAGENETIC EQUATION

Given all of the observations in this section, if we re-examine eqns. (6.2)–(6.5), we can write the following general steady-state diagenetic equation for sediment solids:

$$0 = \frac{\partial}{\partial z}\left(\varphi_s D_b \frac{\partial S}{\partial z}\right) - \varphi_s \omega \frac{\partial S}{\partial z} + \varphi_s \Sigma R_i \qquad (6.22)$$

Similarly for pore waters we obtain

$$0 = \frac{\partial}{\partial z}\left(\varphi[D_b + D_s]\frac{\partial C}{\partial z}\right) - \varphi v \frac{\partial C}{\partial z} + \varphi \Sigma R_i \qquad (6.23)$$

In deriving this latter equation it is also important to note that because molecular diffusion occurs only in pore waters, diffusive fluxes cannot result from porosity gradients (Berner, 1980). Therefore molecular diffusive fluxes are expressed as $\varphi D_s \dfrac{\partial C}{\partial z}$ and not $D_s \dfrac{\partial (\varphi C)}{\partial z}$.

Finally, how the derivatives are expanded in the diffusive and bioturbational flux terms in equations such as eqns. (6.22) and (6.23) will depend on whether porosity and/or D_b are functions of depth. Similarly if φ is a function of depth then so is D_s (e.g., see eqn. 6.11). Depending on how the depth dependences of these three terms are parameterized, this often leads to a differential equation that cannot be solved analytically, but must be solved numerically (see the beginning of section 6.3 for further details).

6.1.3 Adsorption

For an ion in the pore waters that undergoes reversible, equilibrium adsorption (see section 2.4.3 and eqn. 2.7), Berner (1976, 1980) discusses how adsorption can be included in diagenetic equations such as eqn. (6.23). If we start with his approach and incorporate the treatment above of bioturbative fluxes into his derivation, we obtain a similar, but slightly different, equation than the one he originally derived:

$$0 = \frac{1}{\varphi(1+K)}\frac{\partial}{\partial z}\left([\varphi D_b(1+K) + \varphi D_s]\frac{\partial C}{\partial z}\right) - \left(\frac{v + K\omega}{1+K}\right)\frac{\partial C}{\partial z} \\ + \frac{1}{1+K}\left(\Sigma R_i + \frac{\varphi - 1}{\varphi}\rho_s \Sigma \overline{R}_i\right) \qquad (6.24)$$

(compare this equation to eqn. 4-51 in Berner, 1980). In this equation $\Sigma \bar{R}_i$ represents surface reactions involving C, e.g., radioactive decay of an adsorbing ion (Berner, 1976), and K is the dimensionless adsorption coefficient,

$$K = \frac{\rho_{ds}(1-\varphi)}{\varphi} K' \tag{6.25}$$

where K' is defined in eqn. (2.7)

Assuming constant porosity, then K becomes a depth-independent constant, and $\omega = v$ (as discussed above). If we also assume that the sediments are not bioturbated, nor are there surface reactions affecting C, then eqn. (6.24) can be rewritten as

$$0 = \frac{D_s}{1+K}\frac{\partial^2 C}{\partial z^2} - \omega \frac{\partial C}{\partial z} + \frac{1}{1+K}\Sigma R_i \tag{6.26}$$

As this equation indicates, adsorption decreases the importance of molecular diffusion and biogeochemical reactions relative to pore water advection. However, if ω is also very small then eqn. (6.26) becomes,

$$0 = D_s \frac{\partial^2 C}{\partial z^2} + \Sigma R_i \tag{6.27}$$

Under these conditions, then, reversible equilibrium adsorption has no apparent effect on sediment pore-water profiles, regardless of the strength of adsorption (see Berner, 1976, for further details).

6.2 Solutions to the Diagenetic Equations

Although eqns. (6.22)–(6.24) look complicated, simplifying assumptions can often be made that do not apparently compromise the accuracy of the resulting solution. For example, if one assumes constant porosity in sediments, then eqn. (6.23) can be simplified to

$$0 = (D_b + D_s)\frac{\partial^2 C}{\partial z^2} - \omega \frac{\partial C}{\partial z} + \Sigma R_i \tag{6.28}$$

SOLUTIONS TO THE DIAGENETIC EQUATIONS

Eqn. (6.22) for solid phase constituents (S) can also be similarly simplified. If we also neglect bioturbation (which is likely to be a reasonable assumption in organic-rich, sulfidic sediments), then eqn. (6.28) can be written as

$$0 = D_s \frac{\partial^2 C}{\partial z^2} - \omega \frac{\partial C}{\partial z} + \Sigma R_i \tag{6.29}$$

As discussed in section 5.4, the Peclet number comparing sedimentation and diffusion generally indicates that diffusion is the predominant process (also see table 5.1), allowing us to further simplify eqn. (6.29) as

$$0 = D_s \frac{\partial^2 C}{\partial z^2} + \Sigma R_i \tag{6.30}$$

This equation therefore demonstrates that steady-state pore water profiles are often controlled by the balance between diffusion and biogeochemical reactions. Furthermore, depending on the functional relationships used for ΣR_i, eqns. (6.28)–(6.30) may also be simple second-order linear differential equations that can easily be solved analytically (see discussions below for further details).

At the same time, if φ (and therefore D_s) is not constant with depth, both terms can sometimes be expressed as fairly straightforward functions of depth. Under such conditions these equations can be expressed as

$$0 = f(z)\frac{\partial^2 C}{\partial z^2} - g(z)\frac{\partial C}{\partial z} + \Sigma R_i \tag{6.31}$$

which again may be solved analytically depending on the degree of complexity of $f(z)$, $g(z)$, and ΣR_i.

One important aspect of these solutions is the way in which the reaction term ΣR_i is handled. In one case, ΣR_i can be mathematically expressed as a function of C or z and then substituted into the diagenetic equation. Again, depending on the complexity of the resulting differential equation, the equation can often be solved analytically. Examples of this approach will be discussed in later sections of this chapter.

Another approach to this problem is to first fit the concentration data to some arbitrary function of depth (e.g., a numerical spline fit to the data or an analytical exponential or polynomial function). The first and second derivatives of this function can then be calculated and used to determine ΣR_i versus depth by substitution into eqns. (6.27)–(6.30). This approach is particularly useful if one does not wish to make any assumptions about the depth or concentration dependence of ΣR_i (see further discussions of this in Goldhaber et al., 1977; Alperin et al., 1994; Berg et al., 1998). Furthermore, in conjunction with other sediment biogeochemical data, the resulting depth profile of ΣR_i can then be used to further understand the kinetics of the process of interest.

Generally, solutions to time-dependent sediment diagenesis models ($\partial C/\partial t \neq 0$) require numerical techniques (e.g., Klump and Martens, 1989; Soetaert et al., 1996b). However, exact analytical solutions can be obtained for some models (e.g., Aller, 1980a; Boudreau, 1997a). Additional details about such models are discussed below and in later chapters.

6.2.1 Boundary Conditions

Obtaining a specific solution to any of these steady-state diagenetic models for a given sediment requires a set of boundary conditions. Mathematically, the boundary conditions are used to obtain specific values for the constants of integration in the solution to the differential equation of the model. Since many of these diagenetic models are second-order differential equations they need two boundary conditions.[2]

The lower boundary condition used in many sediment diagenesis models assumes that at great depth ($z \to \infty$) concentration gradients ($\partial C/\partial z$ or $\partial S/\partial z$) go to zero; i.e., concentrations no longer change with depth. Other models may, however, specify a particular concentration value at a specific depth (e.g., see discussions in section 8.4 of layered diagenesis model). For pore-water equations the upper

---

[2] As is discussed in any text on differential equations the number of boundary conditions needed to solve a differential equation depends on the order of the equation. Similarly for a time-dependent differential equation one also needs an initial condition (e.g., concentration versus depth at $t = 0$) as well as the appropriate number of boundary conditions (see, e.g., Boudreau, 1997a, for further details).

boundary condition generally sets the pore-water concentration at z = 0 to the bottom water value. However, some models (e.g., Boudreau, 1996; Hammond et al., 1996) specify the flux at z = 0, based on the assumption that transport through the diffusive sublayer above the sediments controls exchange between the sediments and bottom waters (see section 12.2). In diagenetic models of solid phase constituents either concentration or flux boundary conditions are also used at the upper boundary, and this flux is often prescribed with sediment trap measurements made close to the sediment surface (e.g., Jahnke et al., 1982; Emerson et al., 1985). Note that in many papers this flux is often referred to as the *rain* or *rain rate* of material to the sediments. Flux boundary conditions are used more often in models of bioturbated sediments. This approach is generally taken because in bioturbated sediments the relationship between the rain rate of a constituent and its concentrations at the sediment surface is complicated by the bioturbation process (see section 6.3.2 for further details).

6.3 Solutions to Specific Diagenetic Equations

In this last section I will discuss analytical solutions to several common forms of steady-state diagenetic equations (i.e., $\partial C/\partial t = \partial S/\partial t = 0$). The assumption of steady state is one that is often made but generally not explicitly discussed, although in many cases this assumption may indeed be reasonably valid. A further discussion of these considerations is postponed until chapter 14.

As we will also see, several of the simplifying assumptions discussed earlier in this chapter are incorporated into these model equations. In recent years, a number of workers have also presented numerical solutions of more complex forms of these equations that do not make many of these assumptions. Because of the way these models are developed, they consist of sets of coupled differential equations that then allow one to model several pore-water and solid-phase species simultaneously (Boudreau, 1996; Dhakar and Burdige, 1996; Soetaert et al., 1996c; Wang and Van Capellen, 1996; Archer et al., 2002; also see section 8.4 and general discussions of these numerical models in Boudreau, 2000). The level of complexity of these models varies from relatively complex computer codes to more simple models that can be solved using an Excel spreadsheet (Burdige, 2002).

6.3.1 Organic Matter Remineralization without Bioturbation

In this model we assume steady-state conditions and constant porosity with depth. We also assume that there is only one "type" of reactive organic matter (G) in the sediments, and its degradation is a first-order process with rate constant k (see discussion in section 8.1 for further details). Given these assumptions, eqn. (6.22) can be rewritten as

$$-\omega \frac{dG}{dz} - kG = 0 \tag{6.32}$$

Since we have eliminated the time derivative here, partial derivatives are now written as full derivatives. Similarly, since we have neglected bioturbation, this is now a first-order differential equation, requiring only one boundary condition. This equation can be re-expressed in a way such that its form is similar to the radioactive decay equation,

$$\frac{dG}{dz} = -\gamma G \tag{6.33}$$

where $\gamma = k/\omega$. With the boundary condition $G = G_o$ at $z = 0$ the solution to this equation is

$$G = G_o e^{-\gamma z} \tag{6.34}$$

This equation then predicts that both the total organic carbon concentration and its rate of remineralization ($= -kG$) should decrease exponentially with depth, as is often seen in many sediments. It also predicts that the TOC concentration goes to zero at depth in the sediments. This, however, is not generally observed, since almost all marine sediments contain some organic carbon at depth. The incorporation of this latter observation into sediment diagenesis models is discussed in section 8.1. Furthermore, this long-term carbon burial (or preservation) is an important part of the global carbon cycle and will be discussed in further detail in later sections (e.g., section 8.3).

SOLUTIONS TO SPECIFIC DIAGENETIC EQUATIONS

6.3.2 Organic Matter Remineralization with Bioturbation

Here we assume that there is constant bioturbation with depth in the sediments, leading to the equation,

$$D_b \frac{d^2G}{dz^2} - \omega \frac{dG}{dz} - kG = 0 \tag{6.35}$$

where D_b is the bioturbation (or biodiffusion) coefficient as defined in section 12.2. Since this is a second-order homogeneous linear differential equation (with constant coefficients), solving this equation requires two boundary conditions. These will be a flux condition at the upper boundary and a concentration boundary condition at depth. If we specify the flux to the sediment surface (rain rate) as J, then based on eqns. (6.3) and (6.6) this implies that

$$J = \left(-D_b \frac{dG}{dz} + wG \right)_{z=0} \tag{6.36}$$

We will also assume here that at the lower boundary $G = 0$ as $z \to \infty$.

Overall, the assumptions in this model are somewhat unrealistic since they imply that bioturbation occurs to an infinite depth in the sediments. A more realistic approach to the inclusion of bioturbation in sediment diagenesis models is discussed in sections 8.4 and 12.4. Nevertheless, because the solution to this model does illustrate several interesting points, its discussion is included here.

Eqn. (6.36) can be solved using standard procedures (see Boudreau, 1997a, or any text on differential equations). Doing so we obtain an equation whose form is similar to eqn. (6.34),

$$G = G_o e^{-\beta z} \tag{6.37}$$

where

$$\beta = \frac{-\omega + \sqrt{\omega^2 + 4kD_b}}{2D_b} \tag{6.38}$$

and

$$G_o = \frac{J}{D_b \beta + \omega} \tag{6.39}$$

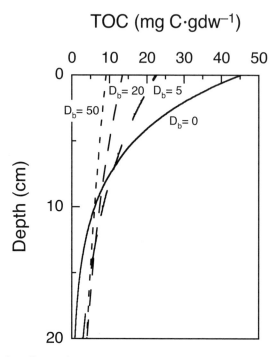

Figure 6.3 The effects of varying D_b on TOC depth profiles predicted by eqns. (6.37)–(6.39).

As noted above, this solution illustrates several interesting points. First, it can be seen that when $D_b = 0$, $G_o = J/\omega$. This makes intuitive sense since in the absence of bioturbation the concentration of organic carbon at the sediment-water interface should be related to its flux to the sediments or rain rate (J) and the rate at which it is diluted out by other sedimentary material (parameterized here by ω). This point can also be seen in eqn. (6.36) by setting $D_b = 0$ and recognizing that in the absence of bioturbation, G at $z = 0$ is G_o. Second, eqn. (6.39) also shows that G_o decreases with increasing bioturbation (i.e., as D_b increases). Again, this makes sense since bioturbation mixes surface sediments that are more carbon rich with deeper sediments that have been depleted in organic carbon as a result of remineralization. At the same time, increasing D_b values also lead to an increasingly smaller depth gradient in G (i.e., β also decreases; see fig. 6.3).

SOLUTIONS TO SPECIFIC DIAGENETIC EQUATIONS

6.3.3 Organic Matter Remineralization Coupled to Sulfate Reduction

Expressing sulfate reduction as

$$2CH_2O + SO_4^{-2} \rightarrow H_2S + 2HCO_3^- \tag{6.40}$$

(see chapter 7) the diagenetic equation for sulfate (C) is often written as,

$$D_s \frac{d^2C}{dz^2} - \omega \frac{dC}{dz} - \mathcal{L}\widetilde{\mathfrak{F}} k_s G = 0 \tag{6.41}$$

where we neglect bioturbation, assume constant porosity, and also assume that sulfate reduction is independent of sulfate concentration and simply proportional to the amount of organic carbon present (see section 8.1 for further details). The parameter $\mathcal{L}$ is the stoichiometric ratio between the moles of sulfate reduced and the moles of organic carbon oxidized, and is assumed here to be 1/2 (see section 11.2.1 for a further discussion of this value). The parameter $\widetilde{\mathfrak{F}}$ converts solid phase concentration units (e.g., mol g_{dw}^{-1}) to pore-water units (mol l^{-1}), and it is defined as

$$\widetilde{\mathfrak{F}} = 10^3 \rho_{dw} \frac{(1-\varphi)}{\varphi} \tag{6.42}$$

$\widetilde{\mathfrak{F}}$ has units of $g_{dw}\, l^{-1}$. For completeness I have also included the advective term in this equation; as discussed above, under many circumstances it can be neglected.

In solving eqn. (6.41) we need first to define the depth dependence of G. To do this the diagenetic equation for organic carbon given by eqn. (6.32) is used, and substituting its solution (eqn. 6.34) into eqn. (6.41) yields,

$$D_s \frac{d^2C}{dz^2} - \omega \frac{dC}{dz} - \mathcal{L}\widetilde{\mathfrak{F}} k_s G_o e^{-\gamma_s z} = 0 \tag{6.43}$$

where $\gamma_s = k_s/\omega$. This equation is a second-order linear nonhomogeneous differential equation, the general form of which is

$$a \frac{d^2y}{dx^2} + b \frac{dy}{dx} + cy = f(x) \tag{6.44}$$

where a, b, and c are constants. The approach to solving this equation involves first obtaining a solution to the homogenous form of the equation (i.e., where $f(x) = 0$) to get the general solution to the equation, and then using $f(x)$ to guess at the particular solution to this equation (see any text on differential equations or Boudreau, 1997a, for further details). The complete solution to this equation is then a linear combination of the general and particular solutions. For the boundary conditions $C = C_o$ at $z = 0$ and $C = C_\infty$ as $z \to \infty$ the solution to eqn. (6.43) is then

$$C = C_\infty + \Delta C e^{-\gamma_s z} \tag{6.45}$$

where

$$\Delta C = C_o - C_\infty = \frac{\omega^2 \tilde{\mathfrak{F}} \mathcal{L} G_o}{\omega^2 + k_s D_s} \tag{6.46}$$

This equation then predicts that pore-water sulfate concentrations decrease with depth in an exponential-like fashion, as is often seen in many anoxic marine sediments (see fig. 6.4 and section 17.1 for further details). A positive value of C_∞ (i.e., $\Delta C < C_o$) implies that organic carbon, rather than sulfate, limits sulfate reduction. However, this model can also predict physically unrealistic negative C_∞ values for sediments in which sulfate, rather than organic matter, limits sulfate reduction (Berner, 1980; Boudreau and Westrich, 1984; Martens and Klump, 1984). This point can be illustrated by examining eqn. (6.46) and noting that large values of G_o will eventually lead to ΔC values that exceed C_o and thus yield negative C_∞ values. The modeling approach typically used to overcome this problem is described in section 8.4 (although also see discussions in Boudreau and Westrich, 1984, for other solutions to this problem).

6.3.4 Ammonium Production in Anoxic Sediments

A very similar approach can be used to examine ammonium production from organic matter during sulfate reduction. Here we assume that ammonium production is proportional to the amount of metabolizable organic nitrogen in the sediments, and because ammonium is known to adsorb to sediment particles, the diagenetic

SOLUTIONS TO SPECIFIC DIAGENETIC EQUATIONS

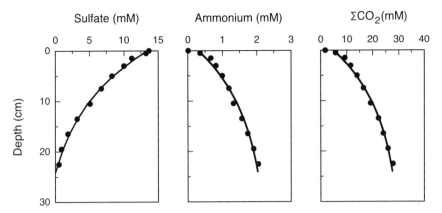

Figure 6.4 Pore-water profiles of sulfate, ammonium, and ΣCO_2 collected in July 1995 from a site in the mesohaline portion of Chesapeake Bay (Burdige unpub. data; see Burdige, 2001, for a site description). The sulfate data have been fit to eqn. (6.45), and the ammonium data to eqn. (6.49). The ΣCO_2 data have also been fit to a modified form of this latter equation (i.e., with no adsorption coefficient). These three best-fit profiles yield very similar values of γ (≈ 0.07 cm^{-1}) and therefore similar values of k_s, k_n, and k_c (≈ 0.05 yr^{-1}). This observation agrees with discussions in the text (sections 7.6.1 and 11.2.1) regarding the tight coupling of the processes responsible for either the consumption or production of these pore-water constituents during organic matter remineralization.

equation must also take into account this process. This then leads to the following diagentic equations for particulate (organic) nitrogen and dissolved ammonium (N and A, respectively),

$$-\omega \frac{dN}{dz} - k_n N = 0 \qquad (6.47)$$

$$\frac{D_s}{1+K_n} \frac{d^2 A}{dz^2} - \omega \frac{dA}{dz} + \frac{\tilde{\delta} k_n N}{1+K_n} = 0 \qquad (6.48)$$

where K_n is the dimensionless reversible ammonium adsorption coefficient (see Rosenfeld, 1979b, or Mackin and Aller, 1984a, for values of K_n). The boundary conditions for these equations are analogous to those used for the organic carbon and sulfate models on which these equations are based. The solutions to eqns. (6.47) and (6.48) are then,

$$N = N_o e^{-\gamma_n z} \tag{6.49}$$

$$A = A_o + \Delta A(1 - e^{-\gamma_n z}) \tag{6.50}$$

where

$$\Delta A = A_\infty - A_o = \frac{\omega^2 \tilde{\gamma} N_o}{(1 + K_n)\omega^2 + k_n D_s} \tag{6.51}$$

and $\gamma_n = k_n/\omega$. This equation for pore-water ammonium predicts that ammonium concentrations increase with depth in an asymptotic fashion, again as is seen in many anoxic marine sediments (see fig. 6.4 and section 16.2 for further details). Also note that similar asymptotic profiles are seen for other pore-water constituents whose production rates decrease with sediment depth (e.g., silica or ΣCO_2; also see section 13.4).

A similar approach can, in principle, be used to model phosphate production in anoxic sediments (e.g., Krom and Berner, 1981; Klump and Martens, 1987). However, in this case, authigenic phosphate mineral formation and adsorption of phosphate onto sediment iron oxides often lead to a more complex picture of phosphorus diagenesis (see section 16.3 for further details).

Although the equations for particulate organic carbon and nitrogen and dissolved ammonium and sulfate appear to be independent of one another, they do have certain interrelationships. The first is that if the C/N ratio of the organic matter undergoing remineralization (r_n) is constant with sediment depth (see section 11.2), then this ratio is given by

$$r_n = G_o/N_o \tag{6.52}$$

At the same time, under many circumstances we can assume that $k = k_s = k_n$, such that the various γ parameters described above are all equal to one another (see fig. 6.4). In general, the tight coupling of the reactions that lead to the overall remineralization of sediment organic matter likely also leads to this apparent equality of k values (see discussions in sections 7.6.1 and 11.2.1 for further details).

6.3.5 Determination of Sediment Accumulation Rates

Most often the determination of sediment accumulation rates involves the examination of depth profiles of a radioisotope (radiogeochemical tracer) whose half-life matches the time scales of sediment accumulation (see section 3.8.1). At the same time, the determination of sedimentation rates is also often strongly tied to the determination of bioturbation coefficients in sediments. Here I will briefly discuss how sedimentation rates are determined in the absence of bioturbation, saving the discussion of the calculation of sedimentation rates in bioturbated sediments for section 12.4.

The steady-state diagenetic equation for a radioisotope in nonbioturbated sediments with constant porosity is written as

$$0 = -\omega \frac{\partial A}{\partial z} - \lambda A \qquad (6.53)$$

where A is the activity (dpm gr^{-1}) and λ is the radioactive decay constant (see section 3.8.1). Implicit in this approach are also the assumptions that there is constant input of this radioisotope to the sediments with time, sediment accumulation is constant, and the isotope undergoes no postdepositional migration or reaction beyond radioactive decay. Given these assumptions the solution to eqn. (6.53) is:

$$A = A_o e^{-\lambda z/\omega} \qquad (6.54)$$

where A_o is the activity at the sediment-water interface. It is interesting to note the similarity of this equation to the classic radioactive decay equation (eqn. 3.15), and to recognize that here the sedimentation rate transforms a time coordinate system to a depth coordinate system.

As discussed previously (section 6.1.2), porosity variations with depth due to compaction complicate the interpretation and modeling of sediment profiles because of the effect of compaction on the velocities of both sediments and pore waters relative to the fixed sediment-water interface. However, Robbins (1978) has shown that if depth is replaced with a term defined as cumulative mass-depth, then an equation analogous to eqn. (6.54) can be obtained that relates sediment activity (A) to mass-depth (also see similar discussions in Mulsow et al., 1998). In this approach cumulative mass-depth (m) is defined as,

$$m(z) = \int_0^z \rho_s [1-\varphi] dz \approx \bar{\rho}_s \int_0^z [1-\varphi] dz \quad (6.55)$$

assuming an average dry sediment density ($\bar{\rho}_s$) for the sediments. This transformation to mass-depth units then leads to the equation

$$A = A_o e^{-\lambda m/r} \quad (6.56)$$

where r (the mass accumulation rate in units of, e.g., g m^{-2} yr^{-1}) is assumed to be constant over time (as was ω above). Note that r is identical to F^s_{sed} discussed in the context of eqn. (6.17).

Using either approach, eqns. (6.64) or (6.66) can be rewritten as

$$\ln(A) = \ln(A_o) - (\lambda/\omega)z \quad (6.57a)$$

$$\ln(A) = \ln(A_o) - (\lambda/r)m \quad (6.57b)$$

and semilogarithmic plots of A versus z or m should both be linear; the slopes of these lines can then be used to calculate either ω or r. Also note that in the case of variable porosity $\omega(z)$ and r are related by,

$$\omega(z) = \frac{r}{\bar{\rho}_s [1-\varphi]} \quad (6.58)$$

(see similar discussions in section 6.1.2 and also compare with eqn. 6.17). Examples of such plots for excess ^{210}Pb versus sediment depth in coastal sediments can be seen in section 12.4.1 (also see Chanton et al., 1983; Cornwell et al., 1996; Nie et al., 2001; or see Robbins, 1978, for an earlier review).

≈ **CHAPTER SEVEN** ≈

Biogeochemical Processes in Sediments

A CENTRAL THEME of this book is that the remineralization of sediment organic matter plays a key role, either directly or indirectly, in many of the geochemical transformations that occur in marine sediments. To understand the diagenesis of sediment organic matter we need to know several things about these processes, including the potential reactions that may occur and the relationships between these reactions (i.e., thermodynamic vs. kinetic considerations). These discussions focus largely on bacterial metabolism, in part because there is a much greater diversity of metabolic processes in prokaryotes[1] than in eukaryotes. However, where appropriate the relevant aspects of the metabolic processes associated with higher organisms that inhabit marine sediments are discussed.

Bacteria have relatively large surface-to-volume ratios as compared to higher, eukaryotic organisms. This then implies that their metabolic processes (on a per organism volume basis) can have a much greater impact on overall sediment metabolism, even if they represent only a few percent of the total benthic biomass (see discussions in Nealson, 1997; Fenchel et al., 1998). At the same time, bacterial biomass dominates the total benthic biomass in many, but not all, marine sediments (see references above and also Aller, 1998). This is most pronounced in anaerobic (oxygen-deficient) sediments, since only a few protozoans and fungi are capable of anaerobic metabolism. Thus in most sediments, bacterial metabolism dominates organic matter remineralization, providing further justification for the need to understand the wide range of metabolic processes in bacteria.

---

[1] The term *bacteria* will be taken here to be synonymous with prokaryotes. Except where specifically noted, it therefore includes archaea (i.e., organisms in the group *Archaea*, formerly referred to as archaebacteria), cyanobacteria, and true bacteria or eubacteria (i.e., organisms in the group *Bacteria*; see Wose et al., 1990, for further details).

7.1 Bacterial Metabolism: General Considerations

Bacterial metabolism occurs in order for cells to acquire energy (*dissimilatory metabolism* or *catabolism*) or to produce new cellular material (*assimilatory metabolism* or *anabolism*). These processes require an external energy source as well as the materials needed for growth. In most cases the two processes are tightly coupled since bacteria generally use the vast majority of the energy they obtain during catabolism on growth (Nealson, 1997; Fenchel et al., 1998).

In further describing these processes we also distinguish between *autotrophic* and *heterotrophic* metabolism. Heterotrophs require an external organic carbon source for growth energy and as a precursor for cellular biosynthesis. In contrast, autotrophs use C-1 compounds (e.g., CO_2 or methane) as their carbon source. These organisms then obtain their growth energy, i.e., "fix" C-1 compounds into new biomass, from either sunlight (*photoautotrophs*) or from redox (chemical) reactions involving inorganic substrates (*chemolithotrophs*; see section 7.4 for further details).

Most bacterial processes that impact sediment geochemistry result from dissimilatory metabolism, although assimilatory processes, i.e., production of new bacterial biomass, may be important in some aspects of sediment carbon metabolism and sediment carbon preservation (see sections 7.5.2 and 11.4 for further details). Broadly, the acquisition of energy during dissimilatory metabolism involves either respiration or fermentation, with the basic difference between these processes being the mechanism by which ATP is generated (ATP, or adenosine triphosphate, is the molecule used by all cells to internally store and transfer energy). During fermentation, ATP is generated by substrate-level phosphorylation, and there is a direct stoichiometric relationship between substrate catabolism and ATP production from ADP (adensosine diphosphate). In contrast, oxidative phosphorylation, or electron transport, does not involve a strict stoichiometric coupling between catabolism and ATP production. Rather, substrate catabolism feeds electrons into an electron transport chain consisting of membrane-bound enzymes, and the process generates a "proton motive force" that produces ATP using a membrane-bound ATPase enzyme (see Nealson, 1997, Fenchel et al., 1998, or any microbiology text for further details).

There are also other important differences between fermentation and respiration. Fermentation is an anaerobic process in which an organic carbon substrate is partially oxidized and partially reduced, with no external electron acceptor involved. Thus during fermentation there is a redox balance between the initial substrate and all of its products. Examples include glucose (alcohol) fermentation,

$$C_6H_{12}O_6 \rightarrow 2CH_3CH_2OH + 2CO_2 \quad (7.1)$$
$$\text{(glucose)} \quad \text{(ethanol)}$$

propionate fermentation,

$$CH_3CH_2COOH + 2H_2O \rightarrow CH_3COOH + CO_2 + H_2 \quad (7.2)$$
$$\text{(propionate)} \quad \text{(acetate)}$$

or amino acid fermentation (Stickland reaction), e.g.,

$$CH_3CHNH_2COOH + 2CH_2NH_2COOH \rightarrow 3CH_3COOH$$
$$\text{(alanine)} \quad \text{(glycine)} \quad \text{(acetate)} \quad (7.3)$$
$$+ 3NH_3 + CO_2$$

Fermentation therefore produces a mixture of oxidized and reduced products, including H_2, CO_2, organic acids, and alcohols. Fermentation is also a relatively inefficient process in terms of ATP production per mole of substrate consumed, and much of the chemical energy in the starting substrate is retained in the fermentation end-products.

In contrast, respiration involves the oxidation of organic compounds coupled to the reduction of an external inorganic electron acceptor. Respiration also generally oxidizes the carbon substrate completely to CO_2 and is a much more efficient process. For example, aerobic respiration of glucose yields ~32 mol ATP per mol glucose consumed, while glucose fermentation yields only 2–4 mol ATP per mol glucose consumed.

7.2 Bacterial Respiration and Biogeochemical Zonation in Sediments

In any natural system (including marine sediments), the occurrence of specific respiratory processes appears to be controlled by the free energy yield per mole of organic carbon oxidized by each of these electron acceptors (Claypool and Kaplan, 1974; Froelich et al., 1979). Based on this model, the oxidant that provides the greatest amount of free energy will be utilized first, and when this oxidant is depleted the next most efficient oxidant will be used. In principle, this will continue until either all available oxidants are consumed or all oxidizable (metabolizable) organic matter is utilized.

On much of the Earth's surface and in the oceans, the amount of oxygen is sufficient relative to the amount of reduced (oxidizable) carbon, and aerobic respiration predominates. If we approximate this organic matter as CH_2O,[2] then we can describe this process as

$$CH_2O + O_2 \rightarrow CO_2 + H_2O \tag{7.4}$$

Although clearly not correct biochemically, we can simplistically think about aerobic respiration as a process that runs photosynthesis "in reverse," oxidizing CH_2O back to CO_2 to acquire the energy that was originally fixed into this organic matter by phototosynthesis.

Naturally produced marine organic matter is actually a complex mixture of proteins, carbohydrates, lipids, and other biopolymers (see chapter 9 for further details) and one representation of the chemical composition of this material that takes into account the N and P containing compounds in this organic matter is expressed by the traditional Redfield-Ketchum-Richards (RKR) equation,

$$(CH_2O)_{106}(NH_3)_{16}(H_3PO_4) + 138O_2 \rightarrow 106CO_2 + 16HNO_3 \\ + H_3PO_4 + 122H_2O \tag{7.5}$$

This equation then predicts that the average oxidation state of carbon in marine organic matter is approximately zero and that the atomic C:N:P ratio of this material is 106:16:1 (also referred to as the Redfield ratio). This equation also predicts that the aerobic remineralization of this material is tightly coupled to organic nitrogen oxidation, present largely in the amide form as proteins. This latter process is referred to as *nitrification* and is carried out by chemolithotrophic bacteria (see section 7.4.1 for further details). Since the formulation of the RKR equation in the 1960s several studies have re-examined the chemical composition of marine organic matter using a wide range of analytical and oceanographic techniques. These will be discussed in further detail in section 9.3.

[2] Note that CH_2O is not formaldehyde, but a shorthand way to describe naturally produced organic matter. It can be thought of as carbohydrate material (e.g., $C_6H_{12}O_6$) "divided by 6" (compare eqn. 7.4 with the equation for aerobic respiration in table 7.1)

TABLE 7.1
Standard-state Free Energy Changes for the Major Organic Matter
Remineralization Processes*

Process	Chemical Reaction	$\Delta G°$ (kJ mol Glucose^{-1})
Aerobic respiration	$C_6H_{12}O_6 + 6O_2 \rightarrow 6CO_2 + 6H_2O$	−2.82
Denitrification	$5C_6H_{12}O_6 + 24NO_3^- \rightarrow 12N_2 + 24HCO_3^- + 6CO_2 + 18H_2O$	−2.66
Manganese reduction	$C_6H_{12}O_6 + 18CO_2 + 6H_2O + 12\,\delta\text{-MnO}_2 \rightarrow 12Mn^{2+} + 24HCO_3^-$	−2.38
Iron reduction	$C_6H_{12}O_6 + 42CO_2 + 24Fe(OH)_3 \rightarrow 24Fe^{2+} + 48HCO_3^- + 18H_2O$	−0.79
Sulfate reduction	$2C_6H_{12}O_6 + 6SO_4^{2-} \rightarrow 6H_2S + 12HCO_3^-$	−0.45
Methanogenesis	$2C_6H_{12}O_6 \rightarrow 6CH_4 + 6CO_2$	−0.30

* Data for $C_6H_{12}O_6$, δ-MnO$_2$, and Fe(OH)$_3$ are for glucose, fine-grained vernadite, and amorphous ferric oxide, respectively. Values of standard free energies of formation are from Stumm and Morgan (1996).

In addition to the aerobic remineralization of organic matter, anaerobic remineralization occurs in oxygen-deficient environments. Such environments are common in many marine sediments, and bacterial metabolism dominates here, utilizing alternate electron acceptors other than O_2 for the oxidation of organic matter. Table 7.1 lists these anaerobic remineralization reactions; the predicted pore-water profiles based on the sequential occurrence of these reactions are shown in fig. 7.1. The organisms responsible for these processes are generally referred to as either *facultative* or *obligate anaerobes*, depending on whether they can use O_2 and these other electron acceptors for catabolism (facultative anaerobes) or whether they are limited to using only one or more of these alternate electron acceptors (obligate anaerobes).

The occurrence of these processes is defined in table 7.1 by their free energy values. However, other formalisms look at this "biogeochemical zonation" from the standpoint of the redox or half-cell

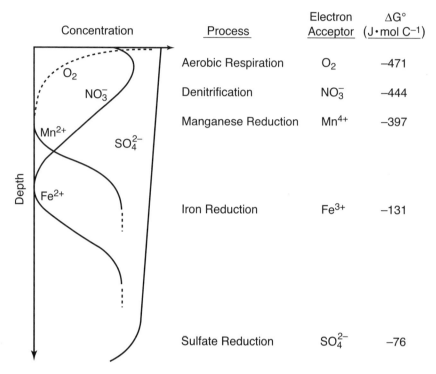

Figure 7.1 Pore-water profiles predicted by the successive utilization of inorganic terminal electron acceptors during the remineralization of organic matter in marine sediments (modified after Froelich et al., 1979, and Burdige, 1993). ΔG° values based on those in table 7.1.

potential (E_H) of the reduction of these terminal electron acceptors, in spite of the many difficulties encountered in the interpretation of measured E_H values in natural systems (Berner, 1981a; Stumm and Morgan, 1996; Fenchel et al., 1998). This approach then leads to a gradient from positive E_H values (oxidizing conditions) in oxic zones where aerobic respiration occurs to negative E_H values (reducing conditions) in deeper, anoxic sediments where sulfate reduction occurs (see fig. 7.2).

Below the oxic zone of marine sediments and above the deeper, more reducing (and often sulfidic sediments), the sediments are referred to as either *suboxic* (Froelich et al., 1979) or *postoxic* (Berner, 1981a). In this region of the sediments, the processes of denitrification and manganese and iron reduction occur. Below the suboxic

BACTERIAL RESPIRATION

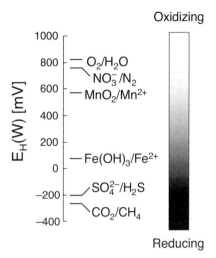

Figure 7.2 The half cell potential at pH 7 [$E_H(W)$] for the reduction of terminal electron acceptors used in organic matter remineralization (compare with fig. 7.1). E_H is related to other thermodynamic parameters such as ΔG or $p\varepsilon$ (the hypothetical electron activity) according to

$$E_H = \frac{-\Delta G}{nF} = 2.303 \frac{RT}{F} p\varepsilon,$$

where R is the gas constant, F is Faraday's constant and n is the number of electrons transferred in the half reaction. Note that under standard conditions E_H^o, ΔG^o and $p\varepsilon^o$ are similarly related, and are also equal to $(RT/nF)\ln K$ (see Stumm and Morgan, 1996, for further details). In this formalism $p\varepsilon$ (or E_H) indicates the relative tendency of a solution to accept (or transfer) electrons. While $p\varepsilon$ and pH are somewhat analogous, it is also important to recognize that free or hydrated electrons (unlike protons) do *not* exist in water. Based on data in Stumm and Morgan (1996).

zone, the sediments are generally referred to as *anoxic*, and in this portion of the sediments sulfate reduction and methanogenesis occur.[3] Highly reducing, anoxic sediments also often contain free sulfide in the pore waters, depending on the availability of reactive iron to form iron sulfide minerals (see section 17.1).

---

[3] Strictly speaking suboxic/postoxic sediments are also anoxic since they too are devoid of oxygen. However, under most, but not necessarily all, circumstances, anoxic sediments are defined as those in which sulfate reduction and methanogenesis occur.

To a first approximation, many pore-water profiles appear to be consistent with this biogeochemical zonation model, particularly in deep-sea (pelagic and hemipelagic) sediments (Burdige, 1993). At the same time, in other sediments, particularly those that are bioturbated and/or bioirrigated, the distribution of biogeochemical processes in the sediments is more complex than that predicted by this relatively simple conceptual model (e.g., see discussions in sections 7.4 and 7.5.2 and chapter 12).

One factor complicating the use of this model to understand sediment biogeochemistry is that the free energies listed in table 7.1 are those at standard state; i.e., they are $\Delta G°$ values at 25°C, 1M concentration for aqueous species and 1 atm pressure for gasses. In actuality, though, under natural *in situ* conditions one needs to consider the ΔG at these specific conditions,

$$\Delta G = \Delta G° + RT \ln \frac{\Pi_i [product]^{ni}}{\Pi_j [reactant]^{nj}} \tag{7.6}$$

where R is the gas constant, T is temperature (°K), and the symbol Π indicates the product of either the product or reactant concentrations, with each concentration raised to its appropriate stoichiometric coefficient. As will be discussed below, this observation can have some importance in an understanding of the occurrence of certain biogeochemical processes in sediments.

Finally, it is important to recognize that the biogeochemical zonation model is not necessarily meant to provide specific information about the controls on carbon remineralization and preservation in sediments (e.g., see Postma and Jakobsen, 1996). Rather, it provides information on the controls and occurrence of the various terminal electron acceptor (or respiratory) processes that are coupled to organic carbon oxidation (e.g., see table 7.1). Such information is of some interest since the occurrence of these specific reductive processes leads both directly and indirectly to the occurrence of a whole host of other processes that play important roles in the overall geochemistry of marine sediments. Therefore, in spite of its limitations, the biogeochemical zonation model has proven quite useful for understanding many aspects of marine sediment biogeochemistry.

7.3 Bacterial Respiration: Specific Processes

7.3.1 Aerobic Respiration

This process uses oxygen as the electron acceptor, and is the most efficient respiratory process in terms of energy acquired per mole of carbon oxidized. It dominates the organic matter remineralization process in most pelagic sediments (section 7.5.2), because of the relatively low carbon input into these sediments as compared to the oxygen input (see chapter 13 for further details). In addition to oxidizing organic matter, other aerobic bacteria obtain growth energy from the oxidation of methane and reduced inorganic substances such as ammonium, hydrogen, sulfide, and reduced iron and manganese (see section 7.4 for further details). Most of these chemoautotropic, or chemolithotrophic, bacteria then use the energy gained from these reactions to fix carbon, i.e., to produce new cellular material from CO_2.

7.3.2 Denitrification

When oxygen is present in sediments, it is generally the thermodynamically preferred electron acceptor for respiration. As a result, the concentration of O_2 usually decreases with sediment depth, and the concentration of nitrate increases as a result of nitrification accompanying aerobic respiration (see section 7.4.1 for further details). However, as oxygen concentrations approach zero at depth, there reaches a point where nitrate becomes the preferred electron acceptor.

From a thermodynamic standpoint, this transition of respiratory processes occurs when the ΔG values for these reactions, i.e., their free energies at *in situ* conditions, become such that denitrification provides a greater amount of free energy per mole of carbon oxidized than does aerobic respiration. The discussion above of eqn. (7.6) therefore indicates that, all other factors held constant, as the oxygen concentration decreases the ΔG for aerobic respiration will become less negative, and the ΔG for denitrification will become more negative (since nitrate levels increase as a result of nitrification coupled to aerobic respiration). At some point, then, these ΔG values will cross over, and nitrate will become the thermodynamically favored electron acceptor, even though some oxygen may still be present. However, studies with marine bacteria isolated from oxygen-deficient environments and analyses of water column data from similar regions suggest that this crossover occurs at O_2 levels very close to zero, between ~2 and 8 μM (Devol, 1978).

When oxygen levels fall below this critical level, denitrification occurs. This remineralization process uses nitrate as an electron acceptor and produces N_2 gas as its predominant end-product. The reductive pathway for nitrate can be written as

$$NO_3^- \to NO_2^- \to NO \to N_2O \to N_2 \qquad (7.7)$$

and all of these intermediates are known to accumulate to some extent during denitrification. Furthermore, the production of nitrous oxide by denitrification may represent the major process by which N_2O is added to the atmosphere from the oceans (Codispoti et al., 2001). Denitrifying bacteria are widespread throughout the groups *Archaea* and *Bacteria* and many are facultative anaerobes. Denitrifying bacteria can also use a wide range of organic carbon substrates. In most sediments there is often a tight coupling between denitrification and nitrification, and much of the nitrate used in sediment denitrification appears to come from nitrate produced in the sediments by nitrification (see section 16.2).

Recent work has also shown that the reduction of nitrate coupled to the oxidation of reduced inorganic substrates, such as ammonium, sulfide, or reduced iron and manganese, may be important in some marine sediments; these reactions will be discussed in section 7.4.2. Finally, under some anaerobic conditions nitrate can be reduced to ammonium by a process referred to as *nitrate fermentation* or *nitrate ammonification* (Sørensen, 1987; Koike and Sørensen, 1988; Fenchel et al., 1998). This process differs from denitrification in that the fermentative organisms that reduce nitrate to ammonium use nitrate as an electron sink to dispose of excess reducing equivalents without directly coupling this reductive process to ATP production. Nitrate ammonification can be thought of as part of a broader group of processes referred to as *dissimilatory nitrate reduction to ammonium* (DNRA), which also includes nitrate reduction to ammonium at the expense of sulfide oxidation by organisms such as *Thioploca* (Schulz and Jorgensen, 2001). The controls on denitrification versus dissimilatory nitrate reduction are poorly understood, and in shallow estuarine and tidal flat sediments the rates of the two processes may be comparable (Sørensen, 1987; Koike and Sørensen, 1988; Tobias et al., 2001; An and Gardner, 2002). An and Gardner (2002) also suggest that the controls on the occurrence of these processes may be related to sulfide levels in these sediments, with high sulfide level inhibiting denitrification and enhancing DNRA. This observation would then be consistent with the observation that much of the DNRA in such sediments appears to occur via sulfide oxidation, as opposed to nitrate fermentation. However, this remains to be demonstrated definitively.

7.3.3 Manganese and Iron Reduction

Unlike the other respiratory processes discussed here, reduction of manganese and iron oxides uses solid, rather than dissolved, electron acceptors. As a result, the form of the solid oxide (e.g., mineralogy, degree of crystallinity, available oxide surface area) affects its reactivity, i.e., the energetics (ΔG) of reduction. The most reactive metal oxides (with respect to reductive dissolution) are generally poorly crystalline, amorphous phases (Burdige et al., 1992b; Thamdrup, 2000). In sediments they are often found as coatings on inorganic or biogenic particles (e.g., clays or siliceous tests). Their small crystal size and lack of long-range ordering further adds to their apparent amorphous nature when analyzed by conventional X-ray diffraction techniques.

7.3.3.1 MANGANESE AND IRON REDOX CHEMISTRY AND MINERALOGY

An important aspect of the biogeochemistry of these metals is that they undergo redox transformations at Earth surface conditions that change the solubility of the metals: reduced forms are generally soluble, while oxidized forms (solid oxides) are highly insoluble. Iron oxidation by O_2 is generally faster than manganese oxidation and likely not biologically catalyzed except under acidic conditions (see section 7.4.1 and Stumm and Morgan, 1996). Aerobic manganese oxidation also appears to be an autocatalytic process, in that Mn (and Fe) oxide surfaces catalyze further Mn^{2+} oxidation. In contrast, iron oxides are more resistant to reduction than are manganese oxides (Nealson, 1997) for thermodynamic and/or kinetic reasons that will be discussed below.

Under most circumstances, manganese oxidation produces an amorphous to poorly crystalline form of vernadite (or δ-MnO_2), although the mineralogy of these oxidation products is often not well defined (also see section 13.3). The oxidation of Mn^{2+} appears to be nonstoichiometric, producing a compound whose chemical formula is sometimes written as MnO_x, where x generally varies from ~1.3 to

1.9 (Burdige, 1993; Stumm and Morgan, 1996). Whether this is caused solely by mixtures of Mn^{2+} and Mn^{4+} in the oxide or involves the occurrence of Mn^{3+} in the solid is uncertain (Burdige, 1993). A more detailed discussion of the mineralogy of manganese oxides is beyond the scope of this discussion and can be found in several reviews (Burns and Burns, 1979, 1981; Tebo et al., 1997; Post, 1999).

The predominant product of iron oxidation appears to be amorphous ferrihydrite (hydrous ferric oxide or $Fe(OH)_3$), although amorphous and more crystalline forms of FeOOH (lepidochrosite or goethite) and Fe_2O_3 (hematite) are also found in marine sediments (Murray, 1979; Canfield and Raiswell, 1991b; Canfield et al., 1992; Haese et al., 1997; Thamdrup, 2000). Given the many difficulties in the quantitative separation and identification of these iron oxides in sediments, a number of selective leaching techniques have been developed in recent years to examine these mineral phases (see Canfield, 1989a; Kostka and Luther, 1994; Raiswell et al., 1994, and summaries in Haese, 2000; Thamdrup, 2000). Based on these studies, it has been estimated that ~25–28% (±10%) of the total iron found in sediments is iron oxides that are reactive with respect to either microbial or chemical reduction (by, c.g., sulfide; see below and section 17.1) on early diagenetic time scales, i.e., with half-lives of less than 1 month with respect to reduction by dissolved sulfide (Canfield et al., 1992; Raiswell and Canfield, 1998; Poulton et al., 2004). Recent work, however, using ^{57}Fe Mössbauer spectroscopy suggests that nanophase goethite (nanogoethite) may be the dominant reactive iron oxide in both marine and lacustrine sediments (Van der Zee et al., 2003).

Over time, the amorphous iron and manganese oxides that form in sediments from either chemical or biological Mn^{2+} or Fe^{2+} oxidation should slowly transform into more stable crystalline forms. However, the active diagenetic (redox) cycling that these metals undergo in sediments (see sections 7.4.3 and 13.2) likely plays a role in inhibiting such transformations on early diagenetic time scales.

The predominant products of metal oxide reduction, whether it occurs chemically or biologically, are soluble Mn^{2+} and Fe^{2+}. Both ions are known to strongly adsorb to sediment surfaces, including their own oxides, and in some sediments their adsorptive behavior may strongly influence the geochemistry of these metals (Slomp et al., 1997; Thamdrup, 2000). In some cases the soluble reduced iron and manganese produced by these reactions can precipitate out as sulfide and carbonate minerals (see sections 13.1, 13.2, 17.1, and 17.3). Magnetite (Fe_3O_4), a magnetic phase containing both Fe^{2+} and Fe^{3+}, can also be a product of iron reduction, although its production appears to be of minor importance in marine sediments (see discussions in Thamdrup, 2000).

7.3.3.2 MICROBIAL MANGANESE AND IRON REDUCTION

The occurrence of iron and manganese reduction in marine sediments is generally consistent with the concept of biogeochemical zonation discussed above. However, Mn and Fe oxides can also be reduced by a number of organic acids and inorganic compounds, such as sulfide, and Fe^{2+} itself can reduce manganese oxides (Sunda and Kieber, 1994; Nealson, 1997; Thamdrup, 2000; also see section 7.4.3). Thus, metal oxide reduction not directly coupled to bacterial respiration may also occur in these regions of sediments. Under such circumstances organisms may indirectly reduce metal oxides by producing these compounds and then excreting them into the pore waters, where they chemically react with the metal oxides.

Since metal oxides are solid particles, bacteria cannot utilize them in the same way that they can use dissolved substances such as O_2, nitrate, or sulfate. Thus if metal oxide reduction is used in respiration, then the responsible organisms need some way to couple electron transport to metal oxide reduction. In many cases this is accomplished by direct contact with the solid oxides (Nealson and Saffarini, 1994). In other cases, electron-shuttling compounds and/or Fe^{3+} chelators (see the discussion above) may overcome the need for this direct contact (Nevin and Lovley, 2002). Many organisms capable of reducing metal oxides during respiration, e.g., the genera *Geobacter* or *Shewanella*, are able to use both metal oxides and nitrate as electron acceptors (Nealson and Saffarini, 1994).

In addition to the microbial reduction of iron oxides, some of the structural Fe^{3+} in clays, e.g., in octahedral sites of smectites (see section 2.4), can be reduced by bacteria and coupled to organic carbon oxidation and respiration (Kostka et al., 1996, 2002). In sediments, the reduction of structural iron in clays generally results in a tan-green color change that marks the upper boundary of the apparent zone of iron reduction (Lyle, 1983; König et al., 1997, 1999; also see sections 7.5.1 and 14.4). Estimates of the reducible iron in clays range from ~12 to 26% of the total sediment iron (see Thamdrup, 2000, for a summary). Since all structural (silicate-bound) iron is ~60% or more of the total sediment iron (Habicht and Canfield, 1997;

Chester, 2000; Thamdrup, 2000), this suggests that only a fraction of the silicate-bound iron can be reduced (also see discussions in Raiswell and Canfield, 1998). In many cases the rates of microbial iron reduction using structural iron are comparable to those observed with iron oxides (Kostka et al., 2002), and under some circumstances the reduction of structural iron in clays is reversible (Lyle, 1983; Heller-Kallai, 1997; König et al., 1997, 1999). However, in other cases the process results in the production of soluble Fe^{2+} and the partial dissolution of the original clay mineral (Kostka et al., 1999a; Dong et al., 2003a,b).

7.3.4 Sulfate Reduction

Bacterial sulfate reduction has a very low $\Delta G°$ as compared to oxic and suboxic processes (table 7.1), and based on the biogeochemical zonation model is therefore predicted to occur in sediments when the organic carbon input is in excess relative to the input of oxidants such as O_2, nitrate, and metal oxides. From the standpoint of sediment redox potentials this process generally occurs only when the sediments become very reduced and have a low E_H (see fig. 7.2). Because of these factors, under most circumstances sulfate reduction occurs in sediments devoid of oxygen, although it also has been observed to occur both in the lab and in certain natural settings in the presence of oxygen (see Fauque, 1995, for a review). In general, sulfate reduction is not important in pelagic sediments (see section 7.5.2), although it is the predominate process by which organic matter undergoes remineralization in coastal sediments and in many continental margin sediments. In part this occurs because of the large quantity of sulfate found in seawater compared to other more energetically favored oxidants (Reeburgh, 1983; Canfield, 1989b; also see discussions in sections 7.5.2 and 15.4).

Compared to aerobic bacteria, sulfate reducing bacteria are able to use a somewhat limited number of substrates in catabolism (also see discussions in 7.6). These include low molecular weight fatty acids such as lactate, acetate, propionate, and butyrate, some alchohols and amino acids, and H_2. Some groups of sulfate reducers oxidize these carbon compounds only to acetate, e.g., the genus *Desulfovibrio*, while others can completely oxidize these compounds to CO_2, e.g., the genus *Desulfobacter*. Because of the limitations of their metabolism, other microbial processes (e.g., fermentation) are important in producing the specific substrates used by sulfate reducers from the larger, more complex organic matter deposited in sediments. These microbial processes are, however, tightly coupled such that the net reaction for these processes can be expressed as the equation in table 7.1, in spite of the fact that sulfate reduction is responsible only for the last portion of the process. This point will be discussed further in section 7.6.2.

During bacterial sulfate reduction, sulfate is reduced to sulfide, which in many sediments reacts with iron to form iron monosulfides (FeS) and eventually pyrite (FeS_2). At earth surface temperatures and pressures the reduction of sulfate to sulfide requires bacterial catalysis, although chemical (abiotic) sulfate reduction occurs at the elevated temperatures and pressures found at hydrothermal (mid-ocean ridge) spreading centers (Elderfield and Schultz, 1996). The reduction of sulfate to sulfide is an important part of the global sulfur cycle and the cycling of many trace metals (see section 17.3). During oxidative portions of the sulfur cycle, sulfur intermediates such as thiosulfate or elemental S can be produced, and their subsequent reduction can also be a source of sulfide in some sedimentary environments (Kasten and Jørgensen, 2000). For example, the genus *Desulfuromona* reduces S°, rather than sulfate, to sulfide during the oxidation of actetate to CO_2 (see section 17.1 for further details).

7.3.5 Methanogenesis

Methane production occurs in marine sediments only when they become highly reducing. Significant amounts of methane are not produced in marine systems until sulfate is almost completely exhausted (less than ~1 mM, versus a normal seawater concentration of ~28 mM; e.g., Alperin et al., 1994; Hoehler et al., 1994), consistent with the thermodynamics arguments (biogeochemical zonation) discussed earlier (see table 7.1). Methane production is more important in freshwater sediments, given the much lower sulfate concentrations in freshwaters. As is the case for sulfate reduction, methane production does not occur at Earth's surface conditions in the absence of biological catalysis. Methane is, however, produced at high temperatures and pressures when organic matter is buried to great depths, producing what is referred to as *thermogenic* or *catagenic methane*, as opposed to the *biogenic methane* produced by the microbial processes being discussed here (see Hunt, 1996, for further details).

Methanogens are able to use an even smaller number of substrates than sulfate reducers, and there are two major types of methanogenesis. The first is acetate disproportionation,

$$CH_3COOH \rightarrow CO_2 + CH_4 \tag{7.8}$$

performed by, e.g., microbes of the genus *Methanosarcina*. The second is CO_2 reduction,

$$4H_2 + CO_2 \rightarrow CH_4 + 2H_2O \tag{7.9}$$

and is performed by microbes of the genera *Methanobacterium* or *Methanococcus*, among others. The latter process can also use CO and formate as the carbon substrate. Methane production by some organisms also occurs using methanol and certain methyl amines. This type of methanogenesis appears to be most important as the source of small amounts of methane that are produced in anoxic sediments in the presence of sulfate. Here, these methanogens use noncompetitive substrates that sulfate reducers apparently cannot metabolize (King et al., 1983).

However, as discussed above, significant amounts of methane production occur only in the absence of sulfate. Based on lab and field studies, this biogeochemical zonation appears to occur because sulfate reducers outcompete methanogens for acetate and H_2, key substrates that are used by both groups of organisms (Crill and Martens, 1986; Fauque, 1995; Hoehler et al., 1998). As we will see in section 7.5.1, these molecules are also important end-products in many fermentative reactions.

To further describe the nature of this competition, we can look at this problem using the Michaelis-Menten (or Monod) equation for an enzyme-mediated process:

$$R = \frac{V_{max}C}{K_m + C} \tag{7.10}$$

where R is the rate of the process, C is the concentration of the substrate being utilized (e.g., here, acetate or H_2), V_{max} is the maximum uptake rate, and K_m is the half-saturation constant, i.e., the concentration at which R equals $1/2\ V_{max}$. At the low acetate or H_2 concentrations typical of most marine sediments (see sections 7.6.2 and 10.3.1), eqn. (7.10) simplifies to

$$R = (V_{max}/K_m)C \qquad (7.11)$$

which implies that for similar values of V_{max} that lower K_m values will lead to faster uptake rates. Several studies have shown that the K_m values for acetate or H_2 uptake by sulfate reducers are indeed lower than the same values for methanogens (Widdel, 1988; Fauque, 1995), consistent with this explanation. Therefore, when sulfate is present, this competition for acetate or H_2 implies that sulfate reduction will predominate.

7.3.5.1 ANAEROBIC METHANE OXIDATION

In many marine sediments much of the methane produced during methanogenesis is oxidized anaerobically, apparently using sulfate as the electron acceptor. This reaction can be written as,

$$CH_4 + SO_4^{2-} \rightarrow HCO_3^- + HS^- + H_2O \qquad (7.12)$$

(Hoehler et al., 1994; Valentine and Reeburgh, 2000), and thermodynamically it should be a favorable reaction ($\Delta G = -25$ kJ mol^{-1} under typical *in situ* conditions; Martens and Berner, 1977). While geochemical evidence since the 1970s has been consistent with the occurrence of this process (Reeburgh, 1982; Alperin and Reeburgh, 1985), isolating the organism(s) responsible for anaerobic methane oxidation has proven difficult. Furthermore, most of the well-studied methanotrophs, i.e., aerobic methane-oxidizing bacteria (see section 7.4.1) specifically require O_2 to oxidize methane.

As an explanation for these geochemical observations (described in section 17.2), it has been suggested that anaerobic methane oxidation (AMO) is mediated by a consortium of sulfate-reducing bacteria and methanogens, through a process termed *reverse methanogenesis* (Hoehler et al., 1994). Additional evidence for this comes from Boetius et al. (2000), who presented microscopic evidence of a consortium of sulfate-reducing bacteria and methane-consuming archaea that apparently mediates anaerobic methane oxidation; related evidence in support of this mechanism of AMO is also presented in Orphan et al. (2001).

According to this model of reverse methanogenesis, the methanogen in the consortium oxidizes methane with water by essentially running reaction (7.9) in reverse; the H_2 produced by this reaction is then subsequently consumed by the sulfate-reducing bacteria. If the H_2 partial pressure is kept sufficiently low because of a tight coupling between its production and consumption, then this H_2 production reaction becomes thermodynamically favorable, and the sum of these processes yields the apparent methane oxidation by sulfate according to rxn. (7.12).

While this suggestion is consistent with many of the studies of anaerobic methane oxidation, it has also been suggested that reverse methanogenesis involves the production of acetate instead of (or in addition to) H_2 from methane in the first step in anaerobic methane oxidation, i.e., by running rxn. (7.8) in reverse (Valentine and Reeburgh, 2000). Again the tight coupling of this acetate production to sulfate reduction would yield rxn. (7.12).

7.4 CHEMOLITHOTROPHIC REACTIONS

In addition to redox reactions involving reduced (organic) carbon oxidation, redox reactions involving only inorganic substances also occur in sediments. Many of these reactions are mediated by microbes generally referred to as *chemoautotrophs* or *chemolithotrophs*. As their name implies, these organisms are generally autotrophic, using the chemical energy obtained from these reactions to fix carbon. However, some of these organisms can also assimilate fixed (reduced) carbon compounds to produce new biomass.

7.4.1 Aerobic Processes

During anaerobic respiration, a number of reduced end-products are produced whose oxidation by O_2 are thermodynamically favored; i.e., the reaction has a negative $\Delta G°$. These compounds include ammonium, methane, reduced iron and manganese, and several reduced forms of sulfur, e.g., sulfide and elemental sulfur. Microbial hydrogen (H_2) oxidation by O_2 is also known to occur, but because of the tight coupling of H_2 production and consumption in anoxic sediments its aerobic oxidation in nature is apparently limited (Fenchel et al., 1998).

Aerobic methane oxidation (methanotrophy) is carried out by several groups of bacteria (e.g., the genera *Methylomonas* or *Methylococcus*). Furthermore, the balance between methanotrophy and methanogenesis plays a key role in regulating the methane flux to the atmosphere from many aquatic environments (e.g., see discussions in Fenchel et al., 1998).

Ammonium oxidation, or nitrification, is an extremely common process in nature, both in sediments and in the water column. It occurs in two steps, with ammonium first being oxidized to nitrite by one group of organisms (the genera *Nitrosomonas* or *Nitrocystis*), and nitrite to nitrate by another group (the genera *Nitrobacter*, *Nitrococcus*, or *Nitrospirina*). As is the case for denitrification (section 7.3.2), during nitrification there is also production of the trace gases NO and N_2O (e.g., Anderson et al., 1993). The ammonium used in nitrification generally comes from particulate organic nitrogen that is regenerated during aerobic respiration (see section 7.2). However, in many sediments ammonium produced at depth in the sediments by anoxic remineralization diffuses upward to the oxic portion of the sediments to be nitrified (also see sections 12.5.2 and 16.2).

The bacterial oxidation of reduced sulfur species involves several oxidation reactions that eventually produce sulfate:

$$HS^- \text{ (sulfide)} \rightarrow S° \text{ (elemental sulfur)} \rightarrow S_2O_3^{2-} \text{ (thiosulfate)} \rightarrow SO_3^{2-} \text{ (sulfite)} \rightarrow SO_4^{2-} \text{ (sulfate)} \quad (7.13)$$

This process is carried out by some species of *Thiobacillus*, as well as by mat-forming organisms such as *Beggiatoa*, *Thioploca*, and *Thiothrix*. This complete oxidative process may actually involve several groups of organisms, each mediating only a portion of the sequence; as a result, there is often the transient production of partially oxidized intermediates such as elemental sulfur or thiosulfate. The cycling of many of these intermediates is an important part of sulfur dynamics in some environments (see section 17.1 for further details). However, in most cases, the net oxidation of sulfide to sulfate is of most geochemical significance.

The oxidation of reduced iron and manganese (Fe^{2+} and Mn^{2+}) can also be biologically mediated. For iron, this likely occurs only under acidic conditions, mediated by organisms such as *Thiobacillus ferrooxidans* (Fenchel et al., 1998). At pH values typical of most marine sediments, though, abiotic (chemical) iron oxidation is extremely rapid and likely outcompetes biological oxidation (Nealson, 1997). In contrast, at natural pH levels reduced manganese is more kinetically stable in the presence of oxygen, and its oxidation using O_2 is often biologically catalyzed. It has been suggested (Tebo et al., 1997) that the majority of the manganese oxidation occurring in sediments is biologically catalyzed (also see Thamdrup et al., 1994), although more work is clearly needed to definitively examine this suggestion. At the same time, though, unlike the other chemolithotrophic processes discussed in this section, evidence that manganese-oxidizing bacteria gain energy for growth from this oxidation reaction remains equivocal (see discussions in Tebo et al., 1997).

7.4.2 Anaerobic Processes

In recent years, field observations have suggested that a wide range of other inorganic oxidative reactions may occur in marine sediments using nitrate and manganese oxides, rather than O_2, as the oxidant (Sørensen, 1987; Aller and Rude, 1988; Bender et al., 1989; Aller, 1994b; Luther et al., 1997; Hulth et al., 1999; Thamdrup, 2000; Mortimer et al., 2002; Thamdrup and Dalsgaard, 2002; also see additional discussions in chapters 16 and 17). Under natural conditions these reactions have negative ΔG values (see table 7.2), and therefore chemolithotrophic bacteria should be able to couple these reactions to carbon fixation. Although many of these geochemical studies are indeed consistent with biological mediation of these processes, many of these reactions can also occur abiotically. Furthermore, in most cases the organisms responsible for these processes have not yet been isolated and/or studied in pure culture. One likely exception to this biological mediation is manganese reduction coupled to iron oxidation (rxn. 9 in table 7.2; Myers and Nealson, 1988; Burdige et al., 1992b). In spite of the fact that the reaction itself may be closely coupled to other biological processes (e.g., see section 7.5.1), its kinetics appear to be sufficiently rapid, and the process likely occurs abiotically.

7.4.3 Linkages between Chemolithotrophic and Organic Matter Remineralization Processes

Many of the chemolithotrophic reactions described above involve the oxidation of reduced end-products of anoxic and suboxic respiration. In many marine sediments this leads to the internal redox cycling of

TABLE 7.2
Some Possible Anaerobic (Non-O_2 Utilizing) Oxidation Reactions that May Occur in Marine Sediments*

Process	Chemical Reaction	ΔG (pH = 7)	ΔG (pH = 8)
Reactions coupled to NO_3^- reduction			
1. Anammox	$5/3NH_4^+ + NO_3^- \rightarrow 4/3N_2 + 3H_2O + 2/3H^+$	−421	−426
2. Denitrification coupled to sulfide oxidation	$NO_3^- + 5/8FeS + H^+ \rightarrow 1/2N_2 + 5/8SO_4^{2-} + 5/8Fe^{2+} + 1/2H_2O$	−419	−413
3. Denitrification coupled to Fe^{+2} oxidation	$NO_3^- + 5Fe^{2+} + 12H_2O \rightarrow 5Fe(OH)_3 + 1/2N_2 + 4H^+$	−319	−370
4. Denitrification coupled to Mn^{2+} oxidation	$5/2Mn^{2+} + NO_3^- + 2H_2O \rightarrow 5/2MnO_2 + 1/2N_2 + 4H^+$	−49	−71
Reactions coupled to manganese reduction			
5. Anoxic nitrification	$4MnO_2 + NH_4^+ + 6H^+ \rightarrow 4Mn^{2+} + NO_3^- + 5H_2O$	−175	−141
6. Mn reduction coupled to oxidative N_2 production	$3/2MnO_2 + NH_4^+ + 2H^+ \rightarrow 3/2Mn^{2+} + 1/2N_2 + 3H_2O$	−224	−212
7. Mn reduction coupled to complete sulfide oxidation	$2H^+ + MnO_2 + 1/4FeS \rightarrow Mn^{2+} + 1/4SO_4^{2-} + 1/4Fe^{2+} + H_2O$	−148	−137
8. Mn reduction coupled to partial sulfide oxidation	$3H^+ + MnO_2 + HS^- \rightarrow Mn^{2+} + S^0 + 2H_2O$	−136	−119
9. Mn reduction coupled to iron oxidation	$MnO_2 + 2Fe^{2+} + 4H_2O \rightarrow Mn^{2+} + 2Fe(OH)_3 + 2H^+$	−42	−127

*Modified from Hulth et al. (1999). The following conditions were used to calculate ΔG values: $[NO_3^-] = 5$ μM; $[NH_4^+] = 25$ μM; $pN_2 = 0.791$ atm; $[Mn^{2+}] = 100$ μM; $[Fe^{2+}] = 5$ μM; $[SO_4^{2-}] = 24$ mM; $[H_2S] = 25$ μM; $[HS^-] = 5$ μM. Activity coefficients were assumed to be 1. Values for the standard free energies of formation ($\Delta G°$) were taken from Berner (1980) and Stumm and Morgan (1996).

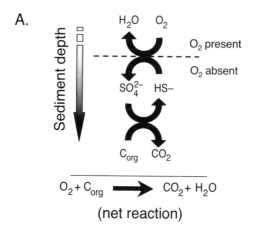

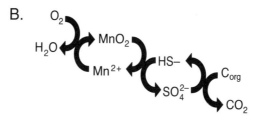

Figure 7.3 A. A schematic representation of the coupling of bacterial sulfate reduction and sulfide oxidation, such that in a net geochemical sense O_2 appears to be the electron acceptor used in the remineralization of sedimentary organic matter. B. A more complex coupling of oxygen, manganese, sulfur, and carbon redox cycling that may occur in some marine sediments (modified after Aller, 1994a, and Burdige, 1993).

these elements, such that aerobic chemolithotropic or abiotic oxidative processes are coupled with suboxic, or anoxic, organic matter remineralization (fig. 7.3A). These processes can then be thought of as an electron "shuttle" that brings O_2 oxidation equivalents into deeper anoxic (O_2-devoid) portions of the sediments (e.g., Burdige, 1993). Thus in a net sense O_2 "appears" to be the oxidant of organic matter in the sediments, in spite of the fact that the actual remineralization occurs by anoxic, or suboxic, metabolism (also see related discussions in Aller, 1994a). While this coupling can occur through pore water diffusive processes, more often it is mediated by macrofaunal activity, i.e., bioturbation and bioirrigation (see recent discussions in

Hulth et al., 1999, and references therein). This coupling implies that certain redox-sensitive elements, e.g., Mn or S, can undergo repetitive oxidation-reduction cycles before ultimately being buried in sediments (see sections 13.2, 16.1). Furthermore, as will be discussed in section 17.1 in more detail, this type of sulfur redox cycling has important implications in terms of understanding the sulfur isotope record in sediments and sedimentary rocks.

Similarly, when one examines the occurrence of anaerobic chemolithotrophic processes in marine sediments, it can also be shown that these reactions may be coupled with aerobic chemolithotrophic processes, as well as suboxic or anoxic organic matter remineralization (fig. 7.3B). As a result, elements such as manganese and sulfur may actually undergo more complex internal redox cycling than that described above; i.e., compare figs. 7.3A and 7.3B (also see section 7.5.1). The occurrence of such interactions suggests that the sediment biogeochemical cycles of Mn, Fe, N, S, C, and O_2 may be closely coupled and more complex than previously thought. Furthermore, these processes also have important implications for the internal dynamics of sediment processes, as well as net sediment (and perhaps global) budgets of these elements.

An important consequence of these interactions is that in some cases sediment oxygen uptake (also sometimes referred to as sediment oxygen demand) results not only from aerobic respiration, but also from oxidation of these reduced compounds (Mackin and Swider, 1989; Sampou and Oviatt, 1991). However, these processes may often be tightly coupled, such that there is no net loss of reduced intermediates, e.g., sulfide burial or benthic fluxes of dissolved Mn^{2+}, during this redox cycling. Under these circumstances sediment oxygen uptake occurring by aerobic respiration, chemolithotrophic reactions, and/or abiotic chemical oxidation still provides a reasonably good estimate of the overall (depth-integrated) rate of sediment carbon oxidation (i.e., oxic plus anoxic remineralization), even though aerobic respiration is only a fraction of the total sediment oxygen uptake (Sampou and Oviatt, 1991; Canfield et al., 1993a; Soetaert et al., 1996b; also see section 8.3.1). In contrast, the nonsteady-state conditions occurring in energetic margin sediments such as those on the Amazon shelf (section 5.5) may result in an uncoupling of sediment O_2 uptake from the cycling of these reduced metabolites, and here estimates of benthic metabolism based on sediment oxygen uptake are likely to be lower limits of their true value (Aller et al., 1996; McKee et al., 2004). Such sediment systems, with complex transport processes and nonsteady-state diagenetic conditions, therefore require the independent determination of the rates of sediment processes to ascertain estimates of depth-integrated rates of sediment carbon oxidation.

7.5 THE DISTRIBUTION OF ORGANIC MATTER REMINERALIZATION PROCESSES IN MARINE SEDIMENTS

7.5.1 Depth Scales of Biogeochemical Zonation

In the absence of significant bioturbation and/or bioirrigation, the reaction sequence outlined above (fig. 7.1) might be expected to be everywhere similar, given an elastic (or variable) depth scale. As will be discussed below, this depth scale is roughly controlled by the relative inputs of metabolizable organic matter and these various oxidants (see chapters 8 and 9 for a more detailed discussion of the concept of metabolizable organic matter).

As was discussed in section 7.3.3, the energetics of metal oxide reduction are a strong function of the form of the specific oxide undergoing reductive dissolution. For such reasons, the energy difference between manganese reduction and denitrification is often difficult to determine, and the zones of denitrification and manganese reduction apparently coexist in some sediments (fig. 7.4; also see Burdige, 1993). In contrast, iron reduction more often appears to occur below this zone, at least in pelagic and hemipelagic sediments, and one generally does not see the occurrence of net iron reduction until there is complete (or near complete) depletion of pore water nitrate. The onset of iron reduction also results in a tan-green color change in these sediments (see section 7.3.3.2; Lyle, 1983; König et al., 1997, 1999).

The causes of this dichotomy are not well understood, although one possible explanation is that Fe^{2+} produced by microbial iron reduction in suboxic sediments is reoxidized at the expense of either manganese reduction or denitrification (reactions 3 and 9 in table 7.2). The occurrence of these reactions then implies that if nitrate or manganese oxides are present, then one will not see a distinct (apparent) zone of iron reduction, i.e., Fe^{2+} accumulation in pore waters, until nitrate and manganese oxides are depleted. As discussed above, this observation is consistent with field data from many sediments. This suggestion is also consistent with the observation that in many

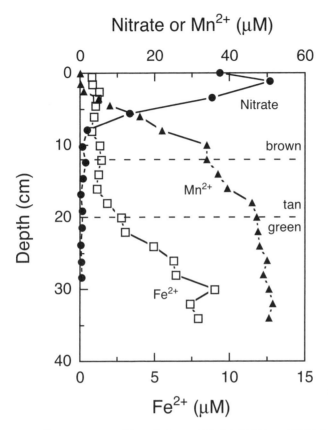

Figure 7.4 Pore-water profiles of nitrate (●), Mn^{2+} (▲) and Fe^{2+} (□) at MANOP site M. As discussed in the text, there is some degree of overlap between the zones of denitrification and manganese reduction above ~15 cm. Also note that the zone of iron reduction (as inferred from the accumulation of Fe^{2+} in the pore waters) does not appear to occur until nitrate is depleted, in association with the tan-green color change in the sediments. Data sources: Lyle, 1983 (Mn^{2+}, Fe^{2+}, sediment color); Klinkhammer, 1980 (nitrate). These data (along with other data from MANOP site M) are also shown in fig. 8.4.

of these sediments oxidants other than O_2 appear to be responsible for oxidizing pore water ferrous iron produced by iron reduction (fig. 7.5; Burdige, 1993). Model results support the occurrence of these coupled reactions involving anaerobic iron oxidation (Dhakar and Burdige, 1996; Wang and Van Capellen, 1996), although more work is still needed to verify these interactions.

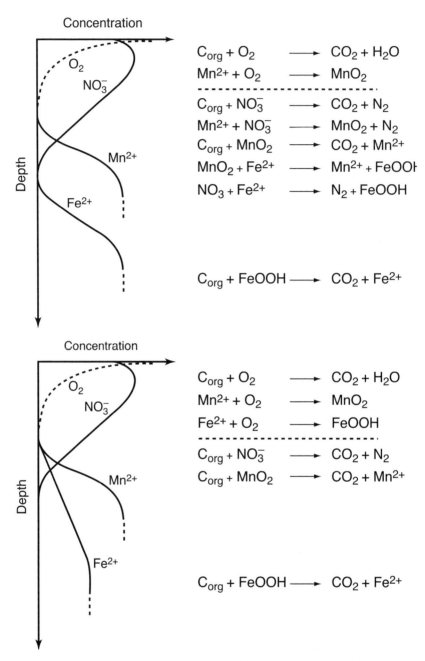

Figure 7.5 A. (Upper panel) Pore-water profiles in which the oxidation of iron in the pore waters is coupled to the reduction of either manganese oxides or

DISTRIBUTION OF ORGANIC MATTER

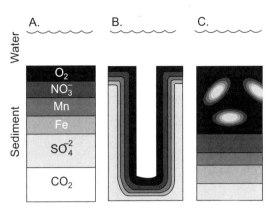

Figure 7.6 A. The classical biogeochemical zonation that occurs in many marine sediments (e.g., see fig. 7.1). Note the vertical redox zonation of electron acceptors used in these sediments. B. A representation of the zonation of redox reactions in the microenvironment surrounding an irrigated macrofaunal burrow. C. A representation of the zonation of redox reactions in the microenvironment surrounding a fecal pellet in a marine sediment (modified after Aller, 1982b).

In sediments that are actively bioturbated and/or bioirrigated, the concept of biogeochemical zonation as illustrated in fig. 7.1 is generally much more complicated than that discussed above. Rather than occurring as simple vertical zonation (also see fig. 7.6A), this zonation may occur radially around cylindrical animal burrows (fig. 7.6B), or near-spherically around organic matter "hotspots" such as fecal pellets in the sediments (fig. 7.6C). This then leads to a complex "three-dimensional mosaic of biogenic microenvironments" in these sediments (Aller, 1982b, 1988). Even in sediments that are not bioturbated, organic matter remineralization may still be concentrated in discrete, highly reactive microsites (Brandes and Devol, 1995).

---

nitrate (see equations to the right). As discussed in the text similar profiles are generally seen in many pelagic sediments (e.g., see fig. 7.4). B. (Lower panel) Hypothetical pore-water profiles that would exist if O_2 were the primary oxidant of pore-water iron (see equations to the right). Here it is assumed that iron oxidation is not coupled to the reduction of either manganese oxides or nitrate. Therefore iron must diffuse through the zones of manganese reduction and denitrification to a depth in the sediments where the upward flux of Fe^{2+} reaches the downward oxygen flux (modified after Burdige, 1993).

In bioirrigated or bioturbated sediments, particles and pore waters can also be exposed to alternating oxic and anoxic conditions as a result of these macrofaunal activities. In the simplest sense we think of these sediments as being periodically, or episodically, oxidized (or oxygenated), although the details of just how this occurs will vary (Aller, 1994a). The time scales over which these redox oscillations occur are quite variable, ranging from minutes for bioirrigation (Aller, 1994a) to longer time scales characteristic of sediment mixing (see table 5.1 and section 12.4). These redox oscillations are also generally asymmetrical in length, with anoxic conditions occurring for substantially longer times (10x or greater) than oxic conditions (Aller, 1994a; Aller et al., 2001). These oscillating (or mixed) oxic/anoxic sediment redox conditions occur, for example, when mobile deposit feeders oxygenate the sediments as they crawl through the sediments, or when burrowing organisms pump fresh (oxygenated) seawater through their burrows (also see Kristensen, 2001). Physical mixing (reworking) of sediments (section 5.5) can also lead to the occurrence of such mixed redox conditions. The importance of these redox oscillations on sediment processes will be discussed further in section 7.6.3 as well as in sections 12.5, 15.3, 16.2, and 17.1.1.

7.5.2 General Trends with Water Column Depth or Sediment Type

Broadly speaking, the occurrence of specific remineralization processes is controlled by the flux of reactive, or metabolizable, organic matter to the sediments. Since fluxes of organic matter to marine sediments generally decrease with water column depth (Fowler and Knauer, 1986; Wakeham et al., 1997b), this fact implies that there should be some regular variation of sediment carbon oxidation rates and organic matter remineralization processes with water column depth. Although factors other than water column depth also affect carbon fluxes to sediments, the use of water column depth as the master variable to examine organic matter remineralization processes in sediments appears to adequately describe the basic trends seen in the data (see an analogous approach in Middelburg et al., 1997).

Figure 7.7 shows a plot of depth-integrated sediment carbon oxidation rates (R_{cox}) versus water column depth, along with the partitioning of sediment carbon oxidation into the remineralization processes

DISTRIBUTION OF ORGANIC MATTER

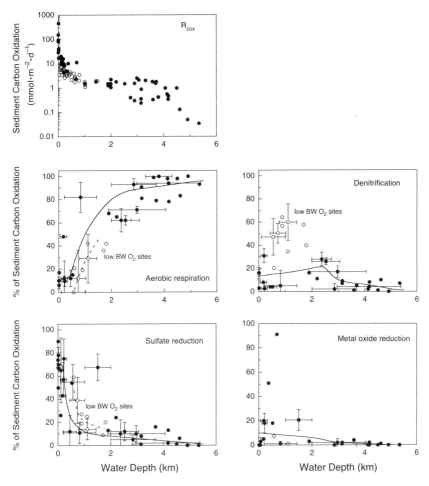

Figure 7.7 The depth-integrated rate of sediment carbon oxidation (R_{cox}) in selected marine sediments as a function of water column depth of the sediment. Also shown is the partitioning of this carbon oxidation into aerobic respiration, dentrification, metal oxide (Mn + Fe) reduction, and sulfate reduction (plus methanogenesis), again as a function of water column depth (see the references cited below for details on how these values were calculated or determined). Open symbols represent samples from sediment underlying low (<50 μM) bottom-water oxygen concentrations along the Washington, central California, and northwest Mexican continental margins (because of the strong oxygen minimum zone in the mid-waters of the eastern Pacific Ocean). For ease of comparison, in all panels except the R_{cox} panel individual data points represent results either from individual sites or average values

aerobic respiration, denitrification, metal oxide reduction (Mn + Fe reduction), and sulfate reduction plus methanogenesis.[4] The results shown here have been obtained using a number of analytical and modeling techniques, including benthic lander measurements and modeling of pore-water profiles, described in other sections of this book (e.g., sections 8.3.1 or 12.1). As such, a detailed comparison of all of these data is difficult. Furthermore, this figure obscures, in general, interocean-basin differences in sediment remineralization processes that may exist at similar water column depths. An additional uncertainty in these results is that some of the calculations shown here do not necessarily incorporate recently described anoxic chemolithotrophic reactions (section 7.4.2) in the mass balance equations used to estimate the values in this figure (also see discussions in section 15.3). Nevertheless, in spite of these potential problems, fig. 7.7 illustrates several important general trends.

◄──

from several sites within a given sediment setting. The curves drawn through the data are meant to help illustrate the trends in the data, and do not imply any known functional relationships. Sample locations and references: Equatorial Pacific between 2°S and 2°N (Hammond et al., 1996); MANOP sites M, H, S, and C and eastern Equatorial Atlantic (Bender and Heggie, 1984); Goban Spur (northeast Atlantic [European] continental margin; Lohse et al., 1998); Canadian continental margin (Boudreau et al., 1998); northwest Atlantic continental margin (Heggie et al., 1987); California Borderlands (Bender et al., 1989; Nealson and Berelson, 2003); Santa Barbara Basin (Reimers et al., 1996; Meysmann et al., 2003b); Patton Escarpment (Nealson and Berelson, 2003); central California (northeast Pacific) continental margin (Reimers et al., 1992); Washington state (northeast Pacific) and northwest Mexican (eastern tropical Pacific) continental margin (Hartnett and Devol, 2003); Chilean continental margin (Thamdrup and Canfield, 1996); Gulf of Maine (Christensen, 1989); Skagerrak (Canfield et al., 1993a); North Sea continental margin (Slomp et al., 1997); Danish coastal sediments (Jørgensen, 1982); Aarhus Bay, Denmark (Moeslund et al., 1994); Arctic coastal sediments (Kostka et al., 1999b); Long Island Sound (Mackin and Swider, 1989); mesohaline Chesapeake Bay (Roden et al., 1995; Marvin-DiPasquale and Capone, 1998); Cape Lookout Bight (Chanton et al., 1983; Martens et al., 1992).

[4] In most, but not all, highly reducing, anoxic sediments, sulfate reduction dominates organic matter remineralization, as compared to methanogenesis (see section 16.2). Therefore, for ease of presentation the two processes have been summed here.

Starting with deep-sea, pelagic sediments it is a general observation that aerobic respiration dominates organic matter remineralization in these sediments (Bender and Heggie, 1984; Emerson et al., 1985; Emerson and Hedges, 2003; Martin and Sayles, 2003). A recent global model of sediment diagenesis (Archer et al., 2002) suggests that at water depths below 1,000 m more than 95% of organic matter remineralization occurs via aerobic respiration, consistent with the results shown in fig. 7.7. In general, aerobic respiration dominates organic matter remineralization in deep-sea sediments because there is a relatively small flux of organic matter to the sediments that is also of low reactivity (see section 8.1). As a result, oxygen transport into the sediments (primarily by diffusion) is sufficiently rapid to "keep pace" with this organic matter input.

In examining remineralization processes in deep-sea sediments, Emerson et al. (1985) identified two end-members cases (fig. 7.8): a carbon-limited case, where detectable oxygen persists down to sediment depths of ∼1 meter or more, and an oxygen-limited case, in which oxygen in the pore waters decreases with depth in an exponential-like fashion to zero, often within the surficial zone of bioturbation. Because aerobic respiration dominates carbon oxidation in most pelagic sediments, virtually all sediment oxygen uptake is used in organic matter oxidation; that is, aerobic respiration coupled to nitrification. Oxidation of reduced solutes such as Fe^{2+}, Mn^{2+}, or HS^- either chemically or by chemoautotrophic reactions is, therefore, of minimal importance because there is limited production of these reduced solutes in deep-sea sediments by either suboxic or anoxic remineralization processes.

The net result of this is that estimates of deep-sea sediment oxygen uptake, based on either benthic lander or pore water measurements, can also be used to reasonably estimate rates of deep-sea sediment carbon oxidation/respiration. For these reasons, a number of empirical relationships have been proposed that relate sediment oxygen uptake to quantities such as bottom water oxygen and core top TOC concentrations (Cai and Reimers, 1995) or surface water primary productivity and water column depth (Christensen, 2000; Wenzhöfer and Glud, 2002). Such relationships can then be used to determine integrated rates of sediment oxygen uptake over larger regions; these predictions can then be compared to integrated rates of deep ocean oxygen utilization, which incorporate both water column and sediment respiration. Such comparisons suggest that ∼30–60% of the aerobic respiration in the deep sea occurs in the sediments (Jahnke, 1996; Christensen, 2000;

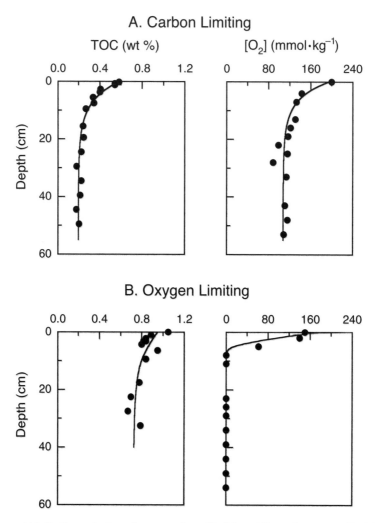

Figure 7.8 Sediment data from carbon limiting (A) and oxygen limiting (B) pelagic sediments (modified after Emerson and Hedges, 2003). Carbon limiting data are for a site in the central equatorial Pacific (Grundmanis and Murray, 1982) while oxygen limiting data are for a site in the northeast Pacific (Murray and Kuivala, 1990).

Schlüter et al., 2000; Wenzhöfer and Glud, 2002). In part, the large range appears to be a function of potential problems associated with applying relationships defined with regional data sets, e.g., the central and south Atlantic, to larger oceanic regions, e.g., the entire Atlantic or all deep oceans globally (see discussions most recently in Wenzhöfer and Glud, 2002).

In their studies of organic matter remineralization in deep-sea sediments Emerson et al. (1985) also suggested that the carbon content of marine sediments, and hence sediment carbon burial and preservation, is controlled by the carbon rain rate to the sediments and the oxygen content of the bottom waters. The extent to which oxygen availability controls carbon burial in sediments in general (or in deep-sea sediments in particular) has been a controversial topic (e.g., see discussions in Calvert and Pedersen, 1992; Hedges et al., 1999a). However, recent results do appear to indicate that oxygen availability may indeed play a role in sediment carbon preservation, at least in some marine sediments (see sections 7.6.3 and chapter 15).

As one moves out of pelagic sediments into hemipelagic (lower continental margin) sediments water column depths obviously decrease and fluxes of organic matter to the sediments increase. Sediment carbon oxidation rates similarly increase and eventually lead to a situation in which oxygen consumption exceeds its input. In such sediments pore-water O_2 will be completely consumed within several centimeters to millimeters below the sediment-water interface (e.g., Cai and Sayles, 1996), and the oxygen penetration depth in the sediments, i.e., the sediment depth of zero oxygen, decreases with decreasing water column depth (fig. 7.9). The apparent difference in the trends shown here for Atlantic versus Pacific sediments will be reexamined in the context of discussions in section 14.3.

As a result of these trends, as water column depth decreases, suboxic and anoxic remineralization processes become increasingly important, in a relative sense, in overall sediment carbon oxidation[5].

---

[5] While the *relative* importance of aerobic respiration decreases with decreasing water column depth, this trend is countered by the fact that the rate of total sediment carbon oxidation increases by more than 2 orders of magnitude as one moves from the deep-sea to nearshore sediments (see the upper left panel in fig. 7.7). When taken together, this then leads to *absolute* rates of sediment aerobic respiration actually increasing with decreasing water column depth. Overall, then, aerobic respiration accounts for ~70% of organic matter oxidation in all marine sediments (Emerson and Hedges, 2003).

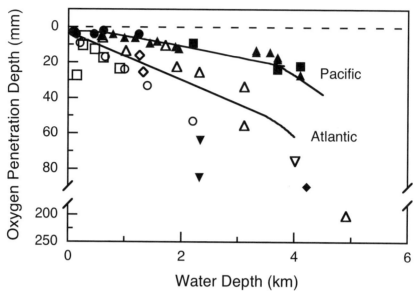

Figure 7.9 The sediment oxygen penetration depth as a function of water column depth in the Atlantic Ocean (open symbols) and Pacific Ocean (closed symbols). Data are from several sources (Murray and Kuivala, 1990; Archer and Devol, 1992; Reimers et al., 1992; Glud et al., 1994; Cai et al., 1995; Hales and Emerson, 1996, 1997a; Lohse et al., 1998; Martin and Sayles, 2003; Adler et al., 2001; Nealson and Berelson, 2003). Martin and Sayles (2003) suggest that these interocean differences are related to the presence of the intense oxygen minimum in the northeast Pacific. However, differences in the reactivity of sediment organic matter at equivalent depths in the two oceans may also play a role here (see discussions in section 14.3).

At the same time, the occurrence here of suboxic and anoxic remineralization processes increases the production of reduced metabolic end-products such as ammonium, sulfide, and reduced metals in these sediments. With this, some fraction of the total sediment oxygen uptake begins to be used for processes other than aerobic respiration, e.g., chemolithotrophic reactions such as those described in section 7.4.1 (for a discussion of this in continental margin sediments see Reimers et al., 1992; Ferdelman et al., 1999; Hartnett and Devol, 2003). This trend continues as one moves to nearshore, coastal sediments where ~50% or more of the sediment oxygen uptake is used (either directly or indirectly) in such nonrespiratory, oxidation reactions (Jørgensen, 1982; Sampou and Oviatt, 1991). However, in spite of this, under some circumstances sediment oxygen uptake measurements may still allow for a reasonable estimate of the total sediment organic matter oxidation rate, as was discussed in section 7.4.3.

Examining fig. 7.7 we see that the relative importance of denitrification increases with decreasing water column depth. In deep-sea sediments denitrification accounts for ~10% or less of the total organic matter remineralization, with this value increasing to ~20% in nearshore sediments. The general shape of the curve shown here is very similar to that published by Middelburg et al. (1996) in their synthesis modeling study of dentrification in marine sediments. The obvious exception to this general trend is in continental margin sediments underlying low bottom-water oxygen concentrations; for example, in the intense oxygen minimum zone of the northeast Pacific. Globally, denitrification accounts for ~7–11% of the total organic matter oxidation in all marine sediments (Middelburg et al., 1996; Brandes and Devol, 2002).

Figure 7.7 also indicates that the relative roles of sulfate reduction and aerobic respiration in sediment organic matter remineralization roughly mirror one another. With decreasing water depth, the importance of sulfate reduction increases from less than a few percent in deep-sea sediments to 25–50% (Jørgensen, 1982) or more as one moves to continental margin and nearshore, coastal sediments. In addition to the increasing carbon rain rate with decreasing water depth, the large concentration of sulfate in seawater (~28 mM) versus that of nitrate (less than ~40 μM) or reactive Mn and iron oxides (Reeburgh, 1983; Thamdrup, 2000) also plays a role here.

Finally, an examination of the results in fig. 7.7 suggests that under most circumstances metal oxide reduction accounts for at most ~10–20% of sediment organic matter remineralization. There is, however, some uncertainty in this estimate because of limited data on metal oxide reduction rates in sediments. In part this is due to the fact that, in many of the results summarized here, the calculations assume that aerobic respiration, denitrification, and sulfate reduction are the only remineralization process that occur in the sediments being examined. However, where metal oxide reduction processes have been explicitly examined in the context of sediment organic matter remineralization, under most circumstances manganese reduction indeed appears to be of minor importance. Therefore, in all but a few exceptions discussed below, iron reduction is the predominant metal oxide reduction process coupled to organic matter oxidation.

In a recent summary of the available data on manganese and iron reduction rates in sediments, Thamdrup (2000) observed that high rates of metal oxide reduction require physical and/or biological mixing of the sediments to continually supply the needed reactive metal oxides, which are generally more poorly crystalline, amorphous forms (also see Burdige et al., 1992b). Since this needed supply of metal oxides generally exceeds its input by sedimentation, this implies that there must be significant internal redox cycling between the reduced and oxidized forms of these metals to sustain these rates of metal oxide reduction (Burdige, 1993; Canfield et al., 1993b; Thamdrup, 2000; also see discussions in section 13.2). Therefore as discussed in section 7.4.3, metal oxide reduction, as a part of this internal metal redox cycle, is serving here as an electron shuttle that essentially couples oxygen reduction in surface sediments to organic matter remineralization in deeper suboxic sediments that are devoid of oxygen (also see fig. 7.3).

Clearly, more work is needed to better quantify the role of metal oxide reduction in sediment organic matter remineralization. Nevertheless, it is also important to remember that metal oxide reduction, and Mn and Fe redox cycling in general, is an important component of other aspects of sediment biogeochemistry, as is discussed in sections 7.4.3 and 13.2 and chapter 15.

At the same time, though, the relatively minor role that these processes play in organic matter remineralization likely occurs for the following reasons. In low carbon flux (deep-sea) sediments, aerobic respiration dominates (as discussed above), thus limiting the importance of metal oxide reduction. As the carbon flux increases—for example, in continental margin and nearshore, coastal sediments— the importance of suboxic and anoxic remineralization processes, including metal oxide reduction, should therefore increase, and this indeed appears to be the case (fig. 7.7). However, in such sediments increasing carbon fluxes also lead to a "compression" of the oxic and suboxic zones closer to the sediment surface (as in fig. 7.9) along with the increased relative role of sulfate reduction. As such, there will also be increasing production of the reduced end-products of anoxic diagenesis (e.g., sulfide, ammonium, methane) near the sediment surface, along with an increase in their upward flux. Processes such as bioturbation can also physically mix reduced iron sulfide minerals upward into surface sediments where manganese oxides predominate (see Hulth et al., 1999, and references therein). As a result, a large range of either abiotic reduction processes such as the reduction of iron oxides by sulfide, or chemolithotrophic processes such as those described in section 7.4.2 can (apparently) outcompete metal oxide reduction that is coupled to organic matter oxidation in these sediments (e.g., Aller, 1994b). This interplay between these processes may therefore "conspire" to minimize the importance of suboxic metal oxide reduction in directly oxidizing organic matter in many sediments. In their model study of denitrification in marine sediments (another suboxic remineralization process) Middelburg et al. (1996) come to very similar conclusions in terms of the controls on the importance of denitrification in marine sediments (also see section 16.2).

Despite these general observations, there are environments where metal oxide reduction is apparently an important component of sediment organic matter remineralization. These include sediments where the Mn levels are substantially elevated over typical sediment values, such as the hemipelagic sediments in the Panama Basin, where elevated Mn levels in sediment result from nearby hydrothermal sources (Aller, 1990; Aller et al., 1998), and some sites in the Skagerrak (Canfield et al., 1993a).[6] In such sediments, these elevated manganese levels apparently allow microbial manganese reduction coupled to organic matter oxidation to outcompete other suboxic and anoxic remineralization processes.

Perhaps the best documented sediment systems in which iron reduction clearly dominates organic matter remineralization are mobile tropical mud belts and deltas, such as the Amazon continental shelf (Aller et al., 1986) or the Gulf of Papua/Fly River (New Guinea) delta complex (Alongi, 1995; Aller et al., 2004). Several factors appear to be at play here to create this situation. The first is that these sediments are relatively iron rich as a result of the tropical weathering regimes that supply sediments to these continental margin settings (see chapter 18). At the same time, the frequency and intensity of the physical reworking of these sediments, coupled with the reoxidation of the sediments associated with this disturbance (section 5.5), poise the sediments at a redox state such that iron reduction, rather than sulfate reduction, dominates organic matter remineralization (see sections 7.6.3 and 11.2.3).

---

[6] Note that the value in fig. 7.7 where metal oxide reduction accounts for ~90% of the total sediment organic matter remineralization is from such a site in the Skagerrak.

7.6 Dynamics of Organic Matter Decomposition in Sediments

7.6.1 General Considerations

In very broad terms, the equations in table 7.1 describe the oxidation of complex organic compounds, e.g., biopolymers such as proteins and carbohydrates, by the various respiratory processes listed in this table. While these reactions do indeed occur in a net sense, the internal dynamics of these processes in sediments are generally more complex. This is particularly true under anoxic and suboxic conditions, where no single group of anaerobic bacteria is generally able to completely degrade any of these complex biopolymers.

In contrast, aerobic bacteria, as well as other higher organisms that carry out aerobic metabolism (e.g., benthic macrofauna), are generally capable of the complete remineralization (oxidation) of a wide range of complex organic compounds (see discussions in Capone and Kiene, 1988, and Fenchel et al., 1998). Aerobic bacteria also differ from anaerobes in that aerobic metabolism is a relatively efficient process, and a large fraction of the metabolized organic material is used to produce new cell material. In contrast, anaerobic metabolism is less efficient than aerobic metabolism and a larger fraction of the metabolized organic material ends up as dissolved inorganic nutrients, as opposed to new cell material (Fenchel et al., 1998).

The presence of oxygen also has another significant impact on the dynamics of organic matter decomposition in sediments. In addition to serving as an electron acceptor, enzymes such as oxygenases or peroxidases can use the O_2 molecule as a cofactor in breaking nonhydrolyzable bonds in materials such as lignins, hydrocarbons, and other more refractory organic compounds. Often this occurs through the production of strong oxidants such as peroxide (H_2O_2) and other reactive oxygen-containing radicals (see discussions in Canfield, 1994; Fenchel et al., 1998; Emerson and Hedges, 2003). Since many of these enzymes are nonspecific, i.e., they do not target specific compounds or types of bonds, they are useful in degrading materials such as lignins, which are randomly polymerized upon formation (see section 9.6 for details). Such oxidants may also play a role in decomposing the molecularly uncharacterized component of sedimentary organic matter that may consist of refractory (nonhydrolyzable) biopolymers, or may have a similar random structure as a result of abiotic recondensation reactions (Hedges and Keil, 1995; Hedges et al., 2000; also see sections 9.8 and 15.6).

7.6.2 Anaerobic "Foodchains"

Anaerobic decomposition of organic matter in sediments occurs stepwise, involving a consortium of several different functional groups of bacteria (Laanbroek and Veldkamp, 1982; Capone and Kiene, 1988; Fenchel et al., 1998). During this sequence the end-product of one process becomes the reactant for the next process, such that in an overall net sense decomposition can be described by equations such as those in table 7.1. In each reaction in this sequence the responsible organism acquires only a portion of the total energy available from the substrate being utilized, passing on the remaining energy in the end-products of the reaction.

A conceptual model describing this process is shown in fig. 7.10. This model has its roots in the field of ruminant microbiology (Wolin, 1979); in biogeochemical studies it was initially developed for anoxic sediments in which sulfate reduction and methanogenesis occur (e.g., see references cited above). In particular, we will see that it is useful in understanding the kinetic factors controlling the occurrence of these two processes (as was discussed in section 7.3.5).

A key feature of this model is that the various processes shown here produce and consume dissolved organic matter (DOM) compounds of increasingly smaller molecular weights, eventually leading to a limited number of relatively low molecular weight (LMW) compounds that are then used by the terminal respiratory organisms. Because bacteria can transport only compounds smaller than ~600 daltons across their cell walls (Weiss et al., 1991), many of these processes are mediated by extracellular ectoenzymes that are attached to cell surfaces, or exoenzymes that are released into solution.

The first step in this sequence, the depolymerization of complex sedimentary organic matter, produces high molecular weight (HMW) compounds (e.g., oligomers) such as dissolved polysaccharides, proteins, and fatty acids by the hydrolysis of biopolymers such as proteins, carbohydrates, and some lipids (Colberg, 1988; McInerney, 1988). In contrast, the depolymerization of nonhydrolyzable materials such as lignins and hydrocarbons is more limited under strict anaerobic conditions than when oxygen is present (Schink, 1988), as was discussed in the previous section. As an example of this, studies in freshwater

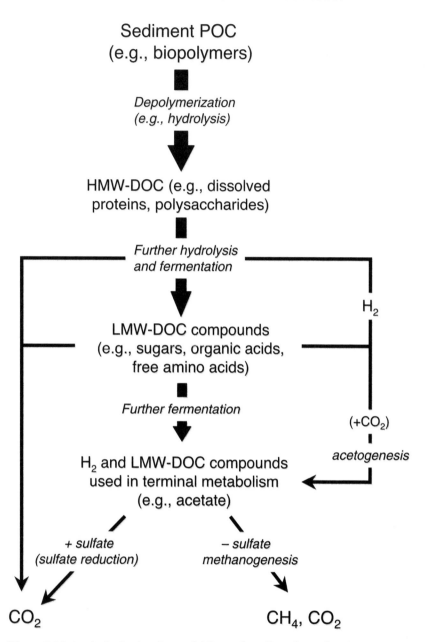

Figure 7.10 A relatively simple model for carbon flow through DOC intermediates during anoxic remineralization (sulfate reduction and methanogenesis; compiled from information in several sources).

peatlands demonstrate that the absence of oxygen inhibits the activity of the enzyme phenol oxidase, which then greatly inhibits overall anaerobic peat decomposition (Freeman et al., 2001).

While the nonhydrolytic decomposition of complex organic matter is generally limited under anaerobic conditions, it can occur for some types of this material. For example, studies of anaerobic lignin degradation indicate that the anaerobic metabolism of both the aromatic subunits and the complete lignin polymer can occur (Benner et al., 1984; Young and Frazer, 1987; Colberg, 1988). Nevertheless, the decomposition of such types of hydrolysis-resistant, or recalcitrant (refractory) organic matter is often less efficient in the absence of oxygen than in its presence (Benner et al., 1984; Canfield, 1994; Kristensen et al., 1995; Sun et al., 2002). Such effects may therefore play some role in controlling sediment carbon preservation under anoxic conditions, as will be discussed in chapter 15 in further detail.

The next group of steps in the sequence shown in fig. 7.10 involves the continued hydrolysis and fermentation of these initial DOM intermediates, eventually producing monomeric low molecular weight DOM compounds such as simple sugars, amino acids, and fatty acids. Because fermentation requires redox balance between the starting substrate and all of its products, reduced H_2 is often an important fermentation end-product, along with CO_2 and organic acids and alcohols of intermediate oxidation state (see section 7.1). While much of this hydrogen is consumed by terminal respiratory processes such as sulfate reduction or methanogenesis, it can also be consumed by homoacetogens, a group of organisms that live autotrophically by reducing CO_2 to acetate using H_2 (Fenchel et al., 1998).

Regardless of the mechanisms of consumption, H_2 concentrations are generally quite low in anoxic sediments, i.e., nM levels, or less than $\sim 10^{-3}$–10^{-4} atm (Lovley and Goodwin, 1988; Novelli et al., 1988; Hoehler et al., 1998). This occurs because of a tight coupling between H_2 production and consumption by a process referred to as *interspecies hydrogen transfer* or *syntrophy*. The need for this tight coupling appears to be related to the fact that many H_2- producing fermentation reactions actually have a positive $\Delta G°$ under standard conditions (i.e., $p_{H_2} = 1$ atm). However, if hydrogen is kept at the low levels found in most sediments, then these reactions have negative ΔG values under these nonstandard, *in situ* conditions.

Continued fermentation leads to the production of a limited number of LMW-DOM compounds, primarily short-chain fatty acids such as acetate and formate.[7] These compounds, along with H_2, are then consumed by the terminal respiratory organisms in the sediment. As discussed above, competition for these substrates is thought to control the relative importance of sulfate reduction and methanogenesis in many anoxic sediments. Consistent with this model, studies have shown that at high substrate concentrations these kinetic barriers can be overcome, leading to the simultaneous occurrence of sulfate reduction and methanogenesis (Winfrey and Zeikus, 1977; Crill and Martens, 1986).

In examining the kinetic controls on organic matter remineralization in a multistep scheme such as that shown in fig. 7.10, it is often assumed that the initial steps of extracellular hydrolysis or oxidative cleavage of the sediment organic matter is the overall rate-determining step in the process (e.g., see Brüchert and Arnosti, 2003, and references cited therein). Such an assumption is often used to explain differences in the observed remineralization rates of "fresh" versus "aged" sedimentary organic matter, and also to explain the role of oxygen in enhancing the degradation of certain types of refractory organic materials (see discussions above, in Emerson and Hedges, 2003, and in section 15.2). In contrast, though, other studies (e.g., Arnosti et al., 1994; Brüchert and Arnosti, 2003; Arnosti, 2004) have demonstrated that under some conditions the consumption of "downstream" fermentative intermediates and not the initial enzymatic hydrolysis can be the slow step in the degradation of polysaccharides in sediments.

At the same time Westrich and Berner (1988) have observed that the apparent activation energy[8] for sulfate reduction in Long Island Sound sediments is not a constant (see section 14.3), but appears to vary inversely with the reactivity of the organic matter being decomposed; less reactive organic matter has a higher apparent activation energy for overall decomposition. Although Westrich and Berner (1988) discuss several possible explanations for these observations, they

---

[7] These compounds are also sometimes referred to as volatile fatty acids or low molecular weight organic acid.

[8] The temperature dependence of the rates of sediment organic matter remineralization is often quantified by assuming that the rate (or its rate constant) can be described by the Arrhenius rate law (R or $k = Ae^{-Ea/RT}$). The apparent activation energy (Ea) for these sediment processes appears to vary between ~15 and 25 kcal mol^{-1} (Aller, 1980a; Klump and Martens, 1989). Expressing this temperature dependence in terms of Q_{10} (the increase in a rate for a 10°C temperature increase) these Ea values lead to Q_{10} values of ~3–4.

suggest that the most straightforward explanation for these results is that the initial degradative steps in the remineralization of the sediment organic matter by the microbial community (e.g., hydrolysis or oxidative cleavage) limits the overall remineralization process. This result is not necessarily in disagreement with the results of Arnosti and coworkers that are discussed above, because differences in the activation energies of the different steps in the overall remineralization process could lead to changes in the rate-limiting step of remineralization for different types of organic matter, or for the same type of organic matter at different temperatures.

This anaerobic foodchain model also appears to explain the zonation of iron reduction, sulfate reduction, and methanogenesis in freshwater sediments (Lovley and Phillips, 1987), and may describe, in general, the zonation of all suboxic and anoxic processes in marine sediments (see discussions in Thamdrup, 2000). However, the significance of the model in suboxic marine sediments has not been examined in nearly the detail that it has been in anoxic sediments. In particular, less is known about the specific carbon substrates used in suboxic terminal respiration, as compared to sulfate reduction and methanogenesis, particularly under *in situ* conditions.

7.6.3 Dynamics of Organic Matter Decomposition under Mixed Redox Conditions

The discussion above illustrates some of the fundamental differences between the decomposition of organic matter under strictly oxic versus anoxic conditions, particularly with regard to the depolymerization of refractory organic compounds. In many organic-rich sediments the absence of physical reworking (section 5.5) or bioirrigation (section 12.5) greatly limits oxygen penetration into the sediments by diffusive transport, and anoxic remineralization processes predominate. However, in sediments that undergo redox oscillations, mixed oxic/suboxic [anoxic] redox conditions (as defined in section 7.5.1) occur in association with these physical or biological transport processes. Such redox oscillations have the potential to greatly enhance organic carbon remineralization in sediments over that which might occur under more strict anoxic conditions, in spite of the general asymmetry between cycles or periods of anoxic versus oxic conditions.

The specific mechanisms by which this enhanced organic matter remineralization occurs under mixed redox conditions are not well characterized, although there are at least four likely possibilities. First, the oxidation of refractory components of the sediment organic mater pool may be catalyzed by the input of fresh organic matter through a process referred to as *cometabolism* or *co-oxidation* (Schink, 1988; Aller, 1994a; Canfield, 1994). In this situation a reactive carbon input "primes" the sediments, and the resulting microbial decomposition stimulated by the addition of this reactive organic material then catalyzes the degradation of more refractory material. The input of this fresh/reactive organic matter may occur in several ways, including physical or biological mixing of the sediments, as well as the rapid death of a component of the microbial population due to changes in sediment redox conditions (Aller, 1994a; Aller et al., 1996). For example, in Amazon margin sediments as well as other muddy tropical sediments, physical mixing of the sediments adds relatively fresh, planktonic organic matter to sediments that contain a significant amount of more refractory terrestrial organic matter (much of which is derived from soils) to stimulate the overall remineralization of both types of organic matter (also see section 15.5 and Aller et al., 2004, as well as discussions of terrestrial organic matter in sediments in sections 9.8.4, 9.8.5, 11.1.2, and 11.2.3).

Second, the more direct activity of benthic macrofauna in mixed redox sediments plays an important role in the dynamics of carbon remineralization in these sediments. The grazing of bacterial biomass in these sediments by higher organisms provides a benthic microbial loop that enhances sediment carbon remineralization (Lee, 1992; also see section 11.1.5), as does the direct macrofaunal degradation of sediment organic matter by chemical, mechanical, and/or enzymatic digestion (see table 15.1 and general discussions in Mayer et al., 1997). Additional aspects of this problem will be discussed in chapter 15.

Third, the periodic introduction of oxygen into these sediments may initiate the subsequent anaerobic microbial decomposition of certain types of organic matter that would otherwise be refractory under continuous anaerobic conditions (Aller, 1994a). This may occur, for example, as a result of the initial depolymerization (oxidative cleavage) of these compounds by O_2-requiring enzymes or reactive O_2 products (e.g., H_2O_2; see section 7.6.1).

Finally, oxygen input to such mixed redox sediments also leads to the production of Mn and Fe oxides by the oxidation of reduced iron and manganese. Iron oxidation by O_2 at oxide surfaces can also lead to the production of many of the same reactive oxidants, such as peroxide, that are produced by the aerobic enzymes discussed above (see discussions in Hedges and Keil, 1995). Furthermore, the reaction between Fe^{2+} and H_2O_2 produces Fenton's reagant, an extremely powerful oxidizing agent for degrading organic substances. At the same time a wide range of refractory organic compounds, including many aromatic compounds, as well as fulvic acid extracts, can be partially oxidized by manganese and iron oxides (see discussions in Stone et al., 1994; Mayer, 1995). The ability of these oxides to cleave a wide range of organic bonds suggests that these oxides could play a role similar to that of oxygen and associated enzymes in depolymerizing many refractory, and nonhydrolyzable, organic compounds (see discussions in Hedges and Keil, 1995, and section 7.6.1). Some of these same refractory organic compounds can apparently be used by metal oxide reducing organisms (Nealson, 1997), although these organisms may simply take advantage of abiotic reactions between metal oxides and refractory organic compounds (Sunda and Kieber, 1994) and utilize the end-products of these reactions.

Recent studies suggest that some types of organic matter degradation in sediments are strongly enhanced by oxygen exposure prior to and during deposition (Hartnett et al., 1998; Hedges et al., 1999a; Hedges, 2002). However, as noted in these papers, this does not imply that the specific mechanism by which this occurs directly involves molecular oxygen as the sole, causative electron acceptor during organic matter remineralization. Rather, it is likely that oxygen exposure leads to some set of unknown processes/conditions that then enhance remineralization over that which occurs under more strict anoxic conditions. The discussion here of processes occurring in mixed redox sediments suggests several possible ways this may occur, and their significance will be discussed in further detail in later chapters (e.g., chapter 15).

≈ CHAPTER EIGHT ≈

Quantifying Carbon and Nutrient Remineralization in Sediments

THIS CHAPTER WILL continue the discussion begun in previous chapters of the techniques that can be used to quantify sediment biogeochemical processes.

8.1 MODELS OF ORGANIC MATTER DECOMPOSITION IN SEDIMENTS

Despite the complex multistep mechanisms for organic matter remineralization in sediments (e.g., see fig. 7.10), the rate equations used for modeling these processes are often relatively simple. Such equations generally neglect the inherent complexity of the process and begin with the initial assumption that the rate of organic matter remineralization is related to the amount of organic matter and the concentration of the terminal oxidant in the sediments (although see Burdige, 2002, and discussions in section 10.2 for an alternate approach to this problem). Assuming that oxidant uptake is controlled by Michaelis–Menten (or Monod) kinetics (eqn. 7.10), this then implies that

$$\frac{dG}{dt} = -kG\frac{[Ox]}{K_m + [Ox]} \tag{8.1}$$

where k is the rate constant, G is the amount of organic carbon (TOC) present, and $[Ox]$ is the oxidant concentration (e.g., O_2 or sulfate).

For many oxidants, K_m is much smaller than most *in situ* concentrations (see summaries in Rabouille and Gaillard, 1991b; Boudreau, 1996; Dhakar and Burdige, 1996). This then implies that organic matter remineralization is essentially independent of $[Ox]$, since if $K_m \ll [Ox]$ then $(K_m + [Ox]) \approx [Ox]$ and the fraction on the right side of eqn. (8.1) is ~1. Consequently, the rate expression for organic matter remineralization in sediments is simply first-order decay:

$$\frac{dG}{dt} = -kG \qquad (8.2)$$

(see Berner, 1964, 1980, or Westrich and Berner, 1984, for additional details regarding this model). Note that this point will be also be revisited in section 8.4.

In spite of the simplicity of this approach, under many circumstances this kinetic formulation has proven successful in describing early diagenesis in sediments. This suggests that at a first-order level, eqns. (8.1) and (8.2) are likely "correct" in terms of their basic assumptions regarding the quantification of these processes in sediments (although also see discussions at the end of section 12.5 that describe sediment systems in which this formulation may break down).

In relatively simple, anoxic, nonbioturbated sediments the inclusion of eqn. (8.2) in diagenetic models predicts that organic carbon will decrease with sediment depth in an exponential-like manner (see section 6.3.1). However, such a model also predicts that sediment TOC concentrations will go to zero with depth. This observation stands in contrast with what is observed in almost all sediments: namely, that while TOC concentrations decrease with depth they also approach asymptotic, but nonzero, concentrations. At the same time, the overall rate of organic matter remineralization in sediments generally decreases more slowly with depth than that predicted by such simple models (e.g., Jørgensen, 1978).

To explain these observations a variant of the model described above, the *multi-G model*, was proposed by Jørgensen (1978; also see Berner, 1981b). This model assumes that sediment organic matter is composed of various groups, or fractions, of organic matter (G_i below) that have different degrees of reactivity, and that the sum of these individual fractions (G) then comprises the total amount of organic matter in the sediments. If each individual fraction undergoes first-order decomposition according to eqn. (8.2) then the overall rate of decomposition can be expressed by,

$$\frac{dG}{dt} = -\sum k_i G_i \qquad (8.3)$$

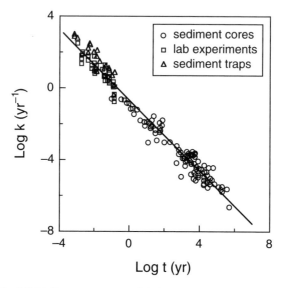

Figure 8.1 The Middelburg power model showing the inverse relationship between the reactivity of organic carbon (k) and the age (t) of the material. Solid squares, open circles, and open triangles represent, respectively, results from organic matter decomposition experiments in the lab, organic carbon depth profiles from dated sediment cores, and sediment trap organic carbon versus water column depth profiles. The original references for the data used in this compilation can be found in Middelburg (1989) and Middelburg et al. (1993).

where successively "higher" fractions of organic matter have lower degrees of reactivity (i.e., $k_1 > k_2 > k_3$ etc.). Differences in the reactivity of each fraction then lead to the near sequential utilization of each fraction, and the model predicts that over time, or with sediment depth, there will be selective utilization of reactive fractions and preservation of refractory fractions.[1] In general, this observation is consistent with past discussions about how the amounts and reactivity of the bulk sediment organic matter change over time (also see fig. 8.1 and associated discussions, and section 9.1).

[1] This concept of selective utilization is a slight oversimplification since all organic matter fractions actually undergo some amount of decomposition simultaneously (e.g., Burdige, 1991). However, if the inequalities among the k's are sufficiently large, then in any given time window the reactivity of the most reactive fraction will generally predominate. As a result, less reactive fractions "appear" to be inert until the more reactive fraction has undergone near complete utilization.

MODELS OF ORGANIC MATTER DECOMPOSITION IN SEDIMENTS

These models, however, do not account for the synthesis of new bacterial biomass in sediments (Blackburn, 1988) or the role that the amount of viable *in situ* biomass plays in determining the apparent reactivity of sediment organic matter. In the former case, synthetic reactions producing new biomass may produce material that is either more reactive or more refractory than the original sediment organic matter substrate (e.g., see discussions in sections 11.1.5 and 12.5.1). Thus while models such as the ones used here imply the occurrence of unidirectional remineralization, in actuality remineralization is a net process that includes both degradative and synthetic reactions.[2]

In the latter case, it is not unrealistic to assume that k_i values might be proportional somehow to this amount of living sediment biomass (e.g., $k \sim k' \cdot \mathbf{B}$, where $\mathbf{B}$ is defined here as the viable bacterial biomass in sediments; Aller et al., 2001). In contrast, though, experimental results and theoretical calculations both suggest that k is actually independent of bacterial biomass (see Boudreau, 1992, for details; also see discussions in section 11.1.5).

The simplest version of the multi-G model assumes that there are two fractions of organic matter in sediments: one that is reactive, or metabolizable (G_1), and one that is refractory, or nonreactive (G_2 or more often G_{nr}). Re-examining the organic carbon diagenetic model discussed in section 6.3.1 in this light then indicates that the amount of the reactive fraction decreases with sediment depth according to eqn. (6.33) and the inert fraction is simply buried in the sediments (this latter point can be seen by examining eqn. 6.31 and setting k_2, or k_{nr}, equal to zero). The depth profile for total organic carbon in the sediments is then given by

$$G = G_{nr} + (G_o - G_{nr})e^{-\alpha z} \qquad (8.4)$$

where $(G_o - G_{nr})$ is the concentration of metabolizable organic matter at the sediment surface (often referred as G_m^o, and the total amount of organic matter in the surface sediments is defined as G_o. Such a model then explains nonzero asymptotic TOC concentrations in sedi-

---

[2] Expressed another way, sediment organic matter remineralization is a net heterotrophic process that is a balance between gross sediment heterotrophy and new biomass production (autotrophy or biosynthesis by heterotrophs).

ments, and has been used in many studies examining organic matter diagenesis in sediments.

Initial "tests" of the multi G model (Westrich and Berner, 1984) supported the basic assumptions of the model. These results also suggested that sulfate reduction and organic matter decomposition in coastal sediments could be described by assuming there are two reactive organic matter fractions (G_1 and G_2) and one fraction (G_{nr}) that is nonreactive on early diagenetic time scales. The rate constants for the degradation of these two reactive fractions (T = 20–22°C) differ by roughly an order of magnitude ($k_1 \approx 8$ yr^{-1} and $k_2 \approx 1$ yr^{-1}). Later work by Burdige (1991) yielded similar rate constants and also demonstrated that these different apparent fractions had different C:N ratios, with the G_1 fraction having a Redfield-like ratio (~6) and the G_2 material being depleted in nitrogen (C/N ≈ 30–40).

In spite of the success of this approach, organic matter in sediments is likely not "quantized" into such discrete fractions with distinct reactivities, but more likely has a continuum of reactivities. In contrast, though, methods of data analysis generally used to extract information from field or laboratory studies of sediment organic matter reactivity can determine, at most, only two or three fractions of material, regardless of the number of fractions that may actually exist (see discussions in Van Liew, 1962; Middelburg, 1989). Therefore, the definition of these metabolizable and nonreactive fractions is strongly dependent on the "time window" or relevant time scale of the sediment system being examined. This is of some importance since material that appears to be refractory on the short time scales of early diagenesis in a coastal sediment, generally less than a few hundred years, may indeed undergo remineralization on longer time scales (e.g., kyr or greater), if it is eventually transported to a deep-sea sediment.

Perhaps the best indication of this continuum of organic matter reactivity comes from the compilations presented by Middelburg (1989) and Middelburg et al. (1993), where it can be seen that as marine organic matter ages there is a continual decrease in the average reactivity of the bulk material (fig. 8.1). This result also suggests that organic matter reactivity in sediment diagenesis models might be better described by the equation,

$$\frac{dG}{dt} = -k(t)G \qquad (8.5)$$

where the function $k(t)$ might be described by the best fit line shown in fig. 8.1. However, this approach is generally not used in sediment diagenesis models for a number of reasons that are discussed elsewhere (Soetaert et al., 1996c; Boudreau, 1997a), including the fact that this approach severs the direct link that exists between the properties and composition of the sediment organic matter and its reactivity (Boudreau and Ruddick, 1991).

Nevertheless, an interesting aspect of the Middelburg power model (fig. 8.1) is that the slope of this line is essentially -1. This then implies that if one were to plot log t (the age of the organic matter) versus log τ ($= 1/k$; the mean lifetime of the decomposing organic matter with respect to its oxidation), one would observe a nearly 1:1 relationship (e.g., see Emerson and Hedges, 1988, for such a plot with a smaller data set). This then implies that decomposition rates scale with the time window of observation, such that the degradation of only some particular fraction of the sediment TOC will be observed within this time window (also see Burdige, 2002, for a similar discussion relating the reactivity of pore water DOM and time and depth scales of observation). This result also implies that the attenuation of organic matter remineralization rates with sediment depth is a function of the time window of the sediment being examined.

Given these observations, what has evolved in many sediment diagenesis models is the assumption that in any given sediment there are three organic matter fractions: two that undergo remineralization, and one that is inert on the time scale of the particular sediment system (e.g., Soetaert et al., 1996c; Boudreau, 1997a; Archer et al., 2002). The "reactive" (or G_1) fraction appears to undergo remineralization on depth scales ranging from <1 cm in the deep sea to 10–20 cm in continental margin and coastal sediments. The less reactive (or G_2) component decomposes on correspondingly longer depth scales (from several centimeters in the deep sea to greater than $\sim$1 m in shallower sediments; Hammond et al., 1996; Boudreau, 1997a; Hales and Emerson, 1997a).

This approach is an outgrowth of both the multi-G model and the results discussed above regarding the ability (or difficulty) of extracting information on the reactivity of multiple fractions of organic matter from field or laboratory observations. At the same time, in many deep-sea sediments where aerobic respiration dominates carbon remineralization, it is often observed that a double exponential function provides the best fit to pore water oxygen profiles (Hammond et al., 1996; Hales and Emerson, 1997a). This observation is also roughly consistent with this formalism.

In a given sediment, then, one can apparently adequately describe organic matter remineralization using such operationally defined G_1, G_2, and G_{nr} fractions and appropriate rate constants. However, it is also important to keep in mind that this does not also imply that the reactive (G_1) fraction of organic matter in a deep-sea sediment has the same reactivity, or is even the same type of material, as that found in a coastal sediment. Rather, a series of empirical relationships have been developed that relate the reactivity of these two operationally defined organic matter fractions in a given sediment to external parameters, such as the carbon rain rate to the sediments (F_c) or the bulk sediment accumulation rate (ω). Since both F_c and ω vary systematically (i.e., generally decrease) from coastal to deep-sea sediments (e.g., Middelburg et al., 1997), then so do the particular values of k_1 and k_2 in the sediments along this continuum.

For the G_2 fraction, k_2 (units of yr^{-1}) can apparently be parameterized according to

$$k_2 = 0.057\, \omega^{1.97} \tag{8.6}$$

where ω has units of cm yr^{-1} (e.g., Tromp et al., 1995). In contrast, several different relationships have been proposed for k_1 (again units of yr^{-1}). A data compilation presented in Boudreau (1997a) suggests the following relationship,

$$k_1 = 0.38\, \omega^{0.59} \tag{8.7}$$

while a compilation by Tromp et al. (1995) suggests

$$k_1 = 2.97\, \omega^{0.62} \tag{8.8}$$

Although both relationships show significant scatter among the limited data used in the compilations, it is interesting to note that only eqn. (8.8) predicts the existence of highly reactive material (k_1 on the order of ~1–10 yr^{-1}) for sediments with high (but not unrealistic) sediment accumulation rates (i.e., ω between ~0.2 and 7 cm yr^{-1}). The significance of this will be discussed in chapter 14.

At the same time, others (Emerson et al., 1985; Murray and Kuivala, 1990) have argued that k_1 shows a better correlation with F_c, and Boudreau's (1987) summary of the relevant data yields the following equation,

$$k_1 \approx 2.2 \times 10^{-5} F_c^{2.1} \tag{8.9}$$

where here F_c has units of μmol cm^{-2} yr^{-1}. However, because empirical relationships also exist between F_c and ω (Henrichs and Reeburgh, 1987; Tromp et al., 1995; Middelburg et al., 1997) these two approaches are not necessarily fundamentally different.

The diagenetic model of Soetart et al (1996c) uses an alternate approach to the examination of this problem. Their model defines two end-member types of organic matter with rate constants that differ by two orders of magnitude (26 yr^{-1} and 0.26 yr^{-1}), and that also have different C/N ratios; specifically, the reactive end-member is nitrogen rich and has a Redfield C/N stoichiometry, and the less reactive fraction is depleted in nitrogen. They then use an algorithm to determine how the fluxes of these two fractions attenuate with water column depth, to estimate the rain rate of each fraction to the sediments. They also assume that these two rate constants decrease with decreasing temperature, which itself varies with the water column depth of a sediment. Thus for a sediment at a given water column depth, this approach defines both the rain rate and the reactivity of organic matter deposited in the sediments.

While one can indeed envision why such relationships between organic matter reactivity and water depth, carbon flux, etc. should or could exist, it is also important to note that for now these equations are simply empirical observations that come out of existing data compilations. However, to look at these observations in a slightly more unified context I note that Middelburg (1989) also suggests that the rate constant for organic matter remineralization is a function of the apparent "initial age" of the material (roughly defined as the time period between its production/synthesis and its deposition in a sediment). In a sense, then, what this argues is that predepositional degradation moves organic matter along the power curve in fig. 8.1, and that this apparent initial age represents the point on the curve at which postdepositional diagenesis begins. At the same time, Middelburg et al. (1997) have also observed, based on a fairly extensive compilation of sediment parameters, that water column depth may be an appropriate master variable to predict quantities such as F_c and ω for a given sediment. Therefore if one assumes that the apparent initial age of sediment organic matter is itself some function of water depth, which does not appear to be an unreasonable suggestion, then one might imagine how relationships between k_i values and F_c and ω could be observed, through relationships that all of these parameters may have with water column depth.

8.2 SEDIMENT BUDGETS FOR REACTIVE COMPONENTS

One approach to quantifying sediment carbon remineralization and preservation involves determining sediment nutrient budgets. In its simplest formulation, this approach starts by considering the surface sediments where early diagenesis occurs as a "box" in steady-state with respect to inputs and outputs of material. Under steady-state conditions this then leads to

$$J_{in} = J_{out} + J_{bur} \qquad (8.10)$$

where J values represent fluxes of material into or out of this sediment box (see fig. 8.2a).

Defining the size, i.e., depth, of this sediment box can be done in several ways. One approach uses sediment profiles of a solid phase constituent (e.g., TOC) undergoing remineralization. At the depth in the sediments where an asymptotic, nonzero TOC concentration is reached (e.g., see eqn. 8.4 and associated discussions), it can generally be assumed that organic matter remineralization has ceased and that we are thereafter only burying "refractory" organic carbon out of the surface sediment box (also see fig. 8.2B).

We can also define the size of this box using pore-water profiles or sediment reaction rates, which are either directly determined or estimated from the pore-water profiles (see fig. 8.2C). For example, in a deep-sea sediment where aerobic respiration generally dominates organic matter remineralization, the depth at which oxygen either goes to zero or approaches a nonzero asymptotic concentration (because

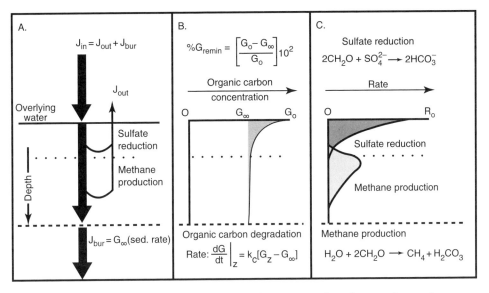

Figure 8.2 Illustrations of three independent approaches that can be used to quantify organic matter remineralization in Cape Lookout Bight sediments (modified after Martens and Klump, 1984). A. A carbon budget approach based on direct benthic flux measurements (ΣCO_2 and CH_4; $= J_{out}$) and calculated organic carbon burial rates ($= J_{bur}$). B. Modeling of sediment organic carbon (TOC) depth profiles. C. Calculation of organic carbon remineralization rates based on measured rates of sediment organic matter remineralization processes (sulfate reduction and methanogenesis).

reactive organic carbon rather than O_2 is the limiting reactant in these sediments) can be used to define the lower boundary of the sediment box (see Emerson, 1985, and section 13.1).

8.2.1 Theoretical Considerations

The diagenetic equations discussed in chapter 6 can be used to define some of the relationships that are used in sediment nutrient budget calculations. Starting with eqn. (6.2) and assuming steady-state conditions, this equation can be rewritten as

$$\frac{dJ}{dz} = \Sigma R_i \qquad (8.11)$$

where J represents fluxes associated with advection, diffusion, bioturbation, and/or bioirrigation (see eqn. 6.3). When integrated over the depth of the sediment box this yields,

$$J_L - J_0 = \int_0^{z_L} \Sigma R_i dz \qquad (8.12)$$

where z_L is the bottom of the sediment box. The term on the right side of this equation is the depth-integrated rate of the processes producing or consuming the solute of interest (= DIR_i). Under many circumstances the term J_L (the flux at the bottom of the box) is equal to zero because, for example, all reactive organic carbon is consumed within the surface sediments and none is therefore buried with depth. The same is also generally true for pore-water solutes (again see eqn. 6.3) because: (a) fluxes associated with bioturbation or bioirrigation are almost always zero at depth (i.e., below the surface mixed or burrowed layer), (b) pore-water gradients (dC/dz) go to zero with depth, leading to no diffusive fluxes, and (c) pore-water burial fluxes are, in most cases, generally a very small term that can be safely neglected in such calculations. Thus J_L can generally be neglected and eqn. (8.13) can be rewritten as

$$J_o \approx -DIR_i \qquad (8.13)$$

which shows that under steady-state conditions fluxes of a solute across the sediment-water interface (J_o) are balanced by its depth-integrated sediment production or consumption rate. This is a very important observation that we will come back to many times, since it illustrates the importance of benthic flux measurements in quantitatively examining sediment remineralization rates. Also recall, however, discussions in section 7.4.3, which indicate that under nonsteady-state conditions there can be a breakdown in this relationship between benthic fluxes and depth-integrated remineralization rates.

For a compound undergoing decomposition in sediments, another important quantity of interest is the mean residence time of this material with respect to decomposition (τ_r). As with the Damkohler number (eqn. 5.5), determining τ_r allows one to examine the time scale over which removal occurs relative to the time scales of other relevant processes or properties of a sediment. In a general sense, τ_r is defined as the integrated amount of this reactive material in the surface sediments divided by its integrated removal rate. Using organic carbon as an example, τ_r is then given by

$$\tau_r = \frac{\Sigma G_m}{J_{out}} \quad (8.14)$$

where J_{out} equals ωG_m^o (see the next section) and ΣG_m is defined as

$$\Sigma G_m = \int_0^{z_L} G_m dz = \int_0^{z_L} G_m^o e^{-\alpha z} dz = \frac{G_m^o}{\alpha}(1 - e^{-\alpha z_L}) \approx \frac{G_m^o}{\alpha} \quad (8.15)$$

Note that the approximation at the end of this derivation is based on the fact that, because metabolizable organic matter in the sediments goes to zero at $z \approx z_L$, so does the term $e^{-\alpha z_L}$. With these equations we now see that

$$\tau_r = \frac{G_m^o/\alpha}{\omega G_m^o} = \frac{1}{\omega \alpha} = \frac{1}{k} \quad (8.16)$$

since $\alpha = k/\omega$ (see section 6.3.1).

8.2.2 Sediment Nutrient Budgets Using Cape Lookout Bight as an Example

In this section we will show how the approaches discussed above can be used to calculate sediment nutrient budgets, in part by using data from the sediments of Cape Lookout Bight, NC (CLB) as an example. An important aspect of this exercise is that these calculations use directly measured *in situ* benthic flux and rate measurements, along with diagenetic models of pore-water and solid-phase profiles, in the calculations. With this approach one then can make independent estimates of many of the terms in the budget, and this provides additional confidence in the calculations because of the built-in overdetermination of many of the budget terms.

The approach described here was initially used to examine the organic carbon budget in Cape Lookout Bight sediments (Martens and Klump, 1984), and was subsequently used to examine CLB sediment budgets of nitrogen, phosphorus, and amino acids (Klump and Martens, 1987; Burdige and Martens, 1988). Similar approaches have also been used to examine sediment nutrient budgets in other estuar-

TABLE 8.1
An Organic Carbon Budget for the Sediments of Cape Lookout Bight, NC[a]

Flux	mol C m^{-2}yr^{-1}
1. Sediment-water exchange (benthic flux)	
ΣCO_2	29.9 ± 4.5
Methane	8.5 ± 2.9
DOC	2.2 ± 1.7
2. Pore water burial	
ΣCO_2	6.5 ± 1.1
Methane	0.14 ± 0.02
DOC	0.35 ± 0.07
3. Total organic carbon remineralized (J_{out} [= 1 + 2])	47.6 ± 5.7
J_{out} based on modeling TOC profiles[b]	22.4–50.5
4. Particulate organic carbon burial (J_{bur})	117 ± 19
5. Particulate organic carbon input (J_{in} [= 3 + 4])	165 ± 20
6. Burial Efficiency (eqn. 8.19)	71 ± 14%

a. All values are mean annual average (from Martens and Klump, 1984, and Martens et al., 1992).
b. As discussed in Martens and Klump (1984), this relatively large range is caused in part by problems estimating the surface TOC concentration from field measurements (because of, for example, seasonal inputs of organic matter to these sediments).

ine and coastal sediments (e.g., Bender et al., 1989; Alperin et al., 1992b; Aller et al., 1996; Burdige and Zheng, 1998).

The three approaches that will be used here are shown in fig. 8.2A. In the first of these, a carbon balance based on fluxes into and out of the surface sediments is determined using eqn. (8.10). In most sediments (e.g., see the discussion in section 8.3.1) the term J_{out} is generally based on either directly measured or calculated benthic fluxes (see section 12.1 for further details). Since sulfate reduction and methanogenesis dominate organic matter remineralization in CLB sediments, here J_{out} is largely determined from measured benthic fluxes of both ΣCO_2 and CH_4 (see table 8.1). However, as this table also indicates, benthic fluxes of dissolved organic carbon (DOC) are also a minor component of the loss of organic carbon from these sediments (Alperin et al., 1994; also see section 10.4).

Because these benthic fluxes are balanced by depth-integrated remineralization rates, J_{out} is also often referred to as J_{remin}.[3] At the same time, though, because of the rapid sedimentation rate in CLB sediments (~10 cm yr^{-1}) and the high rates of sediment organic matter remineralization, the burial of pore-water ΣCO_2 and methane (and DOC) represents a minor, but not insignificant, component of carbon remineralization in these sediments. While this burial flux is not explicitly shown in fig. 8.2A (Martens et al., 1992) it is included in table 8.1. In most sediments the term J_{bur} is determined by modeling sediment organic carbon profiles (see below), and J_{in} is then obtained by summing J_{out} and J_{bur}. The term J_{in} can also be obtained by modeling sediment TOC profiles (in some sediments; see below), using near-bottom sediment trap measurements (section 6.2.1), or using algorithms that predict organic carbon rain rate as a function of sediment water column depth and surface water primary productivity (e.g., Jahnke, 1996).

The second approach shown in fig. 8.2B involves modeling CLB organic carbon profiles using eqn. (8.4) to calculate remineralization and burial fluxes. In the absence of bioturbation J_{in} will be given by ωG_o (see section 6.3.2), J_{bur} will be given by ωG_∞, and J_{out} (or J_{remin}) then equals $\omega(G_o - G_\infty)$ or ωG_m^o (again note that G_∞ here is the same as G_{nr} in eqn. 8.4).[4] As we will discuss below, J_{out} also represents the flux of metabolizable organic matter to the sediments. Since it was also estimated above using benthic fluxes plus rates of pore water nutrient burial, this approach now provides two independent estimates of J_{out} in these sediments (table 8.1). Looked at slightly differently, this comparison tells us that under steady-state conditions the flux of inorganic nutrients out of the sediments must be balanced by the input of reactive organic matter, because the latter is the source of the former. It will also be shown below (eqn. 8.22) that the depth-integrated rate

---

[3] Throughout the discussion I have attempted to use the most common abbreviations found in the literature for rate terms, fluxes, etc. In some cases I have also included other abbreviations that one may also come across in the literature, to hopefully avoid any confusion on the part of the reader. For example, when specifically referring to organic carbon remineralization, J_{out} is also sometimes defined as C_{ox} or R_{cox}.

[4] If ω has units of cm yr^{-1} and these G valves have units mmol g$_{dw}^{-1}$, then this flux term must also be multiplied by the term $(1 - \varphi_\infty)\rho_{ds}$ to convert this value to more conventional flux units (mmol cm^{-2} yr^{-1}). However for simplicity in this discussion I have not included this unit conversion in the equations discussed in the text. Also note that this is effectively the same as converting ω (the sediment accumulation rate) to r (the mass accumulation rate; see section 6.3.5 and eqn. 6.58 for further details).

of carbon remineralization based on eqn. (8.4) is similarly given by ωG_m^o. Since the model being used here assumes that we are not burying any metabolizable (reactive) organic carbon, its input via sedimentation has to be balanced by its depth-integrated rate of consumption—that is, loss via remineralization.

With this approach to modeling organic carbon profiles we can also estimate the fraction of the organic matter input that is remineralized:

$$\% G_{remin} = \frac{G_o - G_\infty}{G_o} \tag{8.17}$$

In some sediments, however, $\% G_{remin}$ may be more appropriately expressed in terms of sediment organic carbon fluxes as

$$\% G_{remin} = \frac{J_{out}}{J_{out} + J_{bur}} = \frac{J_{out}}{J_{in}} \tag{8.18}$$

For example, in bioturbated sediments the magnitude of G_o is controlled by both sediment deposition/carbon rain rate and bioturbation (see section 6.3.2). Thus eqn. (8.17) is not a valid estimate of $\% G_{remin}$ because here $G_o - G_\infty \neq G_m^o$. For similar reasons, in bioturbated sediments a slightly different approach is also taken in estimating these J terms since J_{in} is also not equal to ωG_o. Here, if one can estimate J_{in} using sediment trap measurements, then J_{out} will be given by $J_{in} - J_{bur}$, where again J_{bur} equals ωG_∞, which is determined at some depth below the sediment mixed layer where diagenesis "ceases."

The term $\% G_{remin}$ is also more commonly referred to as the *sediment oxidation*, or *remineralization, efficiency* (OE). A related term is then the *carbon burial*, or *preservation, efficiency* (BE), which is defined as,

$$BE = J_{bur}/J_{in} \tag{8.19}$$

A great deal of effort has gone into examining relationships between different environmental parameters and carbon burial efficiency, as a means of unraveling the factors that ultimately control sediment carbon preservation (e.g., Hedges and Keil, 1995). This will be discussed in later chapters. Also note that the concept of burial efficiency is not limited to use in the examination of sediment carbon budgets, but is also used in the context of sediment budgets of other biogenic components such as opaline silica or calcium carbonate.

SEDIMENT BUDGETS FOR REACTIVE COMPONENTS

The final approach taken in fig. 8.2 to examine sediment nutrient budgets uses measured or modeled rates of sediment processes to estimate J_{out}. Because sulfate reduction and methanogenesis dominate organic matter remineralization in Cape Lookout Bight sediments, depth profiles of the rates of these processes can be integrated to estimate depth-integrated rates (DIR) of methane and ΣCO_2 production, thus providing yet another independent estimate of J_{out}.

Modeled rates can also be used to estimate DIR values. If, for example, one assumes a specific functional relationship for the kinetics of the process of interest, such as ΣR_i in any of the diagenetic equations in chapter 6, then one can model solid phase and/or pore-water depth profiles and use the resulting best fit of the equation to the actual data to calculate the depth integrated rate of the process of interest. Using this approach with the solution to the diagenetic equation for metabolizable organic carbon (e.g., see sections 6.3.1 and 4.5.1) yields the following equation for the depth-integrated rate of organic carbon remineralization (Martens and Klump, 1984):

$$DIR = \int_0^{z_L} kG_m^o e^{-\alpha z} dz = \frac{k}{\alpha}(1 - e^{-\alpha z_L})G_m^o \approx \omega G_m^o \qquad (8.20)$$

again because $\alpha = k/\omega$ (also see footnote 4 in this chapter).

Alternately, in modeling sediment profiles to estimate DIR values one may choose not to specify a functional relationship for ΣR_i, but rather simply fit the data to an arbitrary function of depth (e.g., see section 6.2). Then using a general diagenetic equation of the form of eqn. (6.29), for example, DIR values can be determined as,

$$\int_0^{z_L} \Sigma R_i dz = DIR = \int_0^{z_L} \left(-D_s \frac{\partial^2 C}{\partial z^2} + \omega \frac{\partial C}{\partial z} \right) dz \qquad (8.21)$$

While the right side of this equation may look complicated, because C can often be defined as a fairly straightforward function of z, differentiating this function and performing the integration on the right side of eqn. (8.22) is usually not difficult.

Again returning to results from CLB sediments, we can independently estimate depth-integrated rates of sulfate reduction using these different techniques. One approach (shown in table 8.2) involves the depth integration of directly measured sulfate reduction rates (Crill

TABLE 8.2
Estimates of Annual Average, Depth-integrated Rates of Sulfate Reduction in Cape Lookout Bight Sediments (All Values mol m^{-2} yr^{-1})

Method	Calculated with D_s^a	Calculated with D_s and $D_s'^b$	Directly Measured
Depth-integrated rates based on eqns. (8.21) and (8.22)[c]	13.1		
Benthic flux calculation (linear gradient at the sediment-water interface)[c,d]	12.1	16.2	
Benthic flux calculation (gradient predicted by the best-fit values to eqn. 8.22)[c,e]	12.1	15.5	
Sediment incubation[f]			15.6
^{35}S-sulfate radiotracer techniques[g]			16–24

a. Bulk sediment diffusion coefficients corrected for seasonal temperature variations.

b. The term D_s' is an apparent diffusion coefficient that takes into account enhanced transport across the sediment-water interface at this site for pore-water solutes in general, as a result of seasonal (summer to early fall) methane gas ebullition from the sediments. During periods when ebullition does not occur (late fall to spring) molecular diffusion dominates sediment-water exchange at this site, and standard D_s values (see section 6.1.1) are used in these calculations. Ebullition occurs during warmer months as a result of high rates of sediment methanogenesis and pore waters that reach saturation with respect to dissolved methane (see section 12.6 for further details). During these times it has been shown (Martens and Klump, 1980; Kipput and Martens, 1982) that an apparent diffusion coefficient D_s', which takes into account enhanced transport across the sediment-water interface as a result of gas ebullition, can be estimated as described in section 12.6. With this approach, during times of active ebullition D_s' values can be up to ~3 times greater than D_s values.

c. From Martens and Klump (1984).

d. In this approach to calculating dC/dz_o it was assumed that there is a linear pore-water gradient in the upper few centimeters of sediment, and the surface sulfate data were fit to a straight line.

e. In this approach to calculating dC/dz_o the gradient was based on eqn. 8.22 (= $-\Delta C\beta$), and was determined using best-fit parameters obtained from fitting sulfate data to eqn. (8.22).

f. From Crill and Martens (1983).

g. From Crill and Martens (1987).

and Martens, 1983, 1987). However, Martens and Klump (1984) also used two independent approaches to calculate these DIR values with pore water sulfate data. First they fit individual pore-water sulfate profiles collected for approximately 2.5 year, to the equation

$$C = C_\infty + \Delta C e^{-\beta z} \tag{8.22}$$

where $\Delta C = C_o - C_\infty$. Although this equation is essentially the same as the solution derived in section 6.3.3 for a pore-water sulfate diagenetic model, here Martens and Klump (1984) did not explicitly define the kinetics of sulfate reduction and thus their parameter β is not directly related to other sediment or rate parameters.

They then used each of these individual solutions to eqn. (8.22) with eqn. (8.21) to obtain "instantaneous" depth-integrated sulfate reduction rates based on each profile. These were finally used to calculate the mean annual average, depth-integrated sulfate reduction rate in CLB sediments (see table 8.2). Because of seasonal variations in the rates of processes in these sediments (and most estuarine and coastal sediments in general) any given instantaneous rate is not necessarily a good rate estimate on more appropriate seasonal or annual time scales (see discussion in section 14.3). Hence one needs to calculate this mean annual average.

Since depth-integrated rates are balanced by benthic fluxes, Martens and Klump (1984) were also able to use sulfate pore-water profiles to estimate the sulfate flux across the sediment-water interface (J_o) using the equation

$$J_o = -\varphi_o D_s \left(\frac{dC}{dz}\right)_0 + \varphi_\infty \omega_\infty C_o \tag{8.23}$$

(see eqns. 6.3 and 6.20). Although such calculated benthic fluxes generally do not include an advective term (e.g., see discussions in section 12.2), this term is included here because of the high sedimentation rate in these sediments. Again, individual estimates of J_o for specific cores were used to calculate a mean, annual average benthic sulfate flux. The results of these calculations are also shown in table 8.2. As can be seen, there is good agreement among these three independent estimates of the annual average, depth-integrated rate of sulfate reduction in CLB sediments, providing additional confidence in each of these values.

Although the discussion here has used Cape Lookout Bight sediments as an example of how independent techniques can be used to crosscheck calculations in a sediment budget calculation, the general approach outlined here is clearly applicable to other sediment systems. At the same time, though, before leaving this discussion it is of some interest to examine the specific results from these calculation to see what they tell us about biogeochemical processes occurring in CLB sediments.

Based on the CLB carbon budget in table 8.1, we see that the carbon burial efficiency of these sediments is quite high (~70%). High burial efficiencies are common in organic-rich sediments such as these (see compilations in Henrichs and Reeburgh, 1987; Canfield, 1994; Aller, 1998), and the significance of this in terms of sediment carbon burial and preservation will be discussed in chapter 15. The results of these calculations also indicate that sulfate reduction accounts for 68% of the carbon remineralization and methanogenesis 32%. Sulfate reduction accounts for similar amounts of carbon remineralization in other coastal marine sediments (see section 7.5.2), although in most of these sediments the remainder of the carbon oxidation occurs by oxic or suboxic remineralization rather than by methanogenesis.

The mean residence time (τ_r) of metabolizable organic carbon in CLB sediments calculated with eqn. (8.15) is also relatively short (~4.3 months) as compared to other coastal sediments such as Long Island Sound where τ_r is on the order of 2 years (Turekian et al., 1980). This short τ_r in CLB sediments, coupled with the high sedimentation rate at this site, implies that >98% of the remineralization in these sediments occurs in the upper 25 cm of sediment. This short τ for organic matter in these sediments, which is also equivalent to a large k value based on eqn. 8.18, also plays a role in the occurrence of seasonal changes in both pore-water profiles and rates of biogeochemical processes in these sediments (see section 14.3).

In Cape Lookout Bight sediments similar approaches have also been used to construct sediment nitrogen and phosphorus budgets (Klump and Martens, 1987), as well as budgets for sedimentary (total hydrolyzable) amino acids (THAAs; Burdige and Martens, 1988). A comparison of these results with the carbon budget results indicates that the τ_r for metabolizable organic carbon in Cape Lookout Bight sediments is roughly similar to that for total nitrogen (~4–6 months) and THAAs (~8 months).

In contrast, the mean residence time for phosphorus in CLB sediments is substantially longer (~19–23 months; Klump and Martens, 1987). This enhanced retention of sediment phosphorus relative to nitrogen and carbon may be related in part to precipitation of phosphate minerals at depth. At the same time, it may be related to adsorption of dissolved phosphate to iron oxides that form near the sediment-water interface, that is, dissolved phosphate produced at depth that diffuses upward toward the sediment-water interface, where it encounters a thin veneer of iron oxides that form when the upward diffusion of Fe^{2+} reacts with downwardly diffusing O_2. These points will be discussed further in section 16.3.

8.3 Carbon Burial in Sediments

Sediment organic matter that is not remineralized is considered to be buried or preserved in the sediment. In its most simple sense, the problem can be thought of in the context of the "reversibility" of the following reaction,

$$CO_2 + H_2O \rightleftharpoons CH_2O + O_2 \tag{8.24}$$

where the balance between the forward reaction (photosynthesis) and the reverse reaction (respiration) results in both carbon burial in marine sediments and the net addition of O_2 to the atmosphere. The burial of organic carbon in marine sediments represents the major link between "active" surface pools of carbon in the oceans, in the atmosphere, on land, and in marine sediment, and carbon pools that cycle on much longer (geological) time scales, i.e., carbon in sedimentary rock, coal, and petroleum deposits (also see Berner, 1989). The balance of rxn. (8.24) can also be considered in the context of a larger-scale global redox balance and controls on atmospheric O_2, as will be discussed in sections 15.4.1 and 17.1.2.

Given these considerations, understanding the controls on organic matter burial and preservation in marine sediments has been one of the most important areas of research in marine biogeochemistry. Numerous controlling factors have been proposed (see Hedges et al., 1999a, for a recent summary), ranging from primary productivity in the surface ocean (e.g., Calvert and Pedersen, 1992) to processes occurring in or near the sediments (e.g., see the beginning of chapter 15 and references cited therein). These will be discussed in more detail in chapter 15.

Most of the organic matter buried in marine sediments is eventually transformed into kerogen during later stages of diagenesis (e.g., Hunt, 1996) when sediments are subject to elevated temperatures and pressures during their transformation into sedimentary rocks (a small amount of this organic matter is also transformed into fossil fuels during this late diagenesis). Kerogen is operationally defined by solubility considerations (see section 9.8.3 and Whelan and Thompson-Rizer, 1993), and represents the largest repository of organic carbon on the Earth's surface (Berner, 1989). Kerogen is generally thought of as being extremely refractory, in part because it represents the very small fraction of organic matter that escapes remineralization in surface carbon cycles. Kerogen oxidation on land, after uplift of sedimentary rocks, is generally thought to balance carbon burial in marine sediments, and to a lesser extent the associated net addition of O_2 to the atmosphere (see discussions in Berner, 1999; Hedges, 2002). Additional information about the geochemistry of kerogen in marine sediments is also presented in section 11.1.4.

Another important point about carbon burial and preservation in marine sediments is that >80% of this preservation, and remineralization as well, occurs in nearshore deltaic and continental margin (shelf and upper slope) sediments (see table 8.3). This occurs despite the fact that these environments constitute less than 20% of the total area of all marine sediments. Thus understanding biogeochemical processes occurring in these sediments plays an important role in understanding the overall controls on sediment carbon preservation (e.g., Hedges and Keil, 1995; Hedges et al., 1999a), as will be discussed in greater detail in chapter 15.

8.4 Layered and Coupled Models of Sediment Diagenesis

Discussions presented throughout earlier sections have demonstrated that biogeochemical processes occurring in sediments vary with sediment depth. The most obvious example of this is the biogeochemical zonation of organic matter remineralization processes that occurs in nonbioturbated/nonbioirrigated marine sediments (see section 7.2). At the same time, the redox cycling of elements such as S, Mn, Fe, and

TABLE 8.3
Organic Carbon Burial Rates in Different Marine Sediment Regimes

Sediment Type	Carbon Burial Rate[a]	% of Total Burial
Data from Berner (1989)		
Terrigenous deltaic-shelf sediments	104	82
Biogenous sediments (high-productivity zones)	10	8
Shallow-water carbonates	6	5
Pelagic sediments (low-productivity zones)	5	4
Anoxic basins (e.g., Black Sea)	1	1
Total	126	
Recalculation of Berner's (1989) data by Hedges and Keil (1995)		
Deltaic sediments[b]	70	44
Shelf and upper slope sediments	68	42
Biogenous sediments (high-productivity zones)	10	6
Shallow-water carbonates	6	4
Pelagic sediments (low-productivity zones)	5	3
Anoxic basins (e.g., Black Sea)	1	0.5
Total	160	
Calculations presented by Middelburg et al. (1997)[c,d]		
0–200 m water depth (coastal sediments; 9%)		68
200–2,000 m water depths (continental margin sediments; 7%)		27
>2,000 m water depths (pelagic sediments; 84%)		3

a. Units of 10^{12} gC yr^{-1}. For comparison, dividing these numbers by 1,000 expresses them in units of Gt C yr^{-1} (=10^{15} gC yr^{-1}).

b. As discussed in Hedges and Keil (1995) deltaic-shelf sediments in the Berner (1989) calculations were reapportioned here by assuming that 33% of the sediment discharge from rivers is deposited in shelf and upper slope sediments outside of river deltas, and that these sediments contain 1.5% TOC (versus 0.7% TOC in strict deltaic sediments). As discussed in section 5.5 such deltaic sediments are found largely in river-dominated continental margins.

c. The total carbon burial rate calculated by Middelburg et al. (1997) is much higher than these other estimates (590 to 2600×10^{12} gC yr^{-1}), and may be biased toward higher values for reasons discussed by these workers. However, in spite of this possible problem with their estimate of the rate of total carbon burial, their partitioning of this burial is very similar to the other estimates listed here.

d. Values in parentheses are the percentages of the total sediment surface area found in each sediment regime.

N can also occur in such a way as to lead to a spatial separation of the different parts of each cycle. In its simplest sense, this occurs when reduced end-products of suboxic or anoxic remineralization are produced at depth and then transported upward into oxygen- or nitrate-containing sediments, where they are subsequently reoxidized (see discussions in section 7.4.3). Thus, for example, this can lead to a situation in which there is a surficial zone where Mn oxidation occurs and a deeper zone where Mn reduction occurs (see section 13.2 and figs. 13.2 and 13.3).

Modeling such "layered" systems can be carried out in two different ways. One approach involves independently defining the depth zonation of sediment processes and then writing separate diagenetic equations for each compound being modeled in each of these different sediment regions. For the Mn example discussed above, one would thus write equations for particulate (solid-phase) and dissolved (porewater) manganese in each of these two regions (four equations in total). Solutions of these equations for the same forms of manganese (e.g., dissolved or solid) are then be coupled at the boundary between the two regions. This is done by requiring that each solution predict the same concentrations and fluxes at this boundary, because there can be no concentration and flux discontinuities at this interface. Models of this type have been developed to examine manganese redox cycling in sediments (Burdige and Gieskes, 1983; Gratton et al., 1990), as well as nitrate profiles in pelagic sediments where both nitrification and denitrification occur (Goloway and Bender, 1982). This approach has also been used to model bioturbation in sediments in which it is assumed that there is a mixed (bioturbated) layer of fixed depth overlying deeper sediments in which bioturbation does not occur (e.g., Aller, 1982b). This latter model is discussed in further detail in section 12.4.2.

Advantages of this modeling approach are that the model equations can often be solved analytically rather than numerically. However, to achieve these analytical solutions assumptions are often made in the models to simplify the mathematical expressions for certain sediment processes. These models also require that the depth of the boundary between different sediment regions be prescribed a priori, rather than allowing the boundary to be determined directly by chemical distributions in the sediments. This approach is then difficult to incorporate into a nonsteady-state diagenetic model.

To overcome these difficulties, another modeling approach uses a single diagenetic equation for each chemical constituent over the entire sediment column (see eqn. 8.27 below); functionally it then turns "off" and "on" various biogeochemical processes in the model equations. Since the occurrence of particular processes (i.e., the turning off and on of these processes) is generally defined by the geochemistry of the sediments, these models make no initial assumptions about the depths of different sediment boundaries. Thus while these models must be solved with numerical techniques, they are amenable to use in time-dependent models of sediment diagenesis. Several approaches to this type of diagenetic modeling have been described in the literature (Boudreau, 1996; Dhakar and Burdige, 1996; Soetaert et al., 1996c; Wang and Van Capellen, 1996; Boudreau, 1997a; Martens et al., 1998; Archer et al., 2002; Burdige, 2002).

To examine how these models control the occurrence of specific sediment processes we first note that in some cases, the occurrence of one process can be thought of as being inhibited by the presence of another compound. For example, we generally think of oxygen as inhibiting the occurrence of suboxic and anoxic remineralization processes (see section 7.3.2) or of sulfate inhibiting the occurrence of methanogenesis (see section 7.3.5). Although this inhibition is not necessarily directly related to the presence/absence of these oxidants per se (e.g., see discussions in the sections cited above), this approach appears to serve as a reasonable proxy for the factor controlling the occurrence of these processes.

As a result, models of these processes generally include some sort of concentration-dependent function that, for example, inhibits methanogenesis in the presence of sulfate, or inhibits suboxic remineralization in the presence of O_2. Several different types of inhibition functions have been used in such models, although they all have the basic property that above some critical concentration the inhibition function equals zero and below this critical concentration the function quickly goes to 1. Thus, for example, when a rate expression for denitrification is multiplied by such an O_2-dependent inhibition function the denitrification rate goes from zero to some finite value below a threshold oxygen concentration (fig. 8.3).

Controlling the occurrence of other processes relies on using more explicit kinetic expressions that let the chemistry of the system carry out this task. For processes such as manganese oxidation or nitrification (ammonium oxidation), the kinetic expression for such reactions is written such that the rate depends on both oxidant and reductant

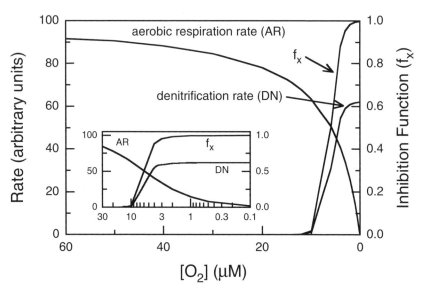

Figure 8.3 The oxygen inhibition function f_x defined by Dhakar and Burdige (1996) as a function of oxygen concentration. This function is given by $f_x = 1/(1 + e^{([O_2] - y)})$ where $y = 6$ μM. The rate of aerobic respiration (AR) shown here as a function of $[O_2]$ is based on eqn. (8.1). Similarly, the rate of denitrification (DN) is a function of nitrate and also based on eqn. (8.1) and f_x. Shown as an insert are f_x, AR, and DN on an expanded log scale of $[O_2]$, to better illustrate the sharp transition of f_x at low values of $[O_2]$.

concentrations. This may be either a simple product of the two reactants, for example,

$$R_{Mn\ oxid} = k_{Mnox}[Mn^{2+}][O_2] \quad (8.25)$$

or as a more complex hyperbolic (or Monod-type) function of both the oxidant and reductant (see eqn. 7.10 and recent discussions in Gilbert et al., 2003). Therefore the rates of these processes go to zero as oxygen concentrations go to zero. In contrast, the Mn models discussed above assume that manganese oxidation is only a function of pore-water manganese concentrations, with no explicit oxygen dependence (e.g., Burdige and Gieskes, 1983). As noted above, such models must therefore independently define a sediment redox boundary (i.e., the zero oxygen level) to "stop" the occurrence of manganese oxidation.

Similarly, kinetic expressions for organic matter remineralization processes also explicitly include some dependence on oxidant concentration (e.g., eqn. 8.1). At high oxidant concentrations these expressions are zero order with respect to oxidant concentration, and equivalent to eqns. (8.2) or (8.3). However, as oxidant concentration decreases, the kinetics of the process eventually become first order with respect to oxidant concentration, and eqn. (8.1) then approximately becomes

$$\frac{dG}{dt} \approx -\left(\frac{k}{K_m}\right) G[Ox] \qquad (8.26)$$

The use of equations such as eqn. (8.1) rather than eqns. (8.2) or (8.3) therefore leads to a situation in which the rate of the relevant organic matter remineralization process goes to zero as the oxidant concentration also goes to zero.

At the same time, these K_m values are generally small in comparison to *in situ* concentrations (see references in section 8.1). Therefore this transition from zero-order to first-order kinetics will have a minimal impact on the overall rates of sediment organic matter remineralization by a given oxidant, since most remineralization will occur by zero-order kinetics (see similar discussions at the beginning of section 8.1). Furthermore, in a layered sediment this implies that sediment organic matter remineralization will largely look as if it has a weak dependence (at best) on oxidant concentrations, and eqns. (8.2) or (8.3) will still roughly describe the kinetics of remineralization (also see discussions in Boudreau, 1999, for further details).

At the same time, this approach is of importance for modeling oxidant concentration profiles. In conjunction with the use of the inhibition functions described above, this allows chemical distributions in the sediments to produce smooth transitions between biogeochemical processes in different sediment layers (see the discussions below for further details).

When all of these approaches are incorporated into a diagenetic model, they then lead to fairly complex differential equations for each of the various solutes and solids being modeled. For example, in the steady-state model of Dhakar and Burdige (1996) the equation for pore-water nitrate (N) is given by

$$0 = \frac{\partial}{\partial z}\left[-(D_n + K_D)\frac{\partial N}{\partial z} + \omega N\right] - \frac{f_x F k_{dn} G N}{K_n + N} - k_{an} A_m N \\ - k_{fn} F_e N + k_{ao} A_m O_x + \frac{l_{an} F G O_x}{K_{ox} + O_x} \qquad (8.27)$$

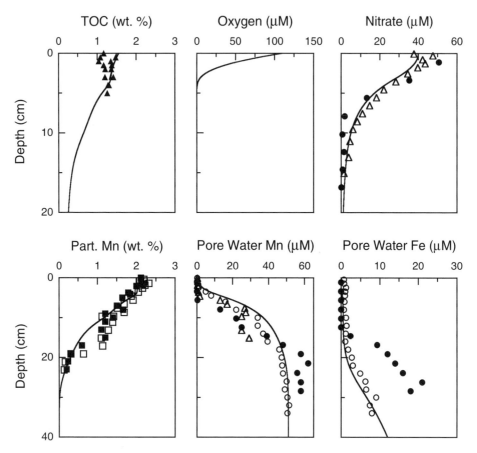

Figure 8.4 The fit of a steady-state, coupled nonlinear sediment diagenesis model to data from MANOP site M in the eastern equatorial Pacific (model results from Dhakar and Burdige, 1996). Data sources: ▲ = Emerson et al. (1985); ● = Klinkhammer (1980); △ = Heggie et al. (1986); ■ = Kalhorn and Emerson (1984); □ = Lyle et al. (1984); ○ = Lyle (1983).

where, moving from left to right, the terms in this equation represent: nitrate transport (depth-dependent diffusion, bioturbation, and advection), nitrate consumption via denitrification (G = organic carbon), nitrate consumption via anammox (A_m = ammonium), nitrate reduction coupled to iron (F_e) oxidation, nitrate production via nitrification of upwardly diffusing ammonium (see section 15.2; O_x = oxygen), and nitrate production via nitrification coupled to aerobic respiration. Furthermore, in an equation such as this, as the oxygen concentration goes

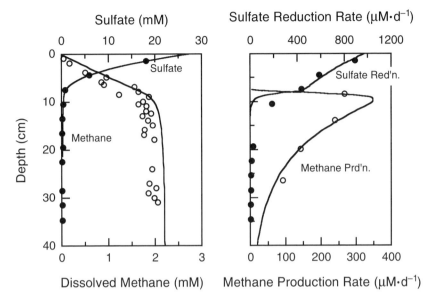

Figure 8.5 The fit of a steady-state, coupled nonlinear sediment diagenesis model to data from Cape Lookout Bight sediments (redrawn from results presented in Martens et al. 1998).

to zero with sediment depth aerobic respiration also goes to zero, the inhibition function (f_x) described above goes from zero to 1, and denitrification begins. Therefore this approach allows one to model the biogeochemical zonation (e.g., oxic and suboxic respiration) that occurs in many deep-sea sediments (fig. 8.4).

In organic-rich sediments where sulfate reduction and methanogenesis occur, this approach also implies that for situations in which sulfate, rather than organic matter, limits sulfate reduction (see section 6.3.3), the rate of sulfate reduction, expressed in the form of eqn. (8.1), decreases to zero as sulfate goes to zero. This approach therefore does not allow model sulfate concentrations to become negative (as discussed in section 6.3.3). Instead, the model transitions smoothly from sulfate reduction to methanogenesis below the zero-sulfate depth, consistent with field observations in such sediments (fig. 8.5).

Models of this latter type generally lead to a set of coupled differential equations for some number of pore water solutes and solid phase constituents. For example, in Boudreau's (1996) CANDI model there are equations for ten pore-water solutes and five solid-phase constituents. Furthemore, because the kinetic expressions for the biogeochemical reactions in such models are often nonlinear functions (e.g., see eqns. 8.1, 8.25, and 8.27), the resulting differential equations are also generally nonlinear. A variety of numerical techniques have been described in recent years to solve such coupled nonlinear diagenetic models (Boudreau, 1996; Dhakar and Burdige, 1996; Soetaert et al., 1996c; Wang and 0Van Capellen, 1996; Boudreau, 1997a; Archer et al., 2002; Meysman et al., 2003b).

≈ **CHAPTER NINE** ≈

An Introduction to the Organic Geochemistry of Marine Sediments

IN PREVIOUS CHAPTERS we began to examine the composition of sediment organic matter as it relates to sediment processes such as organic matter remineralization. In this chapter we will look at the organic geochemistry of sediments in greater detail, setting the stage for discussions in later chapters that more directly link organic geochemistry to quantitative studies of early diagenetic processes in sediments.[1]

In discussing the organic geochemistry of sediments we will begin by examining some of the bulk properties of sediment organic matter and then look at the major biochemical classes of this material. In our examination of the biochemical composition of sediment organic matter we will consider four broad biochemical classes. Amino acids (proteins) and carbohydrates (sugars) represent the bulk of the organic matter produced by most marine and terrestrial organisms and also play a major role in organic matter remineralization in the water column and in sediments. Lignins are a class of phenolic compounds found exclusively in vascular plants and represent important tracers of terrestrial organic matter. Lipids, and fatty acids in particular, represent a diverse class of compounds that have a wide range of functionalities in marine organisms; they are also used extensively as source indicators of specific types of organic matter. Finally, a large component of sedimentary organic matter falls into the category of molecularly uncharacterized organic matter (Hedges et al., 2000). The mechanisms by which this material "forms," and its possible role in carbon burial and preservation, will also be discussed.

To understand the controls on carbon preservation versus remineralization in sediments it is also important to have information not

---

[1] This chapter is dedicated to the memory of John Hedges, who died in July 2002 as I was beginning to work on this chapter. An examination of this chapter and others in this book clearly indicates the major impact of John's work on the development of many of the ideas presented here. John was also a person of special charm and wit, who always seemed to be tremendously enjoying whatever he was doing. He is greatly missed by family, friends and colleagues.

only on organic matter composition, but also about the "matrix" in which the organic matter is contained. This "packaging" of sedimentary organic matter has significance in terms of the reactivity of the material as well as in terms of our ability to chemically analyze the material. Work in recent years (discussed below) has shown that physical form is at least as important as chemical composition in determining the reactivity of organic matter in many natural settings (including sediments).

9.1 General Considerations

Different types of biologically produced organic matter have different reactivities, and selective preservation and/or remineralization of these classes of organic matter may occur during diagenesis. This organic matter fractionation appears to begin in the oceanic water column, where sediment trap studies show that as organic matter sinks through the water column, preferential decomposition of reactive components appears to occur (Wakeham et al., 1997b). This process also results in "production" of organic matter that cannot be characterized at the molecular level by conventional analytical techniques such as gas or liquid chromatography, thus leading to an increase in a molecularly uncharacterized component of this organic matter (Hedges et al., 2001).

Historically, this molecularly uncharacterized organic matter (MU-OM) was thought to form through abiotic *heteropolycondensation*, or geopolymerization, reactions involving simple organic matter intermediates such as monomeric sugars, amino acids, or fatty acids. This classical model of humification is discussed in section 9.8.3 and in, for example, Tissot and Welte (1978) and Hedges (1988). However, recent studies suggest that this model may not be entirely correct, as will be discussed in section 9.8.

Regardless of the mechanism(s) by which MU-OM forms, this "fractionation" of organic matter continues during early diagenesis in sediments (fig. 9.1; Cowie and Hedges, 1994). As is shown here, not only is there a decrease in the absolute amount of organic carbon in natural samples as a result of this diagenetic "maturity," but the relative amounts of presumably reactive components of the total organic matter pool, i.e., total amino acids and carbohydrates, also appear to decrease. Note that if these components of the total organic matter

GENERAL CONSIDERATIONS

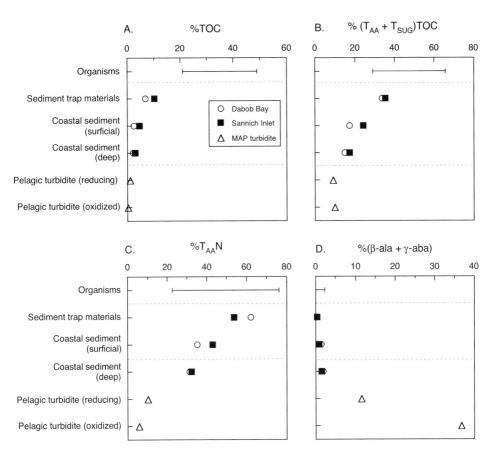

Figure 9.1 The composition of organic matter in natural marine samples as a function of the diagenetic maturity of the material (redrawn from data presented in Cowie and Hedges, 1994). As discussed in the text, diagenetic maturity can be thought of as a process by which increasing remineralization/alteration of organic matter, starting with fresh organic matter in marine organisms, leads to increasingly degraded material. The range for *Organisms* represents a range of values for typical marine organisms and vascular plants. The values labeled *Sediment trap materials* are annual averages of monthly deployments at 30 m (Dabob Bay) and 50 m (Sannich Inlet). *Surficial coastal sediments* are from the 0–1 cm sediment depth at these two sites while *Deep coastal sediments* are from sediment depths of 48–50 cm (Dabob Bay) and 78–80 (Saanich Inlet), and represent sediment ages of ~100–120 yrs. The values labeled *Pelagic turbidite* are from sediments above (*reducing*) and below (*oxidized*) the oxidation front in a turbidite from the Madeira Abyssal Plain

pool did not undergo selective remineralization during diagenesis, then the percentages in figs. 9.1B and C would be roughly constant across this gradient of diagenesis. At the same time, this loss of reactive components may not necessarily be a real loss, but may be related to changes in the nature of the organic matter that occur during diagenetic maturity such that these reactive compounds no longer are recognizable by conventional analytical techniques (see sections 9.8 and 9.8.5 for further details).

At the same time, the results in fig. 9.1D indicate that the mole percentages of two (presumably) refractory, nonprotein amino acids increase through this diagenetic sequence. If diagenetic processes selectively utilize reactive components, then the remaining material could simply become increasingly enriched in refractory material, as is seen in this figure. In a more general sense, this explanation is an example of possible mechanism of carbon preservation, namely the selective preservation of inherently refractory biologically produced macromolecules (Hatcher and Spiker, 1988; de Leeuw and Largeau, 1993). However, the increase in the relative amounts of refractory components during diagenetic maturity may also result from their *in situ* production during organic matter diagenesis. These two general models of carbon preservation will be discussed in greater detail later in this chapter.

9.2 Concentrations and Sources of Organic Matter in Marine Sediments

Globally, the total organic carbon (TOC) content of marine sediments ranges from <2.5 mg C g_{dw}^{-1} to >20 mg C g_{dw}^{-1}, while the total nitrogen (TN) content ranges from <0.3 mg N g_{dw}^{-1} to >4 mg N g_{dw}^{-1} (Premuzic et al., 1982; Romankevich, 1984). For the most part, concentrations

---

◂

(see section 14.4 for further details). A. Weight percent total organic carbon [%TOC]. B. The percentage of TOC in the form of hydrolyzable amino acids and aldoses (sugars) [%($T_{AA} + T_{SUG}$)TOC] C. The percentage of total nitrogen in the form of hydrolyzable amino acids [%T_{AA}N]. D. The mole percentage of hydrolyzable amino acids found as the nonprotein amino acids β-alanine and γ-aminobutyric acid [%(β-ala + γ-aba)].

are lowest in open ocean (pelagic) sediments and highest in coastal and continental margin sediments. A more recent compilation by Keil et al. (2000) supports these observations and provides additional information on organic matter concentrations in organic-rich coastal and continental margin sediments (e.g., sediments within oxygen deficient zones in regions of intense upwelling). In such sediments, TOC concentrations can be as high as 200 mg C g_{dw}^{-1}, while TN concentrations can be as high as 12 m N g_{dw}^{-1}.

In a very broad sense, we can think of organic matter in sediments as being either marine derived or terrestrial derived. The "endmember" for marine organic matter is generally considered to be phytoplankton debris, or detritus, whose chemical components are predominantly proteins, carbohydrates, and lipids. Terrestrial organic matter consists of living biomass, plant litter, and soil organic matter, the latter being largely composed of highly altered and degraded remains of this living terrrestrial biomass, e.g., soil humus (see discussions in Stevenson, 1994; Hedges et al., 1997). Terrestrial organic matter is brought to the oceans largely by rivers, in either a dissolved or particulate form, although atmospheric inputs could be as large as 25% of the total river flux (Romankevich, 1984). Additional details on the differences between marine and terrestrial organic matter will be discussed in chapter 11. Other possible sources of the organic matter in marine sediments include the remains of *in situ* microbial biomass (section 11.1.5), black carbon (or products of incomplete biomass burning; section 9.8.1), and weathered kerogen transported back to oceans after its uplift in sedimentary rocks (section 11.1.4).

9.3 THE BULK CHEMICAL COMPOSITION OF MARINE SEDIMENT ORGANIC MATTER

Since the formulation of the Redfield-Ketchum-Richards equation (eqn. 7.5) several studies have re-examined the elemental composition of marine organic matter using a range of analytical and oceanographic techniques. In general, the results of these studies suggest that marine planktonic organic matter (collected by surface-water plankton tows) is less rich in hydrogen and oxygen than that proposed by the RKR equation, and its composition is more consistent with a chemical formula in the range $C_{106}H_{175-180}O_{35-44}N_{15-20}S_{0.3-0.5}$ (see recent discussions in Hedges et al., 2002). This material then requires 150–155 moles of O_2 for its complete oxidation, versus 138 for RKR material (rxn. 7.5). The average carbon oxidation state in this material is also slightly more reduced than that in RKR material (between approx. −0.3 and −0.7 vs. 0; see table 9.1).

TABLE 9.1
The Oxidation State of Carbon in Naturally Occurring Organic Matter

Material	Carbon Oxidation State
"RKR" marine organic matter	~0
Marine plankton[a]	−0.3 to −0.7
Marine organic matter undergoing water column remineralization[b]	−1.3 ± 0.7
Average bacteria[c]	−0.8 to −0.9
Bulk algae and bacteria[d]	−0.6 to −1.2
Reactive POC in Cape Lookout Bight (NC) sediments[e]	−0.7, −0.1 ± 0.4
Reactive organic matter in Chesapeake Bay sediments[f]	−0.5 ± 0.1, 0.3 ± 0.3
Surficial marine sediments[g]	−0.4 to −0.8
Reactive POC in pelagic red clay sediments[h]	0.2 ± 1.3
Reactive POC in Equatorial Pacific sediments[i]	−0.2

a. Based on the chemical formula listed in the text (section 9.3) from Hedges et al. (2002), and the following assumptions about this material: H has an oxidation state of +1, O has an oxidation state of −2, N has an oxidation state of −3 (i.e., it is reduced amide nitrogen), and S has an oxidation state that is either −2 (i.e., sulfide) or +6 (i.e., sulfate).

b. From Anderson and Sarmiento (1994) in which water column nutrient data were used to estimate the following remineralization ratios: $P:N:C_{org}:-O_2 = 1:16(\pm 1):117(\pm 14):170(\pm 10)$. These ratios were then used with eqns. (11.7) and (11.8) to calculate the value shown here.

c. Based on elemental composition data presented in Fenchel et al. (1998) assuming C, H, O, N, and S have the oxidation states listed above, and that P has an oxidation state of +5, i.e., phosphate.

d. From Gujer and Zehnder (1983).

e. The first value is from Alperin et al. (1994), based on a plot of $\Delta\Sigma CO_2$ vs. ΔSO_4^{2-} in a laboratory sediment decomposition experiment. The second value is based on stoichiometric modeling of pore-water sulfate and ΣCO_2 data (see section 11.2.1 and eqns. 11.4–11.6) from six cores reported in Klump and Martens (1987).

f. The first value is based on stoichiometric modeling of pore-water sulfate and ΣCO_2 data (see eqns. 11.4–6 and fig. 11.2) from 18 cores collected at station M3 in the meso

THE BULK CHEMICAL COMPOSITION

Other results also suggest that the carbon in sediment particulate organic matter (POM), or more specifically the sediment POM undergoing remineralization, is slightly more reduced than that predicted by the RKR ratio (see table 9.1). In particular, the NMR results presented by Gélinas et al. (2001), when used to estimate sediment POM carbon oxidation states, suggest that the observed range in carbon oxidation states is a function of both the degree of organic matter degradation and its oxygen exposure time (fig. 9.2). Increasingly degraded POM is more oxidized (less negative carbon oxidation states), and longer periods of oxygen exposure also lead to more oxidized POM (see section 9.4 for a discussion of the amino acid degradation index used here, and see Hedges et al., 1999a, and discussions in chapter 15 on the relationship between oxygen exposure and organic matter degradation).

---

haline portion of Chesapeake Bay, collected between 8/91 and 11/97 (Burdige, unpub. data; see Burdige, 2001, for additional information on the sediments at this site). The second value is based on plots of $\Delta\Sigma CO_2$ vs. ΔSO_4^{2} in laboratory sediment decomposition experiments using sediments collected in the lower Chesapeake Bay and tributaries (Burdige, 1991).

g. Based on results in Gélinas et al. (2001); here they examined demineralized sediments using ^{13}C NMR, and applied a carbon mixing model to the NMR spectra to estimate the amount of carbon in the following pools: amino acid carbon; carbohydrate carbon; black carbon; alkyl nonprotein carbon. For these calculations amino acid carbon was assumed to have an average oxidation state of either –0.13 (based on the average amino acid composition of marine plankton; Hedges et al., 2002) or –0.48 (based on the average amino acid composition of marine sediments; Keil et al., 2000). Both carbohydrate carbon and black carbon are assumed to have an average oxidation state of 0 (Hedges et al., 2002). Alkyl non-protein carbon was assumed to have an average oxidation state of either –1.67 (assuming that the average "lipid" is roughly equivalent to oleic acid; Hedges et al., 2002) or –2.1 (assuming that this material is slightly less oxidized). The samples examined by Gélinas et al. (2001) are all in the upper 20 cm of sediment from sites that range from coastal (continental margin) to deep-sea sediments. The TOC content of these sediments ranges from ~7 to 0.3 wt%.

h. Calculated from data in Grundmanis and Murray (1982) with pore-water stoichiometric models (see section 11.2.1) using O_2, nitrate and ΣCO_2 data from equatorial Pacific pelagic red clay sediments (water column depths ~5,000 m, sediment TOC values generally less than 0.5–0.7 wt%).

i. Calculated with benthic flux data from Hammond et al. (1996), as discussed in section 11.2. These sites are in the equatorial Pacific (locations that range from ~105°W to 140°E between 2°S and 2°N) in water depths between 3,300 and 5,000 m. Sediment TOC contents are generally less than ~0.5 wt%.

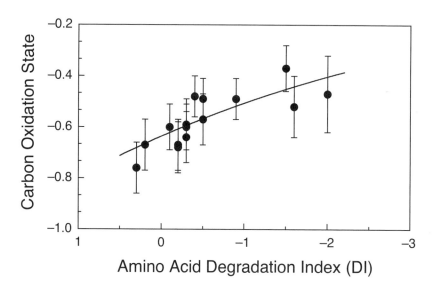

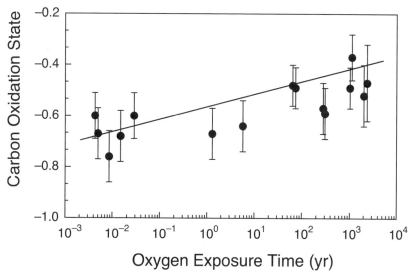

Figure 9.2 (Top) The carbon oxidation state of organic matter in surface marine sediments as a function of the amino acid degradation index (DI). Carbon oxidation states were calculated with data in Gélinas et al. (2001) as described in note g of table 9.1, while DI values were taken from the original work. The sediments examined here range from continental margin to deep-sea sediments (see table 9.1 and the original work for additional details about these sediments). As discussed in the text, this figure indicates that increasing

9.4 Amino Acids

Amino acids are the building blocks of proteins and peptides and make up much of the organic nitrogen in almost all organisms. The 20 protein amino acids are α-amino carboxylic acids (fig. 9.3A) where the functional group R can be as simple as a proton (the amino acid glycine), an alkyl group (e.g., alanine or leucine), or a more complex functional group such as a phenyl group (e.g., phenylalanine or tryptophan) or a heterocyclic nitrogen ring structure (e.g, tryptophan or histidine).

Protein amino acids can be further classified as acidic, neutral, or basic amino acids (table 9.2), depending on the chemical properties of these functional groups. In proteins and peptides individual amino acids are joined together by peptide linkages (fig. 9.3B) that form through a dehydration reaction involving the carboxyl group of one amino acid and the amino group of the other. Hydrolysis, either chemical or enzymatic, then results in the addition of water across this bond, producing either free amino acids or smaller peptides.

Proteins produced by marine organisms have several different functions. Some occur intracellularly and are used, for example, as hormones (chemical messengers) or antibodies, or to control intracellular trace metal concentrations, e.g., metallothioneins. Most frequently, though, they are used in metabolic processes as enzymes. In some phytoplankton and zooplankton they are a part of the organic matrix that aids in the precipitation of carbonate and silica skeletons (see section 2.2 for further details). Some are also found in compounds referred to as peptidoglycans, the primary structural components of bacterial cell walls. Peptidoglycans are also generally pre-

---

alteration or diagenetic maturity, i.e., increasingly negative DI values, leads to increasingly oxidized sediment organic matter. The curves here, and in the next panel, are included to better illustrate the trends in the data, and not meant to imply any known functional relationships between these quantities. (Bottom) The carbon oxidation state of organic matter in surface marine sediments as a function of organic matter oxygen exposure time (OET). OET values were determined as described in Gélinas et al. (2001) using procedures described in section 15.4. Again note that longer periods of oxygen exposure lead to more oxidized sediment organic matter (again see related discussions in section 15.4).

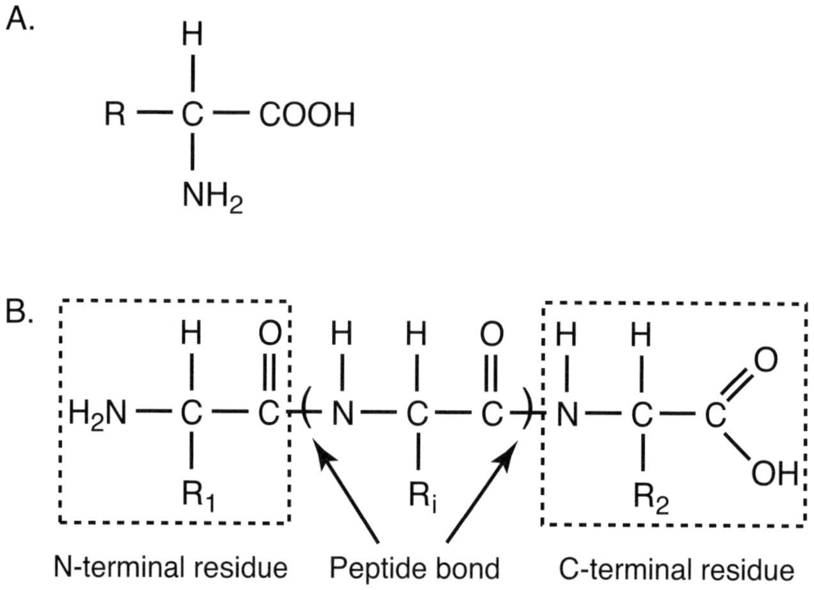

ORGANIC GEOCHEMISTRY OF SEDIMENTS

A.

$$R - \underset{\underset{NH_2}{|}}{\overset{\overset{H}{|}}{C}} - COOH$$

B.

$$H_2N - \underset{\underset{R_1}{|}}{\overset{\overset{H}{|}}{C}} - \overset{\overset{O}{\|}}{C} + N - \underset{\underset{R_i}{|}}{\overset{\overset{H}{|}}{C}} - \overset{\overset{O}{\|}}{C} + N - \underset{\underset{R_2}{|}}{\overset{\overset{H}{|}}{C}} - C\underset{OH}{\overset{O}{\diagup}}$$

N-terminal residue   Peptide bond   C-terminal residue

Figure 9.3 A. The basic structure of most protein, and many nonprotein, amino acids. Note, however, that in proline and hydroxyproline the amide group is part of a 5-member heterocyclic ring. B. A schematic representation of a protein or peptide chain indicating the N-terminal and C-terminal amino acid residues and the remaining amino acids in the structure joined together by peptide linkages.

sumed to be more resistant to degradation than other nonstructural proteins (de Leeuw and Largeau, 1993; McCarthy et al., 1998; Pedersen et al., 2001; Grutters et al., 2002). In spite of this wide range of uses, however, the bulk amino acid composition of marine organisms is very similar (fig. 9.4), generally making it difficult to use sediment amino acid compositional data to determine sources of organic matter.

All protein amino acids except glycine have a chiral carbon atom adjacent to the carboxyl group and therefore form optical isomers or enantiomers. Virtually all amino acids produced by organisms are of the L-form; with the death of an organism, the L-form will eventually transform into an equilibrium, racemic mixture of D- and L-amino acids. With some knowledge of the kinetics of this reaction, amino acid D/L ratios can be used to date Quaternary materials such as sediments, shells, and bones (Wehmiller, 1993).

Table 9.2
General Classification of Protein Amino Acids*

Acidic amino acids	**aspartic acid**[a], **glutamic acid**[b]
Neutral amino acids	alanine, asparagine[a], cysteine[c], **glutamine**[b], **glycine, isoleucine, leucine, methionine, phenylalanine**, proline[d], **serine, threonine**, tryptophan[e], **tyrosine, valine**
Basic amino acids	**arginine, histidine, lysine**

*Amino acids that are commonly quantified after acid hydrolysis are listed in **bold**.

a. In samples that have undergone acid hydrolysis, asparagine is converted to aspartic acid; therefore, in such samples the measured aspartic acid is the sum of asparagine and aspartic acid.

b. In samples that have undergone acid hydrolysis, glutamine is converted to glutamic acid; therefore, in such samples the measured glutamic acid is the sum of glutamine and glutamic acid.

c. In samples that have undergone acid hydrolysis, cysteine is converted to cysteic acid.

d. Cannot be analyzed by HPLC and OPA derivitization.

e. Destroyed during acid hydrolysis.

In addition to the standard protein amino acids, there are numerous nonprotein amino acids. These nonprotein amino acids are known to occur in free or combined forms, but not in proteins, and are generally derived from protein amino acids. In nonprotein amino acids, the amino group is often found in the β-, γ- or δ-position (relative to the carboxyl carbon), as compared to the α-position in protein amino acids. The most common nonprotein amino acids in (particulate) marine sediments are β-alanine, γ-aminobutyric acid, α-aminobutyric acid, and ornithine (Keil et al., 2000). In sediment pore waters other nonprotein amino acids, including β-aminoglutaric acid (β-glutamic acid), can be important (Burdige, 2002).

The analysis of combined amino acids in sediments or dissolved in seawater/pore waters generally involves acid hydrolysis of the sample, which produces free amino acids that can then be analyzed by liquid or gas chromatography. Additional details about these analytical procedures can be found elsewhere (Lee, 1988; Cowie and Hedges, 1992a; Keil et al., 2000). The analysis of sedimentary amino acids usually quantifies 15–16 protein amino acids, with the number

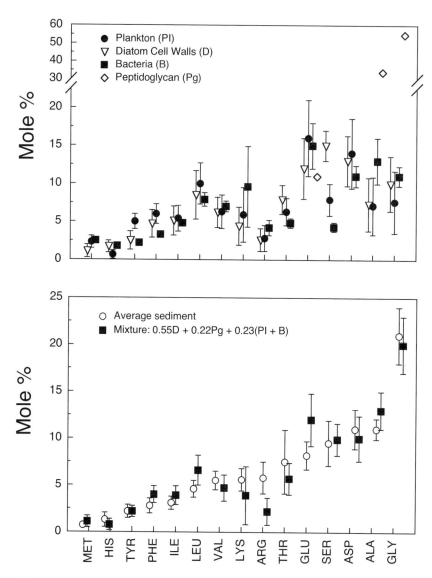

Figure 9.4 (Top) Mole percentages (±1 s.d.) of amino acids in various marine organic matter sources. Note that the plankton results are for cell-wall free diatoms and that the bacteria results include the associated peptidoglycan. (Bottom) The average amino acid model percentages of shallow marine sediments (±1 s.d.) and a mixture containing 55% diatom cell wall material (D), 22% peptidoglycan (Pg) and 23% plankton and bacterial proteins (Pl + B). Redrawn from data presented in Keil et al. (2000) and references cited therein.

of nonprotein amino acids that are determined varying among individual investigators.

Based on a recent literature summary, Keil et al. (2000) observed that amino acids represent ~50% of the organic carbon in marine organisms (range: ~30–70%) and ~75% of the nitrogen (range: ~45–100%). These percentages decrease as particles sink through the water column, possibly as a result of their reactivity (see the discussions below and in sections 9.8.5 and 9.9). Amino acid carbon represents ~10–30% of the organic carbon and ~20–70% of the total nitrogen found in shallow water sediments. The limited data on amino acids in pelagic (deep-sea) sediments further indicate that in these low-carbon and -nitrogen sediments, amino acids are generally <10% of the organic carbon and <20% of the total nitrogen (also note that trends similar to those described here can be seen in fig. 9.1C).

In shallow-water sediments, the relative amounts of total amino acids do not appear to vary systematically with sediment TOC or TN content. One might expect to see some relationship here if bulk amino acids were preferentially degraded as compared to total organic carbon and nitrogen, as appears to be the case for sinking particles in the water column. However, amino acid compositional data do suggest that across a spectrum of sedimentary environments that there is a relationship between sediment TOC content and the inferred degradation state of the organic matter based on its amino acid composition (see the discussion below for further details).

In sediments where amino acid remineralization has been quantified in terms of overall sedimentary carbon and nitrogen cycling, we see that amino acids represent a major fraction of both carbon and nitrogen undergoing remineralization (table 9.3). These percentages appear to vary more for nitrogen (~20–80%) than for carbon (~10–30% for all but Peru Upwelling sediments). These observations will be discussed in further detail in section 11.2.4.

In general, surficial marine sediments tend to be depleted in acidic and basic amino acids relative to biological sources. There are no systematic trends seen for the enrichment or depletion of neutral amino acids in sediments (relative to sources). However, some individual neutral amino acids do show some relative changes between sources and sediments. Glycine, serine, and threonine are preferentially preserved in sediments relative to sources, whereas leucine and isoleucine are preferentially depleted. Since glycine, serine, and threonine are components of the organic matrix associated with diatom cell

TABLE 9.3
The Importance of Amino Acids in Sediment Carbon
and Nitrogen Remineralization*

Site	Relative Amino Acid Remineralization	
	(% of TOC remin.)	(% of TN remin.)
Coastal sediments		
Cape Lookout Bight[a]	27 ± 11%	82 ± 43%
Long Island Sound[b]	~20%	~20–30%
Florida Bay[b]	~30%	~20%
Buzzards Bay[c]	11–23%	23–42%
Saanich Inlet[d]	22%	49%
Inner Oslofjord, Norway[e]	4–12%	13–40%
Laurentian Trough[f]	21%	67–80%
Continental margin sediments		
Peru Upwelling region[g]	35–60%	60–70%
Okhotsk Sea[h]	~7–17%	~30–80%
Goban Spur (northeast Atlantic continental slope)[i]		
651 m water depth	31%	
1,296 m water depth	17%	
3,650 m water depth	14%	
Deep sea sediments		
Madeira Abyssal Plain turbidite[j]	~3%	nc

*See chapter 8 (e.g., section 8.2.2) for additional general information on how such calculations are carried out. Also see the appendix for a brief discussion of some of the sites discussed here.

  a. From Burdige and Martens (1988).

  b. Calculated from data in Rosenfeld (1979a).

  c. Carbon results from Henrichs and Farrington (1987); nitrogen results calculated from data in this paper.

  d. From Cowie et al. (1992).

  e. From Haugen and Lichtentaler (1991).

  f. From Colombo et al. (1998). The Laurentian Trough is in the St. Lawrence River estuary (Canada) at a water depth of 300–450 m.

  g. Carbon results from Henrichs et al. (1984). Nitrogen results calculated from data in this paper and in Henrichs and Farrington (1984).

  h. Calculated from data in Maita et al. (1982).

  i. From Grutters et al. (2001).

  j. Calculated from data in Cowie et al. (1995). Results for nitrogen remineralization were not calculated (nc) because of difficulties quantifying nitrogen remineralization during the diagenesis of turbidite organic matter.

walls, or frustules (Hecky et al., 1973; also see section 2.2), preferential preservation of these amino acids may occur because of their association with these biominerals (Ingalls et al., 2003). Preferential preservation of these amino acids is also apparent in some, but not all, studies of amino acid diagenesis in sediments, expressed as an increase in the mole% of these amino acids with sediment depth (compare Burdige and Martens, 1988; Cowie et al., 1992; Colombo et al., 1998).

In many cases though (Henrichs et al., 1984; Henrichs and Farrington, 1987; Cowie and Hedges, 1992c; Cowie et al., 1995; Dauwe and Middelburg, 1998), amino acid mole% shows no systematic change with sediment depth, suggesting that the remineralization of protein amino acids can be fairly nonselective during early diagenesis in sediments. However, discussions below of the Dauwe degradation index suggest that on different time and space scales systematic changes in the amino acid composition of marine organic matter can occur during diagenesis. As an extreme example of this, Whelan et al. (1992) present results indicating the fractionation of amino acids during the long-term (million-year) diagenesis of organic matter in deeply buried marine sediments (sediment depths of hundreds of meters).

Building on some of these observations, Keil et al. (2000) recently used the results in fig. 9.4 to estimate the relative sources of amino acids to the "average" marine sediment. Their approach used a simple mixing model that assumes the predominant end-member amino acid sources are intracellular planktonic or bacterial proteins, diatom frustules, and bacterial peptidoglycan. The best fit to this three-component mixing model is shown in fig. 9.4, although, as discussed by these authors, some degree of caution must be taken in the interpretation of this calculation. Nonetheless, these calculations do provide some interesting insights into several problems in marine organic geochemistry, including the roles of sediment–amino acid interactions and *in situ* bacterial biomass production in sediment carbon preservation.

With specific regard to this latter issue, peptidoglycans represent one of the few classes of biologically produced materials that contain D-enantiomers of certain amino acids (alanine, glutamic acid, and aspartic acid). The nonprotein amino acid muramic acid is also found in bacterial cell walls, in the form of N-acetylmuramic acid. Therefore, these D-enantiomers, along with muramic acid and the $\delta^{13}C$ of other amino acids such as valine (Keil and Fogel, 2001) all have the potential to be useful as tracers of bacterial biomass in sediments.

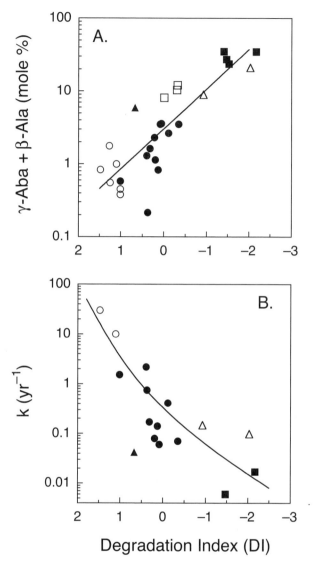

Figure 9.5 The mole percentage of the sum of the nonprotein amino acids γ-aba and β-ala (A) and the first order rate constant for organic matter degradation (B), as a function of the amino acid degradation index (redrawn from results presented in Dauwe et al., 1999, and references cited therein). Note that lines through these data are included to better illustrate the trends in the data and not meant to imply any known functional relationships between these quantities. Symbols: ○ = source materials (phytoplankton, zooplankton, or

Although much of this discussion has focused on protein amino acids, nonprotein amino acids are an important component of the amino acids found in low organic carbon, deep-sea sediments (e.g., see fig. 9.1). In particular, β-alanine, γ- and α-aminobutyric acid, and ornithine can represent more than 20 mole% (and up to 70 mole%) of the amino acids in these sediments (Whelan, 1977; Whelan and Emeis, 1992; Cowie et al., 1995; Keil et al., 2000). These amino acids are found at low levels in common sources of sediment organic matter (less than ~5 mole%) and the increasing mole% of β-alanine plus γ-aminobutyric acid has proven useful as a qualitative indicator of the degradation state of marine organic matter during its diagenetic alteration or maturity (Cowie and Hedges, 1994; Hedges et al., 1999a; also see fig. 9.5A). Selective preservation of these nonprotein amino acids during organic matter diagenesis represents a possible mechanism by which they achieve high concentrations in deep-sea sediments (e.g., see discussions in Cowie et al., 1995). These amino acids can also be produced during amino acid diagenesis, since the partial decarboxylation of aspartic or glutamic acid produces β-alanine and γ-aminobutyric acid, respectively; thus, *in situ* production could represent another source of these amino acids in deep-sea sediments. Keil et al. (2000) suggest that this latter mechanism dominates in environments low in organic matter, although the relative importance of these two possible mechanisms is not clearly established.

To more quantitatively use amino acid compositional data to examine the reactivity and degradation of sediment POM, Dauwe et al. (1998, 1999) used a principal component analysis of protein amino acid mole percentages in natural samples, i.e., source materials and marine sediments, to develop a degradation index (DI) for sediment POM. Keil et al. (2000) also used this index to examine amino acid compositional data from a wider range of marine settings.

In general, there is a decrease in the Dauwe index with increasing diagenetic maturity of marine organic matter, and DI values decrease as one moves from source materials (DI values generally >1) to sinking marine particles (DI between approx. −1 and 1) and then to marine

---

bacterial cultures); ● = coastal sediments; △ = hemipelagic sediments from the eastern Mediterranean (water depth = 2.539 m); ▲ = sapropel sediments from the eastern Mediterranean; ■ = reduced sediments from the MAP f-turbidite; □ = oxidized sediments from the MAP f-turbidite.

sediments (DI values generally <0). A positive relationship also exists between sediment organic matter DI values and both sediment TOC content and its first-order degradation rate constant (fig. 9.5A). Conversely, an inverse relationship exists between DI values and the mole % of nonprotein amino acids (fig. 9.5B).

An examination of the approach taken in defining the Dauwe index indicates that it is strongly related to the selective preservation of structural amino acids, such as those found in diatom frustules, versus those found more predominantly in intracellular material (also see related discussions in Ingalls et al., 2003). Furthermore, the index also appears to be most sensitive to diagenetic changes in organic matter prior to sediment deposition (Keil et al., 2000). This latter observation may then explain why DI values can vary widely among sediments that are of a broad similar "type"; for example, within shallow marine sediments DI values vary between almost −2 and +1, yet once the material is deposited in many of these sediments, the remineralization of amino acids appears to be much more nonspecific, that is, amino acid mole percentages generally do not change with sediment depth.

At the same time, Keil et al. (2000) found no organic-rich sediments with evidence of highly degraded amino acids, on the basis of either the mole% of nonprotein amino acids or the Dauwe DI. In these sediments large-scale compositional changes in the sediment organic matter may not occur prior to deposition, and the sediment amino acid composition may, therefore, better reflect the organic matter sources of the sediment rather than the impact of predepositional diagenesis. Such an explanation could then partially explain why in some of these sediments amino acid degradation exhibits some degree of selectivity, i.e., changes in mole percentages of some amino acids with sediment depth, similar to that seen during amino acid remineralization in sinking marine, water column particles (Lee and Wakeham, 1988; Keil et al., 2000).

As will be discussed throughout this chapter and in later chapters, the factors controlling carbon preservation versus remineralization appear to be a function of both the chemical composition of the organic matter as well as the "matrix" in which the organic matter is contained (be it some sort of organic matrix or a physical association with a sediment particle). The amino acid data discussed here support this general observation. For example, in the coastal sediments of Cape Lookout Bight and Buzzards Bay only ~40–50% of the amino acids deposited in these sediments that can be chemically analyzed are remineralized on early diagenetic time scales (Henrichs and Farrington, 1987; Burdige and Martens, 1988). Since amino acids are presumably a relatively reactive component of the marine organic-matter pool, some aspect of pre- or postdepositional diagenesis may therefore protect sediment amino acids, such that their complete remineralization is significantly impeded and/or prevented (also see recent discussions of this problem in Ingalls et al., 2003, 2004). Similar factors also appear to operate on longer time scales, since the preservation of proteins/amino acids is seen in fossilized marine organisms and in early Pleistocene sediments (~1 million years before present; Whelan and Emeis, 1992). Incorporation of amino acids into structural components such as peptidoglycans or the organic matrix associated with calcareous and siliceous shells may play a role here (as was discussed above regarding the Dauwe degradation index). Other factors that may be important here will be discussed in sections 9.8 and 9.9.

9.5 Carbohydrates

Carbohydrates, or polysaccharides, are major biochemical components of all living organisms. They are composed of monomeric sugars, or monosaccharides, produced by condensation or dehydration reactions, similar to those that form peptide linkages, across hydroxyl groups of adjoining sugars (fig. 9.6). Again, either chemical or enzymatic hydrolysis of polysaccharides produces monomeric sugars and/or polysaccharides of smaller sizes. Generally the term *polysaccharide* is reserved for carbohydrates with greater than 10 monomeric sugar units, while the term *oligosaccharide* is used to describe carbohydrates with 2–10 sugar units.

Simple monomeric sugars are either 5- or 6-carbon polyhydroxy aldehydes or ketones, i.e., pentoses and hexoses, and have the chemical formulas $C_5H_{11}O_5$ or $C_6H_{12}O_6$. Although their structures can be written in simple linear forms that clearly indicate the aldehyde or ketone functional groups, sugars are actually 5- and 6-member hemiacetal or hemiketal rings, in which a reaction between the aldehyde or ketone group and one of the hydroxyl groups forms the ring structure (fig. 9.6). The chemical modification of monomeric sugars produces deoxy sugars (through the deoxidization of a sugar hydroxyl group), amino sugars (through the substitution of an amine group for a hydroxyl group), uronic acids (through the oxidation of a sugar alcohol

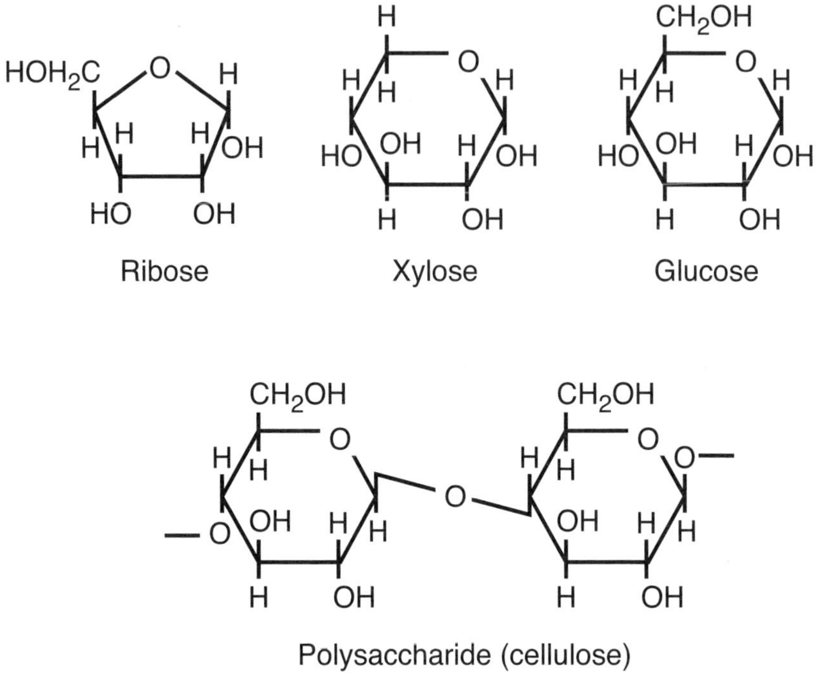

Figure 9.6 The structures of selected monosaccharides and a polysaccharide, cellulose.

group to an acid carboxyl group), and sugar alcohols (through the reduction of an aldehyde group).

Carbohydrates are used by marine and terrestrial organisms for both structural purposes (see section 2.2) and carbon and energy storage. The polysaccharide cellulose, a long-chain linear polymer of glucose, is a major structural component of vascular plants and many algae. Other important carbohydrates include hemicellulose, starch (e.g., amylose), fructans, alginic acids, and chitin (see de Leeuw and Largeau, 1993, for further details).

Chitin, a polymer of N-acetylglucosamine, is a structural polysaccharide found in a wide variety of marine organisms, e.g., arthropods and crustaceans. Chitin production in the biosphere is estimated to be second only to cellulose production (Stankiewicz and van Bergen, 1998). In chitinous organisms polysaccharides can be thought of as analogous to cellulose in vascular plants. Chitin is generally cross-linked to proteins, and these protein-chitin complexes can themselves be mineralized to inorganic materials, usually $CaCO_3$. In sediments, chitin degradation may lead to the production of the amino sugar glucosamine (e.g., Steinberg et al., 1987); however, N-acetylglucosamine is also a component of bacterial peptidoglycan, so this too represents a potential source for this amino sugar in sediments.

Carbohydrates in sediments or sediment pore waters are generally analyzed by either gas or liquid chromatography, after acid hydrolysis, sample clean-up, and derivitization (in some techniques). These procedures determine up to nine neutral sugars, again depending on the analytical techniques being used: the 5-carbon sugars lyxose, arabanose, xylose, and ribose and the 6-carbon sugars rhamnose, fucose, mannose, galactose, and glucose. Additional details about these methods can be found elsewhere (Steinberg et al., 1987; Mopper et al., 1992; Cowie et al., 1995; Skoog and Benner, 1998). Depending on the nature of the analytical scheme that is used (e.g., differences in the clean-up steps of the hydrolysates), many of these same techniques can be also be used to quantify modified sugars such as amino sugars and uronic acids.

Total carbohydrates, generally expressed as glucose-equivalent concentrations, can also be determined by spectrophotometric techniques (Arnosti and Holmer, 1999; Burdige et al., 2000). In coastal and continental margin sediments total sediment carbohydrates determined by these spectrophotometric methods are similar to values determined using the chromatographic techniques described above, in which individual neutral sugars are determined in sediment hydrolysates and then summed to obtain total sediment carbohydrates (see discussions in Burdige et al., 2000). This observation suggests that neutral sugars may represent the majority of the particulate carbohydrates in sediments. Consistent with this suggestion, total carbohydrate concentrations in two coastal sediments were dominated by eight neutral sugars—all of the sugars discussed above except lyxose, with uronic acids and O-methyl sugars accounting for <10% of the total carbohydrate yield (Bergamaschi et al., 1999).

Particulate carbohydrates are a major component of vascular plants (~75%) and marine plankton (~20–40%) and represent ~5–20% of the POC in sediments (see Hamilton and Hedges, 1988; Burdige et al., 2000, and references therein). Carbohydrate remineralization in sediments appears to be a similar percentage of total organic carbon remineralization (~8 ± 1%; range 5–15%) in a variety of sediments whose carbon oxidation rates span a range of more than 3–4 orders of magnitude (i.e., Cape lookout Bight sediments to the MAP-f turbidite; Burdige et al., 2000).

There appear to be some compositional differences among different types of biologically produced carbohydrates. For example, marine versus terrestrial source materials have different neutral sugar compositions, as do various types of terrestrial materials (Cowie and Hedges, 1984). Among marine sources, e.g., plankton vs. bacteria, attempts to use neutral sugars as organic matter source indicators have proven equivocal (e.g., Cowie and Hedges, 1984), although a detailed analysis of this problem, in a fashion analogous to that undertaken by Keil et al. (2000) for amino acids, has not been carried out. Minor sugars such as O-methyl aldoses and uronic acids can have a variety of marine and terrestrial sources, although marine bacteria appear to be their primary source in sediments (Bergamaschi et al., 1999).

In contrast, neutral sugar concentrations in natural organic-matter mixtures, e.g., sediment material, sediment trap samples, or riverine suspended particulate organic matter, do appear to be useful in qualitatively ascertaining the degradation state of this material. For example, the weight percentage of glucose decreases as organic matter degradation increases, while the weight percentage of the sum of fucose plus rhamnose increases (Hedges et al., 1999a). The causes of this may be related to where and how these neutral sugars are used by organisms, since glucose tends to predominate in intracellular (energy plus carbon storage) polysaccharides, while rhamnose and fucose are generally more abundant in structural (cell wall) polysaccharides (Cowie et al., 1992). As was discussed above for amino acids, these observations suggest that physical protection/matrix effects may also affect the reactivity (or preservation) of carbohydrates during early diagenesis.

In spite of these observations, carbohydrate diagenesis in a given sediment appears to be largely nonselective (Hamilton and Hedges, 1988; Cowie et al., 1992; Cowie and Hedges, 1994; Hedges et al., 1999a), and the weight percentages of individual neutral sugars generally do not change with sediment depth. These observations suggest that external (predepositional) factors are more important than internal factors associated with sediment diagenesis in controlling sediment carbohydrate distributions. Again, this trend is broadly consistent with that discussed above for amino acid diagenesis.

9.6 LIGNINS

Lignins are a group of polymeric materials based on phenylpropanoic monomers rich in methoxyl substituents. They occur uniquely in vascular plant tissues and are generally associated with cellulose and hemicellulose, forming a material collectively referred to as *lignocellulose*. Terrestrial plant material (on a carbon percentage basis) is ~42% cellulose, ~30% lignin and ~25% hemicellulose, with proteins, lipids, and cutin making up the remaining ~3% (Hedges et al., 1997; also see table 11.3). In lignocellulose, lignin appears to act as the "glue" that binds together the hemicellulose and cellulose microfibrils, although the exact details of the chemical interactions between these components of lignocellulose are not entirely resolved (see discussions in Colberg, 1988).

In contrast to many other biopolymers, lignin is composed of non-repeating units linked together in a random network by carbon-carbon and ether bonds (Young and Frazer, 1987; de Leeuw and Largeau, 1993). Because lignin degrades relatively slowly under both aerobic and anaerobic conditions and represents a major fraction of terrestrial plant material, lignin in sediments represents an important tracer not only of vascular plant debris but also of terrestrial organic matter in general (see discussions in Hedges et al., 1997, and section 11.1).

Alkaline CuO oxidation of lignin-containing materials produces 11 simple phenols that can be used to characterize different types of vascular plants, i.e., angiosperm vs. gymnosperm, and different tissue types, i.e., woody vs. nonwoody (Hedges and Mann, 1979; Haddad and Martens, 1987; Hamilton and Hedges, 1988). Some degree of caution must, however, be taken in interpreting such results because under some circumstances selective degradation of some individual lignin phenols can occur. Evidence to date suggests that while anaerobic degradation of lignin is nonselective, aerobic lignin degradation by white rot fungi is selective, altering the relative amounts of acidic and aldehydic lignin phenols (Ertel and Hedges, 1985; Hamilton and Hedges, 1988; Hedges and Weliky, 1989).

Lignin phenols also appear to represent a significant fraction of the humic and fulvic acids that can be extracted from marine sediments (e.g., Ertel and Hedges, 1985), and there are at least two possible explanations for these observations. One is that these lignin phenols may actually be incorporated *in situ* into humic or fulvic acid, if these humic "compounds" indeed form by geopolymerization reactions (see discussions in section 9.8). At the same time, fresh vascular plant material contains components that can be extracted by the procedures used to isolate humic substances (Ertel and Hedges, 1985). This then suggests that in addition to, or perhaps instead of, geopolymers, unaltered plant fragments found in marine sediments could represent some component of sedimentary humic substances. This point will be discussed in further detail below.

9.7 Lipids

Lipids are an extremely diverse group of compounds that are insoluble in water but soluble in non- or semipolar solvents. They include hydrocarbons, wax esters, triacylglycerols, sterols, glycolipids, phospholipids, and pigments. Lipids have a wide range of uses, including energy storage, structural purposes in cell walls and membranes, and protective tissues.

In comparison to monomeric sugars, amino acids, and lignin phenols, the number of lipid compounds is extremely large and diverse. These compounds often have a high degree of structural complexity, and many individual lipids, or classes of lipid compounds, often have unique biological sources (table 9.4). Such lipids are therefore potential biomarkers for these organisms in both modern and ancient sediments (Cranwell, 1982; Volkman et al., 1998). At the same time, understanding the early diagenesis of lipid biomarkers is important for their accurate use and interpretation as paleoenvironmental indicators (see discussions below and in Hedges and Prahl, 1993).

Fats or oils (liquid fats) are lipids used for energy storage, as are wax esters in some organisms. Chemically, fats are triacylglycerols (fig. 9.7A) composed of fatty (carboxylic) acids esterified to a glycerol backbone. The chain lengths of these fatty acids range from ~C_{12} to C_{36} (Killops and Killops, 1993). Saponification (base hydrolysis) of lipid extracts[2] breaks these lipids into glycerol and their constituent

---

[2] All sediment lipid analyses begin with an extraction using an organic solvent (or solvent "cocktail") to isolate the sediment lipids. An examination of the literature indicates that a range of solvents can be used here, with different extraction procedures varying in their recovery of polar lipids and in the preservation of other more labile lipids.

LIPIDS

TABLE 9.4
Selected lipid biomarkers*

Compound(s)[a]	Source
14:0, 16:1ω7 fatty acids	algae
branched and normal C_{13}, C_{15}, and C_{17} fatty acids	bacteria
17:1 fatty acids	bacteria
polyunsaturated C_{16}–C_{20} (even C numbers) fatty acids	algae
22:0–32:0 (even C numbers) fatty acids	terrestrial vascular plants
Hopanoids (in general)	bacteria (including cyanobacteria)
4-methyl sterols (e.g., dinosterol)	dinoflagellates
Archaeol (a glycerol ether)	archaea (especially methanogens)
C_{15}–C_{19} (odd C numbers) n-alkanes	algae
C_{23}–C_{35} (odd C numbers) n-alkanes	terrestrial vascular plants
C_{25} and C_{30} highly branched isoprenoid alkenes	algae (diatoms)
Crocetane (a C_{20}-isoprenoid; 2,6,11, 15-tetramethylhexadecane)	archaea

*Compiled from various sources (Canuel and Martens, 1993; 1996; Eglinton et al., 1997; Volkman et al., 1998; Rullkötter, 2000; Pancost et al., 2001; Thiel et al., 2001). Note that this list is far from a comprehensive list of the known lipid biomarkers for these, and other, types of organisms.

a. For fatty acids defined as x:yωz, x denotes the length of the carbon chain, y the number of double bonds, and z is the inclusive number of carbon atoms from the terminal methyl carbon to the carbon atom of the first double bond (note that y = 0 compounds are saturated fatty acids).

fatty acids; the latter can then be analyzed by gas chromatography (see references cited in this section for further details).

In recent years, studies have tended to focus on free lipids that can be extracted from sediments by simple solvent extraction. However, there is also a bound lipid fraction that is released from the sediments only by combined saponification (base hydrolysis) and solvent extraction (Farrington et al., 1977; Lee et al., 1977; Van Vleet and Quinn, 1979; Prahl et al., 2003). The source of these bound lipids is not well understood (see discussions in Haddad et al., 1992; Killops and Killops, 2005); they may result from strong adsorption to sediment surfaces, as well as from esterification of free lipids with other forms of sedimentary organic matter. There may also be a direct bacterial

A. Triacylglycerol

$$\begin{bmatrix} O-C-R_1 \\ \| \\ O \\ O-C-R_2 \\ \| \\ O \\ O-C-R_3 \\ \| \\ O \end{bmatrix}$$

B. Isoprene

$$\begin{array}{l} CH_2 \\ \| \\ C-CH_3 \\ \| \\ CH \\ \| \\ CH_2 \end{array}$$

C. Squalene (acrylic triterpene)

D. Phytol (acrylic diterpene)

E. Diploptene (hopane series)

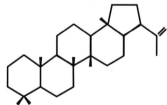

F. Sterol base unit

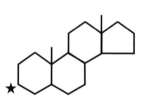

Figure 9.7 The structures of selected lipids and lipid subunits. A. Triacylglycerols (note that R_1, R_2, and R_3 represent fatty acids of various chain lengths). B. The isoprene subunit that forms the basis for many lipid compounds, including squalene (C), phytol (D), and diploptene (E). F. The base structure of steroids. ★ = the position at which a hydroxyl group is attached in sterols (steroidal alcohols).

contribution to the bound lipid pool, e.g., lipids associated with bacterial membranes (see section 11.1.5). Studies of Madeira Abyssal Plain turbidites (see section 14.4) have shown that such bound lipids, regardless of their apparent source, are preferentially preserved relative to their free counterparts during organic matter remineralization in these sediments (Hoefs et al., 2002; Prahl et al., 2003).

Wax is a general term for the protective tissue, or cuticle, on plant leaves, although waxes are also found in animals, algae, and bacteria. Waxes are a diverse collection of long-chained lipids with wax esters (esters of long-chain fatty acids and a straight-chain saturated alcohol other than glycerol) being particularly important. Other components of wax include hydrocarbons (e.g., branched and straight-chain n-alkanes), ketones, alcohols, and aldehydes, as well as an insoluble matrix.

Many lipids are used for structural purposes. These include phospholipids, glycolipids, and some sterols. Phospholipids are similar to triacylglycerides except that in place of the third acetyl (fatty acid) group there is a phosphate group, or more likely an organophosphate group. Phospholipids are a major component of bacterial and eukaryotic cell membranes (see section 16.3 for a further discussion of organic phosphorus compounds). In contrast, cell membranes in archaea (e.g., methanogens) are dominated by ether-linked, rather than ester-linked, lipids (e.g., glycerol diethers, and diglycerol tetraethers; Woese et al., 1990; Madigan et al., 1997).

The lipids described above are complex lipids in that they can be thought of as containing various fatty acids attached, generally through ester linkages, to some backbone molecule (e.g., glycerol). In contrast, simple lipids do not contain fatty acids. Simple lipids include pigments such as chlorophyll or carotenoids, hydrocarbons, terpenoids, and steroids. Terpenoids, steroids, and hydrocarbons exist not only in their unfunctionalized form, i.e., as basic carbon-hydrogen skeletons, but also in functionalized forms such as alcohols and their esters, aldehydes, and ketones.

Terpenoids are a class of linear or cyclic compounds based on the 5-carbon isoprene unit (fig. 9.7B). Terpenoids include many essential oils found in terrestrial plants, carotenoid pigments, phytol (a side chain of chlorophyll *a*), phytane, pristane, and squalene (figs. 9.7C and D). The cyclization of squalene forms a variety of pentacyclic triterpenes, with the hopanoids (fig. 9.7E) representing the geochemically most important group of triterpenes (Killops and Killops, 2005).

Hopanoids are exclusively biosynthesized by prokaryotes, where they are used to stabilize cell membranes.

Squalene cyclization also produces the 6- and 5-membered ring structures that forms the basis of all steroids (fig. 9.7F). Steroids are ubiquitous in marine and terrestrial organisms and are used both as structural components of membranes and as regulatory hormones. Sterols, in particular, are steroidal alcohols with a hydroxyl group attached to the carbon at position 3. They represent a large class of compounds with different branched side chains attached to the core structure and also have different stereochemistry around asymmetric carbons.

The discussion above presents information on the major lipid classes, and most (but certainly not all) lipids fall within these broad categories. One group of lipids not included here, but of some interest, is sulfur-containing lipids and organic sulfur compounds in general. A wide range of organic sulfur compounds have been found in recent marine sediments and crude oils (Sinninghe Damsté and de Leeuw, 1989), and evidence to date suggests that most of these compounds are likely not directly biosynthesized (although also see discussions in Brüchert and Pratt, 1996). Rather they appear to form through the incorporation of inorganic sulfur (sulfide or polysulfides) into both functionalized lipids and carbohydrates (see Wakeham et al., 1995; Kok et al., 2000; Aycard et al., 2003; Werne et al., 2003, and references therein; also see section 17.1). These natural "vulcanization" reactions may play some role in carbon preservation in certain environments (Tegelaar et al., 1989). They may also help preserve structural information in reactive biomarkers by protecting sulfurized biolipids from diagenetic transformations or remineralization (e.g., Kohnen et al., 1992, and references therein).

Lipids comprise 5–25% of the carbon in phytoplankton and bacteria (Killops and Killops, 2005) and lower percentages of the total organic carbon in marine sediments. In coastal and estuarine sediments lipids are at most ~3–5% of the sediment TOC (e.g., Haddad et al., 1992; Zimmerman and Canuel, 2001) and generally less than ~1% of the sediment TOC in pelagic and hemipelagic sediments (e.g., Wakeham et al., 1997a; Belicka et al., 2002). In terms of their contribution to sediment carbon remineralization, in the anoxic sediments of Cape Lookout Bight lipids account for 17% of the sediment organic carbon undergoing remineralization (Martens et al., 1992). Since total sediment lipids account for ~5–8% of the total pool of organic carbon in these sediments, this suggests that there is preferential remineralization of lipids relative to total organic carbon in these sediments. This observation is also consistent with observations about lipid reactivity in sinking particles in the water column (Wakeham et al., 1997b). Similar results were also obtained in studies of fatty acid diagenesis in Gulf of California sediments (Camacho-Ibar et al., 2003), where total fatty acids accounted for ∼1–2% of the total organic carbon undergoing remineralization, yet comprised only ∼0.2–0.4% of the TOC in the sediments. Since fatty acids are generally the most abundant lipids that are quantifiable in sediments (see below), these results also suggest that total lipid diagenesis in these sediments is likely only a few percent of the total organic matter that is remineralized.

In studies where multiple lipid classes have been examined (e.g., fatty acids, sterols, alchohols, hydrocarbons) fatty acids tend to predominate the quantifiable lipids (Smith and Eglinton, 1983; Canuel and Martens, 1993; Wakeham et al., 1997a; Pearson et al., 2001; Zimmerman and Canuel, 2001; Belicka et al., 2002 and numerous other studies[3]). The ability to further examine this problem is hampered by the fact that in many studies, not all lipid subclasses are analyzed. Furthermore, quantifiable lipids are often a small fraction of the total lipid extract, i.e., the total amount of material extracted from a sediment by solvent extraction (e.g., Haddad et al., 1992; Canuel and Martens, 1993), adding additional uncertainty to such a comparison.

Concentrations of most lipids generally decrease with depth in sediments, and several recent studies have examined aspects of lipid diagenesis in sediments. The results of these studies have led to several general observations. Focusing first on fatty acids, it appears that under almost all conditions unsaturated fatty acids are more reactive than saturated fatty acids (Van Vleet and Quinn, 1979; Haddad et al., 1992; Sun and Wakeham, 1994; Canuel and Martens, 1996; Harvey and Macko, 1997b; Sun et al., 1997; and others; also see table 9.5). Furthermore, within the saturated fatty acid group, short-chain fatty acids are degraded more rapidly than longer chain fatty acids, e.g., <C_{18} in

---

[3] Given the large literature on lipids in sediments, it is impossible to present a comprehensive bibliography on this subject in the present text. The papers cited here illustrate the key points being made in this section, and because many are relatively recent papers, they also provide the interested reader with an entry point into this literature (also see Cranwell, 1982, for a summary of earlier studies).

TABLE 9.5
Fatty Acid Concentrations and Remineralization Rate Parameters in Cape Lookout Bight Sediments[a]

Fraction[b]	Relative Conc. (normalized to TFA conc.) or Absolute Conc. (TFA only)			% Remineralization (5–100 cm)[c]	k (yr^{-1})[d]	τ(yr)[e]
	5–10 cm	45–50 cm	95–100 cm			
TFA (μg gdw^{-1})	1119	221	88	88%	0.76	1.32
TSFA	50%	75%	89%	77%	0.60	1.59
MC	35%		18%	94%	0.98	1.02
LC	15%		67%	42%	0.15	6.90
TunSFA	43%	16%	4%	99%	0.98	0.96
MC				99%	1.02	1.02
LC				>99%	1.43	1.42
TbrFA	8%	9%	7%	88%	0.95	1.05

a. Data from Haddad (1989) and Haddad et al. (1992).
b. Abbreviations: TFA = total free fatty acids; TSFA = total saturated fatty acids; MC = medium chain length fatty acids (C_{12} to C_{19}); LC = long chain length fatty acids (C_{20} to C_{34}); TunSFA = total unsaturated fatty acids ; TbrFA = total branched fatty acids.
c. Loss of fatty acids between 5 and 100 cm sediment depth.
d. Based on fitting fatty acid profiles to an equation similar to eqn. 8.4 (see Haddad et al., 1992, for further details).
e. τ (the turnover time) is $1/k$ (see eqn. 8.16 and section 8.2.1).

the study of Haddad et al. (1992). However, this relationship is not a smooth one, but rather there appears to be an abrupt decrease in reactivity above this threshold carbon chain length (fig. 9.8). Within the unsaturated fatty acid group, different studies have observed different relative reactivies of monounsaturated versus polyunsaturated fatty acids (Haddad et al., 1992; Canuel and Martens, 1996; Harvey and Macko, 1997b; Sun et al., 1997).

Finally, branched-chain fatty acids, i.e., common biomarkers of bacterial sources, show differing degrees of reactivity depending on the time scale over which their diagenesis has been examined. On long time scales the reactivity of branched fatty acids is comparable to that of unsaturated fatty acids in both the upper ~5 cm of Black Sea sediments (age of ~150 yrs; Sun and Wakeham, 1994) and in the upper 100 cm of Cape Lookout Bight, NC, sediments (age of ~10 yrs; Haddad et al., 1992; table 9.5). However, in a study of much more short-term

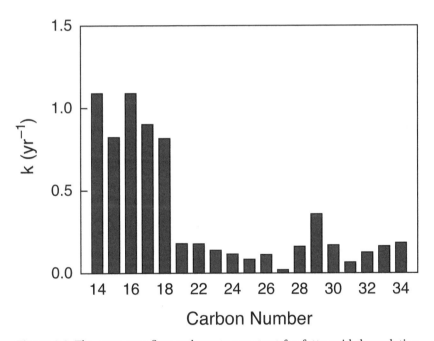

Figure 9.8 The apparent first-order rate constant for fatty acid degradation as a function of fatty acid carbon number (redrawn from results in Haddad et al., 1992). See section 6.3.1 for additional information on the modeling approach used to determine these rate constants.

diagenesis in surficial (<8 cm) Cape Lookout Bight sediments (time scale of <1 yr), Canuel and Martens (1996) observed more variable, but generally lower rates of reactivity for branched fatty acids. These differences may be related to the fact that branched fatty acids are both produced and consumed in anoxic sediments. Therefore, differences in their relative *net* rates of production or consumption on different time scales may affect their apparent reactivity, i.e., enhanced bacterial activity and a tighter balance between bacterial fatty acid production and consumption in "young" surficial sediments, versus lower activities and net consumption of bacterially derived lipids over longer time and depth scales.

Comparisons of the reactivity of different lipid classes are difficult for a number of reasons (e.g., see discussions in Canuel and Martens, 1996). However, in a compilation of sediment data on fatty acid versus sterol degradation, Canuel and Martens (1996) observed that the reactivity of both classes of compounds decreased with increasing age of the material (fig. 9.9; also see section 8.1 for similar results with total organic carbon). Of particular interest here is that sterol rate constants decrease more rapidly with increasing organic matter age than do those for fatty acids. Thus in "young" material, e.g., surface Cape Lookout Bight sediments, sterols are more reactive than fatty acids, while in older material, e.g., Buzzards Bay or Black Sea sediments, fatty acids are more reactive than sterols. Perhaps related to these results is the general observation in many of these studies that the reactivity of lipids is not only a function of their chemical structure/composition, but is also dependent on characteristics of the matrix in which the lipids are incorporated (often referred to as organic matter "packaging"). This common theme was previously discussed for carbohydrates and proteins and will be discussed later in this and other chapters.

Several studies have examined the effects of oxygen on lipid degradation (Harvey et al., 1986; Harvey and Macko, 1997b; Sun et al., 1997). In general, lipid degradation, either of total lipid material or of specific subclasses, occurs more rapidly under oxic conditions, although the extent of these differences can vary among different lipid subclasses. Redox oscillations, which vary the fraction of time the sediment sees oxic conditions (section 7.5.1), fall in with the general trend of an increase in lipid degradation from completely anoxic to completely oxic conditions (Sun et al., 2002). The causes of these differences are not well understood although it has been suggested

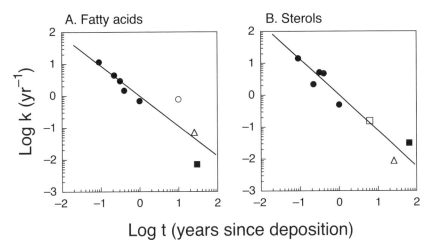

Figure 9.9 The apparent first-order rate constants for fatty acid (A) and sterol (B) degradation as a function of time following sediment deposition (redrawn from Canuel and Martens, 1996). Note that the inverse relationship shown in these two graphs is very similar to that seen for total organic carbon in marine sediments (see section 8.1 and fig. 8.1). ● and ○ = Cape Lookout Bight sediments. The different sediment ages here represent analyses of sediments and sediment cores at this site of different lengths (Haddad et al., 1992; Canuel and Martens, 1996); △ = Black Sea surface sediments (Sun and Wakeham, 1994); ■ = Buzzards Bay surface sediments (Farrington et al., 1977; Lee et al., 1977); □ = surface sediments from the Peru upwelling zone (McCaffrey, 1990).

(Harvey and Macko, 1997b) that lipid structural characteristics, i.e., fewer functionalized oxygen atoms than other biochemicals, may limit their anaerobic degradation (also see discussions in section 7.6). However, this issue remains unresolved (see discussions in Sun et al., 2002, and in section 15.2).

As noted above, lipids are often used as biomarkers for a specific source of organic matter. However, because of differences in the reactivity of different lipids during diagenesis, the distribution of lipids deposited at the sediment surface can often look vastly different than that which is buried at depth. An example of this is seen in table 9.5 for different fatty acid classes in Cape Lookout Bight sediments. As a result, then, of these diagenetic transformations, the biomarker record preserved in sediments, or ancient rocks, may be difficult to interpret in more than a qualitative sense for the assessment of sediment organic matter sources. Hedges and Prahl (1993) summarize the factors that must be considered if one wishes to accurately employ biomarkers in sediments to assess organic matter sources to sediments (also see de Lange et al., 1989; Killops and Killops, 2005, and discussions therein).

Several approaches may, however, help to overcome some of the difficulties in interpreting lipid biomarker data. First, studies of sulfurized lipids may be of use, because sulfurization decreases the reactivity of the resulting compound relative to that of the parent lipid (Kohnen et al., 1992). Second, coupling biomarker concentration studies with compound-specific $\delta^{13}C$ and $\Delta^{14}C$ analyses of the biomarkers (e.g., Hayes et al., 1990; Eglinton et al., 1997; Schouten et al., 2000) can also help resolve some of the source ambiguities of certain lipid compounds. Stable isotopic differences occur because the $\delta^{13}C$ of the carbon assimilated by different organisms varies depending, for example, on whether their carbon source is oceanic bicarbonate or atmospheric CO_2 (see section 11.1.1 for further details). Furthermore, $\Delta^{14}C$ analyses place an age constraint on the biomarker, further helping to resolve biomarker sources, metabolic pathways, and diagenetic fates (also see sections 11.1.1 and 3.8.2).

In addition to their use in the assessment of sediment, or sedimentary rock, organic matter sources, lipid biomarkers have also been used as other types of paleoceanographic tracers. Perhaps the best known, and most successful, of these tracers is based on the relative amounts of two long-chain unsaturated ketones, or alkenones, produced by marine prymesiopytes (coccoliths; Brassell, 1993). This alkenone unsaturation index (U^K_{37}) appears to be well preserved in sediment records (i.e., it is not apparently affected by sediment diagenesis), and has also been shown to have a strong positive correlation with sea surface temperatures over a range of annual mean temperatures between 0°C and ~30°C (Müller et al., 1998).

9.8 Humic Substances and Molecularly Uncharacterized Organic Matter

The results in table 9.6 illustrate that less than ~30–40% of the organic carbon in marine sediments can be identified in the four compound classes discussed above using conventional analytical techniques.

TABLE 9.6
"Identifiable" Biochemicals in Marine Sediments and End-Member Sources

Sediment type/site	Amino acids	Carbohydrates	Lignin	Total lipids	Identified Components
"Typical" modern coastal marine sediments[a]	0–15%	5–10%	3–5%	<5%	<35%
Cape Lookout Bight, NC sediments[b]	<8–13%	6–8%	<1%	5–8%	<30%
Namibian shelf diatomaceous ooze[c]	~11%	~22%	na	~5%	~38%
Equatorial Pacific sediments[d]	16–17%	1–12%	na	<1%	<30%
NE Pacific sediments[e]	11–19%	3–18%	na	2–3%	<40%
"End-member" sources					
Marine organic matter[f]	~50–60%	20–40%	0%	5–30%	75–130%
Vascular plant material[g]	~1–2%	~70%	~30%	~1–2%	~100%

na = not analyzed but assumed to be zero.
a. From Hedges and Oades (1997).
b. From Martens et al. (1992). This range is based on analyses of sediment samples from depths of 0–5 cm and 95–100 cm.
c. From Klok et al. (1984). The water depth of this site was 106 m, and the sample analyzed was from a sediment depth of 40–75 cm.
d. From several sources (Hernes et al., 1996; Wakeham et al., 1997a; Lee et al., 2000) for samples collected from 0–12 cm sediment depth at several sites. Water depths at these sites are all >4,000 m.
e. From Wang et al. (1998). This range is based on samples collected from 0–14 cm sediment depth. The water depth at this site is 4,100 m.
f. Data from a variety of sources (Libes, 1992; Martens et al., 1992; Hedges et al., 2001, 2002).
g. From Hedges et al. (1997).

In contrast, in end-member organic matter sources, surface plankton samples and sinking particles leaving the euphotic zone, greater than 80% of the carbon can be partitioned into the lipid, carbohydrate, and amino acid fractions (Wakeham et al., 1997b; Wang et al., 1998). Understanding what comprises this molecularly uncharacterized organic matter (MU-OM) is of great importance and interest (Hedges et al., 2000), in part because this material may play some role in controlling carbon preservation in marine sediments. This particular problem will be discussed in later sections of this book (e.g., section 15.6); the discussion here will focus on what we know about the chemistry of this uncharacterized material.

In discussing MU-OM, it is important to recognize that its "existence" is ultimately inferred from specific analytical techniques that use some sort of extraction procedure (e.g., acid hydrolysis or solvent extraction) and chromatographic separation and identification of individual compounds in the extract (e.g., free amino acids or neutral sugars). The summation of these individual biochemicals then yields the "total" amount of material in this compound class. Thus, it appears that the majority of the organic matter in sediments escapes the analytical "window" of these techniques. For example, the presence of bound (versus free) lipids not only apparently enhances the preservation of lipids, but also can contribute to the operationally defined MU-OM pool, if the appropriate sediment extractions are not carried out.

This problem is not unique to marine sediments, since similar large fractions of soil organic matter, dissolved organic matter in seawater, and organic matter in wastewater treatment effluent are similarly uncharacterized at this compound class level (see discussions in Hedges et al., 2000). In sections 9.8.4 and 9.8.5 we will discuss analytical techniques that may help to overcome some of these analytical difficulties.

9.8.1 Black Carbon

One potential component of the MU-OM pool is black carbon. Black carbon represents a broad range of heterogeneous, aromatic, and refractory carbon-rich materials that can form during the incomplete combustion of fossil fuels or organic matter (biomass burning). Black carbon includes graphite (elemental carbon), soot, charcoal, and char, and represents both combustion residues and condensates (Goldberg, 1985; Hedges et al., 2000). Recent work using $\delta^{13}C$ and radiocarbon analyses (Dickens et al., 2004a,b) also suggests that a significant fraction of the black carbon in marine sediments can be graphite weathered from continental rocks that is simply reburied in marine sediments (also see related discussions in section 11.1.4). Black carbon is ubiquitous in the atmosphere, cryosphere, soils, oceans, and marine sediments (albeit at very low levels) because of its global production and apparent resistance to degradation (Goldberg, 1985). From the perspective of global climate, black carbon is important as a light-scattering atmospheric aerosol.

In marine sediments, black carbon represents from ~2% to perhaps 30% of the sediment TOC (Gustafsson and Gschwend, 1998; Masiello and Druffel, 1998; Middelburg et al., 1999; Gélinas et al., 2001; Masiello et al., 2002; Dickens et al., 2004a,b). However, because of differences in (and potential problems with) the various methods used to determine black carbon, some caution should be placed in the interpretation of these estimates (also see discussions in Massiello, 2004). While black carbon is generally thought to be extremely resistant to both biological and chemical degradation, studies by Middelburg et al. (1999) suggest that oxygen exposure of black carbon in pelagic turbidites over long time periods (~10–20 kyr) can lead to significant (~64%) degradation of the material.

9.8.2 Molecularly Uncharacterized Organic Matter (MU-OM): General Considerations

The results in table 9.6 combined with estimates of sedimentary black carbon still imply that roughly half of the sediment POM remains uncharacterized. There are at least three possible explanations for just what "causes" or "forms" MU-OM; as we will also see, these explanations are not necessarily mutually exclusive. These possibilities are illustrated schematically in fig. 9.10.

The first explanation assumes that during the decomposition of sedimentary POM reactive intermediates such as monomeric sugars or amino acids recombine in abiotic chemical reactions to form humic substances or *geopolymers* (e.g., Tissot and Welte, 1978). These geopolymers are presumably too complex to be either enzymatically decomposed by organisms or chemically analyzed. In other words, the humification process sufficiently degrades organic matter to the point that conventional analytical procedures used to analyze amino acids, carbohydrates or lipids no longer recognize the precursor compounds in the humic material.

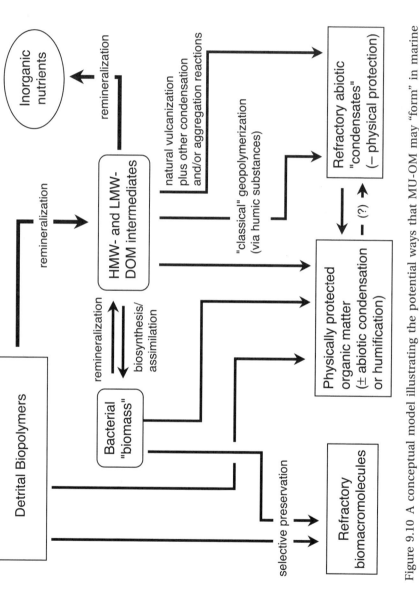

Figure 9.10 A conceptual model illustrating the potential ways that MU-OM may "form" in marine sediments (modified after Tegelaar et al., 1989, and Killops and Killops, 2005).

A second possible explanation is based on the observation that organisms produce hydrolysis-resistant, biologically refractory macromolecules (de Leeuw and Largeau, 1993). Selective preservation of these materials can then can lead to their preservation in sediments (Hatcher et al., 1983; Hatcher and Spiker, 1988; Tegelaar et al., 1989). Finally, reactive organic matter may be shielded from degradation through its interaction with, and/or protection by, refractory inorganic or organic matrices (Knicker and Hatcher, 1997; Hedges et al., 2001).

9.8.3 Geopolymerization: The Formation of Humic Substances

The study of humic substances in the marine environment has grown out of the soil science literature, where there is a long and continuing history of the study of soil humic substances (Aiken et al., 1985; Stevenson, 1994). As noted above, humic substances are generally thought of as chemically complex materials that form via abiotic (chemical) reactions. Humics are also considered amorphous, hydrophilic materials that are refractory with respect to both chemical and biological degradation.

Operationally, however, humic substances are defined by their aqueous solubility at different pH values. Base extraction of sediment (or soil) solubilizes the humic plus fulvic acid fractions, while the remaining insoluble material is defined as the humin fraction. Acidification of this base extract then precipitates out the humic acid fraction, leaving behind the fulvic acid fraction. Fulvic acids are, therefore, defined as being soluble in both base and acid, humic acids as being soluble in base but insoluble in acid, and humin as being insoluble in both acid and base. Humin is also sometimes referred to as protokerogen, in part because its definition is similar to that for the kerogen fraction of organic matter found in lithified sediments or sedimentary rocks (Whelan and Thompson-Rizer, 1993; also see section 11.1.4 for further information on kerogen geochemistry).

In terms of the general properties of humic substances, there appears to be an increase in molecular weights as one moves from fulvic acids to humic acids to humin, along with a decrease in the oxygen content of the material (Hedges and Oades, 1997). There also appear to be broad differences between marine and terrestrial humic substances, presumably due to differences in their source materials; marine plankton versus vascular plants, for example (Aiken et al., 1985; Rashid, 1985; Hedges and Oades, 1997).

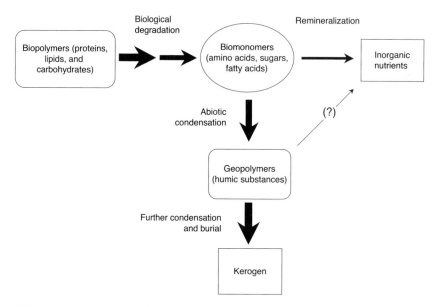

Figure 9.11 An expanded view of the classical model of formation of humic substances (modified after Berner, 1980, and references cited therein).

The classical view of the formation of humic substances is shown in fig. 9.11. As described in chapter 7, the degradation of biological polymers in the sediment POM pool eventually leads to the production of a variety of compounds referred to here as *biological monomers*, e.g., amino acids, simple sugars, fatty acids (see figs. 7.11 and 10.2). While most of these monomers undergo remineralization to inorganic nutrients, in the geopolymerization model some small fraction of the monomers undergo abiotic condensation reactions, forming humic substances. Several such condensation reactions have been proposed (Rashid, 1985; Thurman, 1985) and one well-studied example is the Maillard or "browning" reaction, a sugar-amino acid condensation reaction that forms compounds known as melanoidins. Such reactions occur, for example, during the preparation of foodstuffs, and synthetic melanoidins produced in the laboratory show some similarity to marine humic substances (Hedges, 1988).

In the context of the geopolymerization model, fulvic acids are considered the first products of abiotic condensation, and increased condensation and chemical modification of fulvic acids transforms them into humic acids and eventually into humin or protokerogen (Brown et al., 1972; Nissenbaum and Kaplan, 1972; Rashid, 1985). An alternate model for humic substance formation (Waksman, 1938; Hatcher and Spiker, 1988) assumes that these reactions actually occur in the opposite direction. In this *biodegradation model*, humic and fulvic acids result from the partial oxidative degradation of biologically produced macromolecules, and the following reaction sequence is envisioned:

$$\text{Biopolymers} \rightarrow \text{Humin} \rightarrow \text{Humic acids} \rightarrow \text{Fulvic acids} \quad (9.1)$$

Interestingly, studies of dissolved organic matter in the marine water column (Amon and Benner, 1994) and in marine sediment pore waters (Burdige and Gardner, 1998) tend to support this model of humification. This will be discussed in further detail in chapter 10.

Returning to the geopolymerization model in fig. 9.11, recent work has led to a serious re-examination of the role of these processes (as described here) in the formation of refractory, and uncharacterized, sediment organic matter. First and foremost, there is little direct evidence for the occurrence of these geopolymerization reactions in nature (Hedges, 1988; Henrichs, 1992; Hedges and Oades, 1997). Under most circumstances abiotic condensation reactions involving the commonly discussed monomeric reactants are likely to be quite slow in comparison to their biological uptake or remineralization to inorganic nutrients (Alperin et al., 1994).

Second, the notion that humic substances form via abiotic condensation reactions appears to be somewhat problematic because the extraction procedures used to isolate humic substances almost always coextract known biochemicals such as lignins, carbohydrates, or proteins (Hedges et al., 1997). One interpretation of this observation is that monomers derived from these biomacromolecules are incorporated into humic substances in such a way that they still retain their chemical "signature" (Rashid, 1985; Killops and Killops, 2005). However, as noted in section 9.6, results presented by Ertel and Hedges (1985) demonstrate that unaltered plant fragments found in marine sediments are coextracted along with humic substances, suggesting that they could represent some component of sedimentary humic substances. Furthermore, techniques used to extract protein from marine sediments (Mayer et al., 1986) involve methods very similar to those used to isolate humic substances.

Along these same lines, in Cape Lookout Bight sediments the loss of fulvic acid nitrogen accounts for ~80% of the total nitrogen remineralization, with no concomitant increase in humic acid nitrogen, which actually decreases slightly with sediment depth (Haddad, 1989). At the same time (hydrolyzable) amino acid remineralization in these sediments also accounts for ~80% of the total nitrogen remineralization (Burdige and Martens, 1988). While this similarity may be fortuitous, it is equally probable that reactive proteins are being coextracted into the operationally defined fulvic acid fraction. These observations also suggest that at least in these sediments the fulvic acid fraction is far from a refractory component of the sediment POM.

All of these observations suggest that the geopolymerization model as shown in fig. 9.11 does not significantly contribute to the formation of MU-OM in sediments. However, other related types of condensation reactions may still play a role in preserving sedimentary organic matter and in "forming" MU-OM. Protein in sediments may be preserved and become hydrolysis resistant through processes such as aggregation and cross-linking, both between proteins and perhaps with carbohydrates (Fogel and Tuross, 1999; Nguyen and Harvey, 2001; also see discussions in sections 9.9 and 15.6). Similarly, as discussed in section 9.7, it has been suggested that the sulfurization of lipids and carbohydrates (natural vulcanization reactions) might play a role in carbon preservation in certain environments (Tegelaar et al., 1989). However, recent studies in the highly sulfidic, permanently anoxic sediments of the Cariaco Trench, where such natural sulfurization reactions would be expected to be of greatest importance, suggest that these processes are slow in comparison to other more "classical" modes of protokerogen formation (e.g., fig. 9.11), and that they therefore may play a minor role in the formation of sediment protokerogen (Aycard et al., 2003).

9.8.4 Selective Preservation of Refractory Biomacromolecules

A wide range of organisms produce highly aliphatic, macromolecular material that is insoluble, nonhydrolyzable, and resistant to biological degradation (de Leeuw and Largeau, 1993). These refractory molecules tend to be produced by vascular plants and algae (de Leeuw and Largeau, 1993; Gélin et al., 1999) and include algaenans (algal cell wall components consisting of long-chain aliphatic compounds with hydroxyl or ester functional groups) and cutans (a nonhydrolyzable component of the cuticles of higher plants). Because these compounds are nonhydrolyzable, they would historically be considered MU-OM because they fall outside the conventional analytical window used to determine biomacromolecules such as proteins and polysaccharides. However, techniques such as analytical pyrolysis (Larter and Horsfield, 1993), Curie point pyrolysis (Gélin et al., 1999), and cross polarization/magic angle spinning ^{13}C NMR (Baldock et al., 2004) have proven useful in examining the structure and composition of these materials.

These refractory biomacromolecules likely represent a very small fraction of the initial biomass produced by marine and terrestrial organisms. However, they have the potential to make up a major fraction of the organic matter preserved in sediments and soils because of their extreme recalcitrance, and the fact that >99% of the carbon fixed into biomass is eventually remineralized (e.g., Hedges and Keil, 1995). Consistent with this, kerogens in ancient rocks have also been shown to resemble, both chemically and morphologically, these types of refractory biomacromolecules from ancient plants and bacteria (de Leeuw and Largeau, 1993). Thus, selective preservation of refractory biomolecules represents another mechanism by which this carbon may be initially preserved in marine sediments (Tegelaar et al., 1989).

9.8.5 Physical Protection

Recent work has also suggested that physical protection of organic matter by both organic and/or inorganic matrices may play a role in the apparent formation of MU-OM and in the preservation and burial of sediment organic matter. Under some circumstances, this may involve physical protection of labile compounds such as reactive proteins, in a way that affords these compounds protection from chemical attack (e.g., abiotic acid hydrolysis) or enzymatic degradation. Physical protection may also allow for the subsequent chemical modification of the organic matter that further enhances protection from both degradation and analysis.

In thinking here about the role of physical protection, it is important to recognize that physical protection, which impedes enzymatic degradation of organic matter, does not necessarily also contribute to the "formation" of MU-OM. For example, carbohydrate and protein-rich material associated with calcareous shells, or biominerals (see section 2.2), may be physically protected from biological degradation, yet still be susceptible to acid hydrolysis and subsequent chemical analysis (Ingalls et al., 2003, 2004).

One mechanism by which this physical protection may occur is via encapsulation of reactive molecules within an insoluble (hydrolysis-resistant) organic matrix such as algaenans (Knicker and Hatcher, 1997). While encapsulation may protect reactive proteins from acid hydrolysis or biological degradation, ^{15}N NMR or pyrolysis GC/MS techniques can detect the occurrence of such proteinaceous compounds in hydrolysis-resistant fractions of sediment organic matter (Knicker et al., 1996; Knicker et al., 2001; Zang et al., 2001).

Much of the work examining encapsulation has been carried out either in organic-rich sapropel sediments with very low (<10%) mineral content, or in short-term algal degradation studies of generally less than ~1 yr. Studies of more typical marine sediments from the northwest African upwelling region (Zegouagh et al., 1999) also indicate that what appears to be protein-derived material is found in a hydrolysis-resistant component of these sediments. However, these authors suggest that here this occurs by melanoidin-type condensation reactions. Similar conclusions were reached by Aycard et al. (2003) in studies of recent sediments of the Cariaco Trench.

Inorganic matrices, such as small mesopores on mineral surfaces or between mineral grains, may also provide physical protection of organic matter. Recent work has suggested that such processes may indeed play a role in preserving organic matter in sediments (see section 15.1), in part because this organic matter resides in mesopores or intragranular pores (slits between mineral grains) that are too small for microbial enzymes to enter (see Mayer, 2004, for a recent summary). However, for this process to contribute to organic matter being preserved in sediments and being defined as MU-OM, this attachment must be sufficiently strong, e.g., near irreversible adsorption, such that there is effectively no detachment/desorption of this organic matter, particularly during acid or base hydrolysis, or solvent extraction (see discussions in Mayer, 1994b; Henrichs, 1995). At the same time, though, subsequent chemical changes in the organic material after its initial attachment/adsorption to sediment particles may play a role, in that this could render this attached organic matter chemically nonrecognizable and/or increase the strength of its attachment (also see discussions in section 15.6).

Recent NMR studies of sinking marine particles, however, argue against substantive chemical changes in such physically protected organic matter, at least on the time scales of particle sinking in the water column (Hedges et al., 2001). In this work, the molecularly uncharacterized fraction of organic matter in sinking particles from both the upper (~500–1,000 m) and lower (~3,000–3,500 m) water column has lipid, amino acid, and carbohydrate concentrations that appear to be very similar to that seen in surface water plankton samples, where, in contrast, 80% or more of the organic matter can be classified in terms of known biochemicals. Similar conclusions are also reached based on the results of NMR analyses of sediment organic matter (after sediment demineralization) across a range of sediment types, e.g., organic-rich coastal sediments to organic-poor pelagic sediments (Gélinas et al., 2001). These observations will be further discussed in section 11.2.4.

The results of these NMR studies argue that physical protection alone could lead to this gross compositional similarity between characterized and uncharacterized (≈nonhydrolyzable) natural organic matter. At the same time, though, it is also possible that because of slight alterations of natural biochemicals these compounds may be missed by standard chromatographic techniques, but still detected by NMR techniques (e.g., see discussions in Lee et al., 2000). Thus physical protection in association with some from of chemical modification cannot be unequivocally ruled out.

9.9 Organic Nitrogen Diagenesis in Sediments

In ending this chapter, I would like to briefly re-examine the topics discussed above with specific reference to organic nitrogen diagenesis in sediments. In contrast to carbon, the biogeochemistry of organic nitrogen "starts" out much more simply, since the overwhelming majority of the nitrogen in organisms is amide nitrogen in proteins. Although other forms of nitrogen are produced by organisms—for example, amino sugars in chitin, nucleic acids in DNA or RNA, tetrapyrroles in pigments such as chlorophyll—they are generally of minor importance as compared to proteins and amino acids (Stankiewicz and van Bergen, 1998).

While much of the organic nitrogen produced by organisms is remineralized, results discussed in section 9.4 indicate that with diagenetic maturity of organic matter, amino acids appear to become an increasingly smaller fraction of the total organic nitrogen (also see fig. 9.1C and discussions in Derenne et al., 1998). In an attempt to characterize this "missing" organic nitrogen, a variety of techniques have been used to look for other types of organic nitrogen compounds such as heterocyclic nitrogen compounds, e.g., pyrroles and pyridines. However as discussed above, studies of organic-rich sapropel sediments and the products of short-term algal degradation suggest that all of the organic nitrogen in these materials is in the amide form. This suggests that processes such as encapsulation, aggregation, and/or cross-linking may render reactive proteins in sediments inert to chemical analysis (i.e., abiotic acid hydrolysis) and biological degradation (e.g., see section 9.8.5); this could therefore explain results such as those in fig. 9.1C.

In contrast, studies of recent marine sediments with more typical mineral contents (versus sapropelic sediments with low [<10%] mineral content) using both destructive (e.g., pyrolysis-gas chromatography) and nondestructive (e.g., K-edges XANES [high energy X-ray] spectroscopy) analytical techniques have yielded slightly different results. Although amide nitrogen still generally dominates, significant amounts (up to ~30%) of pyridinic and/or pyrrolic nitrogen are also observed (Patience et al., 1992; Vairavamurthy and Wang, 2002). Studies of fossil sediments such as the Green River Shale or Monterrey Formation, with these same X-ray techniques and ^{15}N NMR, similarly indicate the occurrence of heterocyclic nitrogen along with amide nitrogen in these more ancient materials.

There can be some concern about artifacts associated with the detection of heterocyclic nitrogen in natural samples using all of these techniques (see discussions in Knicker, 2000; Vairavamurthy and Wang, 2002). Nevertheless, these results seem to imply that a change in N functionality, e.g., amide to heterocyclic N, may begin to occur during early diagenesis in sediments (see discussions in the references cited above and in Derenne et al., 1998). In contrast, results discussed above (Gélinas et al., 2001) appear to imply that all of the nitrogen in recent marine sediments is in the form of amino acids/proteins. However, these studies used ^{13}C NMR to characterize the TOC in these sediments, and analyzed their NMR data with a mixing model that implicitly assumes all of the nitrogen in these samples is in the form of amino acids/proteins (also see Hedges et al., 2001). Thus the results of Gélinas et al. (2001) are not necessarily in conflict with the other sediment studies discussed above.

Interest in this problem also stems from the observation that most nitrogen in fossil fuels and coal exists as heterocyclic nitrogen (Patience et al., 1992; Derenne et al., 1998). Since this material presumably begins as amide nitrogen in the source organisms, there is a desire to understand the timing, controlling environmental factors, and mechanisms of these changes in nitrogen functionality in sediments. Based on the discussion above, three possible mechanisms may be at work, although definitive evidence for the occurrence of any of these is lacking. One possibility is simply that selective preservation of refractory heterocyclic nitrogen compounds, relative to more reactive amide nitrogen compounds, concentrates heterocyclic nitrogen compounds during diagenetic maturity. While this suggestion has generally been discounted for a number of reasons (Patience et al., 1992; Knicker et al., 1996; Aycard et al., 2003), in actuality it has not been rigorously tested (see discussions in Derenne et al., 1998).

Another suggestion builds on discussions above regarding processes (e.g., encapsulation or aggregation/cross-linking) that may initially decrease the reactivity of proteins or render them acid-insoluble or nonhydrolyzable. This "stabilization" during early stages of diagenesis may then be followed by subsequent rearrangement reactions such as Maillard-type condensation reactions, which eventually form heterocyclic nitrogen compounds (Rubinsztain et al., 1984; Patience et al., 1992; Nguyen and Harvey, 1998; Mayer, 2004; also see discussions in section 15.6).

Finally, a third process that could be of importance is analogous to the natural vulcanization reactions discussed in section 9.7. This may involve reactions between ammonium and reactive functional groups in non-nitrogen containing organic compounds (e.g., carbonyl groups), followed by autopolymerization to form "new" heterocyclic nitrogen compounds (Knicker et al., 1996; Vairavamurthy and Wang, 2002). Given the large amounts of ammonium produced in anoxic sediments during diagenesis (see section 16.2 and fig. 6.4) such reactions, if indeed they occur in marine sediments at all, might be of most importance in such sediments. Again, careful work is needed to test all of these possibilities.

≈ CHAPTER TEN ≈

Dissolved Organic Matter in Marine Sediments

Prior to this point, the discussion here of sediment organic geochemistry has focused largely on particulate organic matter. However, dissolved organic matter (DOM) found in marine sediment pore waters is also of some importance. Dissolved organic matter in marine sediment pore waters plays a role in sediment carbon remineralization (see section 7.6) and may also be involved in sediment carbon preservation. It is also involved in oceanic and sediment nutrient (N and P) cycling (e.g., Burdige and Zheng, 1998). Finally, DOM in sediment pore waters can form complexes with some trace metals, and therefore affects dissolved metal cycling in sediments and metal fluxes from sediments (Elderfield, 1981; Skrabal et al., 2000)

10.1 General Observations

Pore-water DOM is a heterogeneous collection of organic compounds, ranging in size from relatively large macromolecules, e.g., dissolved proteins or humic substances, to smaller molecules such as individual amino acids or short-chain organic acids (see Burdige, 2002, for a recent summary). In many sediments, concentrations of DOM in pore waters, comprising both dissolved organic carbon (DOC) and dissolved organic nitrogen (DON), are elevated by up to an order of magnitude over bottom water values (fig. 10.1). This implies that there is net production of DOM in sediments as a result of remineralization processes. Based on diffusive arguments alone, this also implies that sediments are a potential source of both DOC and DON to the overlying waters. This will be discussed in further detail in sections 10.4 and 16.2.1.

Much of the total pore-water DOC and DON is of relatively low molecular weight (<3 kDa; see below) and appears to be refractory, at least in a bulk sense. Such observations are consistent with the results of water column studies showing that high molecular weight DOC represents a more reactive and less diagenetically altered fraction of the total

GENERAL OBSERVATIONS

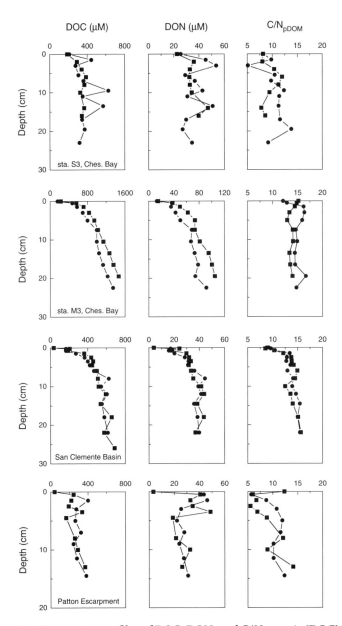

Figure 10.1 Pore-water profiles of DOC, DON, and C/N$_{pDOM}$ (= [DOC]/[DON]) in contrasting marine sediments (modified after Burdige, 2002). Symbols on the x-axes represent concentrations in bottom water samples obtained by hydrocasts. (Top row) Cores collected at site S3 in southern Chesapeake Bay in 10/96 (●) and 8/97 (■). Data from Burdige and Zheng (1998) and Burdige

DOC than the more abundant, and presumably more refractory, low molecular weight DOC (Amon and Benner, 1994; Santschi et al., 1995).

Pore-water DOC and DON profiles have been published from a wide range of surficial marine sediments, in environments ranging from the deep sea to shallow water estuaries, salt marshes, and seagrass environments (see Burdige, 2002, for a recent compilation of these results; also see Krom and Westrich, 1981, for a review of the earlier literature). In an attempt to describe some of the general controls on pore-water DOC depth profiles Starikova (1970) and Krom and Sholkovitz (1977) noted that these profiles tend to fall into two general categories. In anoxic sediments where benthic macrofaunal processes (bioturbation and/or bioirrigation) are insignificant, pore-water DOC (and DON) concentrations generally increase with depth, often approaching asymptotic concentrations at depth. In contrast, in what these authors term "oxic" or "oxidizing" sediments, DOC concentrations are elevated above bottom water values, but are also much more constant with depth in the upper portions of the sediments. While many of these sediments may indeed be oxic, it may actually be more appropriate to recognize that such sediments are often bioturbated and/or bioirrigated and have mixed (or oscillating) redox conditions (see section 7.6.3).

To further understand the production and consumption of DOM in sediment pore waters, Burdige and Gardner (1998) proposed the pore-water size/reactivity (PWSR) model, a model that in part is based on traditional models for carbon cycling in anoxic sediments discussed in chapter 7 (see fig. 7.6). In this model (fig. 10.2) the degradation of sediment particulate organic matter (POM) to inorganic nutrients occurs by a series of processes that produce and consume pore-water DOM intermediates with increasingly smaller molecular weights. Although this process likely leads to a continuum of DOM compounds (in terms of molecular weights), the model assumes that there is an initial class of high molecular weight DOM (HMW-DOM) that contains biological polymers, such as dissolved proteins and

---

(2001). (Second row) Cores collected at site M3 in mesohaline Chespapeake Bay in 7/95 (■) and 10/95 (●). Data from Burdige and Zheng (1998). (Third row) Replicate cores collected in San Clemente Basin (California Borderlands) in 3/94 (Burdige et al., unpub. data). (Fourth row) Replicate cores collected at the Patton Escarpments in 3/94 (Burdige et al., unpub. data).

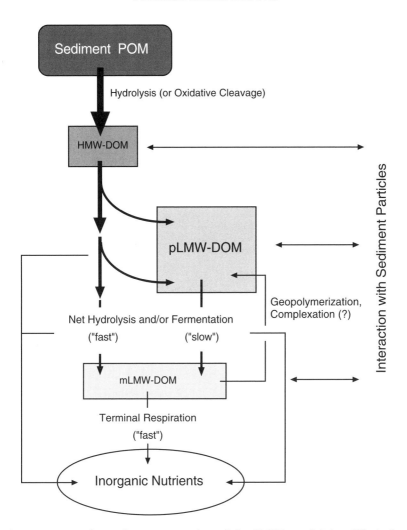

Figure 10.2 A schematic representation of the PWSR model (modified after Burdige and Gardner, 1998). As discussed in the text, most carbon flow in sediments proceeds via POM → HMW-DOM → mLMW-DOM → inorganic nutrients. However, in this model a small fraction of the HMW-DOM is also assumed to be broken down to a group of polymeric low molecular weight compounds (pLMW-DOM) proposed to be much less reactive than HMW-DOM. As is also shown here, this model does not necessarily exclude the possibility of reactions that form pLMW-DOM from mLMW-DOM compounds (e.g., geopolymerization), although this is likely to be of minor significance (see discussions in section 9.8.3 and in Burdige, 2002).

polysaccharides, that result from the initial depolymerization (hydrolysis or oxidative cleavage) of sediment POM.

As fig. 7.6 shows, most HMW-DOM is further hydrolyzed and fermented to monomeric low molecular weight DOM compounds (mLMW-DOM), such as acetate, other short-chain organic acids, and individual amino acids. These are then utilized in terminal respiratory processes in sediments such as sulfate reduction. However, it is also assumed here in the PWSR model that a small fraction of the HMW-DOM is broken down to a group of polymeric low molecular weight compounds (pLMW-DOM) that are proposed to be much less reactive than HMW-DOM.

The existence of these different DOM pools, with different reactivities, is consistent with DOM molecular weight studies showing that the vast majority of the DOC and DON in sediment pore waters has a molecular weight less than ~3 kDa (~80–90% in estuarine sediments and ~60–70% in continental margin sediments; see Burdige et al., 1996; Burdige and Gardner, 1998; Burdige and Zheng, 1998). Since total DOM generally increases with sediment depth (see fig. 10.1), this low molecular weight DOM also accumulates with sediment depth and burial. Therefore, the accumulation of pore-water DOM with depth in sediments results from the net production of refractory low (and not high) molecular weight DOM (also see fig. 10.5). From one perspective this accumulation of DOM in sediment pore waters is somewhat analogous to that which is observed for inorganic nutrients in pore waters (e.g., Krom and Westrich, 1981). However, because many components of the DOM pool are intermediates in the remineralization process (as opposed to inorganic nutrients, which are remineralization end-products), the dynamics of DOM cycling in sediments is likely to be different than that of inorganic nutrients.

Several lines of evidence suggest that the DOM accumulating in sediments is indeed refractory. The first is simply that relatively high concentrations of this material can be found in sediments and that DOM accumulates with depth in sediments as rates of POM remineralization and the reactivity of sediment POM both decrease (Westrich and Berner, 1984; Middelburg, 1989; Burdige, 1991). If the majority of the pore-water DOM had a high degree of reactivity, then one would expect its concentration to eventually decrease with depth on early diagenetic time scales. Second, in many coastal marine sediments, pore-water DOM humiclike fluorescence is strongly correlated with total DOC concentrations (Chen et al., 1993; Burdige et al., 2004). Given that the vast majority of the DOC in sediment pore waters is of low molecular weight (less than ~3 kDa), and that the properties of this low molecular weight DOC appear to be consistent with what is referred to as pLMW-DOC (see Burdige and Gardner, 1998, for further details), these DOM fluorescence observations support the notion that pLMW-DOM is likely what is referred to as dissolved humic substances.

As discussed in section 9.8.3 these observations about DOM cycling in sediment pore waters are consistent with the biodegradation model for the formation of humic substances, which assumes that humification results in increasingly oxidized compounds of lower molecular weights (Waksman, 1938; Hatcher and Spiker, 1988). Studies of dissolved organic matter in the marine water column (Amon and Benner, 1994) also similarly support this model of humification. The relationship between refractory pore-water DOM and sediment organic carbon that is ultimately preserved in sediments will be discussed in chapter 15 (section 15.6).

Consistent with this general view of pore-water DOM cycling is the limited amount of $\delta^{13}C$ data for pore-water DOC. In Santa Monica Basin sediments and sediments on the eastern North Pacific continental rise (water depth 4,100 m), pore water DOC in the upper 10 cm has a $\delta^{13}C$ that is virtually identical with that of the sediment POC (approx. −21 to −22‰; Bauer et al., 1995). This similarity suggests that the source of this DOC is isotopically light marine POC found in these sediments, assuming the DOC is produced with little isotope fractionation (e.g., see section 4.4.2). Similarly, Alperin et al. (2000) observed that seasonal changes in the $\delta^{13}C$ of the DOC in Cape Lookout Bight sediment pore waters are consistent with the known sources of organic matter to these sediments.

Pore-water DOC and DON accumulates with sediment depth as a result of a slight imbalance between production and consumption (Burdige and Gardner, 1998; Alperin et al., 1999). In surficial anoxic sediments these profiles generally approach asymptotic concentrations with depth (fig. 10.1). As shown in fig. 10.3 there is also a positive relationship between maximum pore-water DOC concentrations ($[DOC]_\infty$) in anoxic surficial sediments and depth-integrated sediment carbon oxidation rates (R_{cox}). Bioturbated and bioirrigated sediments do not appear to fall on the trend line for anoxic sediments, as will be discussed below.

At least two possibilities may explain these asymptotic DOC concentrations with depth in anoxic sediments (Burdige, 2002). The first is that

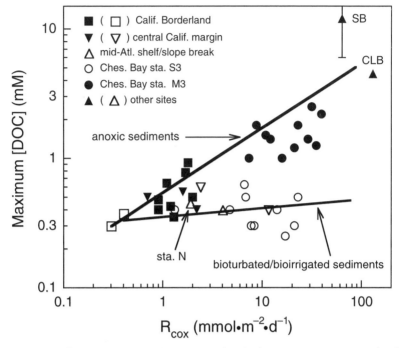

Figure 10.3 The maximum DOC concentration in the upper ~20–30 cm of sediment versus the depth integrated sediment carbon oxidation rate (R_{cox}; modified after Burdige, 2002). Open symbols represent bioturbated/bioirrigated sediments while closed symbols represent more strict anoxic sediments. The two lines are not meant to imply any functional relationships but are presented simply to show the different general trends observed in the two sediment types. Data sources: Chesapeake Bay sites M3 and S3—Burdige and Homstead (1994), Burdige and Zheng (1998), Burdige (2001); California Borderlands and central California margin sites—Berelson et al. (1996) and Burdige et al. (1999, unpub. data); mid-Atlantic shelf/slope break (site WC4)—Burdige et al. (1996, 2000, unpub. data); Cape Lookout Bight, NC (CLB)—Martens et al. (1992) and Alperin et al. (1994); Skan Bay (SB), Alaska—Alperin et al. (1992b); station N—Bauer et al. (1995).

a balance occurs between DOC production from sediment POM and DOC consumption at depth. This possibility will be discussed in greater detail in the next section. A second explanation for these observations is that DOC production rates go to zero with depth, and biotic or abiotic changes in the composition of the refractory pore-water DOC pool ($\approx$ pLMW-DOC) continually decrease its reactivity (fig. 10.4).

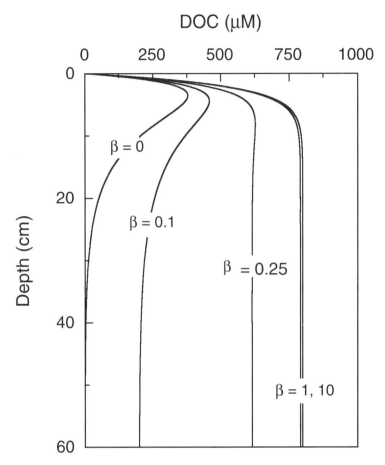

Figure 10.4 Model calculations illustrating a possible explanation for asymptotic DOM concentrations with depth in anoxic sediments. This model is modified from that described in section 10.2 and assumes that DOM production from POM decreases exponentially with depth, as does the first order rate constant for DOM consumption; hence the reactivity of DOM decreases with depth. These assumptions then lead to the following steady-state diagenetic equation:

$$0 = D_s \frac{d^2C}{dz^2} + Pe^{-\lambda z} - (ke^{-\beta z})C$$

As β increases, the consumption of DOM decreases and becomes increasingly localized near the sediment-water interface, because of the greater attenuation of the apparent rate constant (= $ke^{-\beta z}$). At sufficiently large values of β, DOM consumption becomes sufficiently small such that the resulting

This would eventually lead to a situation in which pore-water DOC found at depth is effectively nonreactive on early diagenetic time scales and therefore selectively preserved. In this case, one might think of this DOC at depth much like one thinks of "inert" inorganic remineralization end-products such as ammonium or ΣCO_2, which also show similar asymptotic profiles in anoxic sediments (also see section 6.3.4 and fig. 6.4).

As discussed in Burdige (2002) an examination of pore-water profiles from both shallow, surficial sediments and deeper sediments, i.e., tens to hundreds of meters below the sediment surface, supports the general observation that DOC concentrations at depth are roughly controlled by a balance between production and consumption. However, this analysis also suggests that with time there are also changes in the pLMW-DOC pool (e.g., in its composition and/or structure) that lead to an overall decrease in its bulk reactivity. This latter process then allows this material to continue to accumulate with sediment depth on long (geological) time scales.

In contrast to anoxic sediments, bioturbated and bioirrigated sediments (i.e., mixed redox sediments) generally have lower pore-water DOC and DON concentrations and smaller concentration gradients with sediment depth (figs. 10.1 and 10.3). Results presented in Burdige (2001, 2002) indicate that changes in the kinetics of sediment DOM cycling under mixed redox conditions, in addition to these macrofaunal transport processes, all impact pore-water DOM concentrations and depth profiles in these sediments. Specifically, it appears that sediment redox conditions alter the pathways of sediment DOM remineralization in ways that result in the net build-up of refractory pore-water DOM under anoxic conditions. Whether the production or consumption of this refractory DOM is more affected by redox conditions is uncertain (also see recent discussions in Komada et al., 2004). However, these observations clearly suggest that there is less effective preservation of refractory DOM under mixed redox sediments.

◀

profiles effectively resemble those for inorganic remineralization end-products (e.g., ammonium) in anoxic sediments (compare with fig. 6.4). This model equation was solved numerically (Boudreau, 1997; Burdige, 2002) using the following parameters: $D_s = 50$ cm^2 yr^{-1}; $P = 10$ mM yr^{-1}; $\lambda = 0.5$ cm^{-1}; $k = 1$ yr^{-1}. These parameters are similar to analogous parameters used in the model calculations shown in fig. 10.5.

10.2 Diagenetic Models of Pore-Water DOM Cycling in Sediments

As discussed in chapter 8, diagenetic models of organic matter remineralization in sediments often tend to ignore the complexity of the remineralization process. In many cases remineralization is parameterized as being simply first order with respect to the amount of reactive organic matter present, although some models also include a dependence on the concentration of the terminal oxidant.

To begin to examine the role of DOM cycling in sediment carbon remineralization, Burdige (2002) developed an advection/diffusion/reaction model for these processes. The model is based on the PWSR model shown in fig. 10.2. In this model, it is assumed that the utilization of sediment POM (producing reactive DOC intermediates) decreases exponentially from a value of R_o at the sediment surface to some low, constant value R_∞ at depth (Burdige and Martens, 1988). In some ways, one can think of this formulation as assuming that there is a reactive fraction that degrades by first-order kinetics on early diagenetic time scales (e.g., see eqns. 6.34–6.36), and a less reactive fraction whose k for decomposition is sufficiently small relative to the time scales of early diagenetic processes (e.g., see discussions in section 8.1). Therefore while there is some small amount of remineralization of this less reactive material with depth, its concentration effectively shows no change with depth. As a result, we assume that the rate of remineralization of this material is essentially constant with depth.

Assuming that HMW-DOM (= H) represents the first set of DOM intermediates produced during remineralization, and that its consumption is first order with respect to its concentration (= $k_H H$), the steady-state diagenetic equation for HMW-DOM in nonbioturbated or bioirrigated (i.e., anoxic) sediments can be written as follows:

$$D_s \frac{d^2 H}{dz^2} - \omega \frac{dH}{dz} + (R_o - R_\infty)e^{-\lambda x} + R_\infty - k_H H = 0 \qquad (10.1)$$

where the parameters here are defined above.

Based on the model in fig. 10.2 and arguments presented in Burdige (2001) it is assumed that HMW-DOM consumption then occurs by two pathways, through either pLMW-DOM or mLMW-DOM intermediates. Since mLMW-DOM compounds are presumed to be a relatively small fraction of the total DOC pool that are rapidly oxidized to CO_2, their depth distribution is not modeled here. This then leads to the following steady-state diagenetic equation for pLMW-DOM $(= P)$,

$$D_s \frac{d^2P}{dz^2} - \omega \frac{dP}{dz} + ak_H H - k_p P = 0 \qquad (10.2)$$

where a is the fraction of HMW-DOM consumption that goes through the pLMW-DOM pool, and k_p is a first-order rate constant for pLMW-DOM consumption. The analytical solutions to these equations are described in Burdige (2002), as is the development and solution to modified forms of these equations that include bioturbation and bioirrigation. This latter system of equations can be solved numerically using an Excel spreadsheet available from the author.

Model profiles for anoxic, nonbioturbated sediments are similar in shape to those seen for sediment pore-water profiles of total DOC in such sediments (e.g., compare figs. 10.1 and 10.5). Furthermore, model-derived profiles of pLMW-DOM are consistent with pore-water profiles of humiclike fluorescence, based on the assumption that humiclike fluorescence is a tracer for refractory pLMW-DOM (see Burdige et al., 2004, for further details).

Interestingly, in spite of the fact that model-predicted concentrations of HMW-DOM are significantly lower than those of pLMW-DOM, their pore-water gradients near the sediment-water interface, and presumably their benthic fluxes, are much more similar in magnitude (see the insert to fig. 10.5). These model results and other lines of field evidence suggest that not all of the DOM that escapes sediments as a benthic flux is refractory, and that it does have the potential to be reactive in the water column. This observation will be discussed in further detail in section 10.4.

10.3 Pore-Water DOM Compositional Data

As is the case for DOM in the water column (e.g., Benner, 2002) much of the DOM in pore waters is similarly uncharacterized at the molecular or even compound-class level, e.g., as total carbohydrates, lipids, or amino acids (see Burdige, 2001, 2002, for further details). Lomstein

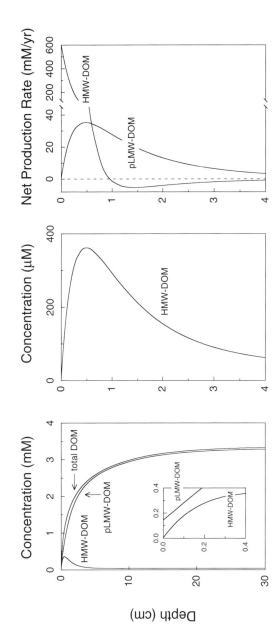

Figure 10.5 Model results obtained with the pore water DOM model described in the text for an anoxic (nonbioturbated or non-bioirrigated) sediment (see eqns. 10.1 and 10.2 and Burdige, 2002). (Left) Pore-water profiles for HMW-DOM, pLMW-DOM, and total DOM (= pLMW − DOM + HMW − DOM) predicted by the model equations. Shown as an insert on an expanded scale are the pLMW-DOM and HMW-DOM profiles, indicating the similarity in concentration gradients for both types of DOM near the sediment-water interface. (Center) The pore-water profile for HLMW-DOM shown in the left panel, again on expanded depth and concentration scales. (Right) Net production rates of HMW-DOM and pLMW-DOM versus depth in these model calculations. Note the scale break for HMW-DOM production in the upper ~0.5 cm, and the fact that there is a small amount of net HMW-DOM *consumption* between ~1 and 12 cm. The net production rate of HMW-DOM is defined as $(\Delta R e^{-\lambda x} + R_\infty) - k_H H$, where $\Delta R = R_0 - R_\infty$ (see eqn. 10.1), while that for pLMW-DOM is $ak_H H - k_p P$ (see eqn. 10.2). The parameters used in these model calculations are listed in Burdige (2002).

et al. (1998) were able to quantify ~40% of the DON in Danish coastal sediment pore waters as dissolved free and combined amino acids, although the detailed nature of these combined amino acids is still uncertain (see the discussion below for further details).

Looking at studies characterizing pore-water DOM in the context of the PWSR model, it can be seen that most efforts have focused on examining concentrations and cycling of compounds that fall into the mLMW-DOM category (see below and Henrichs, 1993). While some work has been carried out examining the dynamics of the HMW-DOM pool (Mayer, 1989; Mayer and Rice, 1992; Arnosti et al., 1994; Boschker et al., 1995; Pantoja and Lee, 1999; Arnosti, 2000) less work has been carried out examining its chemical composition.

10.3.1 Short-Chain Organic Acids

Interest in the study of short-chain organic acids (SCOAs), e.g., volatile fatty acids such as formate, acetate, and propionate, stems from the observation that acetate and other SCOAs are important *in situ* substrates for both sulfate reducing and methanogenic bacteria in anoxic sediments (see section 7.6). Rephrasing this in the context of the PWSR model, this suggests that SCOAs represent major components of the mLMW-DOC pool in anoxic sediments through which much of the carbon flow ultimately occurs (Fig. 10.2).

In most anoxic sediments acetate concentrations range from <1 μM to ~100 μM, and generally increase with depth in the upper 20–30 cm of sediment (see Burdige, 2002, for a recent summary). In general, acetate appears to account for ~5% (at most) of the total pore-water DOC and often less than 1% of the DOC. Concentrations of other SCOAs such as formate or propionate have generally not been determined as frequently as acetate although, based on the available literature, concentrations of these other SCOAs may be comparable to those of acetate (e.g., Barcelona, 1980; Albert and Martens, 1997). Other studies have also concluded that much of the acetate and other SCOAs that can be chemically measured may not be biologically available (see Henrichs, 1993, for a summary). This observation further reinforces the notion that material in the mLMW-DOM pool represents a small fraction of the total sediment pore-water DOM pool whose concentration is held at relatively low levels because of rapid bacterial turnover.

10.3.2 Carbohydrates

Total dissolved carbohydrates (DCHOs) have been determined in a limited number of coastal and continental margin sediments, and concentrations generally range from ~10 to 400 μM C (see Burdige et al., 2000, and references therein). In most cases DCHO concentrations increase with sediment depth (upper ~20–30 cm) and represent ~10–40% of the total DOC. Relative DCHO concentrations generally decrease with sediment depth (again upper ~20–30 cm) although the extent of decrease varies among the few sites that have been examined. Evidence to date also suggests that DCHOs may be preferentially found in the high molecular weight (HMW) pore-water DOC pool, and therefore likely represent some of the initial HMW intermediates produced and consumed during sediment POC remineralization (Arnosti, 2000; Burdige et al., 2000).

10.3.3 Amino Acids

Amino acids in pore waters have been examined in a wide range of marine sediments including salt marsh, estuarine, and coastal sediments. Several studies have also examined amino acid adsorption to marine sediments and dissolved amino acid turnover rates in sediments (see Burdige, 2002, for a summary).

Concentrations of total dissolved free amino acids (TDFAAs) generally decrease with sediment depth, with surface pore-water concentrations ranging from ~20 to 200 μM and concentrations below 10–20 cm being less than 5–10 μM. TDFAAs represent 1–13% of the DON and <4% of the DOC. The predominant amino acids in the free amino acid pool include glutamic acid, alanine, glycine, aspartic acid, and, in some sediments, the nonprotein amino acid β-aminoglutaric acid (β-aga), an isomer of glutamic acid (Henrichs and Farrington, 1979; Burdige, 2002). Other nonprotein amino acids in addition to β-aga can also be enriched in pore waters relative to the sediments (e.g., β-alanine). These and other compositional differences between pore-water and sediment amino acids are likely related to biological and physical processes (e.g., adsorption) that affect pore-water amino acids as they undergo remineralization (see discussions in Henrichs and Farrington, 1987; Burdige and Martens, 1990).

Finally, another pool of dissolved amino acids that has been given little attention is that of dissolved combined amino acids (DCAAs).

These represent amino acids found either in dissolved peptides/proteins, or incorporated into abiotic (humic ?) compounds. In the few studies of DCAAs in marine sediment pore waters, concentrations of total DCAAs appear to be ~1.5–4 times that of total DFAAs (see Burdige, 2002, for a summary). Given the small data set on pore water DCAAs it is difficult to determine whether compositionally the DCAA pool looks more like DFAAs or hydrolyzable sediment amino acids. Such information could be important in determining the extent to which the DCAA pool represents dissolved peptides or proteins, i.e., "reactive" high molecular weight intermediates of sediment organic matter remineralization, or abiotic condensation products of, for example, melanoidin-type reactions.

10.4 Fluxes of DOM from Marine Sediments

In the above discussions the overall remineralization of particulate organic matter deposited in marine sediments has been assumed to result solely in the net production of inorganic end-products. Although remineralization clearly occurs through DOM intermediates (see section 7.6), the implicit assumption here is that there is no net loss of these DOM intermediates from sediments during remineralization.

Because both DOC and DON in sediment pore waters are elevated over bottom waters, the consideration of diffusion alone, as a sediment-water exchange mechanism, implies that sediments are a potential source of DOM to overlying waters. This possibility was discussed in the literature prior to the early 1990s, where it was suggested that sediments might represent a significant source of DOC to the deep ocean (see Burdige, 2002, for a review). In particular, this interest stemmed from suggestions that these fluxes might provide an explanation for the apparent discrepancy between the "old" (~6,000 ybp) ^{14}C age of deep water DOC, the average mixing time of the oceans (~1,000 yr), and other chemical properties of deep water DOC (i.e., $\delta^{13}C$ and lignin concentrations) that suggest this material is primarily of marine origin. However, it was not until recent years that direct measurements of these fluxes were undertaken using both core incubation techniques and *in situ* benthic landers or chambers (again see Burdige, 2002). To date, such benthic DOC fluxes have been determined in a number of estuarine, coastal, and continental margin sediments, with values that range from ~0.1 to 3 mmol m^{-2} d^{-1}.

Based on a compilation of these results, Burdige et al., (1999) observed that there was a positive, but nonlinear, relationship between benthic DOC fluxes (BDF) and depth-integrated sediment carbon oxidation rates (R_{cox}), expressed as

$$BDF = 0.36\, R_{cox}^{0.29} \tag{11.15}$$

(BDF and R_{cox} are both expressed here as mmol m^{-2} d^{-1}). The relationship between BDF and R_{cox} implies that the ratio of BDF to R_{cox} decreases with increasing R_{cox}. With these results Burdige et al. (1999) were able to estimate the global significance of benthic DOC fluxes in oceanic and sediment carbon cycling. This calculation suggests that the integrated DOC flux from coastal and continental margin sediments (0–2,000 m water depth) is ~190 Tg C yr^{-1}.

The implications of these results are several-fold. First, these results imply that DOC fluxes from coastal and margin sediments are generally less than ~10% of sediment carbon oxidation rates. This value is estimated to be 2,570 Tg C yr^{-1} by Middelburg et al. (1997), who determined globally integrated rates of sediment processes in these sediment regimes using an extensive database of published rates of sediment biogeochemical processes. Similar trends also appear to be the case for benthic DON fluxes (see Burdige, 2002, and section 16.2.1).

Both of these observations imply that net sediment DOM production is small in comparison to gross sediment DOM production. This occurs because, under steady-state conditions, net sediment DOM production is balanced by the benthic DOM flux (section 8.2.1), while gross DOM production is approximately equal to gross DOM consumption or inorganic nutrient production, which then results in inorganic nutrient benthic fluxes. Overall, sediments appear to be quite efficient in retaining (oxidizing) DOM produced during remineralization processes, consistent with past discussions here of the tight coupling that generally exists between processes producing and consuming DOM intermediates during the overall remineralization of sediment organic matter (e.g., sections 7.6.1, 10.1, and 11.2.1).

A second implication of these results is that the integrated benthic DOC flux is comparable to estimates of the riverine DOC input (200 Tg C yr^{-1}; Meybeck, 1982). Thus, as has been noted previously (Burdige, 2002), marine sediments may be an important net source of DOC to the oceans. However, the actual impact these fluxes have on the oceanic carbon cycle ultimately depends on the extent to which sediment-derived DOM is reactive or refractory in the water column.

If this material is sufficiently refractory, then these fluxes could represent an important source of DOC to deep waters. Conversely, if sediment-derived DOC is reactive in the water column, that is, if it undergoes remineralization on time scales shorter than deep water residence times, then it will have minimal impacts on deep water DOC properties (Alperin et al., 1999). Several lines of evidence from contrasting marine sediment suggest that not all of the sediment-derived DOM, i.e., from benthic fluxes, is refractory (summarized in Burdige 2002; also see model calculations in section 10.2); thus it does have the potential to be reactive in the water column. This discussion, therefore, suggests that benthic DOC fluxes may not be as important a source of refractory DOC to the water column as was once hypothesized. However, other results suggest another way that sediment porewater DOC dynamics could play a role in explaining the relatively "old" ^{14}C age of deep water DOC. Guo and Santschi (2000) observed that simple desorption of colloidal (>1 kDa) organic matter from continental margin sediments yields DOC that is substantially older than the bulk sediment organic matter (~3,000 vs. 700 yrs, respectively). Desorption of this colloidal organic matter from continental margin sediments in the benthic nepheloid layer, coupled with transport of this material to the deep ocean (e.g., Bauer and Druffel, 1998), could then play an important role in explaining the age of deep ocean DOC. Furthermore, if this old, desorbed colloidal organic matter is in some kind of reversible equilibrium with the sediment pore waters while in the sediments, the relatively refractory nature of the bulk pore-water DOC pool would then further aid in the aging process of this sorbed organic matter while it resides in the sediments.

10.5 DOM Adsorption and Sediment–Organic Matter Interactions

In addition to biological processes affecting DOM in sediment pore waters, physical interactions with sediment particles, e.g., adsorption, are also of some importance. These processes can decrease the availability of DOM to microbial degradation (Gordon and Millero, 1985; Sugai and Henrichs, 1992) and could play some role in carbon preservation (Henrichs, 1995; Mayer, 1995). Cation or anion exchange (ionic interactions) or weaker Van der Waals interactions appear to be the predominant mechanisms by which reversible adsorption occurs (Henrichs and Sugai, 1993; Wang and Lee, 1993; Arnarson and Keil, 2001).

Adsorption coefficients for equilibrium, reversible adsorption (K'; see section 2.4.2 and eqn. 2.7) range from ~0.1 to 200 l kg^{-1} for a wide range of organic molecules, including glucose, monomeric amino acids, and short-chain organic acids (see Sansone et al., 1987; Henrichs, 1992; Henrichs and Sugai, 1993; Henrichs, 1995, and references therein). Larger molecules such as proteins or synthetic melanoidins have K' values that range from 50 to 600 l kg^{-1} (Henrichs and Doyle, 1986; Henrichs, 1995; Ding and Henrichs, 2002). Adsorption studies using natural assemblages of pore-water DOM yield K' values that range from 1 to 100 l kg^{-1} (Thimsen and Keil, 1998; Arnarson and Keil, 2001). However, Thimsen and Keil (1998) also suggest that their adsorption data are consistent with a model in which the pore-water DOM they studied was composed of a roughly 1:2 mixture of two DOM fractions with K' values of 3,200 and 20 l kg^{-1}, respectively.

For most natural sediments with a negative surface charge (see table 2.1), positively charged organic molecules, e.g., basic amino acids or lysine-containing melanoidins, are more strongly adsorbed than neutral or negatively charged organic compounds, e.g., neutral or acidic amino acids. The reverse trend appears to occur on organic-free carbonate surfaces (see Henrichs, 1992, for a summary). Increasing molecular weights and increasing hydrophobicity also enhance adsorption (Henrichs, 1992, 1995).

At the same time, studies of amino acid and protein adsorption have shown that much of this adsorption is effectively irreversible (Henrichs and Sugai, 1993; Ding and Henrichs, 2002). For proteins, this component of irreversible adsorption could result from attachment of the molecule at multiple sites on the sediment surface. However, amino acid adsorption studies are also consistent with melanoidin-type reactions occurring on sediment surfaces.

Interestingly, adsorption studies using naturally occurring pore-water DOM show that much of this adsorption is reversible (Thimsen and Keil, 1998). This observation is not, however, in conflict with the above-discussed studies of irreversible adsorption of proteins and amino acids, since the extraction procedure used by Thimsen and Keil (1998) to isolate natural pore-water DOM only recovers DOM that is presumably in reversible equilibrium with the sediments; therefore, this procedure selects against the extraction and study of irreversibly adsorbed DOM.

Discussions of DOM adsorption in marine sediments have also been presented in terms of broader, and more general, questions regarding the nature of sediment–organic matter interactions and how they may impact sediment carbon preservation (e.g., Hedges and Keil, 1995). A discussion of this topic is, however, reserved for chapter 15.

≈ CHAPTER ELEVEN ≈
Linking Sediment Organic Geochemistry and Sediment Diagenesis

THIS CHAPTER will continue to examine the organic geochemistry of marine sediments, focusing on the sources of organic matter to marine sediments and the composition of organic matter undergoing remineralization in marine sediments. The presentation here will build on discussions in previous chapters, in an attempt to develop some broad generalizations about these topics. A related topic, the composition of organic matter preserved in marine sediments, will be discussed in chapter 15.

11.1 THE SOURCES OF ORGANIC MATTER TO MARINE SEDIMENTS

Broadly speaking, organic matter in marine sediments has a limited number of sources. The vast majority is either marine derived, e.g., phytoplankton debris, or terrestrially derived, e.g., terrestrial biomass, plant litter, or soil organic matter. Other possible allochthonous sources include black carbon (the products of incomplete biomass burning) and recycled, or weathered, kerogen that is transported back to the oceans after its uplift and weathering in sedimentary rocks. At the same time, some fraction of the sediment POM that undergoes remineralization during early diagenesis is reassimilated, generally at the monomer or oligomer level (see fig. 9.10), and repackaged *in situ* as new bacterial[1] "biomass." This bacterial biomass is actually better thought of as bacterially derived organic matter and is not really a new source of sediment POM in the same sense as these other primary sources. However, bacterial production of organic matter in sediments may play a role in sediment carbon preservation.

---

[1] The term *bacteria* is broadly used here (except where specifically indicated) to describe true bacteria or eubacteria, as well as archaea and cyanobacteria. Also see note 1, chapter 7.

TABLE 11.1
End-member Chemical and Isotopic Compositions of Sources
of Organic Matter to Sediments[a]

Organic Matter Type	$\delta^{13}C$	C/N	C/P
Vascular land plants[b]		20–500	830 (300 to >1,300)
C_3 land plants	−25‰ to −28‰		
C_4 land plants	−8‰ to −18‰		
Soil organic matter[c]	−14 to −26‰ (approx.)	~8–15	
Marine phytoplankton	−17‰ to −22‰	5–10	106–117
Bacterial biomass		~3–5	7–80

a. The values listed here are taken from several compilations (Fogel and Cifuentes, 1993; Anderson and Sarmiento, 1994; Prahl et al., 1994; Hedges et al., 1997; Meyers, 1997; Ruttenburg and Goñi, 1997; Goñi et al., 1998; Onstad et al., 2000; Rullkötter, 2000; Hedges et al., 2002).

b. See section 3.4.1 for additional information on C_3 and C_4 land plants.

c. The large range listed here recognizes that the $\delta^{13}C$ of soil organic matter is a function of the organic source materials to the soils (i.e., C_3 or C_4 plants; see, for example, discussions in Prahl et al., 1994; Goñi et al., 1998; Onstad et al., 2000).

11.1.1 Carbon and Nitrogen Isotopic Tracers of Organic Matter Sources

Sediment organic matter is often characterized by examining its elemental composition, i.e., total organic carbon and total nitrogen content, and its isotopic composition, i.e., $\delta^{13}C$ of the TOC and $\delta^{15}N$ of the TN. This information can potentially be used to differentiate sources of organic matter, since the two broad types of organic matter found in sediments (marine and terrestrial organic matter) can have distinctly different bulk elemental and isotopic compositions (table 11.1).

Both marine phytoplankton and most land plants, i.e., those that represent the predominant source of terrestrial organic matter, use the same C_3 carbon fixation pathway, leading to similar amounts of carbon fractionation during photosynthesis. However, organic matter produced by phytoplankton has very different $\delta^{13}C$ values than that produced by land plants (table 11.1) because of differences in the isotopic composition of their carbon sources. The primary carbon source for marine phytoplankton is seawater bicarbonate, with a $\delta^{13}C$

of ~0‰. In contrast, land plants use atmospheric CO_2 as their carbon source, with a $\delta^{13}C$ of around −7‰. Differences in the mechanisms of CO_2 uptake by terrestrial plants (CO_2 diffusion) versus marine plants (active uptake of bicarbonate in most cases) also lead to some additional amounts of carbon fractionation during photosynthesis (see discussions in Fogel and Cifuentes, 1993). As a result of all of these factors, marine organic matter generally has a $\delta^{13}C$ of around −20‰ and terrestrial organic matter a $\delta^{13}C$ of around −26‰, again assuming a C_3 plant source for this material.

Despite these potential differences in end-member carbon isotopic concentrations, the use of carbon isotopes to differentiate between marine and terrestrial organic matter has at least three possible difficulties. First, the $\delta^{13}C$ of marine planktonic material can vary as a function of species composition, growth rate, nutrient availability, and light intensity (Fogel and Cifuentes, 1993; Hayes, 1993, 2001; also see recent discussions in Schubert and Calvert, 2001). Second, $\delta^{13}C$ values of different biochemical compound classes in individual organisms can vary over a range of ~±5–10‰, relative to the $\delta^{13}C$ of the bulk organic matter, with lipids generally being isotopically "light" and amino acids and carbohydrates being isotopically "heavy" (Degens, 1969; Wang et al., 1996; Hayes, 2001). Thus preferential remineralization or preservation of different organic matter classes can potentially bias $\delta^{13}C$ values of bulk sediment TOC. However, in the absence of large selective losses of specific biochemicals, this effect is likely to be small, less than ~2‰ (Meyers, 1997). Finally, carbon fixation by C_4 plants discriminates less strongly against ^{13}C (see section 3.4.1), and organic matter produced by these plants is isotopically heavier than either marine or C_3-terrestrial organic matter (table 11.1). Consequently, based on $\delta^{13}C$ measurements alone, mixtures of organic carbon from C_3 and C_4 terrestrial plant sources can potentially resemble marine-derived organic matter (e.g., Goñi et al., 1998).

In addition to stable carbon isotopes, radiocarbon ($\Delta^{14}C$) can shed light on organic matter production and sources to sediments. In its simplest sense, radiocarbon measurements allow for the differentiation between organic carbon produced either before or after the large influx of bomb radiocarbon into the atmosphere in the 1950s and 1960s as a result of nuclear weapons testing (see section 3.8.1). Most living marine organic matter has a positive $\Delta^{14}C$ value, consistent with the positive $\Delta^{14}C$ values of marine dissolved inorganic carbon (~+60–100‰ in the 1980s for much of the surface ocean, although values as low as $\sim -100‰$ are seen in the Southern Ocean; Levin and Hesshaimer, 2000). For terrestrial organic matter, similar arguments lead to fresh material having $\Delta^{14}C$ values that are greater than zero, since atmospheric CO_2 also has a positive $\Delta^{14}C$ value (+300–500‰ in the 1970s, decreasing to +120–140‰ in the early 1990s; Levin and Hesshaimer, 2000).

However, because of distinct differences in the dynamics of carbon cycling in different soil types (Trumbore, 1993; Trumbore and Druffel, 1995), widely varying $^{14}\Delta C$ values (and ages) occur both within a given soil (i.e., litter-fall and upper soil horizons versus deeper horizons) and among different soil types (i.e., temperate versus tropical soils) (also see discussions in Raymond and Bauer, 2001). Riverine POC has $\Delta^{14}C$ values that range from $\sim +20‰$ to less than $-200‰$, implying that much of this material is old, refractory organic matter eroded from deeper soil horizons (see Raymond and Bauer, 2001, for a recent summary). Consistent with these radiocarbon ages, riverine POM has a C/N ratio of ~ 9–12 (Meybeck, 1982), similar to that of soil organic matter (see section 11.2 and fig. 11.1). At the same time, though, the extent to which these $\Delta^{14}C$ analyses of bulk riverine POC represent degraded soil organic matter is subject to some uncertainty; relatively small contributions from old, radiocarbon "dead" sources, such as recycled or weathered kerogen (see below), can cause the total POC in a sample to be significantly depleted in ^{14}C at the bulk level (see discussions in Raymond and Bauer, 2001). The combined analysis of concentration, $\delta^{13}C$ content, and $\Delta^{14}C$ values of individual organic compounds (biomarkers) does, however, have the potential to differentiate between these terrestrial organic matter sources (e.g., Eglinton et al., 1997; Pearson and Eglinton, 2000), as well as different marine organic matter sources (also see, for example, Pearson et al., 2001).

The determination of the $\delta^{15}N$ of sediment TN can sometimes be useful in determining sources of sedimentary organic matter, because of differences in the $\delta^{15}N$ values of marine and terrestrial organic matter ($>\sim 10‰$ versus $<\sim 5‰$, respectively; Bickert, 2000). As with carbon, such differences occur because of differences in the isotopic composition of the nitrogen utilized by these organisms, and differences in the degree of fractionation during nitrogen uptake. At the same time, $\delta^{15}N$ values within marine and terrestrial organic matter sources also show wide variations, i.e., the "end-member" values are not well constrained. Therefore, some degree of care must also be taken in using nitrogen isotopes to differentiate organic matter sources (see additional discussions in Fogel and Cifuentes, 1993; Hedges et al., 1997).

11.1.2 Elemental Ratios as Tracers of Organic Matter Sources

The use of elemental ratios as organic matter source indicators has tended to focus on the examination of C/N ratios. In contrast, C/P ratios have been used less commonly, because of analytical considerations as well as the need to carefully distinguish between inorganic and organic forms of phosphorus in sediments when making these measurements (Ruttenburg and Goñi, 1997).

Marine organic matter generally has a C/N ratio between ~5 and 10 and fresh terrestrial organic matter has a C/N ratio >~20 (table 11.1). As noted previously, these differences are a result of dissimilarities in the structural components of marine versus terrestrial plants: carbon-rich lignocellulose in the latter and nitrogen-containing proteins in the former. Bacterial biomass, another "source" of organic matter to both marine and terrestrial systems, also has a low C/N ratio, between ~3 and 5, again because of the predominance of protein in their cellular material. The C/P ratios of these end-member organic matter sources show similar trends.

Although the C/N ratio of sediment organic matter has the potential to provide information about organic matter sources to marine sediments, difficulties can again arise in such efforts. In general, these problems are related to how predepositional degradation effects these ratios. Open ocean sediment trap studies show that C/N ratios in sinking particles generally increase with water column depth, presumably as a result of preferential remineralization of N-containing compounds during water column transit and remineralization (Honjo, 1980; Martin et al., 1987; Hedges et al., 2001). In spite of this trend, the C/N ratio of organic matter that is remineralized in the water column appears to be similar to that of marine organic matter (~7.3; Anderson and Sarmiento, 1994). Nevertheless in many open ocean settings (where marine organic matter presumably predominates) C/N ratios in the surface sediments are slightly elevated above the Redfield ratio for planktonic organic matter (i.e., 6.6), with values ranging between ~8 and 12 (Honjo, 1980; Grundmanis and Murray, 1982; Premuzic et al., 1982; Murray and Kuivala, 1990). Similar general trends also appear to exist for the C/P ratios in sinking marine particles, organic matter undergoing remineralization in the water column,

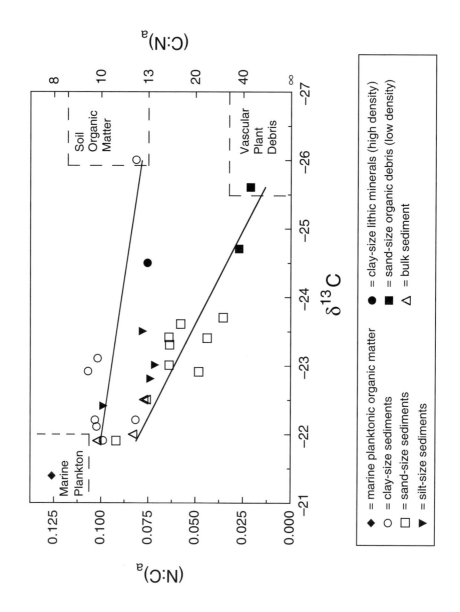

Figure 11.1 Atomic N/C ratio versus $\delta^{13}C$ of organic matter in surface sediments collected on the Washington, USA, continental margin. Redrawn using results presented in Keil et al. (1994b) and references cited therein. ◆ = marine planktonic organic matter (picked from one sediment sample); ○ = clay-size sediment particles; ● = high-density ($\rho > 2.6$ g · cm^{-3}), clay-size sediment particles (i.e., lithic minerals); □ = sand-size sediment particles; ■ = sand-size, low-density ($\rho < 2.6$ g · cm^{-3}) sediment particles (i.e., distinct organic debris); ▼ = silt-size sediment particles; △ = bulk sediments. Also shown are the regions on the plot occupied by the three end-members proposed to make up the organic matter in these sediments. Note that the size-fractionation of these sediments was carried out using hydrodynamic separation (SPLITT-fractionation).

and the C/P ratio in open ocean sediments (see discussions in Suess and Müller, 1981; Ingall and Van Capellen, 1990).

More importantly, though, terrestrial organic matter tends to gain nitrogen during decomposition on land, as lignins and polysaccharides are degraded and leave behind microbial "metabolites" or residues. These materials tend to be acidic, nitrogen rich, aliphatic, often hydrolysis resistant, and associated with mineral grains (Baldock et al., 1992; Knicker et al., 1993; Hedges and Oades, 1997). Because these transformations are associated with the decomposition of coarse vascular plant material, soil C/N ratios also decrease with decreasing particle size. Starting with fresh vascular plant material with a C/N ratio of 20–400, this ratio decreases to 15–45 in the sand size-fraction of soils (avg. ~20), and then further decreases to 7–15 in clay size-fractions (avg. ~11; also see Amazon river POM data in Hedges et al., 1986). The exact nature of these microbial metabolites in soils remains unresolved, although many of the same processes and factors that contribute to the occurrence of molecularly uncharacterized organic matter in marine sediments also appear to operate in soil systems, for example, production of refractory biomacromolecules or physical protection of organic matter (see section 9.8 for further details).

Similar trends in C/N ratio versus grain size are also seen in continental margin sediments (Keil et al., 1994b). These trends have been interpreted in the context of soil organic matter/grain size differences discussed above, coupled with the replacement of high C/N ratio terrestrial organic matter with low C/N ratio marine organic matter as particles move from land to the oceans (see related discussions in section 15.1). As these authors note, the net effect of these processes is a decrease in particle C/N ratio with either decreasing grain size or increasing clay content, particularly in continental margin sediments.

11.1.3 Spatial Trends in the Sources of Organic Matter to Marine Sediments: Marine versus Terrestrial

An examination of the literature suggests that $\delta^{13}C$, $\delta^{15}N$, and the C/N ratio of sediment POM are often used to determine the relative importance of marine versus terrestrial organic matter sources to marine sediments; for example, see de Lange et al. (1994) for a summary. Such estimates are based generally on simple mixing models that use end-member values of these parameters such as those shown in table 11.1. Nevertheless, this approach can be hampered by many of the problems discussed in the previous section, which often render this method more qualitative than quantitative.

In part, this problem stems from difficulties in accurately constraining end-member values of these measures for specific marine and terrestrial sources to a given sediment. In particular, end-member values often are subject to spatial, and perhaps temporal, variability such that values from one region may not be representative of other regions. As a result, such calculations may then require regional validation. Furthermore, degradative processes on land lead to terrestrial organic matter that can be quite heterogeneous, potentially precluding the use of simple mixing models with only two end-members (Keil et al., 1994b; Goñi et al., 1998; Gordon and Goñi, 2003). To overcome some of these difficulties, recent studies have used lipid biomarkers and lignin oxidation products in conjunction with these other bulk tracers, to help better constrain such estimates of sediment organic matter sources (Gough et al., 1993; Jasper and Gagosian, 1993; Keil et al., 1994b; Prahl et al., 1994; Goñi et al., 1998).

One example of this approach is shown in fig. 11.1 (Keil et al., 1994b), where it can be seen that such considerations lead to a situation in which terrestrial organic matter appears to comprise two end-members: a high C/N ratio end-member, containing vascular plant debris, and a much lower C/N ratio end-member, composed of soil organic matter. Furthermore, the mixing between soil organic matter and marine organic matter appears to occur only within the clay size fraction of the sediments (i.e., the upper mixing line in fig. 11.1), consistent with discussions above regarding the mechanism of terrestrial organic matter decomposition on land. Also consistent with these observations is the fact that terrestrial organic matter in clay-sized fractions is, in general, more highly degraded than that found in association with coarser (sand-size) particles (Gough et al., 1993; Keil et al., 1998; Onstad et al., 2000; Bianchi et al., 2002; Gordon and Goñi, 2003, 2004).

The phenomena described above lead to organic geochemical distributions that are a function of sediment particle size; as a result what then occurs in many coastal and continental margin settings is the hydrodynamic sorting of terrestrial organic matter during its offshore transport. In this process, fine-grained particles preferentially transported farther offshore have terrestrial organic matter associated with them that is lignin poor, nitrogen rich, and more highly degraded. In contrast, terrestrial organic matter associated with coarser terrigenous particles is retained in inner shelf deposits and is lignin rich, nitrogen poor, and less degraded (also see Prahl, 1985, in additional to the references cited above).

Based on such observations, quantitative estimates have been made of the relative amounts of terrestrial versus marine organic matter deposited on the Washington (USA) continental shelf near the mouth of the Columbia River (Keil et al., 1994b; Prahl et al., 1994). These studies estimate that ~50–80% of the organic matter deposited on this shelf (0–200 m water depth) is of terrestrial origin, ~10–30% of the slope organic matter is of terrestrial origin (~200–2,000 m water depth), and <15% of the organic matter in offshore Cascadia Basin sediments is of terrestrial origin (~2,000–3,000 m water depth). The ranges in these estimates are a result of differences in the mixing models used to make the calculations, as well as uncertainties in end-member biomarker, isotopic, and elemental concentrations used in the calculations. Gordon and Goñi (2003) also used similar techniques to estimate that terrestrial organic matter accounts for ~70–80% of the organic matter deposited by the Atchafalaya River on the inner continental shelf (0–20 m water depth) of the northern Gulf of Mexico.

Studies of other continental margin regions also provide evidence for terrestrial organic matter transport off the continental shelf (Farrington and Tripp, 1977; Volkman et al., 1987; Gough et al., 1993; Goñi et al., 1998, 2000; Onstad et al., 2000; Bianchi et al., 2002). Such studies have examined sediments associated with significant river discharge systems (e.g, the Mackenzie River/Beaufort Shelf in the Arctic Ocean or the Mississippi River/Gulf of Mexico system), and those removed from major river discharge zones (e.g., the northeast Atlantic Ocean). Work in the northern Gulf of Mexico and lower Mississippi River confirms that hydrodynamic sorting of terrestrial organic matter occurs during offshore transport, as it does on the Washington shelf; there is differential transport of soil organic matter versus vascular plant material based on their association with different particle size classes (also see Gordon and Goñi, 2004).

Historically, it has been assumed that the deposition of terrestrial organic matter is limited to sediments on the inner continental shelf and within major river deltas, with little terrestrial deposition of this material in outer shelf or slope sediments (see Hedges et al., 1997, for a review of this earlier literature). However, the results discussed here suggest that this may not necessarily be the case and that some amount of terrestrial organic matter is indeed transported out of the inner continental shelf region in some margin settings (also see related discussions in section 15.5). Unfortunately, offshore transport of terrestrial organic matter has been quantified for only a limited number of continental margin systems (see related discussions in McKee et al., 2004).

Since the vast majority of carbon remineralization and preservation occurs on the continental margins (see section 8.3), quantifying the contributions of marine versus terrestrial organic matter to continental margin (and deep-sea) sediments is important in terms of understanding global controls on carbon preservation and remineralization processes (see section 15.5 for further details). At the same time, because ~80% of the total productivity in the oceans occurs in the open ocean (e.g., Liu et al., 2000), the potential transport of terrestrial organic matter to deep-sea sediments also has implications for using deep-sea sediment TOC depth profiles as paleoproxies of oceanic productivity (de Lange et al., 1994; Goñi et al., 1998).

Some studies of pelagic sediments suggest that up to ~20% of the sediment TOC could be of terrestrial origin (Gagosian and Peltzer, 1987; Prahl and Muehlhausen, 1989; Hedges, 1992; Prahl et al., 1994). Unfortunately, though, such calculations have large uncertainties; for example, Prahl and Muehlhausen (1989) estimated that the contribution of terrestrial organic carbon to MANOP site C sediments is $27 \pm 23\%$. Atmospheric transport may be an important mechanism by which this terrestrial material is transported to remote pelagic regions, although riverine inputs may also be important (e.g., Volkman et al., 1987). For example, terrigenous materials deposited in continental margin sediments may subsequently be brought to the deep sea by resuspension and lateral transport (see related discussions in sections 5.2 and 15.5).

In the Arctic Ocean, terrestrial organic matter could be up to ~30–50% of the sediment TOC, depending on the particular mixing model applied to the sediment data (Schubert and Calvert, 2001). It is important to note, though, that the Arctic Ocean is perhaps an atypical deep-sea setting for several reasons: it is of relatively small size and has a very broad continental shelf; it is essentially surrounded by land and has significant terrestrial inputs from erosion and several relatively large rivers; and the direct input of terrigenous material to the Arctic by drifting sea-ice is an important means of sediment delivery (also see discussions in Belicka et al., 2002).

While most of the organic matter in deep-sea sediments is likely of marine origin, sediment C/N ratios in many of these sediments are generally elevated above the Redfield ratio. In contrast, in organic-poor pelagic sediments (typically described as oxic red clays) extremely low C/N ratios, below that of typical marine organic matter (<5), have been observed both in the surface sediments and with sediment depth (Müller, 1977; also see results in Grundmanis and Murray, 1982, and Gélinas et al., 2001, which show similar trends). Similar trends are also seen for C/P ratios in such sediments (Ingall and Van Capellen, 1990).

In these sediments some component of this "excess" nitrogen may result from inorganic ammonium produced during sediment organic matter remineralization that is not nitrified but becomes adsorbed to clay minerals and/or fixed in the interlayer sites of clays (see section 2.4.2 and 16.2). However, even when these sediment C/N ratios are corrected for adsorbed and fixed ammonium the true C/N ratios of the sediment organic matter are still relatively low.[2] Müller (1977) originally suggested that these low C/N ratios result from the adsorption of nitrogen-rich organic compounds onto sediments, occurring in such a way as to protect the organic matter from microbial remineralization. Interestingly, such observations are consistent with aspects of the more recently proposed surface adsorption/mesopore protection hypothesis discussed in sections 9.8.5 and 15.1. At the same time, the relatively low C/N ratio of this organic matter suggests that it might represent bacterially derived material that is either inherently refractory, and/or physically protected by inorganic or organic matrices (see discussions in section 11.1.5). Similar suggestions have also been put forth to explain low C/P ratios in such sediments (Suess and Müller, 1981; Ingall and Van Capellen, 1990; also see section 16.3). Thus, an analogy to the refractory microbial metabolites found in terrestrial soils might be relevant in deep-sea sediments (see above for further details). However, further work will be needed to examine this possibility.

---

[2] Sediment organic matter C/N ratios are often determined using TOC and TN (total sediment nitrogen) measurements, based on the implicit assumption that the amount of inorganic sediment nitrogen is small in comparison to organic forms. When this is not the case, though, organic and inorganic nitrogen (e.g., adsorbed or fixed ammonium) must both be determined separately, e.g., see references cited in this discussion and in de Lange (1992).

11.1.4 Other Sources of Organic Matter to Marine Sediments: Black Carbon and Recycled Kerogen

As discussed earlier, black carbon (section 9.8.1) and recycled kerogen represent sources of organic matter to marine sediments that clearly do not fit into the categories of marine or terrestrial organic matter described above. Kerogen oxidation on land, after uplift of sedimentary rocks, is generally thought to balance carbon burial in marine sediments (see discussions in sections 8.3 and 15.4.1). However, many of the details of this process are not well understood. In spite of the fact that kerogen is generally considered an extremely refractory form of natural organic matter, weathering profiles indicate that TOC loss from black shales on land is extensive, i.e., between 60 and ~100% (Petsch et al., 2000). However, such studies have not unequivocally determined the extent to which this TOC loss occurs because of complete kerogen oxidation to CO_2 or partial oxidation/solubilization and subsequent loss of oxidized kerogen by-products (e.g., oxidized fossil DOM or POM) by riverine or steam flow. Thus, some kerogen loss from sedimentary rocks on land may actually result in the transport of "recycled" kerogen to the oceans, where it has the potential to escape remineralization and be reburied on the continental margins (e.g., Dickens et al., 2004a, b).

Kerogen transport from land to continental margin sediments appears to be most important for small, steep, mountainous rivers associated with active continental margins and relatively narrow continental shelves (Masiello and Druffel, 2001; Blair et al., 2003, 2004). In such settings there appears to be little time for kerogen remineralization because of the relatively short time between kerogen exposure, its riverine transport, and rapid deposition in continental margin sediments. This situation contrasts with other river–continental margin settings such as the Amazon river and shelf, where extensive storage and processing of organic matter in upland soils and lowland floodplains leads to the replacement of upland organic matter, i.e., kerogen, with lowland soil organic matter in the riverine suspended matter (Blair et al., 2004). Furthermore, sediments on wide and more energetic margins, such as the Amazon, are exposed to repeated resuspension/redeposition cycles that appear to enhance the degradation of terrestrial-derived organic matter deposited in these sediments (see sections 5.5, 7.6.3, and 11.2.3).

Based on such observations, Blair et al. (2004) estimated that the riverine kerogen flux to the oceans could be on the order of ~40–70 Tg C yr^{-1}. When compared to results in table 8.3, it can be seen that this flux is significant when compared to Holocene rates of carbon burial in marine sediments (~160 Tg C yr^{-1}). On the other hand, these workers also note that recent anthropogenic activity may have perturbed the balance between kerogen export to the oceans and kerogen oxidation on land, and perhaps could also have shifted the locus of kerogen oxidation from the continents to marine sediments (also see Blair et al., 2003).

Because recycled kerogen has already gone through one cycle of sedimentation (burial), uplift, and erosion, any reburial of this material in marine sediments results in no new net O_2 production. Consequently any involvement it has in carbon burial in sediments potentially limits the strength of the feedback between sediment carbon burial and atmospheric O_2 concentrations, and could also play a role in minimizing large-scale swings in atmospheric O_2 (Berner, 1989; Hedges, 2002; also see discussions in section 15.4.1). Kerogen associated with marine bedrock is also likely to have heavy $\delta^{13}C$ values, consistent with that of marine organic matter, yet be essentially depleted in ^{14}C because of its age (Masiello and Druffel, 2001; Blair et al., 2003). Therefore its input to marine sediments could potentially be masked by its marine stable carbon isotopic signature.

Evidence for the occurrence of fossil organic carbon and black carbon in marine sediments has been presented across a wide range of sediment settings (see, e.g., Petsch et al., 2000, and Masiello, 2004, for summaries). Other workers have also begun to examine the quantitative importance of fossil organic carbon input to specific sedimentary environments (e.g., Pearson et al., 2001; Gordon and Goñi, 2004). At the same time, though, studies of the composition of modern marine sediments suggest that recycled kerogen is not a major component of the organic matter that is buried in marine sediments (Hedges and Keil, 1995; Hedges, 2002). However, more work is needed to examine this problem (also see discussions in chapter 15).

11.1.5 Production of Bacterial Biomass in Sediments

In shallow surface sediments bacterial numbers exhibit a relatively small range in values, ~10^9–10^{10} bacterial cells per cm^3 wet sediment (see discussions in Capone and Kiene, 1988). Several empirical and theoretical relationships have been proposed to explain this observation, including the suggestion that these bacterial densities are controlled by sediment TOC concentration or sediment grain-size/surface area (Deming and Baross, 1993; Schmidt et al., 1998; Boudreau, 1999).

Schmidt et al. (1998) have also observed that bacterial numbers across a wide range of marine sediments show a constant proportionality with the volume of sediment pore water, with an average value centering on $\sim 10^9$ bacteria per ml pore water. Re-expressing this relationship in terms of sediment bacterial numbers per cm^3 total (wet) sediment, this implies that sediment bacterial densities scale linearly with sediment porosity. For typical φ values ranging from ~ 0.9 to ~ 0.5 bacterial densities in sediments will then vary by at most a factor of ~ 2, consistent with the discussion above. Also consistent with these observations is the fact that there is a factor of only ~ 4 difference in bacterial biomass (expressed as g C m^{-2}) in Amazon shelf sediments versus that in other continental margin sediments (Aller, 1998). These observations therefore provide additional evidence, in addition to that discussed in section 8.1, that variations in bacterial biomass or numbers do not play a major role in the observed range of sediment organic matter reactivity (e.g., as is shown in fig. 8.1).

The production of bacterial biomass in sediments may, however, be important in terms of its role in the reworking of sediment organic matter. This occurs during sediment bacterial metabolism when some fraction of the sediment POM that undergoes mineralization is reassimilated at the monomer or oligomer level, and repackaged as new *in situ* bacterial biomass.

The importance of autochthonous bacterially derived material as a component of sediment POM, and particularly that which is preserved, has been discussed by numerous authors, and a variety of techniques (e.g., lipid biomarkers, molecular-level isotopic studies, diagenetic and mass balance modeling) all suggest that production of bacterially derived organic matter is important during early diagenesis in sediments (Rice and Hanson, 1984; Burdige and Martens, 1990; Canuel and Martens, 1993; Parkes et al., 1993; Gong and Hollander, 1997; Ruttenburg and Goñi, 1997; Keil et al., 2000; Keil and Fogel, 2001; also see references to earlier organic geochemical studies in Lee, 1992). At the same time, living bacterial biomass is generally thought to be a relatively small component of the total sediment POM pool (Mayer and Rice, 1992; Harvey and Macko, 1997a; Pantoja and Lee, 1999). These observations are not necessarily contradictory, but rather imply that much of the bacterially derived organic matter in sediments is not intact cells, either alive or dead, but rather organic matter derived from living cells—cell exudates, cell lysis products, or remnants of bacterial cell walls. Parkes et al. (1993) refer to this material as bacterial necromass and suggest it can represent a significant component of the molecularly uncharacterized organic matter in marine sediments. While most of these studies assume this bacterial necromass is derived from sediment bacteria, some of this material may actually be input into the sediments from the overlying water column on sinking particles (e.g., see Ingalls et al., 2003, and references therein).

While bacterial necromass may be more or less reactive than the original sediment organic matter (see the references cited above as well as Aller, 1994a), much of the interest in studying bacterially derived organic matter involves its possible contribution to MU-OM and/or carbon preservation in sediments. The mechanisms by which this may occur are not well understood and could include production of inherently refractory materials such as peptidoglycan (see section 9.4) or bacterial membrane lipids (see section 9.7). It may also occur via physical protection of other types of bacterially derived organic matter by encapsulation (section 9.8.5) or through interactions with mineral grains (sections 9.8.5 and 10.4).

Lee (1992) has also suggested that bacterially derived organic matter can be preserved in anoxic sediments as a result of the effective exclusion of all organisms other than bacteria in such sediments (e.g, microbial grazers such as benthic macrofauna). In the absence of a microbial loop to cycle bacterially derived carbon, this material may be less efficiently remineralized and therefore preferentially preserved. However, the results of recent studies (Aller, 1994a; Hartnett et al., 1998; Hedges et al., 1999a) also suggest that bacterially derived organic matter could be an important component of the sediment carbon that is buried/preserved in anoxic sediments not simply because of this lack of microbial grazing by higher organisms. Rather, it may occur because redox conditions in oxic or mixed redox sediments promote the more efficient remineralization of total sediment organic matter in general, and (potentially refractory) sediment bacterial necromass in particular (see section 7.6.3). These suggestions will be discussed further in chapter 15.

At the same time, recent studies also suggest that factors in addition to sediment redox conditions may play a role in controlling the preservation of bacterial necromass in marine sediments. These include: the degree of diffusive openness of mixed redox sediments, independent of sediment redox conditions (Aller and Aller, 1998); physical disturbance/sediment reworking that leads to sediments with relatively low benthic macrofaunal densities and a dominance of bacterial biomass (e.g., on the Amazon shelf or in the Gulf of Papua; Aller, 1998; Aller and Blair, 2004); the preservation of refractory microbial metabolites in very organic-poor, oxic deep-sea sediments (see section 11.1.3). Again, these points will be discussed in chapter 15.

11.2 THE COMPOSITION OF ORGANIC MATTER UNDERGOING REMINERALIZATION IN MARINE SEDIMENTS

Under steady-state conditions the flux of material to the sediment surface (defined as J_{in} in section 8.2) is equal to that preserved or buried in the sediment (J_{bur}) plus that which is remineralized (J_{out} or J_{remin}; also see eqn. 8.10). As has been discussed earlier (e.g., chapter 9), a variety of approaches have been used to directly examine the composition of organic matter deposited in a marine sediment or that which is preserved or buried. In contrast, examining the composition of organic matter undergoing remineralization in marine sediments must be done indirectly because here one is attempting to determine the composition of the fraction of the sediment organic matter that disappears during early diagenesis.

To examine certain compositional characteristics of the organic matter that is remineralized in sediments (e.g., its C/N ratio), inorganic nutrient[3] data are often used. In contrast, far fewer studies have used organic geochemical data to examine this problem. The next section describes techniques used to determine the composition of the organic matter that is remineralized in marine sediments, followed by a discussion of specific results of such analyses. Some of these results were introduced in chapter 9 and here they will be re-examined in light of this broader discussion.

---

[3] Recall that inorganic nutrients are defined here as inorganic remineralization end-products such as ammonium, phosphate or ΣCO_2.

11.2.1 Pore-Water Stoichiometric Models for Nutrient Regeneration/Organic Matter Remineralization

One approach to examining the composition of organic matter undergoing remineralization in sediments uses stoichiometric models of pore-water inorganic nutrient data (Berner, 1977). This approach will be developed here for end-member sediments in which either sulfate reduction or aerobic respiration dominates organic matter remineralization, i.e., anoxic, coastal sediments or oxic, deep-sea sediments. As we will see, in these sediment types this approach is aided by the fact that nitrogen diagenesis is relatively straightforward. In suboxic systems with appreciable denitrification and/or more complex metal/nitrogen interactions, approaches such as those described here are complicated by the fact that, for example, there is both production and consumption of nitrate. Furthermore, nitrogen cycling in these sediments is not necessarily coupled to carbon cycling in a straightforward manner. In such sediments, other techniques must therefore be employed to tease apart nitrogen cycling to ultimately get information on the sediment organic matter that is undergoing remineralization.

For sulfate reducing sediments we start with a modified form of the RKR equation,

$$(C_{red})_x(NH_3)_y(H_3PO_4)_z + w\ SO_4^{2-} \rightarrow x\ CO_2 + w\ S^{2-}$$
$$+y\ NH_3 + z\ H_3PO_4 \quad (11.1)$$

where the C:N:P ratio of the organic matter being remineralized is expressed generically as x:y:z, and using the terminology defined in section 6.3.3, $\mathscr{L} = w/x$. Also note that this equation is not completely balanced because the carbon in the sediment POM is expressed as C_{red}, reduced carbon of some to-be-determined average oxidation state; for the sake of simplicity its H or O content are also not defined. If the carbon in the sediment organic matter undergoing remineralization has an oxidation state of 0, as is assumed in the Redfield equation, then $\mathscr{L} = 1/2$, i.e., the value commonly used in calculations such as these (e.g., see eqn. 6.40). However, if the carbon in the sediment POM is slightly more reduced and has an oxidation state ($= ox$) of -0.7 (see section 9.3 and table 9.1), then slightly more sulfate must be reduced per mole of carbon oxidized. Under these circumstances $\mathscr{L} = 0.59\ (= 1/1.7)$, as will be shown below. Furthermore, since $\mathscr{L}$, ox, w, and x are all interrelated, one can directly estimate $\mathscr{L}$ and ox from pore-water data using the same stoichiometric pore-water models (e.g., see eqns. 11.5 and 11.6).

The approach taken here is based on that presented by Berner (1977), and starts with the steady-state solutions to pore-water diagenetic equations for ammonium and sulfate described in chapter 6. Hammond et al. (1999) have also shown that one can derive similar stoichiometric models in the absence of specific knowledge about the depth dependence of these diagenetic reactions, if one assumes that the reactions involving the two solutes of interest are coupled (see below), and one also has some knowledge of the transport processes affecting these solutes.

The derivation of these stoichiometric models starts with the differential forms of eqns. (6.45) and (6.51) (i.e., dC_s for sulfate and dC_n for ammonium) leading to,

$$\frac{dC_s}{dC_n} = -\frac{k_s \mathscr{L} G_o}{k_n N_o}\left[\frac{(1+K_n)\omega^2 + k_n D_n}{\omega^2 + k_s D_s}\right]\left[\frac{e^{-(k_s/\omega)z}}{e^{-(k_n/\omega)z}}\right] \qquad (11.2)$$

Since plots of pore-water ammonium versus sulfate are often linear (e.g., see fig. 11.2 and Martens et al., 1978; Klump and Martens, 1987), dC_s/dC_n in eqn. 11.2 (the best-fit slope of such a property-property plot) must therefore be a constant. As such, the exponential terms on the right-hand side of this equation must cancel one another out, implying that $k_s = k_n$. This observation is consistent with discussions above and in chapter 7 regarding the tight coupling of the processes that contribute to the overall remineralization of sedimentary organic matter to inorganic nutrients.

If we next recall that the C/N ratio of the organic matter undergoing remineralization (r_n in eqn. 6.52) is G_o/N_o, then eqn. (11.2) can be rewritten as

$$\frac{dC_s}{dC_n} = -\mathscr{L} r_n\left[\frac{(1+K_n)\omega^2 + kD_n}{\omega^2 + kD_s}\right] \approx -\mathscr{L} r_n \frac{D_n}{D_s} \qquad (11.3)$$

where k_n and k_s have been replaced with a single k and the approximation on the far right of eqn. (11.3) assumes that we can neglect advection relative to diffusion in this calculation. An examination of this equation indicates that we can obtain all of the quantities in eqn. (11.3), with the exception of r_n, from either the literature, pore-water nutrient property-property plots, or diagenetic models of pore-water

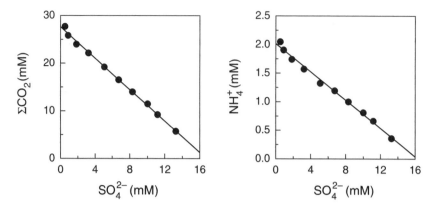

Figure 11.2 Pore-water nutrient property-property plots for a core collected on 7/95 at a site in mesohaline Chesapeake Bay (Burdige, unpub. data). Note that these same data are plotted as depth profiles in fig. 6.5. Using the procedures discussed in the text, the best-fit line in the left panel predicts that $\mathscr{L} = 0.56$ ($= 1/1.79$) and $ox = -0.46$ (also see table 9.1). Using this result and the best-fit line in the right panel predicts that $r_n = 7.7 \pm 1.8$ (also see table 11.2).

profiles (the latter being needed if we cannot make the approximation shown on the right side of eqn. 11.3). This then allows one to use this equation with pore-water data to estimate r_n, the C/N ratio of the organic matter undergoing remineralization (referred to in future discussions as the C/N remineralization ratio). Results based on this approach are listed in table 11.2. In principle, similar approaches can also be used to examine C/P and N/P ratios in sediment POM undergoing remineralization, although factors related to phosphate adsorption on iron oxides and phosphate authigenic mineral formation may complicate such calculations (see section 16.3 for further details).

This approach can also be used to explicitly determine $\mathscr{L}$ in anoxic sediments, and therefore the average oxidation state of carbon in sediment POM, using sulfate and ΣCO_2 pore-water data. Since neither of these species undergoes significant adsorption and since plots of sulfate versus ΣCO_2 are also often linear in such sediments (fig. 11.2), the following equation can be derived using an approach similar to that described above,

$$\frac{dC_s}{dC_c} = -\mathscr{L}\left(\frac{\omega^2 + kD_c}{\omega^2 + kD_s}\right) \approx -\mathscr{L}\frac{D_c}{D_s} \qquad (11.4)$$

TABLE 11.2
The C/N Remineralization Ratio in Marine Sediments

Location	C/N (Atomic)	Method[1]	Source
Coastal sediments			
Chesapeake Bay	~6 (most reactive [G$_1$] fraction)	(s)	2
	~33 (less reactive [G$_2$] fraction)		
Chesapeake Bay (mesohaline)	6.3 ± 1.4, 7.4 ± 0.7	(p, t)	3
Chesapeake Bay (southern)	9.8 ± 0.7	(t)	3
Buzzards Bay, MA	7.7 ± 0.8	(t)	4
Long Island Sound	5.5 to 10.7	(r)	5
Long Island Sound (FOAM site)	7.4 (8.7)	(p)	6
Cape Lookout Bight	6.6, 7.2	(t, p)	7
Saanich, Inlet, British Columbia	7.4, 8.4	(t, f)	8
Skan Bay, Alaska	9.4	(n)	9
Continental margin sediments			
Mexican continental margin		(r)	10
140–510 m	6.4–7.7		
1,020 m	17.1		
Washington continental margin		(r)	10
110–620 m	10.5–24		
1,000 m	10.1		
Amazon continental shelf	~7–9, <~7	(d, r)	11
northwest European continental margin (Goban spur; 200–4,500 m water depth)	~6–7.5	(d)	12
mid-Atlantic shelf/slope break (400–750 m water depths)	5.8–7.9	(t)	13
California continental margin	15.8	(t)	14a
Deep-sea sediments[15]			
central equatorial Pacific	9.1 ± 1.8	(p)	16
	5.6 (6.4)	(p)	17
	9.4 ± 2.0 (11.0 ± 2.4)	(m)	18
	8.6 (10.1)	(p)	19
	6.5 ± 1.6 (7.6 ± 1.9)	(p)	20
subtropical South Pacific	10.2 (12)	(p)	21
equatorial Pacific	8.8 ± 2.2, 7.1 ± 1.2	(m, b)	22
northeast Pacific	10, 10.3 (12.1)	(t, p)	14b
Madeira Abyssal Plain	~17–18	(t, p)	23

1. Methods:
  s = determined in long-term anoxic sediment slurry decomposition experiments;
  p = determined by modeling sediment pore-water data (using property-property plots as discussed in the text);
  t = determined from modeling sediment TOC and TN profiles (property-property plots);
  r = determined from whole sediment rate measurements;
  n = calculated from data from sediment nutrient budgets (carbon and nitrogen) assuming that phytoplankton and kelp are the predominant sources of organic matter undergoing remineralization;
  b = calculated from benthic fluxes estimated from fits to pore-water profiles.
  m = calculated from in situ benthic flux determinations (O$_2$ and nitrate).

where subscript c now represents ΣCO_2. With this approach $\mathscr{L}$ can then be determined from an estimate of dC_s/dC_c, i.e., the slope of the best-fit line through sulfate versus ΣCO_2 data, such as in fig. 11.2, and this diffusion coefficient ratio. Furthermore, $\mathscr{L}$ can then be used to calculate ox, the average oxidation state of carbon in sediment POM. Since the reduction of sulfate to sulfide involves an 8-electron trans-

◄─────────────────────────────────────

    d = based on the roughly constant C/N ratio with depth in the sediment POM
    f = determined using estimated sediment fluxes to the sediment-water interface and that buried at depth.

    2. From Burdige (1991) for sediments in the lower Chesapeake Bay and tributaries.
    3. The TOC/TN plots were determined from data in Burdige and Homstead (1994). The pore-water modeling was carried out with unpublished data (see Burdige and Homstead, 1994, Burdige and Zheng, 1998, and fig. 6.5 for representative depth profiles; also see fig.11.2 for representative nutrient property-property plots).
    4. Calculated from data reported in Henrichs and Farrington (1987).
    5. From Aller (1980a). The observed range increases with increasing water depth at the three sites studied (~8–34 m). The values shown here were calculated with depth-integrated rates of sulfate reduction and ammonium production assuming that $\mathscr{L}$ = 0.5. If as discussed in the text the carbon undergoing remineralization is more reduced (i.e., has an oxidation state of –0.7 versus 0) then $\mathscr{L}$ increases to 0.59 (= 1/1.7) and the estimated C/N ratios decrease by ~15%.
    6. Calculated with results from Berner (1977), assuming $\mathscr{L}$ = 0.59 (value for $\mathscr{L}$ = 0.5 also shown in parentheses). Note that the FOAM site is the shallowest site studied by Aller (1980a).
    7. The first value is from Martens et al. (1992); the second is from Klump and Martens (1987).
    8. The first value is from Hamilton and Hedges (1988); the second is from Cowie et al. (1992).
    9. Calculated with data from Alperin et al. (1992a) as discussed above.
    10. From Kristensen et al. (1999).
    11. From Aller et al. (1996)
    12. From Lohse et al. (1998).
    13. TOC data from Burdige et al. (2000), TN data is from Burdige (unpublished data).
    14a. From Murray and Kuivala (1990) for a site on the California continental margin at a water depth of 3,570 m on the base of the Monterey Fan.
    14b. From Murray and Kuivala (1990) for hemipelagic and red clay sediments in water depths between 4,211 and 5,668 m. The second value has been recalculated assuming that ox equal –0.7, and the value in parentheses is the original value in this paper, which was calculated assuming that ox equals 0.
    15. Note that there is some degree of overlap in the various equatorial Pacific sites discussed here (see the original references for additional details).
    16. From Grundmanis and Murray (1982) for pelagic red clay sediments.
    17. Calculated with data presented in Jahnke et al. (1982) assuming ox equal –0.7 (ox = 0 value in parentheses).
    18. Calculated with nitrate and oxygen fluxes reported in Berelson et al. (1990) assuming ox = –0.7 (ox = 0 value in parentheses).
    19. Calculated with oxygen microelectrode and pore-water nitrate data (Reimers et al., 1984; Emerson et al., 1985) assuming that the slope of the best-fit line obtained by plotting $\Delta[O_2]$ vs. $\Delta[NO_3^-]$ is ~10. As above the value in parentheses assumes ox = 0 while the other values assumes ox = –0.7.
    20. From Martin et al. (1991). The values in parentheses are the original values reported in this paper assuming an ox value of 0. The other values have been recalculated with their results assuming ox = –0.7.
    21. Recalculated with data from Bender et al. (1985/86) assuming that the carbon in the sediment organic matter has an average oxidation state of –0.7 or 0 (the latter in parentheses).
    22. From Hammond et al. (1996) for sites from ~105°W to 140°E between 2°S and 2°N. With their benthic flux results these workers estimated that $\Delta O_2:\Delta \Sigma CO_2$ equals 1.45, which implies that ox equals –0.21 (see table 9.1 for additional details).
    23. From de Lange (1992).

fer and the oxidation of C_{red} to CO_2 requires $(4 - ox)$ electrons, redox balance implies that,

$$\mathcal{L} = (4 - ox)/8 \tag{11.5}$$

or

$$ox = 4 - 8\mathcal{L} \tag{11.6}$$

Results from such calculations using pore-water data from a site in the mesohaline portion of Chesapeake Bay are listed in table 9.1, based on plots such as that shown in fig. 11.2.

At the other extreme, for sediments in which organic matter undergoes remineralization by aerobic respiration (i.e., many deep-sea sediments) we can similarly determine ox and the C/N remineralization ratio using this same approach. Again we start with the following modified form of the RKR equation

$$(C_{red})_x(NH_3)_y(H_3PO_4)_z + w\, O_2 \rightarrow x\, CO_2 + y\, HNO_3 + z\, H_3PO_4 \tag{11.7}$$

where w depends on the oxidation state of C_{red} (ox) according to

$$w = \left(1 - \frac{ox}{4}\right)x + 2y \tag{11.8}$$

and the C/N remineralization ratio ($r_n = x/y$) is then given by

$$r_n = \frac{x}{y} = \frac{(y/w)^{-1} - 2}{1 - (ox/4)} \tag{11.9}$$

Again since plots of pore-water nitrate and O_2 are often linear in such sediments (e.g., Martin et al., 1991), the same general approach outlined above for anoxic sediments can be used here, yielding

$$\frac{dC_{nit}}{dC_{oxy}} = -\frac{y}{w}\frac{D_{oxy}}{D_{nit}} \tag{11.10}$$

where subscripts "nit" and "oxy" represent nitrate and oxygen. Thus to determine the C:N remineralization ratio (r_n or x/y), one first uses the best-fit slope of a plot of pore-water nitrate versus oxygen (dC_{nit}/dC_{oxy}) and values of D_{oxy}/D_{nit} to determine y/w with eqn. (11.10). This value, along with an estimate of ox (see below) is then used in eqn. (11.9) to determine the C/N remineralization ratio. Results from this type of calculation are also presented in table 11.2.

Values of ox are either estimated from the literature, i.e., such as in table 9.1, or determined along with the C/N remineralization ratio using ΣCO_2 pore-water or benthic flux data in conjunction with analogous oxygen and nitrate data. Assuming that there is no calcium carbonate dissolution in the sediments (see the next section for the situation when this is not the case), then one can define the following additional equation from property-property plots of pore-water ΣCO_2 versus O_2:

$$\frac{dC_{tc}}{dC_{oxy}} = -\frac{x}{w}\frac{D_{oxy}}{D_{tc}} \qquad (11.11)$$

where tc represents ΣCO_2. Thus with the addition of this second property-property plot, eqns. (11.7)–(11.11) can be used to determine the ratios y/w and x/w. The C/N remineralization ratio ($= x/y$) is then given by $(x/w)/(y/w)$, while ox is obtained by rewriting eqn. (11.8) as

$$ox = \frac{8(y/w) - 4}{(x/w)} + 4 \qquad (11.12)$$

11.2.2 Benthic Flux and Sediment POM Stoichiometric Models for Nutrient Regeneration

Ratios of benthic fluxes or depth-integrated remineralization rates can similarly be used to determine elemental remineralization ratios, i.e., to directly determine values of the ratios x/w or y/w. Furthermore, with this approach, the contributions from both aerobic respiration and sediment carbonate dissolution to the ΣCO_2 flux can be separated from one another in deep-sea sediments where they co-occur.

This approach takes into account the fact that the alkalinity flux from oxic, deep-sea sediments results from production by carbonate dissolution and consumption by nitrate production during aerobic respiration/nitrificiation. Therefore alkalinity production due to calcium carbonate dissolution alone equals $(J_{Alk} + J_{NO_3})$, where positive fluxes (J values) are *out* of the sediments. If this corrected alkalinity flux is expressed using meq units (e.g., meq m^{-2} d^{-1}) then the depth-integrated rate of sediment carbon oxidation (R_{cox}), or that portion of the ΣCO_2 flux due to respiration, is equal to

$$R_{cox} = J_{\Sigma CO_2} - 0.5(J_{Alk} + J_{NO_3}) \qquad (11.13)$$

(Hammond et al., 1996). Also note that a similar approach can be used with data from anoxic sediments to separate out sediment carbon oxidation via sulfate reduction from carbonate dissolution (Berelson et al., 1996).

Stoichiometric models can also be applied to sediment TOC and TN data to estimate the C/N remineralization ratio. Examining eqn. (6.34) and (6.49), or eqn. (8.4) and its equivalent for TN, it can be shown that

$$\frac{dN}{dG} = \frac{\alpha N_o}{\alpha G_o} \text{ or } \frac{\alpha(N_o - N_{nr})}{\alpha(G_o - G_{nr})} = \frac{N_m}{G_m} = \frac{1}{r_n} \qquad (11.14)$$

Thus plots of TN versus TOC will have a slope equal to the inverse of the C:N remineralization ratio, again assuming that sediment inorganic nitrogen is either a minor or a near constant fraction of the total sediment nitrogen.

11.2.3 The Composition of Organic Matter undergoing Remineralization: Elemental Ratios and Stable Isotopic Composition

Before examining the results in table 11.2 it is important to note that many of the approaches used here to derive these remineralization ratios either explicitly assume that there is only one type of organic matter with a single C:N ratio undergoing remineralization (e.g., eqn. 11.14), or make this assumption implicitly (e.g., eqns. 11.3 or 11.9). Thus these approaches give values that essentially average over all of the organic matter undergoing remineralization in the region of early diagenesis.

However, when this problem is approached using analytical or experimental techniques that do not smooth out these differences, it is seen that these remineralization ratios often increase with sediment depth (Aller and Yingst, 1980; Blackburn, 1981; Burdige, 1991;

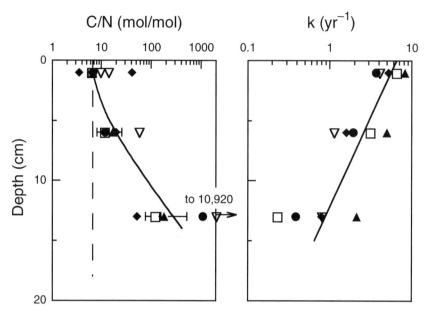

Figure 11.3 The depth-dependence of the C/N remineralization ratio and the first-order organic matter remineralization rate constant for sediments at five sites in southern Chesapeake Bay and tributaries. These results were obtained using long-term anoxic sediment decomposition experiments (redrawn using results presented in Burdige, 1991).

also see fig. 11.3). Since most of these studies were carried out in coastal sediments, one interpretation of these results is that they provide evidence for the increasing importance of terrestrially derived organic matter undergoing remineralization with sediment depth, presumably because more reactive marine organic matter is preferentially remineralized at shallower sediment depths. Such an interpretation is consistent with suggestions that terrestrially derived organic matter is, in general, less reactive than marine-derived organic matter (Cowie et al., 1992; Aller et al., 1996; Prahl et al., 1997; Aller and Blair, 2004). However, another possible explanation, selective utilization of more nitrogen-rich than carbon-rich marine organic matter cannot be ruled out based on remineralization ratio profiles such as those shown in fig. 11.3 (Hansen and Blackburn, 1991; also see data in Ganeshram et al., 1999).

Nevertheless, the results in table 11.2 indicate that, broadly speaking, the POM undergoing remineralization in most marine sediments looks largely like marine organic matter based on its C/N remineralization ratio. In sediments where iron-phosphate interactions do not complicate the interpretation of phosphate benthic fluxes (see section 16.3), sediment C/P remineralization ratios based on benthic fluxes are also generally similar to those for marine organic matter (Klump and Martens, 1987; Hammond et al., 1996; McManus et al., 1997).

Conversely, there are exceptions that can be understood based on the occurrence of well-constrained multiple sources of sediment POM (Skan Bay; see below), or by sediment grain size differences and their effect on sediment C/N ratios—for example, more fine-grained sediments in the mesohaline Chesapeake Bay versus coarser-grained sediments in southern Bay sediments (see the discussion above and in Keil et al., 1994b; Burdige, 2001). Furthermore, an examination of table 11.1 also suggests that based on C/N ratios alone, we cannot rule out the possible contribution of soil organic matter remineralization contributing to these observed C/N remineralization ratios, particularly in coastal or deltaic sediments.

An approach that has the potential to resolve some of this ambiguity involves using pore-water stable isotopes, primarily δ^{13}C of the pore-water ΣCO_2. In Skan Bay sediments this approach was used to determine the relative importance of phytoplankton- and kelp-derived organic matter to the pool of sediment organic matter undergoing remineralization (Alperin et al., 1992b). In Cape Lookout Bight sediments, δ^{13}C analyses of pore-water ΣCO_2 and sediment POC coupled with organic geochemical studies suggest that the vast majority, greater than ~90%, of the organic matter undergoing remineralization is of marine origin (Martens et al., 1992; Boehme et al., 1996; also see related discussions in the next section). In contrast, in Buzzards Bay sediments McNichol et al. (1991) observed that the δ^{13}C of pore-water ΣCO_2 was significantly heavier than the sediment POM (−6 to −15.2‰ vs. −20.6‰). These authors suggest that fractionation during organic matter oxidation or isotope exchange with bottom waters might explain these observations. However, it also seems possible that remineralization of organic matter from isotopically heavy C_4 plants (table 11.1) could play a role in explaining these observations.

In a similar fashion pore-water stable isotopes have been used to examine the sources of organic matter undergoing remineralization in continental margin sediments such as those in the southern California Borderlands (Presley and Kaplan, 1968; Bauer et al., 1995), the central California continental margin (Jahnke et al., 1997), and the North Carolina continental slope (Blair et al., 1994). In these sediments, which are far removed from major river inputs, and where C/N remineralization ratios are consistent with marine values, these isotope studies suggest that there is an organic matter source of ΣCO_2 to the pore waters with a $\delta^{13}C$ value of around −20‰, again consistent with the remineralization of marine-derived organic matter.

In continental margin sediments near large river inputs, e.g., on the Washington continental margin, one often observes C/N remineralization ratios elevated over the range of values for marine organic matter. Such observations are consistent with previous discussions that have shown significant amounts of terrestrial organic matter, including relatively coarse-grained, high C/N ratio material, are deposited in these sediments, and are also apparently remineralized. However, C/N remineralization ratios on the Amazon continental shelf are not as elevated as those seen on the Washington shelf, although mass balance calculations using stable isotope measurements suggest that ∼30% of the total sediment POM remineralized on the Amazon shelf is of terrestrial origin (Aller et al., 1996; see similar results for deltaic sediments in the Gulf of Papua, New Guinea; Aller and Blair, 2004). Much of this terrestrial organic matter is also presumably derived from soils, thus accounting for the observed C/N remineralization ratio (compare tables 11.1 and 11.2, and also see data on Amazon river suspended POM in Hedges et al., 1986). The remineralization efficiency of terrestrial organic matter in Amazon sediments is also quite high (∼60–70%). This efficient remineralization of terrestrial organic matter in muddy tropical sediments appears to be the result of the coupling of physical and biogeochemical sediment processes that were discussed in part in sections 5.5 and 7.6.3. Additional information about the remineralization of terrestrial organic matter in marine sediments is discussed in the next section (11.2.4) and in section 15.5.

For deep-sea sediments, an examination of table 11.2 suggests that, in general, C/N remineralization ratios are within or slightly higher than the range in table 11.1 for marine organic matter. This is particularly true if one assumes that the sediment POM is slightly more reduced than traditional Redfield organic matter ($ox = -0.7$ vs. ∼0). Stable isotope studies of the pore-water ΣCO_2 pool also predict that

TABLE 11.3
The Organic Geochemical Composition of End-member Organic Matter Sources and of Organic Matter that Is Remineralized in Marine Sediments

Sediment Type/ Material	Protein	Carbohydrates	Lipids	Lignin	Unidentified
End-member sources					
Marine organic matter[b]	~50–60%	20–40%	5–30%	0%	
Vascular plant material[c]	~1–2%	~70%	~1–2%	~30%	
Remineralized sediment organic matter					
Cape Lookout Bight, NC sediments[d]	31%	16%	17%		36%
General ranges observed in other marine sediments[e]	~20–30%	5–15% (8 ± 1%)	5–15% (?)		>40%

a. All results are on a % carbon basis.
b. Data from a variety of sources (Libes, 1992; Martens et al., 1992; Hedges et al., 2001; Hedges et al., 2002).
c. From data in Hedges et al. (1997).
d. From Martens et al. (1992).
e. The amino acid results are based on table 9.3, while the carbohydrate results are from a tabulation in Burdige et al. (2000). The value in parentheses for carbohydrates is a best-fit value obtained from a log-log fit of particulate carbohydrate remineralization versus total remineralization. See discussions in section 9.7 for the source of the lipid results.

the POM undergoing remineralization in these sediments has a $\delta^{13}C$ value of around −20‰, again generally consistent with a marine source (McCorkle et al., 1985; Sayles and Curry, 1988; Bauer et al., 1995; Gehlen et al., 1999; Martin et al., 2000).

11.2.4 The Composition of Organic Matter Undergoing Remineralization: Organic Geochemical Composition

An alternate approach to examining the composition of organic matter undergoing remineralization in sediments involves the use of organic geochemical data. As discussed in section 8.3 such information is generally obtained by modeling sediment profiles of organic constituents such as carbohydrates or amino acids. Although there is strikingly little data on this topic, the results shown in table 11.3 suggest a problem similar to that encountered when attempting to determine the molecular composition of total sediment POM (e.g., see table 9.6)—namely, that a large fraction of the remineralized organic matter cannot be identified using conventional analytical techniques. However, if we use C/N remineralization ratios and ox values obtained with inorganic data in conjunction with organic geochemical data, we can place at least some constraints on the lipid, carbohydrate, and amino acid composition of this unidentified material. This then allows us to examine the total composition of the sediment organic matter that is being remineralized, characterized plus uncharacterized.

The results of such calculations are shown in table 11.4. For Cape Lookout Bight sediments it can be seen that the composition of remineralized organic matter in these sediments is ~50% protein and ~25% each lipid and carbohydrate. As compared with results in table 11.3, this composition is broadly consistent with that of marine plankton, thus reinforcing arguments presented in Martens et al. (1992) that marine-derived organic matter dominates the material undergoing remineralization in Cape Lookout Bight sediments.

Similarly, an examination of these calculations for the two hypothetical marine sediments also suggests that the composition of the total organic matter undergoing remineralization is broadly consistent with marine organic matter dominating remineralization. At the same time, these calculations hint at some selective utilization of amino acids, i.e., organic nitrogen, in this material prior to its deposition in sediments (e.g., see discussions in section 11.2). Finally, the high amino acid content and relatively high lipid content strongly argue against a significant terrestrial source (i.e., compare tables 11.3 and 11.4).

The results of calculations presented here suggest that much of the organic matter remineralized in these marine sediments appears to be (roughly speaking) marine organic matter. Nevertheless, it is not in conflict with the observation that most terrestrial organic matter transported to the oceans ultimately is remineralized in the oceans (e.g., Hedges et al., 1997).

First, these calculations are based largely on results from studies of sediments in which terrestrial organic input is likely to be minimal. Second, most terrestrial (riverine) DOC (which is roughly half of the riverine organic carbon flux to the oceans) is presumably remineralized somewhere in the water column (Hedges, 1992). Similarly, terrestrial POC (in riverine suspended particles) may also be remineralized in the water column prior to its deposition in sediments

COMPOSITION OF ORGANIC MATTER

TABLE 11.4

Constraints on the Composition of the Unidentified Component of Organic Matter Undergoing Remineralization in Marine Sediments

	Cape Lookout Bight sediments[a]	"Typical" Marine Sediments[b] I	II
	Fraction of remineralized TOC		
Characterized Material (C_i)			
Amino acids (A)	0.31	0.20	0.30
Sugars (S)	0.16	0.05	0.15
Lipids (L)	0.17	0.05	0.15
Unidentified	0.36	0.7	0.4
Directly measured C/N remineralization ratio (MR)	6.6–7.2	~7–10	~7–10
Uncharacterized Material (U_i)			
Amino acids[c]	0.21–0.25	0.13–0.27	0.03–0.17
Sugars[d]	0.04–0.12	0.19–0.50	0.09–0.40
Lipids[d]	0.02–0.09	0.03–0.28	−0.07–0.18
Total Remineralized Material (= $C_i + U_i$)			
Amino acids	0.52–0.56	0.33–0.47	0.33–0.47
Sugars	0.20–0.28	0.24–0.55	0.24–0.55
Lipids	0.19–0.20	0.08–0.33	0.08–0.33

a. The organic geochemical data is from Martens et al. (1992). The range in measured C/N remineralization ratios is from table 11.2.

b. The organic geochemical data is based on results in table 11.3. The range in measured C/N remineralization ratios is based on results shown in table 11.2.

c. Since carbohydrates and lipids effectively contain no nitrogen, all nitrogen in the remineralized organic matter is assumed to come from the amino acids. Values of U_A (the carbon fraction of uncharacterized material that is amino acids) were determined using the equation

$$MR = \frac{C_A + U_A + C_S + U_S + C_L + U_L}{(C_A + U_A)/R_A} = \frac{1}{(C_A + U_A)/R_A} \quad \text{(see the second equation in note d)}$$

or $\quad U_A = \dfrac{R_A}{MR} - C_A$

where the various Cs are listed above, MR is the directly measured C/N remineralization ratio (listed above) and R_A is the C/N ratio of typical sediment amino acids. In this calculation, R_A was taken to be 3.7 in Cape Lookout Bight sediments (Burdige and Martens, 1988) and 3.3 in other marine sediments (based on results presented in Keil et al., 2000).

(also see Keil et al., 1997; McKee et al., 2004). At the same time, the continental margin sediments that do receive large inputs of terrigenous material, e.g., the Amazon shelf and Washington continental margin, and are very efficient at remineralizing terrestrial organic matter, are also even more efficient at remineralizing marine organic matter (fig. 15.1; Aller, 1998; Aller et al., 2004; McKee et al., 2004). As a result, marine organic matter often still dominates the total pool of sediment organic matter that is remineralized. Therefore in these cases, a terrestrial organic matter signal in the remineralized organic matter may be obscured by the marine signal. Furthermore, as the Amazon margin results indicate, some of the ambiguities of the end-member compositions of marine versus terrestrial organic matter can make such source determinations equivocal when they are based solely on the interpretation of parameters such as the C/N remineralization ratio.

Before leaving this topic it is interesting to look at these results from another standpoint. The sediments listed in table 11.2 have carbon oxidation rates that vary by almost 3 orders of magnitude (a thousand-fold range), yet have TOC values that vary by a factor of only ~10–20. A similar large range in the reactivity of organic matter in such sediments can also be inferred from fig. 8.1 At the same time, the composition of the remineralized organic matter exhibits little variation over this range in remineralization rates, based on either C/N

---

d. These values are based on a calculation that partitions the oxidation state of the remineralized carbon into these three biochemical pools, using the equations

$$ox = (C_A + U_A)X_A + (C_S + U_S)X_S + (C_L + U_L)X_L$$

$$C_A + U_A + C_S + U_S + C_L + U_L = 1$$

where ox, the average oxidation state of the organic matter undergoing remineralization was taken to be –0.7 in Cape Lookout Bight sediments and assumed to range between –0.4 and –0.7 for the two "typical" marine sediments (see table 9.1). Values of U_A were taken from above, as were the three C_i values. The average oxidation states of the amino acid, sugar, and lipid fractions (the various Xs) were assumed to be –0.48, 0, and –1.67 to –2.1 (see note g in table 9.1). These values were then substituted into the two equations above to solve for U_s and U_L. Note that negative values of U_L imply that the assumed amount of lipid in the characterized, remineralized material is greater than that which can be found in the total remineralized material (based on the other parameters used here to describe this material).

remineralization ratios (table 11.2), the oxidation state of the carbon undergoing remineralization (table 9.1), the (identified) carbohydrate content of the organic matter undergoing remineralization (table 11.4), or the (identified) amino acid content of the organic matter undergoing remineralization (tables 11.4 and 9.3).

A possible explanation for this remineralization constancy (Burdige et al., 2000) begins with the observation that as sediment carbon oxidation rates decrease from shallow water to deep-sea sediments, the mode of remineralization changes from almost exclusively anoxic to almost completely oxic. Furthermore, during the early stages of organic matter diagenetic maturity (as defined in section 9.1) reactive (marine) organic matter is apparently utilized equally well under oxic or anoxic conditions (see discussions in section 15.2 and in Cowie and Hedges, 1992b; Cowie et al., 1995). Given these circumstances, it is possible that there is preferential utilization of this reactive material during the early stages of diagenetic maturity that then leaves behind inherently refractory (terrestrial ?) material. However, this possibility would appear to be inconsistent with the remineralization constancy noted above.

Alternately, it is possible that during diagenetic maturity some inherently reactive material escapes remineralization by one of the mechanisms discussed in section 9.8. In this case, if this reactive organic matter is protected from degradation by an organic coating or matrix (encapsulation?) then as organic matter degradation becomes increasingly more oxic with increasing organic matter diagenetic maturity, this would suggest that this protective material can be decomposed only by O_2, or by activated oxygen species such as the hydroxyl radical (Cowie et al., 1995; Hedges and Keil, 1995; also see section 7.6.1). An analogous oxygen effect is also expected if this organic matter is tightly bound to mineral surfaces inside of small mesopores either on these surfaces or between mineral grains (see section 15.1 for further details).

Finally, organic matter associated with calcium carbonate minerals (see section 9.8.5) may also show enhanced degradation under oxic conditions in, for example, deep-sea carbonate oozes. Aerobic respiration produces metabolic CO_2, which dissolves carbonate minerals (see section 13.5) and perhaps then makes mineral-associated organic matter more available for decomposition (e.g., Ingalls et al., 2004). Furthermore, if oxygen input to such sediments does not limit remineralization (see section 13.1), this process also has the potential to operate in a positive feedback mode, since the degradation of carbonate mineral–associated organic matter produces additional CO_2 that then further enhances carbonate dissolution and makes additional mineral-associated organic matter available for decomposition.

≈ CHAPTER TWELVE ≈

Processes at the Sediment-Water Interface

THE TRANSPORT OF DISSOLVED constituents across the sediment-water interface, the benthic flux, links sediment and water column processes. The processes responsible for this transport include simple molecular diffusion, physical processes such as sediment resuspension and water advection, as well as more complex biologically mediated processes such as bioturbation and bioirrigation.

In almost all sediments that are bioturbated, this process is more important than sedimentation as a transport mechanism of material in the upper sediments (see section 4.4 and table 4.1). In deep-sea sediments bioturbation can, therefore, significantly impact the rates and depth-distribution of early diagenetic processes such as organic matter remineralization and silica and carbonate dissolution. In continental margin sediments benthic macrofaunal processes (both bioturbation and bioirrigation) also alter the diagenetic pathways occurring in these sediments, in part because of their ability to bring together reduced species (e.g., Mn^{2+}, NH_4^+, or HS^-) and oxidized species (e.g., O_2, nitrate, or metal oxides) that are found in the sediments (see sections 7.6.3 and 12.5).

Mixing processes such as bioturbation can also "smear" or attenuate (i.e., spread out) downcore variations in sediment depth profiles (e.g., sediment lamination or varves); in some cases these profiles will be completely obliterated by bioturbation. On some time scales this may then limit the paleoceanographic information preserved in such sediment records (e.g., Wheatcroft et al., 1990; Boudreau, 1994; and references therein). In contrast, though, other studies have shown how the presence/absence of bioturbation/sediment lamination may actually be used as a paleoceanographic tracer (Behl and Kennett, 1996; Stott et al., 2000).

In this chapter we will discuss the occurrence of these processes, along with the mathematical models that have been developed to describe them.

12.1 The Determination of Benthic Fluxes

Interest in the quantification of benthic fluxes stems in part from the fact that under steady-state conditions benthic fluxes are equal to depth-integrated remineralization rates (see section 8.3.1). Generally, the total benthic flux (J_{tot}) is the sum of the fluxes due to molecular diffusion (J_{dif}), bioturbation (J_{biot}), bioirrigation (J_{irr}), pore-water advection (J_{adv}), and sediment resuspension or redeposition (J_{rsp}),

$$J_{tot} = J_{dif} + J_{biot} + J_{irr} + J_{adv} + J_{rsp} \qquad (12.1)$$

In principle, pore-water concentration gradients can often be used to estimate either J_{tot} or some of its individual components, assuming that one is able to adequately model and quantify the transport processes affecting pore-water constituents near the sediment surface. However, this latter caveat is not necessarily a trivial one; significant errors in estimated benthic fluxes can result if these transport processes are not properly calculated.

It is also possible in many sediment systems to directly determine the total benthic flux. In the simplest sense, this involves isolating some fixed amount of bottom water over a portion of the sediment surface (either in situ or in a collected sediment core), and examining the change in concentration over time in this water to determine the benthic flux into or out of the sediments (fig. 12.1). Over the years, a variety of techniques have been used in such benthic flux studies; these include the incubation of intact sediment cores, diver-emplacement (and sampling) of in situ benthic chambers, and the in situ determination of benthic fluxes using autonomous, unmanned benthic landers. Additional details about these techniques and their use in different sediment settings can be found in several reviews (Devol, 1987; Tengberg et al., 1995; Reimers et al., 2001). In principle, the direct determination of benthic fluxes using these approaches has the potential to yield more reliable results than fluxes calculated using pore-water gradients. However, this is not always the case, as will be seen, for example, in section 12.3 in the discussion of diagenetic processes in permeable sediments.

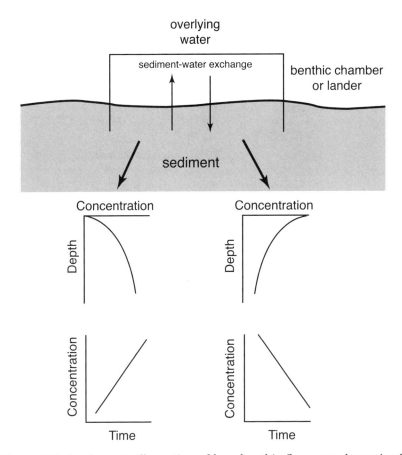

Figure 12.1 A schematic illustration of how benthic fluxes are determined. Note that the example shown here is for a sediment system in which diffusion dominates sediment-water exchange, although similar procedures can also be used to determine benthic fluxes from sediments in which bioturbation and/or bioirrigation occur. By isolating bottom water over a known surface area of sediment, the observed pore-water gradients will lead to changes with time in the isolated bottom waters. The change in concentration versus time is then used to calculate the flux across the sediment-water interface after corrections are made for the geometry of the chamber and for in situ production or consumption of the solute of interest in the bottom waters (see discussions in the text and in Berelson and Hammond, 1986; Devol, 1987; Bender et al., 1989).

12.2 Diffusive Transport and the Benthic Boundary Layer

In the absence of any transport processes other than molecular diffusion and pore-water advection, the flux of a dissolved constituent across the sediment-water interface is given by

$$J_{tot} = -\varphi_o D_s \left(\frac{dC}{dz}\right)_o + \varphi_o v_o C_o \approx J_{diff} = -\varphi_o D_s \left(\frac{dC}{dz}\right)_o \quad (12.2)$$

As was shown in sections 6.1 and 8.2.1, the advective term is generally neglected in this calculation because it is usually very small in comparison to the diffusive term. The interfacial concentration gradient $(dC/dz)_o$ is commonly obtained either from the first derivative of a best-fit curve through the data, or by the approximation $\Delta C/\Delta z$, where ΔC is the difference between the bottom water concentration and that in the first pore-water sample collected and Δz is the depth of this pore-water sample (e.g., see Burdige et al., 1999). Such estimations, however, are subject to some uncertainty because they may not accurately describe pore-water concentration gradients close to the sediment-water interface (see, for example, discussions in Burdige, 2001; Berelson et al., 2003). Such issues can however be addressed by examining this interfacial gradient with high resolution pore-water profiles obtained with either microelectrodes (e.g., Archer and Devol, 1992; Jørgensen, 2001; Wenzhöfer and Glud, 2002), fiber optic microsensors (or "optodes"; Kühl and Revsbeck, 2001), or a whole-core squeezer (e.g., Martin et al., 1991; Hammond et al., 1996).

Often implicit in such flux calculations is the assumption that the water column immediately above the sediment surface is well mixed and homogeneous. However, for many situations this may not be the case. Here, because of steep pore-water gradients near the sediment-water interface, these gradients may extend into what is generally referred to as the diffusive sublayer of the sediment (see below). In these cases fluxes across the sediment-water interface, and therefore rates of sediment processes, can become diffusion, or transport, controlled.

The occurrence of a diffusive sublayer just above the sediment surface can be explained by the physics of flow across the surface (see Boudreau, 1997a, or Boudreau and Jøregensen, 2001). Assuming a flat, immobile surface, flow over this surface creates a velocity profile in

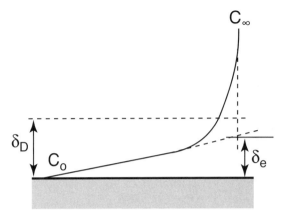

Figure 12.2 An idealized concentration profile in the region of the diffusive sublayer, illustrating the true thickness of the diffusive sublayer (δ_D) and the effective sublayer thickness (δ_e) that produces the same linear gradient at the sediment-water interface (from Boudreau, 1997a, and used with permission of the author).

the overlying waters such that all velocities go to zero at the water-surface interface. This is termed the "no-slip" condition, and is caused by friction between the fluid and this boundary. If the flow over this surface is turbulent in nature, then as one moves closer to the surface this boundary condition also leads to an analogous gradient in eddy diffusivity. At some point the decrease in eddy diffusivity leads to the kinematic viscosity of water becoming more important than eddy diffusivity in transporting momentum through this fluid. The region of flow above the sediment surface where this occurs is generally referred to as the viscous sublayer.

As one continues to move through the viscous sublayer toward the sediment surface, the decrease in the intensity of eddy viscosity leads to a situation in which molecular diffusivity eventually becomes more important than eddy diffusivity. This region where molecular diffusion dominates mass transport in the fluid is referred to as the *diffusive sublayer* or *diffusive boundary layer* (DBL; see fig. 12.2). Within most of the DBL dissolved concentration gradients are linear, consistent with the assumption that there is no net advection or chemical reaction in this region. A variety of experimental techniques, including fine-scale oxygen microelectrode profiles near the sediment-water interface, have been used to verify the occurrence of the DBL and quantify its thickness (e.g., Jørgensen and Revsbech, 1985; Jørgensen, 2001).

In quantifying transport through the DBL, what is generally defined is the effective diffusive sublayer (δ_e in fig. 12.2). This is the region of the diffusive sublayer in which there is a linear gradient that predicts the same flux as that through the real diffusive sublayer of thickness δ_d (Jørgensen and Revsbech, 1985). Given this linear gradient through δ_e, the flux through this portion of the DBL is

$$J_D = \frac{D}{\delta_e}(C_o - C_\infty) = \beta(C_o - C_\infty) \quad (12.3)$$

where C_o is the concentration at the sediment-water interface, C_∞ is that in the overlying region of turbulent flow, and β (= D/δ_e) is often defined as the mass transfer coefficient. As discussed in section 6.2.1, in some cases pore-water diagenetic models use the flux through the DBL as the upper boundary condition. From a practical standpoint, this is equivalent to a coupled diagenetic model (see section 8.4) in which one links a sediment pore-water model to the model shown in fig. 12.2, i.e., diffusive transport through the DBL. In this approach, C_o is now a fitting parameter derived from the boundary condition that requires the pore-water solution to predict a flux at $z = 0$ equal to that predicted by eqn. 12.3 (see table 12.1 for additional details).

Values of δ_e for marine sediments determined *in situ* range from ~300–600 μm at shallow water sites to 500–1,500 μm at continental margin and deep-sea sites (Archer et al., 1989a; Santschi et al., 1991; Jørgensen, 2001). The possible increase in δ_e with water column depth may be the result of lower mean current velocities in the deep sea versus that in shallower waters (since in general δ_e varies inversely with current velocities; see the discussion below). However, as Jørgensen (2001) notes, the database on which these conclusions are based is rather "slim." For comparison, values of the thickness of the viscous sublayer are ~10 greater than that of the diffusive sublayer (Boudreau, 1997a).

Several empirical relationships have been developed that relate β, and therefore δ_e, and u_*, the friction (or shear) velocity of a surface (see discussions in Boudreau, 1997a). One of these (the Shaw and Hanratty equation) is given by

$$\beta = 0.0889 u_* Sc^{-0.704} \quad (12.4)$$

TABLE 12.1
The Effect of the Diffusive Boundary Layer on Sediment Diagenesis

Consider the simple diagenetic model

$$D_s \frac{d^2C}{dz^2} + R = 0$$

for the following three pore water constituents, subject to the following boundary conditions:

1. $C = C_\infty$ at and above the diffusive boundary layer (i.e., above $z = -\delta_e$; see fig. 12.2)
2. $C = C_o$ at $z = 0$ (the sediment-water interface)
3. The flux across the sediment-water interface (J at $z = 0$) is given by $-\varphi_o D_s \dfrac{dC}{dz}\bigg|_{x=0}$

and is equal to the flux through the diffusive boundary layer $\dfrac{D^o}{\delta_e}(C_o - C_\infty)$

4. For ammonium and silica, $dC/dz \to 0$ as $z \to \infty$ (i.e., at depth in the sediments).
5. For oxygen at the lower boundary $z = L$ (the oxygen penetration depth) $C = 0$ and $dC/dz = 0$. Although this appears to involve one too many boundary conditions for the oxygen solution, in actuality this approach allows L to be calculated as a part of the solution (e.g., Hall et al., 1989; Jørgensen and Boudreau, 2001).

	Oxygen	Silica	Ammonium
	$R = -R_o$	$k(C_s - C)$	$R_o e^{-\alpha z}$
	$C = C_0 + \left(\dfrac{R_o}{2D_s}\right)z^2 - \left(\dfrac{2C_o R_o}{D_s}\right)^{1/2} z$ $(0 < z < L)$	$C_S + (C_o - C_S)e^{-\beta z}$	$C_0 + \dfrac{R_o}{D_s \alpha^2}(1 - e^{-\alpha z})$
	$C_0 = C_\infty + \gamma[1 - \sqrt{1 + (2C_\infty/\gamma)}]$	$\dfrac{D\beta\delta_e C_s + C_\infty}{1 + D\beta\delta_e}$	$C_\infty + \dfrac{\varphi_s R_o \delta_e}{\alpha D^o}$
	$J = \varphi_o(2C_o R_o D_s)^{1/2}$ $(\equiv \varphi_o L R_o)$	$\dfrac{-\varphi_o D_s \beta(C_S - C_\infty)}{1 + D\beta\delta_e}$	$\dfrac{-\varphi_o R_o}{\alpha}$
	$L = \sqrt{2C_o D_s/R_o}$		

where $\gamma = \varphi_o^2 R_o D_s(\delta_e/D^o)^2$, $\beta = \sqrt{k/D_s}$, $D = \varphi_o D_s/D^o = \varphi_o^3$ (which is dimensionless here, assuming D_s and D^o are related by eqns. 6.11 and 6.12 with n = 3), and J is positive downward.

277

where Sc, the Schmidt number, is ~1,000 under most circumstances (Boudreau, 1997a). While other equations that relate β and u_* have been developed, predictions of β from all of these equations are quite similar (Boudreau, 1997a). Values of u_* can either be directly determined or estimated with mean velocity measurements made some distance above the sea floor, e.g., at a height of 1 m (see Santschi et al., 1991, for a summary of how these estimates are made). Thus in principle, one can also estimate the thickness of the DBL using current velocity measurements and these empirical relationships. Comparisons of such estimates of DBL thickness with direct determinations of the DBL generally agree to within ~30% or less (Archer et al., 1989a; Santschi et al., 1991; Jørgensen, 2001).

The effect of the DBL on sediment geochemical processes is illustrated in table 12.1 and fig. 12.3 for sediment processes with distinctly different kinetics. The first, sediment oxygen uptake, is a situation in which the water column represents a source to the sediments of the solute of interest. Uptake of Mn^{2+} by manganese nodules (see section 13.3 and Boudreau and Scott, 1978; Santschi et al., 1983) represents another case in which the water column represents a source of reactants for sediment processes, and therefore has the potential to be similarly impacted by transport through the DBL. In the calculations presented here sediment oxygen consumption is assumed to be a zero-order process, independent of oxygen concentration. This kinetic expression is essentially consistent with the assumption that much of this oxygen consumption occurs by aerobic respiration (e.g., Hall et al., 1989; Jørgensen and Boudreau, 2001).

The second case, silica dissolution, represents one in which the sediments are a source of the solute, and its production in the sediments has a first-order dependence on pore-water concentration (assuming that particulate biogenic silica is in excess and therefore not rate limiting). This case may also be representative of calcite dissolution in sediments, since the kinetics of carbonate dissolution may show a first-order dependence similar to that listed in table 12.1 for silica dissolution (see discussions in Hales and Emerson, 1997b, and section 13.5). Finally, the third case, ammonium production, represents a situation in which sediment production is assumed to be zero-order with respect to, or independent of, pore-water ammonium concentrations. However, unlike oxygen consumption, ammonium production has a proscribed depth dependence.

The effect of the DBL on sediment oxygen uptake is examined here over a range of consumption rates that span those observed in deep-sea to coastal sediments (e.g., Archer et al., 1989a; Hall et al., 1989; Glud et al., 1994; Berelson et al., 1996). At low rates of oxygen consumption, typical of deep-sea sediments, transport through the DBL has little impact on sediment oxygen uptake, as is also seen in field data (e.g., Archer et al., 1989a). Specifically, this can be seen in fig. 12.3 by the fact that the normalized sediment O_2 uptake varies by less than ~5% over this range of DBL values; the ratio C_o/C_∞ also deviates only slightly from 1, further indicating that diffusion through the DBL is able to keep up with sediment oxygen uptake. Although oxygen consumption is modeled here as a zero-order process, quantities that impact the oxygen penetration depth L (i.e., δ_e) do therefore impact the depth-integrated rate of oxygen consumption $\left(= \int_o^L R_o dz\right)$, and hence the magnitude of sediment oxygen uptake.

As rates of sediment oxygen consumption increase, transport through the DBL has an increasing impact on the magnitude of sediment oxygen uptake in these model calculations, and therefore may begin to effectively "control" sediment oxygen uptake. These results therefore suggest that the effect of transport through the DBL on sediment oxygen uptake might be most important in near-shore, organic-rich sediments (i.e., high R_o values) or quiescent sedimentary environments (i.e., large δ_e values). In such near-shore settings, low bottom-water oxygen concentrations can also similarly impact sediment oxygen uptake (e.g., Rasmussen and Jørgensen, 1992).

However, as Jørgensen and Boudreau (2001) discuss, sediment oxygen uptake in these environments is likely not controlled by transport through the DBL. This occurs because in near-shore sediments much of the sediment oxygen consumption occurs by the oxidation of reduced metabolites such as sulfide, ammonium, and Fe^{2+}, as opposed to aerobic respiration (see section 7.5.2 for further details). Since these former reactions are generally not zero-order processes (as sediment oxygen consumption has been modeled here), the solutions presented in table 12.1 and illustrated in fig. 12.3 do not accurately represent the relationship between the rates of this type of sediment oxygen consumption, DBL thickness, and sediment O_2 uptake. In fact, calculations presented by Jørgensen and Boudreau (2001) suggest that, for sediment oxygen consumption coupled to the oxidation of these re-

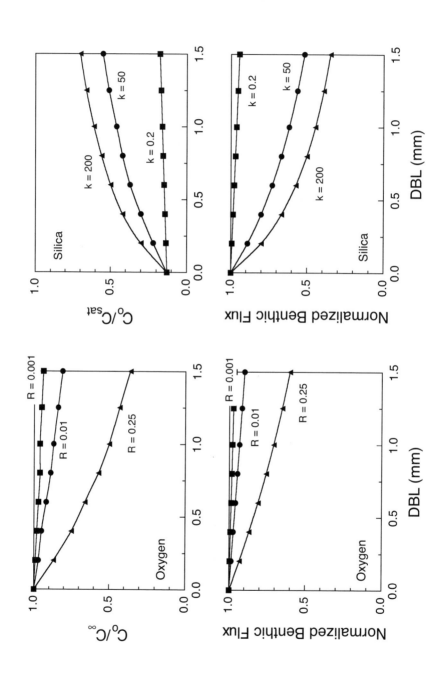

Figure 12.3 (Left) C_o/C_∞ and the normalized sediment oxygen uptake, both versus δ_e, from results obtained with the oxygen model described in table 12.1. For each R_o value, sediment oxygen uptake values have been normalized to the value with $\delta_e = 0$. The three values of R_o (0.001, 0.01, and 0.25 mmol m^{-3} sec^{-1}) span the range of oxygen consumption rates from deep-sea to coastal sediments. For example, with this model $R_o = 0.005$ mmol m^{-3} sec^{-1} predicts a sediment oxygen uptake of ~2 mmol m^{-2} d^{-1}, a value at the upper end of the range for deep-sea sediments. Similarly, with this model $R_o = 0.25$ mmol m^{-3} sec^{-1} predicts a sediment oxygen uptake of ~24–37 mmol m^{-2} d^{-1}, within the range of values observed in coastal sediments, and is now a strong function of δ_e. (Right) C_o/C_s and the normalized benthic silica uptake, both versus δ_e, from results obtained with the silica model described in table 12.1. For each k value the benthic silica fluxes have again been normalized to the value with $\delta_e = 0$. The lower value of k (0.2 d^{-1}) is based on data from a coastal sediment (Aller, 1980c); the other two values (50 and 200 d^{-1}) are used for comparative purposes (see the text for further details). The values of the other parameters used in these calculations are: $\varphi_o = 0.9$; $D^o(O_2) = 2.20 \times 10^{-5}$ cm^2 sec^{-1}; D^o (silica) $= 9.82 \times 10^{-6}$ cm^2 sec^{-1}; $\alpha = 0.1$ cm^{-1}; $C_\infty = 300$ μM (O_2), 74 μM (silica); $C_s = 577$ μM.

duced metabolites, sediment oxygen uptake is actually independent of DBL thickness, in a manner analogous to that discussed below for sediment ammonium production.

In the second example shown here (silica dissolution), the DBL has the potential to become a barrier toward the loss of dissolved silica from the sediments. This then can allow pore-water silica concentrations to build up, thereby decreasing the rate of sediment silica dissolution. This effect is, however, a function of the magnitude of k (or β) relative to δ_e. If $D\beta\delta_e \ll 1$, as is the case shown here for $k = 0.2$ d^{-1} (k value taken from Aller, 1980c, for a coastal sediment) then most silica production occurs sufficiently "far" from the sediment surface (i.e., C_o/C_{sat} is small in Fig. 12.3); transport through the DBL therefore has a relatively small impact on the solute flux (i.e., normalized benthic silica fluxes do not decrease significantly with increasing δ_e; see similar discussions in Jørgensen and Boudreau, 2001).

An alternate way to view this problem is to examine the relationship between the characteristic reaction time of silica production ($\sim 1/k$) and the time scale for diffusion through the DBL (given by eqn. 5.2). Here the characteristic reaction time (5 d) is long as compared to the time scale for diffusion through the DBL (less than ~ 0.02 d, depending on the magnitude of δ_e). Again this demonstrates that mass transport (diffusion) through the DBL is fast relative to dissolution, and therefore will have a minimal impact on the silica benthic flux (and hence the dissolution rate in the sediments).

To generalize these observations to other possible dissolution processes in sediments with similar kinetics, we can carry out the same analysis for larger values of k. As k increases in fig. 12.3 so does β, and this eventually leads to a situation in which $D\beta\delta_e \gg 1$. Now, most dissolution occurs quite close to the sediment-water interface, and there is a much greater build-up of this solute near the sediment-water interface (i.e., C_o/C_s increases and begins to approach 1). As a result, diffusive transport through the DBL now exerts a much greater relative impact on the benthic flux of this solute. If, for example, k is taken to be 200 d^{-1}, the characteristic reaction time is 0.005 d, which is short compared to the time scale for diffusion through the DBL. This analysis similarly shows that the benthic flux of this solute (and hence the rate of the dissolution process in the sediments in general) will be controlled by mass transport (diffusion) through the DBL.

Finally, for ammonium production, changes in the thickness of the DBL have only a very slight effect on pore-water ammonium concentrations (a few percent or less). More importantly, though, the benthic flux of ammonium is independent of the thickness of the DBL, as is shown in table 12.1. Since the rate of ammonium production is assumed here to be independent of pore-water ammonium concentrations and is simply a function of depth, any factors that affect pore water ammonium concentrations (i.e., changing δ_e values) have no effect on the sediment production rate, and therefore the benthic flux.

To further examine this, I note that C_o and δ_e are positively related and therefore covary such that any changes in either value cancel one another out in eqn. (12.3), to maintain a constant value of J_D. Thus if δ_e increases as a result of a drop in current velocity, then C_o similarly increases to maintain a constant flux. While these effects do lead to small changes in pore-water ammonium profiles, in a larger sense the shape of these profiles is actually much more strongly controlled simply by the shape of the reaction rate profile, i.e., the values of R_o and α.

These three cases illustrate some of the ways that transport through the DBL can impact the rates of sediment processes and benthic fluxes. Often times, though, the effects of transport through the DBL can be more complex than that described here. For example, in many situations benthic ammonium fluxes (and net sediment ammonium production) are not simply controlled by production in the anoxic portions of the sediments (as indicated above), but are also impacted by nitrification of this upwardly diffusing ammonium by downwardly transported oxygen (e.g., see section 16.2 and recent discussions in Gilbert et al., 2003). Thus here, if the thickness of the DBL affects oxygen uptake by the sediments (see above), then this will influence sediment nitrifications rates and hence the benthic ammonium flux (see discussions in Alperin, 2003). Jørgensen and Boudreau (2001) discuss other possible indirect effects that transport through the DBL may have on sediment processes. Interestingly, such considerations, along with how differences in zero- and first-order processes respond (in general) to changes in the thickness of the DBL, are analogous to those seen in some models of bioirrigation (see section 12.5 for further details).

12.3 Sediment-Water Exchange Processes in Permeable Sediments

In highly permeable sediments, i.e., sediments in which the average particle size is in the sand category, there is an additional component of pore-water transport near the sediment surface that must be considered. This advective component is driven by sediment pressure gradients (see section 4.3); as indicated by eqn. (4.6) this type of advection requires both high sediment permeability, expressed by the hydraulic conductivity parameter k, and a sustained pressure gradient within the sediments.

This pressure gradient may be generated in several ways. In shallow waters, pressure gradients near the seabed may be induced by wave action (i.e., the passage of gravity waves) because of slight differences in the hydrostatic pressure under wave crests versus troughs. The resulting increase in exchange across the sediment-water interface by this phenomenon is often referred to as the *subtidal pump* (see Shum and Sundby, 1996, and references therein). Equally important, though, are interactions between bottom currents and surface structures such as physical surface roughness, sand ripples, or biogenic structures, that also generate near-seabed pressure gradients in response to flow over these structures. Bottom water flow accelerates as it passes over an elevated structure, leading to a pressure drop (i.e., Bernoulli's principle) that draws bottom water and suspended particles into the sediments upstream and downstream of the physical structure; it also simultaneously forces pore water out of the sediments on the downstream slope of the structure (see discussions in Huettel and Rusch, 2000).

For typical bottom water flows and surface structures, laboratory flume studies suggest that bottom waters can be forced several centimeters into the sediment, simultaneously drawing pore water from depths > ~10 cm up to the sediment surface (Huettel et al., 1996). Field studies of these processes are generally consistent with these observations (e.g., Marinelli et al., 1998; Rasheed et al., 2004; Reimers et al., 2004), and the interaction between these pore-water advective processes and in situ remineralization processes can lead to complex three-dimensional pore-water gradients. As a result, pore water advection is similar to processes such as bioirrigation, which, unlike molecular diffusion or bioturbation, tends to enhance (or produce), rather than dissipate, concentration gradients (Boudreau et al., 2001b).

In addition to their effect on benthic fluxes of solutes, these advective processes can also transport small, suspended particles such as phytoplankton debris and bacteria into the sediments, where they can subsequently undergo remineralization. Evidence for the occurrence of this water column particulate input comes from both laboratory flume studies (Huettel and Rusch, 2000), as well as studies of ^{210}Pb distributions in nonaccumulating relict sands in mid-Atlantic Bight shelf sediments (Bacon et al., 1994).

As a result of these processes, permeable sediments act as efficient sand filters, analogous to those found in sewage treatment plants (Boudreau et al., 2001b). Rates of sediment oxygen uptake are greatly enhanced from increased oxygen input from bottom waters, as well as input of suspended organic matter from the water column (Forster et al., 1996; Ziebis et al., 1996; Huettel and Rusch, 2000). Although these sediments generally have very low organic carbon content, due, in part, to their nondepositional nature, the organic matter that is input into the sediments appears to be relatively reactive, as compared to that found in many fine-grained depositional sediments, with apparent first-order decay constants on the order of ~ 0.1–1 yr^{-1} (Boudreau et al., 2001b; for comparison see the rate constants discussed in section 8.1). This high degree of reactivity results in part because the input of reactive organic matter to these sediments is apparently not diluted by the accumulation of aged, i.e., less reactive, organic matter (Boudreau et al., 2001b). Therefore, the low standing stock of TOC in these sediments, and of other sediment nutrients in general, may be controlled more by rapid turnover rather than by low input.

Pore-water advection in permeable sediments also leads to a situation in which oxygenated bottom waters advected into sediments can come in contact with reduced pore waters (e.g., Huettel et al., 1998). This action potentially stimulates oxic and anoxic chemolithotrophic reactions such as those discussed in section 7.4. At the same time, since many permeable sands occur in shallow waters and therefore exist within the euphotic zone, benthic primary production at such sites can occur. Although estimates of this benthic primary production in shallow water sands are subject to large uncertainties (Boudreau et al., 2001b), in many cases they appear to be comparable to rates of water column primary productivity (Jahnke et al., 2000).

While results to date suggest that sandy, permeable sediments are more biogeochemically dynamic than previously thought, this conclusion is based largely on the extrapolation of results from laboratory flume studies, in part because of methodological difficulties in examining processes in permeable sediments in situ. Although it can be difficult to collect undisturbed sediment cores and pore-water samples from such sediments, some of these difficulties have been overcome in recent years (e.g., Marinelli et al., 1998; Burdige and Zimmerman,

2002). However, estimating benthic fluxes from such sediments using pore-water profiles can be extremely difficult in that it requires not only an accurate estimate of the in situ pore-water velocity, which is difficult to obtain (Reimers et al., 2001), but also information on the three-dimensional pore-water gradients that may exist in these sediments (Huettel et al., 1998) and appropriate three-dimensional sediment diagenesis models for such sediments (Boudreau et al., 2001b).

Similarly, measuring benthic fluxes from permeable sediments using in situ benthic landers or chambers can be problematic because the presence of a benthic flux chamber on the seabed can impede pore-water advection by isolating the sediments from bottom water flow and pressure gradients (Huettel and Gust, 1992b; Jahnke, 2001). Nevertheless, important advances are being made in all of these areas to overcome these methodological difficulties (see recent discussions in Marinelli et al., 1998; Huettel and Rusch, 2000; Boudreau and Jøregensen, 2001; Berg et al., 2004).

Rates of sediment processes in permeable sediments can be comparable to those observed in more fine-grained accumulating sediments (see Huettel and Webster, 2001, and references therein); thus these sediments may be important in regional and perhaps global budgets of carbon and other biologically active elements (also see related discussions in Marinelli et al., 1998; Huettel and Rusch, 2000). Furthermore, interest in understanding the role of diagenetic processes in permeable continental margin sediments is based on the fact that ~70% of continental shelf sediments are relict sands (Emery, 1968). Thus, for example, in the mid-Atlantic Bight, where sandy sediments dominate, such diagenetic processes might help explain why less than ~5–20% of the shelf primary production is apparently exported (see Bacon et al., 1994, and references therein; although also see Liu et al., 2000, for a discussion that suggests these estimates may be low, and that larger percentages of shelf primary production may indeed be exported from continental margins).

12.4 Bioturbation

12.4.1 General Considerations

As discussed in section 5.3, *bioturbation* is defined here as the physical mixing of sediments by benthic macrofauna, primarily in their search for food or construction of burrows/tubes. These processes mix both sediment particles and pore waters and perhaps also sediment porosity, depending on how the sediment mixing occurs (also see discussions in section 6.1.2).

Deposit-feeding organisms responsible for bioturbation act in two basic ways. One group of organisms, generally referred to as conveyor belt deposit feeders, carry out what is thought of as unidirectional feeding. This type of bioturbation will be discussed below (section 12.4.3). A second group of organisms physically mixes sediments by ingesting and defecating sediments and by simply crawling through the sediments. This type of bioturbation, also often referred to as *particle reworking* or *biodiffusion*, has historically been defined as being a random, diffusivelike process analogous to eddy diffusivity (Goldberg and Koide, 1962; Guinasso and Schink, 1975; also see general discussions in Aller, 1982; Boudreau, 2000; Meysman et al., 2003). This leads to the parameterization of bioturbation by a bioturbation, or biodiffusion, coefficient D_b. The possible functional relationship(s) of D_b will be discussed below.

In quantifying bioturbation, Meysman et al. (2003a) have shown that a general framework for bioturbation models leads to two general formalisms: a nonlocal exchange formalism and a local biodiffusion formalism, which is actually a special case of the former. Three fundamental criteria (frequency, symmetry, and scale criteria) differentiate the two formalisms (also see similar discussions in Boudreau, 1986a; Wheatcroft et al., 1990). If all three criteria are met, then the mixing process may be considered a local process; i.e., the biodiffusion model is appropriate. In contrast, nonlocal mixing models generally comply only with the frequency criterion.

The first of these, the frequency criterion, requires that mixing be rapid (relative to other sediment transport processes). Mathematically, this requires that the frequency of mixing events be much shorter than that of tracer disappearance, which is itself the smaller of either the decay time of a radioactive tracer ($1/k$, where k is the radioactive decay constant) or the time of burial of the tracer through the sediment mixed layer ($\approx L/\omega$, here L is the mixed layer depth). As Meysman et al. (2003a) discuss, this criterion is satisfied for many tracers, although it may not necessarily be satisfied for short-lived tracers such as ^{234}Th or ^{7}Be (see table 12.2) at sites with low sediment reworking rates (also see discussions below).

The second of these criteria requires that particle movement be spatially random (on small scales), such that bioturbation has the

TABLE 12.2
Radioisotopes Commonly Used in Studies of Sediment Accumulation and Bioturbation in Marine Sediments

Isotope	Half-life	Input
^{234}Th	24.1 d	continuous
^{7}Be	44 d	continuous
^{228}Th	1.9 yr	continuous
^{14}C	5,730 yr	continuous
^{137}Cs	30.1 yr	pulse (nuclear weapons testing, with peak input ~1963)

same characteristics of Brownian motion and molecular diffusion (see section 5.1 for further detail). Finally, the scale criterion requires that the length scale of individual mixing events (δ^m) be small as compared to the tracer mixing length, which Meysman et al. (2003a) define as

$$\delta^t = 0.708 \frac{v}{k} \qquad (12.5)$$

where v is the particle advection rate ($\approx \omega$). Values of δ^t are tracer dependent and can again lead to situations in which mixing is local with respect to a long-lived tracer, but nonlocal for a short-lived tracer. In contrast, δ^m is a biological parameter that may range from ~0.5 cm to >10 cm (Wheatcroft et al., 1990).

Based on their analysis of δ^t and δ^m values for a range of tracers and sedimentary settings Meysman et al. (2003a) conclude that $\delta^m \geq \delta^t$ in "typical" early diagenetic environments. This therefore implies that the bioturbation mixing length exceeds the tracer length, and the scale criterion is generally not fulfilled.

This observation further suggests that most bioturbation in sediments is nonlocal. In contrast, though, many sediment radiotracer profiles appear to be diffusive in appearance, consistent with the discussion of biodiffusion models in the next section. These biodiffusion models have also proven useful as empirical models for sediment mixing, and this parameterization of bioturbation is used extensively in many general sediment diagenesis models (e.g., Boudreau, 1996; Soetaert et al., 1996c; Van Cappellen and Wang, 1996). Meysman et al. (2003a) term these observations the "biodiffusion paradox" and similarly note that other workers have argued that non-Fickian (i.e., non-random) types of mixing can produce diffusive-like tracer profiles (Boudreau and Imboden, 1987; Wheatcroft et al., 1990). An explanation of this paradox will require further work.

12.4.2 Models of Bioturbation

This section will focus on a discussion of the application of local bioturbation (biodiffusion) models to sediment process and profiles, given their extensive use in many geochemical studies. However, this discussion will also link observations made in such studies with their possible interpretation in terms of nonlocal mixing.

In its simplest sense, the magnitude of the bioturbation coefficient should depend on the abundance and activity of the organisms in the sediments (Aller, 1982b; Wheatcroft et al., 1990). Therefore, the intensity of this process (as a local diffusivelike mixing process) should decrease with depth (also see discussions in Boudreau, 2000). However, this approach of modeling D_b as some decreasing function of sediment depth is not often used; as will be discussed below, the functional relationship(s) of D_b with sediment depth may be even more complex.

Many models of bioturbation assume that there is a single mixed surface layer with a constant D_b value, and that this mixed layer overlays sediments in which bioturbation is absent. Boudreau (1998) has suggested that, worldwide, the mixed layer depth in marine sediments (L) has a mean value of $\sim 10 \pm 4.5$ cm, based in part on a plot of $n = 160$ L values (compiled from the literature) versus water column depth, which shows no obvious trends other than values of L that scatter around this mean value. Potential explanations for the occurrence of such a mean value of L across this wide range of sedimentary environments are discussed by Boudreau (1998). In particular, it may be related to the increasing energy cost to an organism of burrowing deeper than ~ 10–15 cm into the sediment (Jumars and Wheatcroft, 1989).

In contrast, Smith and Rabouille (2002) have shown that in deep-sea sediments the mixed layer depth shows a strong, asymptotic relationship to the sediment POC flux (fig. 12.4). At the same time, though, this asymptotic value agrees reasonably well with the mean value of L suggested by Boudreau (1998). Thus for sediments with higher POC fluxes than those shown here that underlie oxygenated

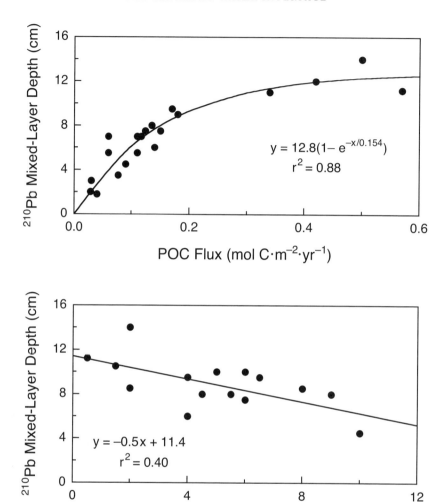

Figure 12.4 (Upper panel) The ^{210}Pb mixed-layer depth as a function of the annual carbon rain rate (sediment POC flux) to deep-sea sediments (>1,200 m water depth). (Lower panel) The ^{210}Pb mixed-layer depth as a function of the oxygen penetration depth in deep-sea sediments. Redrawn using results presented in Smith and Rabouille (2002) and references cited therein.

bottom waters[1] it is possible that the asymptotic value of L in fig. 12.4 may also be valid in these sediments (which again would be broadly consistent with the work of Jumars and Wheatcroft, 1989, cited above). This suggests that within the large range of literature values for L compiled by Boudreau (1998) the relationship in fig. 12.4 could be masked in a broad compilation of published data that simply looks at L versus water column depth (also see related discussions in Boudreau, 2004, and Smith and Rabouille, 2004, on this problem).

Interestingly, Smith and Rabouille (2002) also observed a weak inverse correlation between L and sediment oxygen penetration depth (fig. 12.4), as has also been observed in a wide range of other sediments (Wheatcroft et al., 1990; Smith et al., 1993). This result suggests that L is more strongly controlled by sediment POC flux (as noted above) than by sediment redox conditions, since increasing sediment POC flux generally leads to a decrease in the oxygen penetration depth, particularly in deep-sea sediments lacking significant bioturbation or bioirrigation (see data in Smith and Rabouille, 2002, and model calculations in Rabouille and Gaillard, 1991a). Since benthic macrofauna are aerobic organisms, this also implies that these organisms must maintain "metabolic contact" with the oxic sediment surface. More importantly, the ability of benthic macrofauna to mix suboxic (or even anoxic) sediments with oxic sediments has implications in terms of sediment redox oscillations and their role in controlling sediment carbon remineralization and preservation (see sections 7.6.3, 15.3, and 16.1 for further details).

Bioturbation coefficients can be determined using several approaches. Most often, they are determined by steady-state modeling of naturally occurring radioisotope profiles in a manner analogous to that used to determine sediment accumulation rates (see section 6.4.5 and discussion below). They can also be determined with time-dependent approaches using inert particulate pulse tracers. Perhaps the best example of this is the pulse input of ^{137}Cs to the atmosphere, and subsequently to sediment systems, due to nuclear weapons testing. Since this input peaked in 1963, in nonbioturbated sediments the ^{137}Cs peak can also be used as a stratigraphic marker (e.g., Chanton

[1] Smith and Rabouille (2002) note that sediment POC fluxes exceeding those shown in fig. 12.4 (such as in continental margin or coastal sediments) can sometimes lead to low bottom-water oxygen concentrations. In such settings the mixed layer depth will likely begin to decrease with increasing POC flux (as opposed to remaining constant at the asymptotic value of ~13 cm).

et al., 1983). Similar inert pulse tracers can also be used in experimental studies (Fornes et al., 1999, 2001), as can dissolved tracers (see Berg et al., 2001, for a recent summary).

A radiogeochemical approach (i.e., the use of naturally occurring radioisotopes) to quantify bioturbation, or sediment accumulation, involves several assumptions. The first is that these isotopes are irreversibly fixed to sediment particles and their input to the sediments is constant over time. This implies that once they are deposited in a sediment, their activity (concentration) is affected only by transport processes and radioactive decay. With these assumptions, the steady-state models described below (and in section 6.4.5) can be used to quantify bioturbation and sediment accumulation (in contrast, see DeMaster et al., 1991, for a discussion of the impact of nonsteady-state sediment mixing and/or sedimentation on such radiogeochemical profiles). An implicit assumption with all of these approaches is that the rates of sediment reworking determined by different radiotracers are tracer independent. While this is not always a valid assumption in general (see the discussions below), under some circumstances it may not be that unreasonable.

Models of bioturbation discussed in section 6.1.2 suggest that the process equally mixes sediments and pore waters (interphase mixing) or that the process mixes only sediments and not pore waters (intraphase mixing). In contrast, though, work presented by Berg et al. (2001) suggests that a third case may also exist. In experimental studies with shallow water coastal sediments they observed D_b values for pore waters up to 20 times larger than those for solids, and thus comparable to true bulk sediment diffusion coefficients (i.e., D_s). A possible explanation for these values begins with the observation that at relatively small scales (less than a millimeter), it may be easier to move pore waters than solids. As a result, bioturbation by benthic meiofauna such as nematodes or juvenile polychaetes in addition to, or instead of, benthic macrofauna may therefore differentially affect solids versus solutes (also see earlier discussions in Aller and Aller, 1992). Aller and Aller (1992) further suggest that because the depth scales of sediment reworking by these two groups of benthic infauna differ, meiofaunal bioturbation likely has its largest effect on oxic remineralization reactions occurring near the sediment surface, while macrofaunal bioturbation can impact both oxic and anaerobic (suboxic and anoxic) processes that occur deeper in the sediments. The extent to which the observations of Berg et al. (2001) can be generalized to other marine sediments has not been examined and clearly requires further study, given the potential impact that this meiofaunal bioturbation may have on sediment remineralization rates and benthic fluxes.

In addition to naturally occurring radioisotopes or pulse tracers to quantify bioturbation, sediment distributions of chlorophyll-*a* (chl-*a*) can also be useful in quantifying bioturbation in coastal sediments (Sun et al., 1991; Gerino et al., 1998; Ingalls et al., 2000). Since sediment chl-*a* is a more direct tracer of organic carbon derived from primary production, in some ways its use allows for a more direct link between particle reworking and its effects on sediment carbon remineralization.

As discussed in section 3.8.1, a radiogeochemical approach used to quantify bioturbation or sedimentation rates requires that the time scale of isotope decay match that of the specific process (see table 12.2).[2] Thus in deep-sea sediments, ^{14}C or other long-lived radioisotopes can be used to determine sedimentation rates, while ^{210}Pb can be used to determine the more rapid rates (in a relative sense) of bioturbation. In contrast, in coastal and continental margin sediments where rates of sedimentation and bioturbation are more rapid in general, ^{210}Pb is often used to determine sedimentation rates, and more short-lived isotopes such as ^{234}Th or ^{7}Be are used to quantify bioturbation.

In general, bioturbation tends to destroy gradients of the long-lived radioisotope in the sediment mixed layer because particle reworking is fast relative to radioactive decay. The depth of constant activity near the sediment surface is generally used to determine the depth of the sediment mixed layer (L; see figs. 12.5 and 12.6). In contrast, one generally sees an exponential-like decrease in the activity of the short-lived radioisotope in the mixed layer.

The occurrence of these processes can be shown mathematically using a multilayer model for sediment diagenesis. Starting with eqn. (6.22), and assuming constant porosity and steady-state conditions, the relevant equation for the upper sediment mixed layer ($0 \leq z \leq L$) is

$$0 = D_b \frac{\partial^2 A_u}{\partial z^2} - \omega \frac{\partial A_u}{\partial z} - \lambda A_u \qquad (12.6)$$

[2] In general, five half-lives represents a reasonable estimate for the time scale over which a radioisotope is of use in tracing sediment processes. Thus comparing, for example, the radioisotopes ^{234}Th and ^{210}Pb, the characteristic time scales over which they are of use in studying sediment processes are ~100 d and 100 yr, respectively.

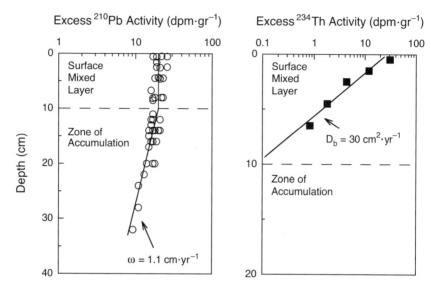

Figure 12.5 (Left panel) Profiles of excess ^{210}Pb from six box cores collected at a site on the North Carolina continental slope. Uniform activity in the upper 10 cm of sediment indicates the surface mixed layer; below this, excess ^{210}Pb decays away exponentially. Fitting the data below the mixed layer to the equations described in the text yields a sedimentation rate of 1.1 cm yr^{-1}. (Right panel) Profiles of excess ^{234}Th in a single box core at this same site. Fitting this data to the equations described in the text yields a bioturbation coefficient of 30 cm^2 yr^{-1}. Redrawn using results presented in DeMaster et al. (1994).

while for the deeper sediments (below the mixed layer: $z \geq L$) eqn. (6.53) is used:

$$0 = -\omega \frac{\partial A_d}{\partial z} - \lambda A_d \qquad (12.7)$$

As discussed in section 8.4 these equations are solved using the following set of boundary conditions,

1. $z = 0$, $J_o = -D_b \left.\frac{\partial A_u}{\partial z}\right|_{z=0} + \omega A_u |_{z=0}$ (upper flux boundary condition) (12.8)

2. $z = L$, $A_u = A_d$ (concentration continuity across the sediment boundary at $z = L$) (12.9)

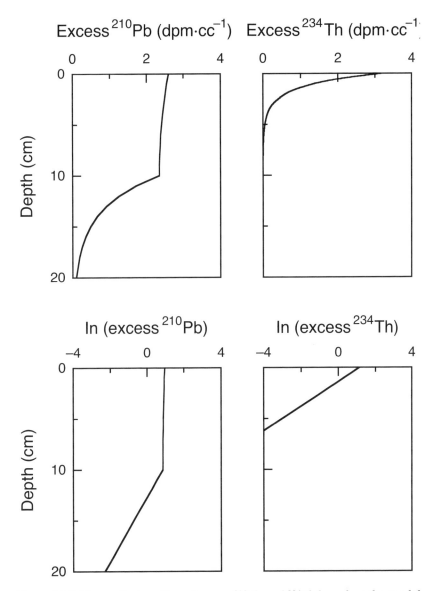

Figure 12.6 Theoretical profiles of excess ^{210}Pb and ^{234}Th based on the model equations (12.11a–g). In these calculations there is a 10 cm mixed layer in which $D_b = 15$ cm^2 yr^{-1}; below this the long-term sedimentation rate (ω) is 0.1 cm yr^{-1}. The fluxes of ^{210}Pb and ^{234}Th to the sediment surface are 1 dpm cm^{-2} yr^{-1} and 40 dpm cm^{-2} yr^{-1} respectively (also see similar plots in Aller, 1982b). As discussed in the text, the slope of the natural log (ln)-transformed ^{234}Th data can be used to estimate D_b while that of the ^{210}Pb data (below the sediment mixed layer) can be used to estimate ω.

3. $z = L$, $\partial A_u/\partial z = 0$ (a modified form of the flux boundary condition (12.10) at this sediment boundary)[3]

Solutions to these equations are (Aller, 1982b),

$$A_u = \beta_1 e^{\eta z} + \beta_2 e^{\gamma z} \quad (12.11a)$$

$$A_d = \beta_3 e^{-\lambda z/\omega} \quad (12.11b)$$

where

$$\eta = \frac{\omega - \sqrt{\omega^2 + 4\lambda D_b}}{2D_b} \quad (12.11c)$$

$$\gamma = \frac{\omega + \sqrt{\omega^2 + 4\lambda D_b}}{2D_b} \quad (12.11d)$$

$$\beta_1 = \frac{J_o \gamma e^{\gamma L}}{\eta(D_b\gamma - \omega)e^{\eta L} - \gamma(D_b\eta - \omega)e^{\gamma L}} \quad (12.11e)$$

$$\beta_2 = \frac{-J_o \eta e^{\eta L}}{\eta(D_b\gamma - \omega)e^{\eta L} - \gamma(D_b\eta - \omega)e^{\gamma L}} \quad (12.11f)$$

$$\beta_3 = (\beta_1 e^{\eta L} + \beta_2 e^{\gamma L})e^{\gamma L/\omega} \quad (12.11g)$$

Model profiles based on these solutions are shown in fig. 12.6. Analogous two-layer models for dissolved constituents in bioturbated sediments can also be developed using similar equations and boundary conditions, and are also discussed in Aller (1982b).

In principle, fitting profiles of a short-lived and long-lived radioisotope to these equations can be used to estimate ω and D_b. However, in practice, the following approach is generally taken. By using the long-lived radioisotope profile to first define the depth of the sediment mixed layer, one can then simply solve eqn. (12.7) in the deeper sediments as described in section 6.3.5, since eqns. (12.7) and (6.53) are identical. The solution to eqn. (12.7) is then a modified form of eqn. (6.54), where $(z - L)$ is simply substituted for z in eqn. (6.54) and the data below the sediment mixed layer are fit to this equation to obtain ω.

[3] The complete form of this flux boundary condition is $-D_b \dfrac{\partial A_u}{\partial z} + \omega A_u = \omega A_d$. However, because at $A_u = A_d$ at $z = L$ (eqn. 12.9) this equation reduces to eqn. (12.10).

Similarly, because the depth of penetration of the short-lived tracer is generally smaller than the mixed layer depth, the short-lived isotope is generally fit only to a solution of eqn. (12.6) subject to the following boundary conditions: eqn. (12.8) and a modified form of eqn. 12.10 (i.e., $z \to \infty$, $A_u = 0$). However, because scaling calculations indicate that bioturbation is generally more important than sedimentation as a transport process in the mixed layer of many sediments (see section 4.4 and table 4.1), the sedimentation term can also be eliminated from eqn. (12.6) under most circumstances. This then yields the following solution for the short-lived radioisotope in the mixed layer:

$$A = \left(\frac{J_o}{\eta D_b}\right) e^{-\eta z} \quad (12.12)$$

where here $\eta = \sqrt{\lambda/D_b}$. Here, the slope of the plot of ln (A) vs. depth can be used to more directly determine D_b. This approach can be seen in the fits to radiogeochemical data from the North Carolina continental slope shown in fig. 12.5.

Values of D_b obtained with the radiogeochemical approaches discussed above vary among marine sediments, with higher values generally seen in coastal versus deep-sea sediments. Values in coastal sediments are often on the order of ~10–100 cm² yr⁻¹ while values in deep-sea sediments may be as low as ~0.01–1 cm² yr⁻¹ (see compilations in Boudreau, 1994; Tromp et al., 1995; Middelburg et al., 1997). These observations appear to be consistent with general observations that both numbers and biomass of benthic macrofauna per unit area decrease with water column depth (Thistle, 2003). Presumably, the decrease in available food for benthic macrofauana with increasing water depth provides a first-order explanation for these observations.

These observations also indicate that D_b values are comparable to D_s values in coastal sediments, notwithstanding the above-discussed possibility of enhanced pore water bioturbation by meiofauna. Thus bioturbation may also affect the shape of pore-water profiles (e.g., see Aller, 1982b). However, strong pore-water gradients can still be maintained in the presence of rapid rates of bioturbation, depending in part on the time scales of the rates of early digenetic processes versus those of sediment reworking (Aller and Cochran, 1976; Burdige, 2002).

Compilations of D_b values have been used to develop empirical relationships between D_b and either ω or water depth. Boudreau (1994) has observed that D_b values (based on ^{210}Pb profiles) show the following empirical relationship with ω:

$$D_b = 15.7\omega^{0.69} \qquad (12.13)$$

(D_b expressed as cm^2 yr^{-1} and ω expressed as cm yr^{-1}). However, the r^2 value of this regression is relatively low ($r^2 = 0.28$ for $n > 130$ data points) and the equation accounts for only ~30% of the variance in the data. Middelburg et al. (1997) have observed a similar relationship between D_b values (again based on ^{210}Pb profiles) and water depth Z (in km),

$$D_b = 5.2 \cdot 10^{(0.762 - 0.3974Z)} \qquad (12.14)$$

although again this relationship explains only a portion of the variance of the data (here $r^2 = 0.43$ for $n = 132$). Although such relationships exhibit a great deal of scatter, they do appear to be of some use in efforts to "scale up" diagenetic models for making global estimates of the rates of sediment processes (Middelburg et al., 1997).

Another important observation that has come out of this radiogeochemical approach to quantifying bioturbation is that measured D_b values can be tracer dependent. In these situations, short-lived tracers (e.g., ^{234}Th) often yield D_b values that are 10–100 times larger than those obtained using longer-lived tracers such as ^{210}Pb (Smith et al., 1993; Pope et al., 1996).

Several explanations have been proposed for these observations. The first is that the known selectivity of deposit feeding organisms leads to particle reworking that is size dependent, with smaller particles being preferentially mixed faster than larger particles (DeMaster and Cochran, 1982; Cochran, 1985; Wheatcroft, 1992). Similarly, since radioisotopes of different half-lives trace processes over different time scales, nonsteady-state sediment mixing or sedimentation may explain these differences (DeMaster et al., 1991; Pope et al., 1996). At the same time, this inverse relationship between D_b values and tracer half-lives may result from a process referred to as *age-dependent mixing*, whereby recently deposited food-rich particles are assumed to be preferentially ingested and therefore mixed faster by deposit feeders than older, food-poor particles (Smith et al., 1993). Additional details of this age-dependent mixing process will be discussed in the next section. Similarly, the fact that different tracers respond differently to the criteria differentiating local versus nonlocal sediment mixing (as defined by Meysman et al., 2003a, and discussed in section 12.4.1) may also play a role in explaining these tracer-dependent D_b values, although again this has not been critically examined.

12.4.3 Nonlocal Sediment Mixing

As noted above, one of the major distinctions between local and nonlocal sediment mixing involves a scale criterion requiring that the length of an individual mixing event be small compared to the tracer mixing length (as defined by eqn. 12.5). In situations where this criterion is not met there is then nonlocal mixing, in which materials are transported between nonadjacent sections of the sediments (Boudreau, 1998).[4] It can also be shown that small-scale sediment resuspension by physical processes can be viewed as a similar type of nonlocal mixing process, and modeled as such (Boudreau, 1997a,b).

Nonlocal transport of sediment particles can occur in several ways. One such process described as *conveyor-belt mixing* is carried out by head-down deposit-feeding organisms that ingest sediment at depth, pass it vertically upward through their gut, and eventually defecate the material at the sediment-water interface (Rhoads, 1974; Robbins, 1986). In this process the sediment void created at depth is also infilled by the settling (compaction) of overlying sediment. As a result, sediment in the vicinity of a head-down deposit feeder is thus moved conveyor belt-like motion between some depth in the sediments and the sediment surface. Although this process is nonlocal (as defined above) the dynamics of particle tracer (e.g., ^{210}Pb) recycling during conveyor-belt mixing tend to retain the tracer in the surficial mixed layer (Robbins, 1986). Steady-state tracer profiles of this type of conveyor-belt activity can, therefore, often resemble those resulting from diffusive-like bioturbation (biodiffusion; Boudreau, 1986b).

Another group of organisms, referred to as *reverse conveyor-belt feeders*, either feed at the sediment surface and then defecate at depth or simply store (cache) food-rich particles at depth in burrows (Rhoads, 1974; Smith et al., 1986/7, 2001). Functionally equivalent to this activity are other processes, including hoeing down of surface sediments

[4] In the literature, this type of bioturbation is sometimes also termed *bioadvection* (as opposed to the type of bioturbation discussed in the previous section referred to as *biodiffusion*).

THE SEDIMENT-WATER INTERFACE

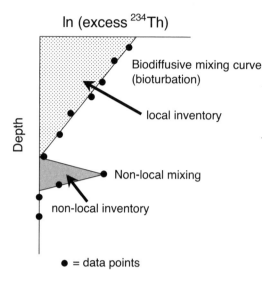

Figure 12.7 The hypothetical profile of a short-lived radioisotope that would result from simple biodiffusive mixing in the upper sediments and nonlocal mixing at depth (modified after Pope et al., 1996).

to some depth in the sediments, head-down deposit feeders, or passive sediment infilling of relict burrows (see discussions in Pope et al., 1996; Boudreau, 1997a). One outcome of this type of mixing is the occurrence of subsurface peaks in short-term radiotracers, superimposed on a more general exponential-like decrease in tracer activity as a result of biodiffusive mixing (see fig. 12.7, Pope et al., 1996, and references therein).

Related to reverse conveyor-belt feeding is the observation that the age-dependent mixing that occurs in some sediments may be best explained by rapid nonlocal mixing of young, i.e., recently deposited, food-rich particles into sediments through feeding and caching activities of benthic organisms, accompanied by relatively slow, and less selective, biodiffusive subsurface mixing (Fornes et al., 2001; Smith et al., 2001). In contrast, though, studies of nonlocal mixing in other sedimentary environments have shown that this type of reverse conveyor-belt activity is much less selective in terms of sediment age and its nutritional quality (Fornes et al., 1999). Factors such as differences in faunal assemblages as well as the quality and quantity of the sediment POC flux likely lead to these differences, although the exact details of how (or why) this occurs are not well understood.

A variety of workers have developed models of these types of nonlocal mixing (Boudreau, 1986b; Robbins, 1986; Smith et al., 1986/87; Boudreau and Imboden, 1987; Smith et al., 1993; Pope et al., 1996; Soetaert et al., 1996a; Fornes et al., 1999). A detailed discussion of these models will not be undertaken here, partly because of the complexity of the models and also because of their current lack of generality (which is itself likely related to an incomplete knowledge of the interplay between the factors controlling this type of sediment mixing).

As an example of one such model, Smith et al. (1986/87) developed a nonlocal mixing model in which material from near the sediment surface ($0 \leq z \leq l$) is transported to some deeper region in the sediments ($z_1 \leq z \leq z_2$; also note that $l < z_1 < z_2 < L$). For a radiotracer being affected by these processes there are thus three diagenetic equations for different portions of the overall sediment mixed layer:

$$0 \leq z \leq l: \quad \frac{\partial A}{\partial t} = D_b \frac{\partial^2 A}{\partial z^2} - \lambda A - r(z) A \qquad (12.15)$$

$$l \leq z \leq z_1 \text{ and } z_2 \leq z \leq L: \quad \frac{\partial A}{\partial t} = D_b \frac{\partial^2 A}{\partial z^2} - \lambda A \qquad (12.16)$$

$$z_1 \leq z \leq z_2: \quad \frac{\partial A}{\partial t} = D_b \frac{\partial^2 A}{\partial z^2} - \lambda A + \frac{1}{l} \int_0^l r(z) A \, dz \qquad (12.17)$$

where $r(z)$ is the nonlocal removal (egestion) coefficient (units of inverse time). For a conservative tracer the amount of its removal from the sediment surface is equal to the amount emplaced at depth, hence the integral term in eqn. (12.17). This latter equation also assumes that tracer removed from the surface is homogenized and distributed equally at depth (i.e., between z_1 and z_2). For a nonconservative tracer, such as chl-a, additional terms would also need to be included in such models to account for tracer degradation during advective transport (Sun et al., 1991).

Models such as this then consist of a series of coupled diagenetic equations, solved using standard upper and lower boundary conditions and the assumption that fluxes and concentrations (activities) are continuous at region boundaries (i.e., at $z = l$, z_1, and z_2; also see section 8.4). Solutions to the steady-state versions of eqns. (12.15)–(12.17) depend on the complexity of $r(z)$, and both analytical and numerical solutions have been presented (Smith et al., 1986/87; Pope et al., 1996).

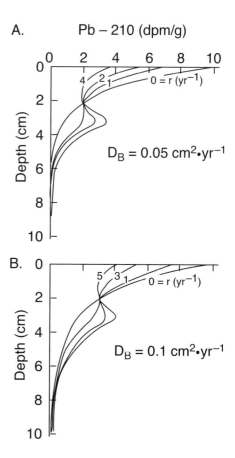

Figure 12.8 ^{210}Pb profiles predicted by a nonlocal mixing model (see eqns. 12.15–17) in which subsurface egestion transports surface sediments to a subsurface zone 3–4 cm below the sediment-water interface (redrawn from Smith et al., 1986/87). The different curves represent different feeding rates of the sediment organisms (see the original reference for further details about the model and these model calculations).

An example of one such model solution is shown in fig. 12.8, where it can be seen that these models predict steady-state subsurface maxima in tracer profiles that are similar to those seen in field data (Smith et al., 1986/87; Pope et al., 1996; and references therein).

12.5 BIOIRRIGATION

Many benthic macrofauna construct sediment burrows or tubes for feeding or dwelling. The time scales over which organisms inhabit these burrows can vary tremendously depending on the mobility of the organisms. Flushing of these burrows (either passively or actively) with overlying bottom water is a common trait of this group of benthic macrofauna, and this process is often referred to as bioirrigation. In the simplest sense, if a burrow or tube wall is permeable to diffusion, then concentration gradients are established between the burrow water and the surrounding sediment pore waters in three dimensions around the burrow or tube. Because burrows and tubes are generally flushed with sufficient regularity, this process enhances benthic fluxes by allowing for an exchange between pore waters and bottom waters across the increased surface area of these macrofaunal structures, in addition to that which occurs across the sediment-water interface (Aller, 1988, 2001; Kristensen, 1988, 2001).

Aside from the (obvious) occurrence of burrowing organisms in a given sediment (e.g., see fig. 5.3), evidence of bioirrigation is often obtained geochemically when measured benthic fluxes of pore-water solutes exceed those that can be predicted by simple molecular diffusion and interfacial pore-water concentration gradients (eqn. 12.2). While bioturbation can, in principle, also enhance solute fluxes, under many circumstances the magnitude of D_b is much less than that of D_s. This then suggests that this type of benthic macrofaunal activity (biodiffusion) does not likely lead to these enhanced solute fluxes (however, also see the discussion in section 12.4 and Berg et al., 2001, for situations in which this assumption may not be appropriate).

At the same time, the impact of bioirrigation on sediment biogeochemistry results from more than simply an increase in the surface area available for exchange between pore waters and bottom waters (see table 15.1). The occurrence of macrofaunal tubes and burrows leads to a complex three-dimensional geometry in the zone of irrigation that significantly alters the spatial (and temporal) distribution of reactions occurring in these sediments. In part, this occurs because macrofaunal irrigation of sediments creates diffusive sources and sinks for pore-water solutes within portions of the sediments that would otherwise be isolated from exchange with the overlying waters (also see section 12.5.1). At the same time, because burrows are flushed with oxygenated bottom water, irrigation enhances sediment oxygen input, thus altering sediment redox conditions, promoting redox oscillations, and allowing for the occurrence of a wide range of coupled oxidative and reductive processes associated with organic matter remineralization (see section 7.6.3 and, e.g., Aller et al., 2001).

The sediments immediately adjacent to many tubes and burrows are often sites of elevated bacterial populations and activity (Aller, 1988) for several possible reasons, including: organic matter (e.g., mucus) enrichment of burrow walls; effects associated with either the direct (e.g., oxygen or nitrate) or indirect (manganese and iron oxides) addition of oxidants for organic matter remineralization; the occurrence of chemolithotrophic reactions due to localized redox boundaries resulting from oxygen diffusion through burrow walls into the surrounding oxygen-deficient sediments.

Various approaches have been taken in the one-dimensional modeling of biorrigation, generally in three broad categories (see summaries in Aller, 1982b; Boudreau, 1997a; Koretsky et al., 2002). Diffusive models assume that bioirrigation can be quantified by an enhanced, apparent diffusion coefficient within a defined (irrigated) sediment layer, while advective models use an apparent pore-water advection velocity to account for bioirrigation. However, the most commonly used approach assumes that bioirrigation can be quantified by a nonlocal bioirrigation coefficient (α) that describes the rate of exchange between bottom waters and pore waters at depth in the irrigated zone of the sediment.

A discussion of these quantitative models of bioirrigation starts with Aller's radial-diffusion tube model (Aller, 1980c), which also forms the basis for many subsequent bioirrigation models presented in the literature. This model approximates the sediment bioirrigated zone as a series of evenly distributed, closely packed hollow cylinders of equal length, with each burrow represented by the hollow portion of the cylinder (fig. 12.9). Although this geometric representation of the irrigated zone of a sediment approximates the true complexity of many of these sediments (e.g., see discussions below), when compared to other bioirrigation models it appears to more appropriately describe the effect of these macrofaunal structures on biogeochemical processes in irrigated sediments.

This model also assumes that these burrows are well flushed with bottom waters and that the burrow and tube linings are diffusively permeable. However, the organic material that makes up this lining is not necessarily freely permeable with respect to diffusion (Aller, 1983; Hannides et al., 2005) and different solutes have different diffusion characteristics through tube walls and burrow linings. In addition, the chemical composition of these linings also varies among different macrofaunal species, and this too can affect solute diffusion. For these reasons the impacts of bioirrigation in the same sediment will vary for different solutes and different burrow/tube types. While this has been discussed in the literature (also see Aller, 1988), most studies of bioirrigation have not considered this variability.

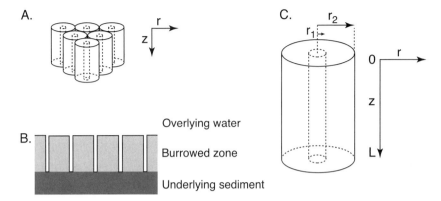

Figure 12.9 (A) The upper region of an idealized bioirrigated sediment with the sediment represented as a series of closely packed uniform hollow cyclinders. (B) A vertical cross-section of this sediment. (C) A more detailed view of the cylindrical microenvironment around a burrow tube. Redrawn from figures originally presented in Aller (1980c).

With the model illustrated in fig.12.9, the three-dimensional nature of an irrigated sediment is reduced to the two dimensions of depth and radial distance, and the following diagenetic equation can be written for each sediment annulus:

$$\frac{\partial C}{\partial t} = D_s \frac{\partial^2 C}{\partial z^2} + \frac{D_s}{r}\frac{\partial}{\partial r}\left(r\frac{\partial C}{\partial r}\right) + \Sigma R(r, z, t) \qquad (12.18)$$

where r is the radial distance away from a given tube axis. In this model porosity is taken to be constant, advection (burial and flow) has been ignored, and a generalized reaction term of the form

$$\Sigma R = R_1 e^{-\gamma z} + R_2 + k(C_{eq} - C) \qquad (12.19)$$

is often assumed (see the discussion below for how R_1, R_2, γ, k, and C_{eq} are determined). This reaction equation has been used in the development of generalized solutions to eqn. (12.18) since it can be used to describe many biogeochemical process that occur in sediments. For example, with $k = 0$ this equation can be employed to describe the depth dependence of sulfate reduction or ammonium production in sediments (as in sections 6.4.3 and 6.4.4), while with $R_1 = R_2 = 0$ this equation can be used to describe silica or carbonate dissolution (see sections 13.4 and 13.5). However, discussions below (in section 12.5.1) will also show that these "idealized" reaction rate equations may not be valid under all circumstances in bioirrigated sediments.

Two of the boundary conditions imposed on eqn. (12.18) specify that the concentrations at the sediment-water interface and at the tube wall equal bottom water values:

$$C(0, r, t) = C_o \qquad (12.20a)$$

$$C(z, r_1, t) = C_o \qquad (12.20b)$$

As discussed above, these boundary conditions implicitly assume that the burrows are continuously well flushed with bottom water. However, macrofauna are also known to ventilate their burrows periodically, and discontinuous burrow irrigation can then lead to boundary condition (12.20b) being a time-dependent function. The third boundary condition specifies that pore-water concentrations go through a maximum or minimum half-way between any two tubes,

$$\left.\frac{\partial C}{\partial r}\right|_{r=r_2} = 0 \qquad (12.20c)$$

while the last boundary condition specifies that the flux at the base of the burrowed sediments matches that from the lower, unburrowed sediments:

$$\left.\frac{\partial C}{\partial z}\right|_{z=L} = B \qquad (12.20d)$$

Both steady-state and transient solutions of this equation have been presented in the literature, where specific details about these solutions can be obtained (Aller, 1980c, 1982b, 1988; Boudreau and Marinelli, 1994; Boudreau, 2000).

In applying these models to actual sediment data, r_1 and L can be determined with field observations while r_2 is often used as an adjustable fitting parameter (Aller, 1980c, 1982b). However, r_2 can also be determined with data on the density of irrigating organisms (N; organisms per m^2); specifically,

$$r_2(cm) \approx 100/\sqrt{\pi N} \qquad (12.21)$$

(Gilbert et al., 2003; also see fig. 12.9A for a schematic view of these systems that can be used to derive this equation). The basal gradient B at $z = L$ is generally calculated from measured pore-water profiles, while the coefficients in ΣR (R_1, R_2, γ, k, and C_{eq} in eqn. 12.19) are determined from either literature data (e.g., Wang and Van Capellen, 1996), parallel sediment incubation experiments (e.g., Aller, 1980a, c), or can be treated as free (fitting) parameters.

Solutions to these equations give radial and vertical pore-water concentrations around the microenvironment of a single burrow. However, pore-water samples from sediment cores are generally obtained over fixed sampling intervals. Assuming then that the sediment core is large enough to sample over several burrow microenvironments, the average pore-water concentration $\bar{C}$ over a finite depth interval (z_1 to z_2) can be calculated using

$$\bar{C} = \frac{V_B C_o}{V_T} + \frac{V_S}{V_T} \frac{2\pi \int_{z_1}^{z_2} \int_{r_1}^{r_2} rC(z,r) dr\, dz}{2\pi \int_{z_1}^{z_2} \int_{r_1}^{r_2} r\, dr\, dz} \qquad (12.22)$$

where $V_T = L\pi r_2^2$ (the total volume of a single burrow microenvironment), $V_S = L\pi(r_2^2 - r_1^2)$ (the sediment volume in a single burrow) microenvironment), and $V_B = L\pi r_1^2$ (the burrow volume in a single burrow microenvironment).

The application of this model to data from a bioirrigated organic-rich coastal sediment is shown in fig. 12.10. Note that such pore-water profiles are similar to those seen in many other bioirrigated sediments (e.g., see discussions in Aller, 1980c). For sulfate, which is consumed by sulfate reduction and has the overlying water column as its ultimate source, we see that biorrigation severely attenuates sulfate depletion with depth because of sulfate replenishment by irrigation. Also note that the one-dimension model shown here for comparison is essentially that discussed in section 6.3.3.

For ammonium, whose source is the sediments and whose overlying water concentration is essentially zero, bioirrigation significantly decreases its pore-water concentration (note that here the 1-D model, such as that described in section 6.3.4, would be off-scale for these sediment rate parameters). Furthermore, one also observes a mid-depth minimum in pore-water ammonium concentrations because

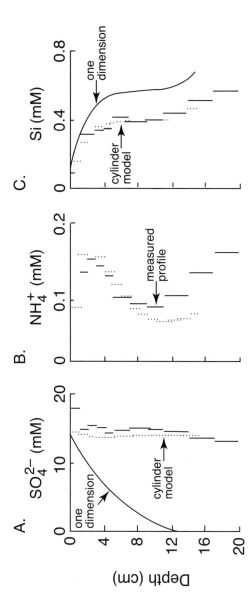

Figure 12.10 The fit of Aller's radial diffusion tube model to sulfate, ammonium, and silica pore-water data from bioirrigated sediments in Mud Bay, South Carolina (redrawn after Aller, 1980c; note that the bottom of the bioirrigation zone, L, is assumed to be at 15 cm). Dashed vertical bars are the best-fit radially averaged concentrations predicted by the model (cylinder model), while the solid vertical bars are the actual pore-water data (measured profile). Also shown as solid lines are profiles from a one-dimension (1-D) diffusion-reaction model (e.g., see section 6.3) that uses the same rate parameters as in the calculations with the radial diffusion model. The 1-D ammonium model profile is off scale and therefore not shown. Note that the tube model captures the major trends in the data (weak sulfate deletion with depth and minima in the pore-water ammonium profiles) that are missed by the 1-D model. Also note the much closer agreement between the 1-D model and the tube model for the silica profile (see the text for further details).

of the interplay between an exponential decrease in ammonium production with sediment depth, a source of ammonium from the deeper unburrowed sediments ($z > 15$ cm), and the removal of dissolved ammonium by bioirrigation in the region $0 < z < 15$ cm.

In contrast, dissolved silica, which also has a sediment source, has profiles that appear to be much less affected by bioirrigation than do those of these other solutes. The reason for this stems from the fact that, unlike ammonium production, silica dissolution is a function of the pore-water dissolved silica concentration (see table 12.1). Therefore, lowering pore-water silica concentrations by irrigation also increases the rate of silica dissolution, i.e., production of dissolved silica. The opposing processes of irrigation and dissolution partially compensate for one another, leading to the observed pore-water profiles.

Benthic fluxes from bioirrigated sediments result from both burrow/tube irrigation and simple molecular diffusion, as a result of vertical pore-water gradients. Thus, the presence of irrigation structures potentially affects the absolute magnitude of the benthic flux as well as the relative importance of bioirrigation versus molecular diffusion in the total benthic flux. To examine this problem, Aller (1980c) examined the impact of changing r_1 and r_2 values on the benthic fluxes of ammonium and silica (note that changing r_2 is equivalent to changing the density of irrigation organisms N, as is indicated in eqn. 12.21). Because ammonium production in sediments is assumed here to be independent of pore-water ammonium concentrations (see table 12.1, section 6.3.4, and the discussion of eqn. 12.19), there is no feedback between ammonium production in the sediments and ammonium loss via benthic fluxes. Consequently, while the size (parameterized by r_1) and number (parameterized by r_2) of tubes or burrows changes the relative importance of irrigation versus diffusion in the total benthic flux, the total benthic flux is independent of any of these parameters; i.e., in this kinetic formulation of ammonium production the benthic flux is fixed by the defined sediment production rate, e.g., the parameters R_1, R_2 and γ in eqn. 12.19. However, as we will see in section 12.5.1, subsequent work has also shown that this assumption of zero-order kinetics for processes such as this may, under some circumstances, break down in bioirrigated sediments.

In contrast, the feedback between silica pore-water concentrations and silica dissolution leads to a situation in which the total benthic flux of silica is very strongly affected by the distribution of irrigating

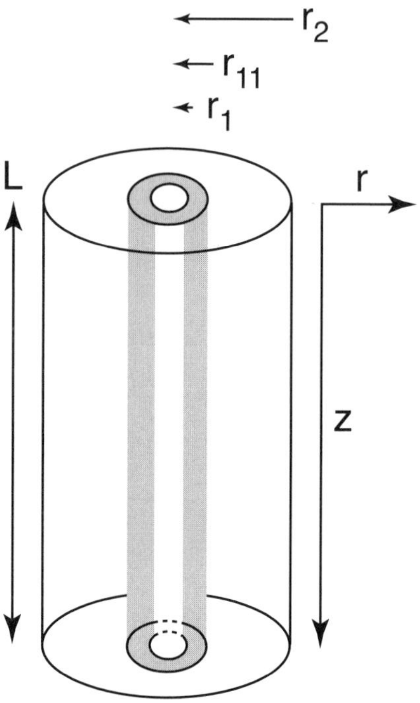

Figure 12.11 The cylindrical microenvironment around a burrow tube in which it is assumed that there is an annulus of oxygen-containing sediment around a burrow tube ($r_{11} < r < r_1$; also indicated by shading). The sediment is assumed to be anoxic for $r > r_{11}$. The impact of such a microenvironment on sediment nitrogen cycling in bioirrigated sediments is also illustrated in fig. 16.5. Redrawn after Aller (1988).

burrows (Aller, 1980c, 1982b). More burrows (decreasing r_2 or increasing N) or larger burrows (increasing r_1) all lead to increasing silica fluxes. Similarly, changing these parameters also leads to changes in the relative importance of irrigation versus vertical molecular diffusion in contributing to the total flux; that is, vertical diffusion becomes more important with decreasing r_2 (i.e., increasing N) or decreasing r_1. The importance of the interplay between such geometrical considerations and rates of sediment biogeochemical processes will be discussed in further detail at the end of this section.

Since the initial presentation of Aller's original radial-diffusion tube model for bioirrigation, there have been modifications/extensions of this model that consider, for example, discontinuous burrow irrigation (Boudreau and Marinelli, 1994), or burrows of various depths and tilt angles (Furukawa et al., 2001). Another version of this model also accounts for the redox zonation that occurs around these burrows (figs. 7.6 and 12.11) by assuming that the annulus of sediment around a given burrow consists of an internal radial zone of oxic sediment surrounded by anoxic sediment (Aller, 1988). For the nitrogen system, this geometry then leads to a situation in which there is a tight coupling between ammonium production (ammonification), nitrification, and denitrification in these burrowed sediments (see section 16.2 for further details).

Another important aspect in the development of bioirrigation models comes from work by Emerson et al. (1984) and Boudreau (1984), who showed that eqn. (12.18) can be re-expressed as a one-dimension diagenetic equation of the form,

$$\frac{\partial \overline{C}}{\partial t} = D_s \frac{\partial^2 \overline{C}}{\partial z^2} - \alpha(\overline{C} - C_o)\Sigma\overline{R} \qquad (12.23)$$

where $\overline{C}$ is the laterally average pore-water concentration and $\Sigma\overline{R}$ is the laterally average reaction rate (as given by, e.g., eqn. 12.19). Here irrigation of sediment burrows is empirically quantified using $\alpha = \alpha(z, t)$, the nonlocal bioirrigation coefficient, which can be thought of as the fraction of pore-water solute exchanged per unit time. Although irrigation actually occurs by a complex network of sediment burrows and tubes, and their three-dimensional geometrical source/sink properties, this approach parameterizes the process by assuming that, at any given sediment depth (and at any given time in a time-dependent model), there is exchange, or mass transfer, between bottom waters and the average pore waters at that depth.

Although α is often taken to be an adjustable parameter obtained by fitting pore-water profiles (see section 12.5.2 for further details), Boudreau (1984) also shows that under some circumstances α can be related to quantities in the original Aller radial diffusion, bioirrigation model according to,

$$\alpha(z) = \frac{2D_s r_1}{(r_2^2 - r_1^2)(\overline{r} - r_1)} \qquad (12.24)$$

where $\overline{r}$ is the radial distance at which C equals its mean value $\overline{C}$. Using this nonlocal irrigation formalism the total benthic flux due to bioirrigation is given by,

$$J_I = \int_0^{l_I} \alpha(z)[C(z) - C_o]dz \qquad (12.25)$$

and if, for now, α is assumed to be constant with depth, then the apparent pore-water exchange velocity due to irrigation (v_I; as defined by Hammond and Fuller, 1979) is given by

$$v_I = \alpha/L \qquad (12.26)$$

(Boudreau, 1997a).

The derivation of eqn. (12.23) demonstrates that under some circumstances, Aller's radial-diffusion model for bioirrigation can be reduced to a one-dimensional nonlocal bioirrigation model (also see recent discussions of this problem in Grigg et al., 2005). This provides some theoretical basis for using this general 1-D approach to model bioirrigation, as well as a certain degree of mathematical simplicity in modeling irrigated sediments.

However, for a given solute $\bar{r}$ is likely to be a function of depth, because of the depth- and radial-dependence of sediment reactions affecting different solutes and the resulting differences in their source-sink reaction geometries in the zone of bioirrigation (Boudreau, 1997a). Thus based on eqn. (12.24) α should likely vary with sediment depth (also see section 12.5.2), and α values may also differ among different solutes because of differing D_s and $\bar{r}$ values (also see recent discussions in Meile et al., 2005). Their ratio will, however, be *approximately* equal to their ratio of diffusion coefficients *if* they have similar reaction geometries (i.e., $\bar{r}$ values; see Boudreau, 1997a, Aller, 2001, also see section 12.5.3.1).

The use of a nonlocal approach to quantifying bioirrigation has some advantages over radial diffusion models and also appears to work reasonably well for describing average pore-water distributions in many bioirrigated sediments. Furthermore, for simple radial geometries of burrow tubes fairly straightforward relationships exist between α and parameters such as r_1, r_2, and N (Aller, 1988). At the same time, benthic macrofauna create burrows of a variety of shapes and sizes; not all burrows are evenly spaced tubes of the same length evenly dispersed in a sediment. Burrow size (e.g., L or r_1), spacing (e.g., r_2), shape (e.g., simple tubes versus U-shaped), and orientation (e.g., vertical or tilted tubes) are therefore often highly variable on local scales, and also can show temporal variability (e.g., see discussions in Aller, 2001; Furukawa, 2001; Koretsky et al., 2002). While some aspects of such complex burrow geometry can be incorporated into radial diffusion models (see above), the paramerization of bioirrigation using this nonlocal approach has certain advantages for such extremely complex burrow patterns and distributions (Aller, 2001; Koretsky et al., 2002).

Modeling bioirrigation using this nonlocal approach has the disadvantage of obscuring the role of burrow geometry on biogeochemical reactions in irrigated sediments. In contrast, radial diffusion models can, for example, explicitly examine the effects of burrow properties (e.g., r_1 and r_2), radial variations in reaction rates (due to, e.g., radial redox zonation), or discontinuous irrigation, on overall sediment reaction kinetics, pore-water transport, and benthic fluxes (e.g., see discussions in Aller, 2001).

12.5.1 The Diffusive Openness of Bioirrigated Sediments

In a broader sense we can think of the impact of burrow geometry on the rates of sediment biogeochemical processes in terms of the extent to which irrigated sediments are open with respect to diffusion, i.e., exchange with bottom waters, and how this diffusive openness impacts sediment processes. For redox reactions, such geometric considerations specifically affect the extent to which oxygen is introduced into otherwise anoxic sediments, as a result of the flushing of burrows with oxygenated bottom water and radial oxygen diffusion (see fig. 12.12). This phenomenon has been studied extensively in terms of its effect on sediment nitrogen cycling (e.g., Aller, 1988; Gilbert et al., 2003), although these factors also impact the cycling of other redox sensitive elements (e.g., Mn^{2+} or I^-) in bioirrigated sediments (also see, for example, Berg et al., 2003).

In particular, burrow spacing and size (i.e., r_1 and r_2) as well as oxic zone thickness around burrows (r_{II}) strongly affect the radial diffusion of oxygen into irrigated sediments (fig. 12.12). These quantities then control the relative amounts of oxidized and reduced sediment around a burrow and, therefore, impact the relative rates of coupled redox reactions such as nitrification/denitrification in bioirrigated sediments (see section 16.2 for further details).

Quantifying the diffusive openness of a bioirrigated sediment, and its impact on sediment processes, is often done in terms of diffusive length scales, which are a function of the burrow spacing geometry in the sediments. For nonredox reactions, this diffusive length scale is roughly given by $(r_2 - r_1)$, while for redox reactions it is $\sim (r_2^2 - r_{II}^2)^{1/2}$ (Aller and Aller, 1998; Aller, 2001; Figs. 12.9 and 12.11).

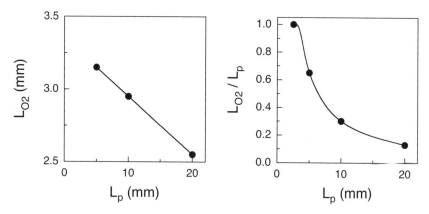

Figure 12.12 (Left) The oxygen penetration depth (L-O_2) in Long Island Sound sediment plugs[5] of varying thickness L_p incubated with identical overlying water oxygen concentration. Note that in comparison to fig. 12.11, L-O_2 is roughly equal to $r_{11} - r_1$ while $L_p \approx r_2$. As L_p increases, diffusion of reduced metabolites from the increasing reservoir of anoxic sediment leads to a decrease in L-O_2. (Right) The fraction of oxic sediment (L-O_2/L_p) as a function of sediment plug thickness (L_p) or, by analogy, burrow spacing (r_2). Redrawn using results presented in Aller et al. (2001).

Geometrical considerations not only impact the relative occurrence of such coupled redox processes, but also appear to affect overall sediment organic matter decomposition rates. In particular, Aller and Aller (1998) have observed that greater diffusive openness, i.e., decreasing diffusive length scales as defined above, enhances net remineralization rates in sediments, independent of effects that can simply be related to enhanced oxygen input into diffusively open bioirrigated sediments (also see general discussions in Aller, 2001). More specifically, they observed that net rates of ammonification, and to a lesser extent phosphate and iodide remineralization, vary inversely with diffusive length scale. In their studies diffusive length scales greater than ~2–3 cm yield ammonification rates that are essentially the same as those obtained in closed sediment incubations;[5] however, as diffusive length scales decrease and sediments become more diffusively open, ammonification rates increase exponentially. At rel-

[5] In these and in other studies cited here (e.g., Gilbert et al., 2003) sediment "plugs" are incubated in open tanks of either oxic or anoxic seawater. These plugs consist of sediment placed in polycarbonate holders of varying thickness (L_p) that are open at the

atively small length scales (<0.5 cm) these rates are more than 5 times greater than those observed in closed incubations.

The causes of these effects are not well understood, though several possible explanations have been proposed. Many build on the general observation that greater openness with respect to diffusion also implies increased overall solute exchange. Thus the observed increase in net remineralization rates with decreasing diffusive length scale could possibly be the result of a decreased efficiency of back reactions (uptake) into new bacterial biomass (see section 11.1.5) or the loss of inhibitory metabolites such as HS^-. Similarly, desorption (or loss from pore waters) of reactive DOM substrates should be enhanced in diffusively open systems and may then also result in enhanced net remineralization rates (Aller and Aller, 1998; also see related discussions in sections 10.2 and 10.4).

In thinking about these observations, it is also interesting to note that in one sense they appear to contradict earlier discussions in this section and in section 12.2 in which it was noted that benthic ammonium fluxes, and therefore rates of sediment ammonification, in bioirrigated sediments should be independent of the geometry of the zone of bioirrigation, i.e., values of r_1 and r_2. However, it is also important to recall that the assumption of a zero-order rate of ammonification used in these earlier calculations is somewhat idealized, and is based partly on results that have been obtained largely in systems that are much more diffusively "closed" than bioirrigated sediments, i.e., anoxic sediments that are nonbioturbated and/or nonbioirrigated. In such systems (e.g., Berner, 1980; Klump and Martens, 1989) this formulation appears to do a reasonably good job at parameterizing ammonium production. At the same time, this kinetic formulation is ultimately based on assumptions discussed in sections 6.3 and 8.1 in which one assumes that the rate of organic matter remineralization (or here, more specifically, ammonification) is not oxidant limited but simply proportional to the amount of reactive organic matter in the sediments. However, the results discussed here indicate that as systems become more diffusively open these assumptions may break down. While transport processes per se are not necessarily

---

top and sealed at the bottom. The geometry of these plugs is such that L_p (e.g.; see fig. 12.12) represents a diffusive transport length roughly equivalent to the half-distance between idealized sediment burrows, i.e., $L_p \approx r_2$ in fig. 12.9 (also see Aller and Mackin, 1989, for further details about this experimental approach).

the sole cause of this, they may apparently lead to changes in local conditions that do cause a breakdown in these assumptions. Furthermore, as discussed above (and in section 16.2), this problem becomes more complex when one considers coupled redox processes such as nitrification-dentrification. More work will be needed to unravel the details of these interactions and their mechanistic basis.

12.5.2 Methods for Quantifying Bioirrigation in Sediments

Interest in the study of bioirrigation stems in part from its impact on the magnitude of benthic fluxes. Quantifying this aspect of bioirrigation in sediments can be done in several ways. One approach involves simply comparing the measured benthic flux to the predicted diffusive flux, to estimate a benthic flux enhancement factor (BFE):

$$\text{BFE} = J_{to}/J_{dif} \qquad (12.27)$$

(see eqns. 12.1 and 12.2). Another approach involves the estimation of the nonlocal irrigation parameter $\alpha(z)$ using either experimental techniques, e.g., benthic chambers or incubated cores, or sediment diagenesis models. In this section we will briefly discuss the various ways that bioirrigation can be quantified in marine sediments. The next section will then discuss specific results from such efforts.

One approach to estimating bioirrigation rates involves the use of nonlocal irrigation models based on eqn. (12.23) to fit either individual pore-water profiles (e.g., Emerson et al., 1984; Martin and Banta, 1992; Schlüter et al., 2000) or multiple pore-water profiles in coupled models of sediment diagenesis, such as those discussed in section 8.4 (Furukawa et al., 2000). In this approach, if all other model parameters are known, particularly those in the ΣR terms, then $\alpha(z)$ can often be obtained as an adjustable fitting parameter.

An important aspect of such efforts is the parameterization of $\alpha(z)$. Often $\alpha(z)$ is assumed to be solute independent, constant over some finite depth interval, and equal to zero below that depth. Although this assumption appears inconsistent with discussions associated with eqn. (12.24), this approach has been used in many studies and appears to provide reasonable results. To address these latter concerns, it has been suggested that $\alpha(z)$ should decrease exponentially with depth:

$$\alpha(z) = \alpha_o e^{-\beta z} \qquad (12.28)$$

where α_o and β are again adjustable fitting parameters. Such an exponential increase may result from the number of burrow tubes decreasing with sediment depth (Martin and Banta, 1992) or from decreasing animal activity with sediment depth (Aller, 2001).

Other approaches to quantify $\alpha(z)$ do not begin with the assumptions of any a priori functional relationship for bioirrigation. Meile et al. (2001) developed an inverse model in which $\alpha(z)$ is determined from measured pore-water concentration and reaction rate depth profiles. A related approach uses an ecological stochastic model in which an irrigated sediment is represented as a three-dimensional region in which a variety of different burrows[6] are distributed based on ecological considerations and field data (Koretsky et al., 2002). Comparisons indicate that bioirrigation coefficients determined by these two approaches agree well with other estimates (see the next section for further details).

Techniques to directly quantify bioirrigation involve the use of purposeful tracer added to either a benthic flux chamber or the water overlying an incubated sediment core. One then examines the loss of this tracer with time either in the chamber or in the water overlying the core, and, in some studies, the temporal evolution of tracer penetration into the sediments. A number of inert tracers have been used for such studies including either deuterated or tritiated water and $^{22}Na^+$, although in recent years Br^- has been the most commonly used tracer (Emerson et al., 1984; Martin and Banta, 1992; Sayles and Martin, 1995; Berelson et al., 1999, 2003; Forster et al., 1999).

Estimating bioirrigation coefficients from such studies also requires the analysis of these results using a sediment model in which assumptions are made regarding the depth-dependence of bioirrigation (e.g., Martin and Banta, 1992; Berelson et al., 1999). Alternately, one can simply calculate the benthic flux enhancement factor (BFE in eqn. 12.27) by comparing the observed tracer loss with that predicted by molecular diffusion alone (Forster et al., 1999).

Determining values of BFE can also be carried out using naturally occurring solutes by comparing measured benthic fluxes (e.g., by using a benthic chamber or lander) with the diffusive flux predicted by the pore-water gradient at the sediment water interface (eqn. 12.2).

---

[6] In the Korestsky et al. (2002) model there are 10 end-member shapes of burrows, including simple vertical tubes, inclined tubes, and L-shaped, U-shaped and Y-shaped tubes.

One natural inert tracer that has proven useful in this regard, and in examining sediment-water exchange processes in general, is ^{222}Rn (Hammond and Fuller, 1979; Smethie et al., 1981; Kipput and Martens, 1982).

Radon has several properties that make it very attractive for such studies. First, radon is chemically inert because it is a noble gas, and its production in sediments is controlled only by the activity of its parent isotope ^{226}Ra. At the same time, because of the large difference in half-lives between ^{222}Rn and ^{226}Ra (3.8 days vs. 1,600 yrs), radon in sediment pore waters should rapidly reach secular equilibrium with its parent isotope; that is, the activity of ^{222}Rn should equal that of ^{226}Ra (see section 3.8.1 for details).

In the absence of transport processes other than diffusion, and neglecting sedimentation, the diagenetic equation for radon in pore waters can be written as

$$\frac{\partial C}{\partial t} = D_s \frac{\partial^2 C}{\partial z^2} - \lambda C + \lambda C^* \tag{12.29}$$

where the concentration of ^{222}Rn (C) is expressed here per unit volume of bulk sediment, λ is the ^{222}Rn radioactive decay constant, and C^* is the ^{222}Rn concentration supported by ^{226}Ra, i.e., the pore water radon concentration at secular equilibrium. Assuming for now that the ^{226}Ra content of the sediment is constant with depth, the steady-state solution to this equation predicts that the diffusive flux of ^{222}Rn is given by,

$$J_{\text{diff}} = -\varphi_o \sqrt{\lambda D_s} (C^* - C_o) \tag{12.30}$$

where C_o is the bottom water ^{222}Rn content and φ_o is the porosity at the sediment surface. Since in general $C^* \gg C_o$ and C^* is controlled by the sediment ^{226}Ra content, this then implies that J_{diff} is essentially a function solely of the sediment radium content. As such the determination of J_{diff} is not subject to uncertainties associated with the accurate assessment of the ^{222}Rn pore-water concentration gradients close to the sediment-water interface (see section 12.2 and discussions in Berelson et al., 2003). Furthermore, even if the radium content of the sediment increases with depth, for example due to sediment compaction, J_{diff} is still only a function of the sediment radium distribution (Smethie et al., 1981; Berelson et al., 1982; Kipput and Martens, 1982). For these reasons, the benthic flux enhancement factor (BFE in eqn. 12.27) may be more accurately defined with radon data than with pore-water and benthic flux data for other solutes. Similarly, deficits in pore water radon concentrations, as compared to those predicted by equations such as eqn. 12.29, can also be used to quantify the sediment depth distribution of nonlocal transport processes such as bioirrigation (Martin and Sayles, 1987) or methane gas ebullition from sediments (see section 12.7).

12.5.3 Rates of Bioirrigation in Marine Sediments

Table 12.3 summarizes data on α and the depth of bioirrigation for several coastal and continental margin sediments. As discussed above, a number of different approaches have been used to estimate α values, some of which assume that α is independent of sediment depth and others that assume α varies with sediment depth. For ease of comparison here I have used these $\alpha(z)$ values to calculate a depth-average value of $\alpha(\bar{\alpha})$ over the zone of bioirrigation (see the notes to this table for additional details about these calculations).

An examination of these results illustrates several interesting trends. The first is that in general, α values decrease with decreasing water depth. This observation is consistent with discussions above regarding the distribution of benthic macrofauna in marine sediments. Second, the results in this table also demonstrate that there is reasonably good agreement between α values calculated by different methods; for example, compare the Buzzards Bay and Washington continental shelf results.

Third, while not obvious from this tabulation, the range in α and BFE values observed by Martin and Sayles (1987) in Buzzards Bay sediments is related to seasonal variations in bioirrigation rates, with enhanced solute transport occurring from early summer to fall. Such seasonality in macrofaunal irrigation rates is consistent with other studies of coastal sediments (e.g., Sun et al., 1991, Gerino et al., 1998), which demonstrate that processes such as particle reworking and macrofaunal abundance in general also vary seasonally. Presumably, this occurs in response to changing temperatures and varying inputs of organic matter to the sediments, e.g., the deposition of spring bloom material to the sediments (also see discussions in section 14.3).

TABLE 12.3
Estimates of the Nonlocal Bioirrigation Coefficient α for Various Marine Sediments

Site	Water Depth (m)	α or $\bar{\alpha}$ (yr^{-1})	α_{max} (yr^{-1})*	Depth of Bioirrigation (cm)	Method of Determination
Sapelo Island, GA (salt marsh)	intertidal	460		4	(a)
Port Phillip Bay, Australia (coastal embayment)	7–24	105–315		20–50	(b)
Quartermaster Harbor, Puget Sound, Washington (coastal embayment)	15	3–16		20	(c)
Buzzards Bay (coastal sediment)	15	12–59	12–152	6–12	(d)
		6–40	7–102	13–43	(e)
		7.7		18	(a)
		3–117	8–300	2–128	(f)
Washington continental shelf	~90	6–43	12–112	30	(g)
		33–41		~24	(a)
Washington continental shelf break	115	1.7	1.7	30	(g)
Eastern North Pacific continental margin	85–160	O$_2$: 6.0	9.4	<15	(h)
		NO$_3$: 2.1	6.0		
		Si: 1.1	2.0		
	225–630	O$_2$: 0.5	0.8	<15	(h)
		NO$_3$: 1.8	2.7		
		Si: 0.6	1.4		
Arctic continental margin	175	294		15	(a)
	330	120		15	

* α_{max} is the maximum value of α obtained from estimates in which α varies with sediment depth. In most cases, this is the value at the sediment-water interface.

a. From Meile et al. (2001). As discussed in the text α(z) is determined by inverse modeling of concentration and rate profiles. Depth profiles of α(z) calculated in this fashion do not necessarily decrease smoothly with sediment depth. These average α values ($\bar{\alpha}$) are based on the irrigation velocities reported by these workers ($= \int \alpha(z) dz$) and their reported values of z_{90} (the depth at which the depth-integrated value of α(z) or irrigation velocity reaches 90% of its total value) according to $\bar{\alpha} = [\int \alpha(z) dz]/z_{90}$. The depth of irrigation reported here is z_{90}. The Buzzards Bay result is a recalculation of the ^{222}Rn data in Martin and Banta (1992), while the Washington shelf result is a recalculation of the results in Christensen et al. (1987).

b. Determined by the loss of deuterated water from benthic flux chambers, modeled by assuming that bioirrigation occurs over a constant sediment depth interval. Both α and the depth of irrigation were used as adjustable fitting parameters in these calculations (Berelson et al., 1999).

c. Determined either by modeling of silica pore-water profiles or by loss of tritiated water from a benthic chamber, modeled by assuming that bioirrigation occurs over a constant sediment depth interval (Emerson et al., 1984).

d. Determined by modeling pore-water ^{222}Rn profiles. Bioirrigation is assumed to decrease exponentially with depth in this model (eqn. 12.28) and the depth-integrated amount of bioirrigation is given by α_o/β. The depth of bioirrigation was estimated here as the depth over which 90% of this irrigation occurs ($= z_{90}$) which equals $-\log(0.1)/\beta$. The $\bar{\alpha}$ value reported here is the average value in the sediment region $0 - z_{90}$ (and therefore equals $0.9(\alpha_o/\beta)/z_{90}$). Also note that the ranges in α_{max} and $\bar{\alpha}$ reported here are due in part to temporal (seasonal) changes in the bioirrigation of these sediments (see the text for further details; Martin and Sayles, 1987).

e. Determined from modeling ^{222}Rn profiles as described in note d (Martin and Banta, 1992).

f. Determined with Br tracer studies using incubated cores. The value of $\bar{\alpha}$ and the depth of irrigation were calculated as described in note d (Martin and Banta, 1992).

g. Determined by modeling pore-water ^{222}Rn and sulfate depth profiles over the depth ranges 2–10 cm, 10–30 cm, and 2–30 cm, assuming a constant α value over each of these intervals. In general, the value for 2–10 cm is greater than that for 10–30 cm. The α value shown here is the 2–30 cm value, while the α_{max} value is the 2–10 cm value (Christensen et al., 1987).

h. Determined with benthic flux and pore water data for oxygen, nitrate and silica, assuming a constant depth of irrigation. Although this depth is not reported, all sediment cores were less than 15 cm in length. The α and α_{max} values shown here are average and maximum values for different sites (Devol and Christensen, 1993).

12.5.3.1 BIOIRRIGATION AND REACTION GEOMETRY

As noted above, differences in reaction geometries for different solutes are likely to impact the values of $\bar{r}$ in eqn. (12.24). Along with differences in D_s values, this may then lead to α values being solute dependent. Given these considerations, this suggests that some care must be taken in applying α values determined for one solute to other solutes. These same considerations also come into play when comparing BFE values determined for different solutes, as will be further discussed in the next section.

For redox sensitive solutes (e.g., Mn^{2+}, Fe^{2+}, or ammonium), this problem stems, in part, from the fact that while there is production of these solutes in the anoxic portion of the sediments away from the tube or burrow (i.e., $r > r_{11}$ in fig. 12.11) there is also consumption, or oxidation, of the solute in the annulus of oxygen-containing sediment around a burrow tube (i.e., $r_{11} < r < r_1$ in fig. 12.11). Therefore, α values determined for solutes that are not redox sensitive will tend to overestimate the magnitude of the bioirrigation coefficient for these redox-sensitive solutes (Berg et al., 2003; also see related discussions in Aller, 2001).

Results in Devol and Christensen (1993) for sediments of the eastern North Pacific continental margin (table 12.3) further indicate that differences in estimates of α appear to occur for different solutes, particularly at the shallower sites they studied. In a first-order attempt to separate out the effects of diffusivity and reaction geometry on α values, one can redefine α as

$$\alpha = \Psi \cdot D_s \qquad (12.31)$$

where Ψ can be thought of as a geometric parameter incorporating all aspects of reaction geometry that ultimately affect solute transport by bioirrigation (Christensen et al., 1984; Aller, 2001). In re-examining the Devol and Christensen (1993) data with this equation, it can be shown that the values of Ψ for nitrate and silicate agree well (within ~30%) while the Ψ value for O_2 is substantially higher (by a factor of ~3). These observations suggest that, roughly speaking, nitrate and silicate have similar reaction geometries in these continental margin sediments, in spite of the different biogeochemical processes that affect them, as well as differences in their source/sink relationships—that is, these sediments are a source of silicate and a sink for nitrate. In contrast, though, these results also indicate that oxygen apparently has a very different reaction geometry than either nitrate or silicate in these continental margin sediments. At the same time, Furukawa et al. (2001) calculated nearly identical $\bar{r}$ values for sulfate and ammonium in mesocosm experiments with coastal sediments to which a subsurface, funnel-feeding hemichordate worm (*Schizocardium* sp.) was added to the sediments.

12.5.3.2 BENTHIC FLUX ENRICHMENT FACTORS

An alternate approach to examining the effects of bioirrigation involves calculation of the benthic flux enrichment factor (BFE in eqn. 12.27). This factor is related to what is often referred to as an *apparent* or *irrigation diffusion coefficient* (defined here as D_I). This definition assumes that bioirrigation can be modeled as a process analogous to molecular diffusion with an enhanced mass transfer coefficient (see summaries in Aller, 1982b, 2001). With this formalism the following relationship can be derived from eqn. (12.27):

$$D_I = \text{BFE} \cdot D_S \qquad (12.32)$$

While this approach is analogous to that in which bioturbation (biodiffusion) is modeled as a random diffusionlike process, it is not necessarily as well grounded either conceptually or theoretically. Enhanced benthic fluxes associated with sediment resuspension-redeposition events have also been modeled using this same approach (Vanderborght et al., 1977; Lohse et al., 1996), although as noted above, it may be more appropriate to view this as a nonlocal exchange process (see section 12.4.3).

Interest in calculating BFE factors is based in part on the assumption that BFE values calculated with one solute or tracer (preferably inert; e.g., ^{222}Rn) at a given site may be applicable to sediment-water exchange in general at this site. However, the extent to which this approach is valid is subject to some uncertainty, as discussed above in section 12.5.3.1. Of particular importance here is the fact that the use of BFE values in quantifying sediment-water exchange is very sensitive to reaction geometry issues such as the depth distribution of reaction rates and kinetics (e.g., Aller, 2001). For solutes whose production or consumption largely occurs in the upper few millimeters of sediment, this approach should yield BFE values at a given site that are similar to one another, and close to 1. In contrast, if the reaction affecting a different solute occurs deeper in the sediments, the BFE

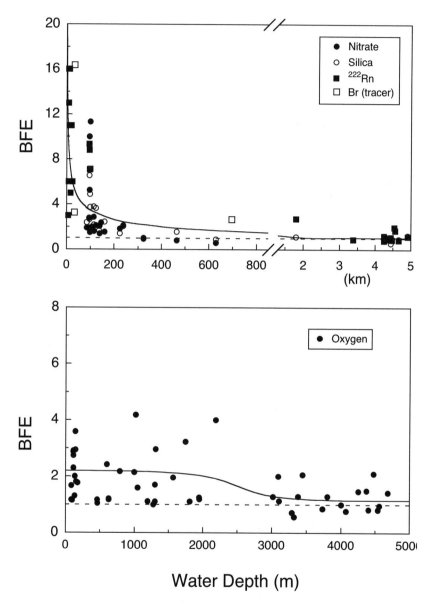

Figure 12.13 The benthic flux enhancement factor (BFE, as defined by eqn. 12.27). (Upper) Results obtained with nitrate, silica, and ^{222}Rn pore-water profiles and benthic flux data as well as Br⁻ tracer experiments. (Lower) Results obtained with O_2 pore-water (or microelectrode) profiles and benthic flux data (see the original references below for further details on the calculations). The two curves are not meant to imply any functional relationships,

value for that solute will likely be greater, as a result of the increasing importance of solute exchange with deeper portions of the irrigated sediment (also see related results in Burdige et al., 2004).

A compilation of BFE values for several solutes in marine sediments is shown in fig. 12.13. The upper panel (fig. 12.13A) shows that BFE values for the solutes nitrate, silicate, ^{222}Rn, and Br⁻ tracer drop off very rapidly with water column depth, consistent with discussions above regarding macrofaunal distribution in marine sediments. As such, in water depths greater than ~200–500 m bioirrigation becomes insignificant as compared to diffusion in terms of controlling benthic fluxes; that is, BFE ≈ 1 (also see similar discussions in Jahnke, 2001). At any given site values of BFE for these different solutes also appear reasonably consistent.

BFE values based on dissolved oxygen results show slightly different trends (fig. 12.13B). Again, BFE values decrease with water depth, as in fig. 12.13A. Data in Glud et al. (1994) for sediments of the eastern South Atlantic Ocean (water depths generally >1,000 m) further indicate that BFE values for O_2 are positively correlated with macrofaunal biomass. However, in contrast to data for other solutes, BFE values for O_2 exhibit a much smaller range, and below water depths of ~200 m do not dramatically increase with decreasing water depth. In light of the discussion above regarding Ψ values for O_2 versus nitrate or silicate in eastern North Pacific continental margin sediments, these observations further support the suggestion that the reaction geometry for O_2 in bioirrigated sediments may be significantly different than that of other solutes.

The details of how or why these differences occur remain to be determined. Diffusion (mass transport) limitations on sediment oxygen uptake at high oxygen uptake rates may play some role in limiting the dynamic range in BFE values obtained with O_2 data (Jørgensen, 2001; also see section 12.2). At the same time, variations in bottom water oxygen concentrations may also be important (e.g., see similar discussions in Gilbert et al., 2003). Nevertheless, these observations suggest that some care must be taken when applying bioirrigation parameters such as Ψ,

---

but are presented simply to aid in the visualization of the general trends in the data. Also shown on each graph is a dashed line for BFE = 1. Data sources: Berelson et al., 1987b, 1990, 1999, 2003; Archer and Devol, 1992; Reimers et al., 1992; Devol and Christensen, 1993; Sayles and Martin, 1995; Hammond et al., 1996; Glud et al., 1998; Forster et al., 1999; Wenzhöfer and Glud, 2002.

α, or BFE obtained with results from other solutes to the quantification of sediment oxygen uptake. The same concern also likely holds when applying parameters defined with oxygen data to other solutes.

Two recent studies have attempted to use results such as those in fig. 12.13 to estimate global parameters for sediment bioirrigation. Using a subset of the oxygen data in this figure, Archer et al. (2002) developed a relationship for irrigation velocity v_I as a function of both carbon rain rate to the sediments and bottom water oxygen levels. This was done as a part of a coupled sediment diagenesis model designed primarily for sediments in water depths >1,000 m. They assumed that irrigation decreased exponentially with depth, and tuning their model to the datasets they used suggested that bioirrigation ceases by ~2 cm in the sediments they looked at. Although the maximum irrigation velocities they observed at the sediment surface (~8–10 cm d^{-1}) are not unreasonable (as compared to values discussed in Boudreau, 1997a, and Meile et al., 2001), the very shallow depth of irrigation is not necessarily consistent with the results reported in table 12.3.

In a similar fashion, Meile and Van Capellan (2003) used a larger database for dissolved oxygen fluxes and sediment profiles (essentially most of the data shown in fig. 12.13B) to develop empirical relationships for D_I (or BFE) and α as a function of total benthic oxygen uptake over a range encompassing coastal to deep-sea sediments. Their choice of benthic oxygen fluxes as the master variable was based in part on the large database of such measurements. Furthermore, their results appear to agree reasonably well with α values in table 12.3 and BFE values in fig. 12.13B. However, it is also important to note that there may be some difficulty in extrapolating these results, as well as those in Archer et al. (2002), to solute transport via bioirrigation in general, based on the discussion above and the results in fig. 12.13.

12.6 OTHER SEDIMENT-WATER INTERFACE PROCESSES: METHANE GAS EBULLITION

In organic-rich sediments where complete sulfate reduction occurs within a meter or so of the sediment surface, large pore-water methane concentrations can build up below the zone of sulfate reduction because of methanogenesis at depth (see sections 7.3.5, 17.2.1 and figs. 8.6, 17.3). In some cases the sediment pore waters become saturated with methane, leading to gas bubble formation. Not only do these bubbles cause what is referred to as acoustic turbidity, but they can also rise through the sediments (ebullition), creating a flux of gas-phase methane from the sediments. Gas ebullition is common in many tidal freshwater and wetland soils and sediments (Chanton, 1989), as well as in some subtidal coastal sediments, such as Cape Lookout Bight and Long Island Sound (Martens and Berner, 1977; Kipput and Martens, 1982; Albert et al., 1998; Martens et al., 1998; Boudreau et al., 2001a). In these latter environments, methane gas ebullition generally occurs seasonally during warmer months, when there are higher rates of sediment organic matter remineralization. More specifically it can also be triggered by hydrostatic pressure release associated with tidal fluctuations, i.e., low tides.

In subtidal sediments methane gas ebullition forms cylindrical bubble tubes that lead to enhanced rates of sediment-water exchange through the creation of additional sediment surface area for benthic exchange (Martens et al., 1980). In Cape Lookout Bight sediments, this process has been successfully modeled for sediment-water exchange in general by using a seasonally varying effective diffusion coefficient D'_s. This value is determined in an analogous way to D_I in eqn. 12.32 (also see eqn. 12.27) using pore-water profiles, i.e., concentration gradients at the sediment-water interface, and directly measured benthic fluxes. During periods of active gas ebullition from these sediments, this process leads to measurable fluxes of, for example, methane, ^{222}Rn, and inorganic nutrients, such as ammonium or phosphate, which are up to 3 times greater than fluxes predicted by molecular diffusion (see eqn. 12.2 and also see Klump and Martens, 1981). Furthermore, on an annual basis methane gas ebullition accounts for >80% of the flux of methane from these sediments (Martens and Klump, 1984). Gas ebullition from methanogenic sediments is also important beyond its effect on sediment-water exchange because it can transport methane, and possibly other reduced gases produced in anoxic sediments, directly to the atmosphere, bypassing their possible oxidation in the sediments or water column (see Kipput and Martens, 1982; Chanton, 1989; Boudreau et al., 2001a; and references therein).

≈ **CHAPTER THIRTEEN** ≈

Biogeochemical Processes in Pelagic (Deep-Sea) Sediments

PELAGIC SEDIMENTS[1] generally have low sediment accumulation rates, typically less than ~5–10 cm per thousand years, and are low in organic matter content. They constitute up to ~80% of all marine sediments, but account for only a small fraction of the total organic matter that is either remineralized or preserved in all marine sediments (see table 8.1). Pelagic sediments are, however, major sites for the deposition and diagenesis of biogenic opal and calcium carbonate. Furthermore, trace metal cycling in deep-sea sediments may play an important role in the formation of ferromanganese nodules commonly found in many of these sediments.

13.1 Organic Matter Remineralization

Aerobic respiration generally dominates organic matter remineralization in deep-sea sediments (section 7.5.2). In defining end-member sediment types in such environments, we previously considered oxygen- and carbon-limiting cases (fig. 7.8), which are implicitly based on the assumption that oxygen availability controls carbon preservation in deep-sea sediments. Since this is not rigorously true, we can also examine analogous end-members for organic matter remineralization based on different pore-water nitrate profiles, shown in fig. 13.1 (additional details about nitrogen diagenesis in pelagic sediments can be found in section 16.2).

The first case, the O_2-consumption model, is identical to the carbon-limiting case in fig. 7.8. and here the asymptotic increase in nitrate with depth results from nitrification coupled to aerobic respiration.

---

[1] As discussed in chapter 2, pelagic sediments are far removed from the continents and often defined as being in water depths >3,000 m (e.g., Berner, 1980; Seibold and Berger, 1996). At the same time, though, some geochemical budget calculations use a shallower water depth to divide the continental margin from the deep sea (Middleburg et al., 1997; Christensen, 2000).

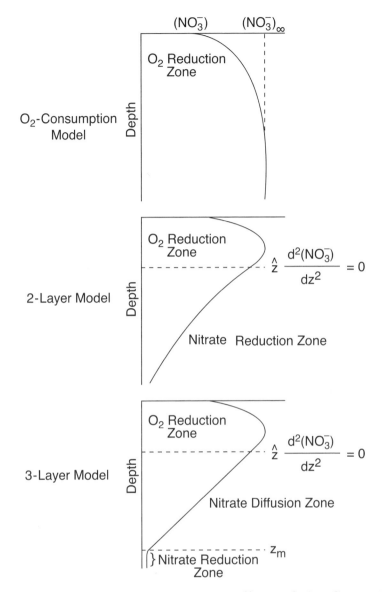

Figure 13.1 Hypothetical pore-water nitrate profiles in pelagic sediments (redrawn after Goloway and Bender, 1982). (Upper) O_2-consumption model, for a carbon-limited sediment (as in fig. 7.8); (Middle) 2-layer model, for an oxygen-limiting sediment in which denitrification occurs (roughly) below the oxygen penetration depth; (Lower) 3-layer model, for a sediment in which the surface zone of aerobic respiration and the deeper denitrification zone are separated by a nonreactive zone in which nitrate simply diffuses downward to be denitrified at depth. Sediments of this type are discussed in section 14.4.

The maximum increase in nitrate, i.e., $[NO_3^-]_\infty - [NO_3^-]_{BW}$, is then roughly controlled by the amount of metabolizable organic matter in the sediments and its C/N ratio. The second case shown in fig. 13.1 is related to the oxygen-limiting case in fig. 7.8, by its upper zone of aerobic respiration and nitrification along with its deeper zone of denitrification, roughly below the sediment oxygen penetration depth. Although in most current models of sediment diagenesis this boundary is defined by the oxygen penetration depth, in the model shown here (Goloway and Bender, 1982) it is defined by the nitrate profile. This can be seen here using a simplified version of the steady-state diagenetic equation for nitrate of the form,

$$0 = D_s \frac{\partial^2 C}{\partial z^2} + R \qquad (13.1)$$

where $R > 0$ for sediments where nitrification occurs and $R < 0$ where denitrification occurs. Based on this equation $\frac{\partial^2 C}{\partial z^2}$ is therefore negative in the upper portion of the sediments where nitrification occurs, and positive in the deeper, denitrifying sediments. As a result, the boundary between these two sediment regions occurs where the second derivative in the nitrate profile equals zero.

In their examination of pore-water nitrate profiles in deep-sea sediments, Goloway and Bender (1982) also observed a third case, in which there is spatial separation between an upper zone of aerobic respiration and a deeper zone of denitrification (also see similar profiles and related discussions in Jahnke et al., 1982; Salwan and Murray, 1983). The separation between these two zones generally averaged 20–30 cm in the cores examined by these workers, with a range of ~5–80 cm. As is shown in fig. 13.1, in the middle region of these sediments nitrate profiles are linear with depth, implying that here diffusion is the only process that affects nitrate.[2] In subsequent work, analogous pore-water oxygen profiles were observed (Wilson et al., 1985, 1986; Thomson et al., 1987) with a linear pore-water oxygen gradient between a surface region where some oxygen consumption

---

[2] This can be seen by examining eqn. (6.30), for example, and setting $\Sigma R_i = 0$. The solution to this equation for any set of boundary conditions is a straight line. At the same time, though, if transport processes other than diffusion are significant and/or porosity is *not* constant with depth then one can obtain nonlinear pore-water profiles for conservative solutes (e.g., McDuff and Gieskes, 1976; Dickens, 2001).

occurs, and a deeper subsurface reaction zone, where additional oxygen consumption occurs and oxygen concentrations go to zero.

A more detailed explanation of these observations is that the subsurface reaction zone in these sediments is associated with buried turbidite sediments (section 5.2). In the overlying carbon-limited sediments, rather than oxygen and nitrate profiles reaching asymptotic, nonzero values with depth (see figs. 7.8 and 13.1), these solutes initially decrease (oxygen) or increase (nitrate) because of surficial aerobic respiration/nitrification, and then diffuse downward, where they oxidize organic matter in the deeper turbidite material. Additional details about the geochemistry of such sediments will be presented in the next chapter.

In the same way that sediment oxygen uptake in deep-sea sediments plays an important role in overall deep-water respiration (see section 7.5.2), so too should sediment remineralization reactions affect nutrient elements such N, P, and Si. Pore-water profiles of phosphate and silicate generally increase with sediment depth, often in an asymptotic fashion, implying there should be a benthic flux of these elements (see typical profiles in, for example, Jahnke et al., 1982; Martin et al., 1991; Zabel et al., 1998). Although nitrate profiles in deep-sea sediments may be more complex (fig. 13.1), they too normally predict a flux of nitrate out of the sediments. There have been few efforts to scale up these benthic fluxes to examine their global significance, and where such calculations have been attempted, they predict fluxes that exceed riverine fluxes of these elements by at least an order of magnitude (Hensen et al., 1998; Colman and Holland, 2000). Such results are not surprising, in that nutrient cycling in the ocean is generally dominated by large internal fluxes such as between deep and surface waters, with small net fluxes into or out of the oceans, e.g., river input or burial in marine sediments (see discussions in Broecker and Peng, 1982, or any other text in chemical oceanography). This then implies that benthic fluxes should contribute significantly to nutrient regeneration in the deep ocean.

To see this more directly, estimates of the benthic silica flux from southern Atlantic Ocean sediments can be scaled up globally (Hensen et al., 1998) and compared to the silica surface water export flux (Nelson et al., 1995). This comparison indicates that ~20% of the silica regeneration in the deep ocean occurs in the sediments. There are a number of reasons why this is a rough estimate, in part because it assumes that benthic fluxes from South Atlantic sediments are representative of those throughout the world's oceans (although also see similar estimates in Tréguer et al., 1995, based on a slightly different approach). Nevertheless, this result further indicates the importance of sediment diagenesis to deep-ocean remineralization processes.

13.2 Trace Metal Diagenesis

The geochemistry of many trace metals and trace elements in surficial marine sediments is strongly influenced by sediment organic matter remineralization processes. In deep-sea as well as some continental margin sediments, the redox cycling of iron and manganese oxides is of most importance, while in coastal sediments the redox cycling of iron sulfide minerals like FeS or pyrite plays an equally important role. The former processes will be discussed here while a discussion of the latter is reserved for section 17.3.

Linkages exist between Fe and Mn redox cycling and trace metal cycling in general because Fe and Mn oxides and sulfides are enriched in a wide range of trace metals (e.g., Cu, Ni, Zn, Cd, Mo, and Co). The processes that lead to these enrichments vary, but include simple adsorption, direct incorporation of these metals into crystal structures, and co precipitation, for example, of trace metal sulfides in conjunction with iron sulfides (Balistrieri and Murray, 1986; Morse and Luther, 1999; Glasby, 2000; Li, 2000). Furthermore, many Mn and Fe oxides occur either as coatings on other particles or as fine-grained amorphous solids, and they therefore tend to have large specific surface areas. Thus for the surface reactions discussed here, small amounts of these oxides can play an important role in controlling the geochemical cycling of trace metals.

Evidence from a wide range of sources suggests that sinking biogenic particles represent the dominant source of most metals to open ocean sediments (Collier and Edmond, 1984; Morel and Hudson, 1985; Fowler and Knauer, 1986). Organic matter itself is the dominant specific carrier phase, with biogenic carbonate and opal being of lesser importance. The mechanisms by which these metals are incorporated into sinking organic matter vary and include direct uptake by living biota (Morel and Hudson, 1985) and scavenging (adsorption) of metals onto organic surfaces in sinking particles (Balistrieri et al., 1981). Another carrier phase in sinking marine particles, lithogenic material, e.g., detrital clays, appears to be most important for iron.

Lithogenic phases have historically also been thought to be the predominant source of Al to sediments. However, recent work by Murray and Leinen (1996) suggests that water column scavenging may also be of some importance in Al transport to sediments.

Upon deposition of these particles at the sediment surface, a variety of processes operate on the associated trace metals. Perhaps the most important of these is the destruction of the organic carrier phases and the subsequent re-uptake of the metals by authigenic sediment phases; in deep-sea sediments these are generally manganese and iron oxides (Bacon and Rosholt, 1982; Thomson et al., 1984a; Chester, 2000). Both pore-water and benthic flux studies suggest that aerobic respiration either within surface sediments or at the sediment-water interface plays a major role in this release of metals associated with organic matter carrier phases (Klinkhammer et al., 1982; Heggie et al., 1986; Johnson et al., 1992). Sediment resuspension in the benthic boundary layer may also contribute to the oxidative degradation of organic matter and the release of associated metals (Walsh et al., 1988). Resuspension similarly allows sediment manganese oxides to scavenge metals remobilized in this fashion as well as scavenge dissolved metals directly from the overlying seawater (Dymond, 1984; Chester, 2000).

Evidence for the transfer of metal from organic carrier phases to metal oxides comes from a variety of studies that use both chemical leaching techniques and normative model calculations. Under most circumstances, these approaches demonstrate that the vast majority of sediment Mn is found in what is generally termed oxide, authigenic, or authigenic+hydrothermal phases[3] (see table 13.1). This is less so for iron, except for sediments that have a significant hydrothermal input (see below). Other trace metals, e.g., Co, Cu, Ni, and Zn, show similar strong associations with these oxide phases (also see Shimmield and Pedersen, 1990, for a review).

Many of these same trace elements, including Mn, also have higher absolute concentrations in deep-sea sediments as compared to either nearshore sediments or average shale (table 13.2). Furthermore, this excess manganese often extends down several meters in these sediments (e.g., Bender, 1971). For manganese, this enrichment appears

---

[3] Although different workers use different terminologies to define these phases, an examination of the various approaches used here indicates that these phases are predominantly Mn and Fe oxides and oxyhydroxides.

TABLE 13.1
The Percentage of Metals in Deep-sea Sediments Found in the "Authigenic" Fraction[a]

Element	Pacific Red Clay[b]	N. Atlantic Deep-sea Sediment[c]	N. Atlantic Hemipelagic Sediment[c]	Mid-Atlantic Ridge Sediment[c]	Nazca Plate Rise Crest Sediments[d]	Nazca Plate Basin Sediments[e]	N. Nazca Plate Sediments[e]
Mn	90%	68%	82%	86%	100%	94%	95%
Fe	<10%	18%	23%	32%	95%	75%	57%
Co	80%	48%					
Ni	80%	45%	36%	83%	79%	82%	63%
Cu	50%	56%	45%	67%	81%	72%	72%
V		29%					
Cr		30%	37%	12%			
Pb			15%	62%			
Zn					83%	68%	43%

a. See the original references for details on the analytical techniques used to make these determinations. All percentages are relative to the total metal content of the sediment.
b. From Glasby (2000).
c. From Chester (2000).
d. These are primarily hydrothermal fractions.
e. From Dymond (1981). The Nazca Plate values are the sum of authigenic + hydrothermal percentages, since both fractions are composed predominantly of metal oxides.

TABLE 13.2
The Elemental Composition of Marine Sediments and Average Shale (all $\mu g\ g^{-1}$ or ppm except Fe)

Element	Average Shale[a]	Nearshore Mud[b]	Nearshore Mud[c]	Pacific Deep-sea Clay[c]	Atlantic Deep-sea Clay[c]	Deep-sea Clay[a]	Deep-sea Clay[c]	Deep-sea Carbonate Ooze[c]
Mn	850	850	850	12,500	4,000	6,700	6,000	1,000
Fe (%)	4.72	6.50	6.99	6.50	8.20	6.50	6.00	0.9
Co	19	13	13	116	38	74	55	7
Ni	68	35	55	293	79	225	200	30
Cu	45	56	48	570	130	250	200	30
V	130	145	130	130	140	120	150	20
Cr	90	60	100	77	86	90	100	11
Pb	20	22	20	162	45	80	200	9
Zn	95	92	95	–	130	165	120	35
Mo	2.6	–	1	–	–	27	8	–

– = not reported
a. From Turekian and Wedepohl (1961). The deep-sea clay value is an average value of Atlantic and Pacific sediments.
b. From Wedepohl (1960)
c. From Chester (2000)

to result from several factors. First, the slow accumulation rates in deep-sea sediments allows Mn oxides at the sediment surface ample time to continually scavenge (oxidize) manganese from seawater, perhaps in association with particle resuspension in the benthic boundary layer (also see similar discussions in Calvert and Pedersen, 1993). In addition, the oxic conditions in most of these sediments minimize diagenetic remobilization of manganese, and its potential loss as a benthic flux (see discussions below). The enrichment of many other trace metals in deep-sea sediments therefore presumably results from their association with manganese oxides as discussed above.

Another important source of metals to some deep-sea sediments is hydrothermal input, through high temperature basalt-seawater interactions at midocean ridges. Hydrothermal vent fluids are extremely enriched in many trace metals such as Mn, Fe, Cu, Co, Zn, with concentrations more than ~3 orders of magnitude greater than typical seawater values (Elderfield and Schultz, 1996). When these vent fluids are injected into the deep ocean, the Fe^{2+} and Mn^{2+} in the vent fluids are oxidized, forming metal oxides. These fresh oxide particles are then very effective at scavenging other trace metals in the vent fluids. The ultimate sink for these oxide particles is the sediments, and metalliferous sediments found in close proximity to mid-ocean ridges, e.g., on mid-ocean ridge crests, are extremely rich in iron and manganese, generally in the form of oxides (Bonatti, 1981; Dymond, 1981). The oxidation of massive sulfide deposits that form at hydrothermal vents (seafloor weathering) can also contribute to the metal oxides found in metalliferous sediments (see discussions in Edmond et al., 1979b; Hannington et al., 1995).

Postdepositional diagenesis also affects sediment metal concentrations as well as their depth distributions. The most important of these processes are associated with Mn/Fe redox cycling, although authigenic clay mineral formation, e.g., of iron-rich nontronite clay, is also important in some sediments (Heath and Dymond, 1977; Hein et al., 1979; Cole, 1985; Kastner, 1999; also see chapter 18).

One view of sediment manganese redox cycling is shown in fig. 13.2. This model is based on early studies that showed that manganese is reduced from the +4 to the +2 oxidation state when sedimentary conditions change from "oxidizing" to "reducing" with depth (Murray and Irvine, 1895; Wangersky, 1962; Lynn and Bonatti, 1965; Li et al., 1969). In this model, manganese oxides that are input from external sources

TRACE METAL DIAGENESIS

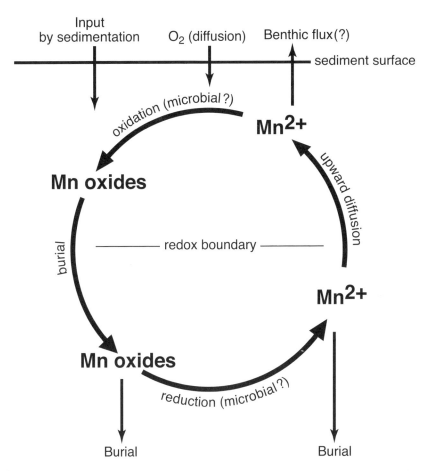

Figure 13.2 A simplified view of the manganese redox cycle in marine sediments. In the absence of oxygen, i.e., below the sediment redox boundary, manganese oxides are reduced to Mn^{2+} (see section 7.3.3), which then diffuses upward as a result of a Mn^{2+} concentration gradient. When this Mn^{2+} reaches oxygen-containing sediments, it is then oxidized back to particulate manganese oxides (see section 7.4.1). Burial of this particulate manganese then completes the cycle. Under steady-state conditions, this cycle results in the concentration of particulate, oxidized manganese in the sediments just above the sediment redox boundary. While manganese cycling in this manner appears to occur in some sediments (Burdige, 1993; Dhakar and Burdige, 1996), processes such as bioturbation (e.g., Aller, 1994b) and the occurrence of manganese oxidation using oxidants other than O_2 (see section 7.4.2) may lead to more complex manganese redox cycling in many (particularly coastal) sediments (see related discussions in section 17.3).

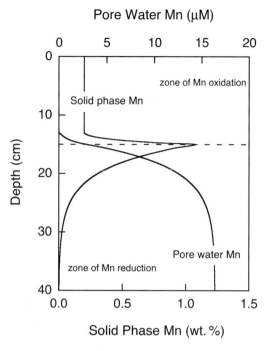

Figure 13.3 Pore-water and solid-phase sediment manganese profiles predicted by the model equations in Burdige and Gieskes (1983). Note that the equations in this diagenetic model are based on the conceptual model in fig. 13.2. Model parameters used in these calculations: $D_s = 71.6$ cm^2 yr^{-1}; $\omega = 0.02$ cm yr^{-1}; $\varphi = 0.8$; depth of the sediment redox boundary = 15 cm; $k_{ox} = 100$ yr^{-1} (first-order rate constant for manganese oxidation); $k_{red} = 0.0005$ yr^{-1} (first-order rate constant for manganese reduction).

are eventually buried below a sediment "redox" boundary, where they undergo reductive dissolution to Mn^{2+}, presumably coupled to organic matter remineralization. This increase in pore-water manganese below the redox boundary then leads to a concentration gradient and the upward diffusion of Mn^{2+} (see fig. 13.3; also see figs. 8.4 and 15.3). Upon diffusion into the upper oxidized sediments, this Mn^{2+} is reoxidized. The depth of the sediment redox boundary is generally assumed to occur where pore-water oxygen concentrations go to near-zero levels (i.e., the oxygen penetration depth), based partly on examination of pore-water manganese and oxygen profiles in pelagic and hemipelagic sediments (Shaw et al., 1990; Reimers et al., 1992). In these settings oxygen is assumed to be the primary oxidant of pore-water manganese, although in other settings such as bioturbated continental margin sediments, nitrate or perhaps iodate may actually serve as the oxidant (see reaction 4 in table 7.2; also see Luther et al., 1997; Aller et al., 1998; Anschutz et al., 2000).

Since upwardly diffusing manganese that is oxidized near the sediment redox boundary is also eventually buried below this boundary, it too is reduced as part of this internal redox cycle. Overall, in this model the net result of these processes is that a peak in particulate (sedimentary, solid phase) manganese occurs in a distinct layer just above the sediment redox boundary (Froelich et al., 1979; Burdige and Gieskes, 1983; Dhakar and Burdige, 1996). Under steady-state conditions, the depth of this peak is controlled by the balance between the upward flux of Mn^{2+} and the downward flux of O_2. Furthermore, because the depth of the redox boundary moves upward at a rate equal to the sediment accumulation rate, over time this manganese peak remains at a fixed depth relative to the sediment-water interface (Burdige, 1993).

The model profiles shown in fig. 13.3 indicate that pore water Mn^{2+} concentrations reach some asymptotic steady-state concentration at depth. The implicit assumption is that this concentration is controlled by the amount of manganese oxide available for reductive dissolution. However, in many settings, particularly in coastal or continental margin settings, the buildup of pore-water Mn^{2+} and alkalinity due to suboxic or anoxic remineralization reactions (see section 16.1) leads to a situation in which the sediments become supersaturated with respect to a Mn carbonate phase, either rhodochrosite ($MnCO_3$) or the mixed-phase kutnahorite ($Mn_xCa_{1-x}CO_3$). At depth, then, pore water Mn^{2+} concentrations can be controlled by the formation of such authigenic phases, although the factors controlling which exact phase forms in a given sediment can be complex (Holdren et al., 1975; Robbins and Callender, 1975; Aller, 1980b; Pedersen and Price, 1982; Middelburg et al., 1987; Shimmield and Pedersen, 1990). Furthermore, in some settings pore-water manganese concentrations also decrease with depth in the reducing portions of the sediments as a result of these precipitation reactions. Presumably this occurs when the sediments become supersaturated with one of these mineral phases and alkalinity production at depth exceeds Mn^{2+} production. Another important aspect of Mn carbonate formation is that it generally occurs some distance away from the manganese oxide peak (usually greater than ~5–10 cm), as a diffuse precipitate dispersed through these deeper sediments (also see Thomson et al., 1986, and discussions in section 14.5.2).

From a paleoceanographic perspective, the presence of manganese carbonates in sediments is taken generally as an indicator of suboxic sediments deposited under oxygenated bottom waters (Calvert and Pedersen, 1996; Calvert et al., 2001). Such settings allow Mn oxides to accumulate in surface sediments because the redox cycling outlined above (e.g., fig. 13.2) acts to "trap" manganese in the sediments. This then allows Mn oxides to eventually be reduced at depth where Mn^{2+} can ultimately precipitate out as a carbonate phase. In contrast, sediments deposited under anoxic bottom waters are very low in solid-phase manganese for two reasons. First, manganese oxides in sinking particles dissolve in the water column before reaching the sediments. And secondly, the absence of a surface oxidized zone in the sediments does not allow for the transfer of any manganese from organic matter carrier phases to authigenic oxide phases (also see results in Shaw et al., 1990). As a result, pore-water Mn^{2+} concentrations never build up to sufficient levels to allow for the formation of authigenic Mn carbonates at depth in these sediments.

While a peak in solid manganese oxide is seen in some marine sediments in conjunction with this sediment redox boundary, other times what is generally observed is a surface oxidized zone in the sediments that is enriched in solid manganese (see Gratton et al., 1990; Rabouille and Gaillard, 1991b; Dhakar and Burdige, 1996; and references therein). In contrast to manganese enrichments seen to depths of several meters in oxic, deep-sea sediments, here this enrichment of manganese generally occurs in the upper ~10–20 cm of sediment (e.g., Bender, 1971). Thus, in contrast to the model described above for the "deep" manganese enrichments, this latter form of manganese enrichment is strongly related to sediment manganese redox cycling.

In particular, bioturbation plays a role here by mixing, or dispersing, manganese oxides produced at the sediment redox boundary throughout the surficial sediment region. Model results in Dhakar and Burdige (1996) demonstrate that the depth zonation of bioturbation versus that of manganese oxidation plays the key role in determining which type of sediment manganese distribution results from this internal manganese redox cycling. When the sediment redox boundary occurs relatively deep in the sediments (greater than ~10 cm), most manganese oxidation occurs below the zone of active bioturbation (see section 12.4), and a well-developed Mn peak forms in the sediments. In contrast, a shallower sediment redox boundary leads to manganese oxidation occurring in the region of active sediment mixing, and a surface Mn-rich sediment layer simply forms instead.

The diagenetic remobilization of manganese in sediments also affects the sediment distributions of many trace metals. Evidence of this can be seen, for example, in pore-water profiles from suboxic sediments in which there are similar increases in the concentrations of Mn, Ni, and Co (Klinkhammer, 1980; Salwan and Murray, 1983; Rutgers van der Loeff, 1990). This covariance implies that these other metals are solubilized along with the reductive dissolution of manganese oxides. Similarly, the surficial enrichment of Mn in many suboxic sediments is also accompanied by enrichments in metals such as Ni, Cu, Zn, Mo, Co (Graybeal and Heath, 1984; Balistrieri and Murray, 1986; Shimmield and Pedersen, 1990). In hemipelagic, and some nearshore, sediments, surficial solid-phase Mn concentrations often exceed those in pelagic sediments (Calvert and Pedersen, 1996). However, because the internal diagenetic processes discussed above eventually reduce these Mn oxides at depth, the amounts of Mn that are ultimately buried in such hemipelagic sediments are much lower than those found in pelagic sediments (fig. 13.4).

Diagenetic remobilization of Mn, Fe, and associated trace metals in sediments also plays a role in controlling their net accumulation in sediments. Since the sediment redox boundary corresponds roughly to the zero-oxygen level in the sediments, increasing carbon rain rate decreases the oxygen penetration depth/sediment redox boundary. One can then think of this as bringing the upper curve of the pore-water Mn^{2+} profile (see fig. 13.3) increasingly closer to the sediment-water interface. Under most circumstances, much (if not all) of this upwardly diffusing manganese is reoxidized in the sediments as a part of the internal redox cycling discussed above. However, as the sediment redox boundary shallows, the relatively slow kinetics of Mn^{2+} oxidation may allow for some of this Mn^{2+} to escape the sediments as a benthic flux (see discussions in Shimmield and Pedersen, 1990, and references therein). For example, in the hemipelagic metalliferous sediments at MANOP site M (see fig. 8.5), steady-state model calculations suggest that >90% of the manganese deposited in these sediments is lost as a benthic flux of Mn^{2+}. Other studies (Dymond, 1984; Kalhorn and Emerson, 1984; Heggie et al., 1986) similarly argue that much of the

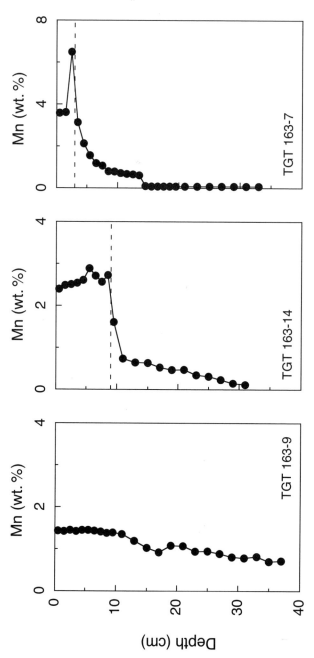

Figure 13.4 Solid-phase sediment manganese profiles for three sites along the Baja California (northwest Mexican) continental margin (redrawn using results presented in Shimmield and Price, 1986). Cores TGT 163-14 and TGT 163-7 are hemipelagic sediments at water depths of 3,168 m and 3,647 m respectively. The dashed line in these two plots represents the approximate depth of the sediment redox boundary (as inferred in this study from the appearance of Mn^{2+} in the pore waters). In these cores, the sediments are red-brown in the upper oxidizing portion of the sediments and gray-green in the lower reduced sediments. Core TGT 163-9 is a more pelagic sediment at a water depth of 3,684 m. This core is more oxidizing and shows no color transition with depth (i.e., it is red-brown throughout).

particulate Mn flux to these hemipelagic sediments does not accumulate in the sediments; however, these works suggest mechanisms that are different from the steady-state explanation presented in fig. 8.5.

In contrast, slightly different results have been obtained in studies of benthic Mn fluxes from sediments subject to less intense Mn input from hydrothermal sources (Johnson et al., 1992). In a transect across the continental shelf and upper slope of the California margin these benthic lander studies suggest that the Mn flux from these sediments can be explained simply by Mn release near the sediment-water interface from organic matter and calcium carbonate carrier phases, that is, as a result of aerobic respiration and release of Mn bound to carbonate particles by sediment carbonate dissolution. These authors argue that Mn^{2+} regenerated by the reduction of sediment Mn oxides plays a minor role in controlling the net benthic flux of Mn^{2+} from these sediments.

For reasons outlined above the upward flux and loss of dissolved manganese from hemipelagic sediments should similarly lead to the loss of other metals associated with sediment manganese oxides. By extension, Glasby (2000) suggests that the reason coastal sediments have metal concentrations that are close to average shale values (table 13.2) is that they do not incorporate this authigenic oxide fraction to nearly the same extent as do deep-sea sediments.

For the trace elements discussed here, their association with Mn and Fe oxides leads to their undergoing indirect (or passive) sediment redox cycling; i.e., changes in redox conditions do not generally change the oxidation state or speciation of these metals, nor directly affect their solubility. In contrast, there are several other trace elements, such as V, Re, and U, that do have multiple oxidation states and also change their solubility with changing (generally reducing) redox conditions. The active redox cycling of these elements in sediments shows some general similarities to that seen for sediment Mn and Fe redox cycling (see discussions in Morford and Emerson, 1999, for further details). At the same time, sediment profiles of these redox-sensitive trace elements have the potential to be used as paleoceanographic indicators of past changes in sediment redox conditions due to changes in either bottom-water oxygen concentrations or the organic matter flux to the sediments. Additional details about the geochemistry of these elements can be found in several recent publications and references therein (Morford and Emerson, 1999; Crusius and Thomson, 2003; also see earlier discussions in Calvert and Pedersen, 1993).

13.3 Manganese Nodules and Crusts

In addition to manganese and iron oxides that form in sediments, there are three other major types of manganese and iron deposits in marine environments. The first are *manganese nodules*, sometimes also referred to as ferromanganese nodules. These are spheroidal or discoidal concretions, generally a few centimeters in diameter, which often contain a central nucleus or core consisting of, for example, skeletal materials such as a shark's tooth, or fragments of an older nodule. Nodule material that grows around this nucleus also often consists of concentric layers or laminae that are traceable around the entire circumference of the nodule. Manganese nodules grow very slowly and are found in almost all of the major ocean basins. Knowledge of deep-sea manganese nodules dates back to results from the *Challenger* expedition (1872–76; Murray and Irvine, 1895).

Second, in some discussions deep-sea manganese nodules are distinguished from *ferromanganese concretions* that form in shallow marine environments like the Baltic Sea, and in many high-latitude temperate lakes. These concretions have much more rapid growth rates than deep-sea nodules and differ from deep-sea nodules in shape, mineralogy, and composition. Finally, *manganese crusts* accumulate on the exposed basalt on seamounts and along mid-ocean ridges. The discussion here will focus on manganese nodules (primarily in the deep sea); additional information on these other types of manganese deposits can be found in several reviews (Glasby, 1977, 1984, 2000; Calvert, 1978).

As noted above, manganese nodules are found in all of the major ocean basins; in some portions of the Pacific and Indian Oceans greater than 50% of the seafloor is covered by nodules (see references above and Chester, 2000). Regions where they are in the greatest abundance (as percent surface coverage of the sea floor) are generally slowly accumulating red clay and siliceous ooze sediments in deep ocean basins (water depths generally greater than ~4,000 m; see map in Glasby, 2000). Relict nodules are also found buried in deep-sea sediments. Deep-sea nodules have extremely slow growth rates, on the order of millimeters per million years; these growth rates are much slower than pelagic sedimentation rates in the regions where nodules are predominantly found (millimeters per thousand years). The fact that not all nodules are completely buried by depositing sediment will be discussed below.

Manganese nodules are composed primarily of amorphous and crystalline forms of iron and manganese oxide. Lesser amounts of detrital, and sometimes authigenic, aluminosilicates and other biogenic phases also occur in nodules. The predominant manganese oxide phases are $\delta\text{-}MnO_2$ (vernadite), birnessite (7 Å manganite), and todorkite (10 Å manganite or buserite; see section 7.3.2 for references discussing manganese oxide mineralogy). The iron oxides tend to be amorphous oxyhydroxides ($FeOOH \cdot nH_2O$).

In addition to scientific curiosity, interest in the study of manganese nodules stems from the fact that they contain high concentrations of trace metals such as Ni, Cu, and Co (see the discussion in the previous section for details). The economic potential of such deposits has been discussed since the mid-1960s (Glasby, 2000), although political issues and technological/economic considerations have precluded the mining of deep-sea nodules.

In light of previous discussions regarding biological involvement in manganese redox cycling (sections 7.3.3 and 7.4.1), the role that bacterial manganese oxidation may play in nodule formation has been a topic of great interest (see Ehrlich, 1996, for a summary). In part, discussions of biological involvement in manganese nodule formation are based on observations of the internal layering of nodules and their possible analogy to biologically produced stromatolites found in shallow-water carbonate environments (see Burnett and Nealson, 1981, for a summary of these arguments). While the discussion below will clearly describe a number of ways that biological processes may play some role in nodule genesis and perhaps in the formation of their internal layering, the exact mechanisms by which this occurs may not necessarily be that suggested by earlier workers.

Studies examining the chemical (trace element) and mineralogical composition of deep-sea manganese nodules have documented their widespread variability on regional scales as well as within individual nodules, e.g., nodule tops versus bottoms. Attempts to explain these differences have focused on possible different sources of metals to nodules, as well as on modes of formation. The most comprehensive discussion of this problem is that presented by Dymond et al. (1984) in which they define three end-member modes of nodule accretion (also see table 13.3). These modes of nodule accretion will be discussed below, followed by a discussion of their overall relative importance in the formation of manganese nodules in the marine environment.

TABLE 13.3
Proposed End-member Modes for the Accretion of Deep-sea Manganese Nodules[a]

Mode of Accretion	Description/Occurrence/Details	Nodule Growth Rate ($mm/10^6\ yr$)	Dominant Mn Oxide Mineralogy	Mn Content	Mn/Fe	Mn/Ni[b]
Hydrogenous precipitation	Direct precipitation or accumulation of metals from seawater. In this mode of accretion the newly-produced material is not in contact with sediment.	1–2	$\delta\text{-MnO}_2$	22%	~1	30–50
Oxic Diagenesis	Accretion occurs in (or in contact with) the oxic zone of the sediments. The most important process here is likely oxic (aerobic) remineralization of sediment organic matter.	10–50	Todorokite (Cu and Ni rich)	32%	5–10	15–20
Suboxic Diagenesis	Accretion driven by reductive dissolution of sediment manganese oxides. This is likely to be an episodic (nonsteady state) process.	100–200	Todorokite (unstable because of low levels of Cu and Ni; transforms to birnessite upon dehydration)	48%	20–70	60–200

a. Modified from Dymond et al. (1984).
b. Note that similar trends are also seen in the ratio Mn/(Cu+Ni+Zn).

The first mode of nodule accretion, *hydrogenous precipitation*, involves direct uptake of metals from seawater into nodules. The second mode, accretion via *oxic diagenesis*, involves processes occurring in oxic sediments that release metals ultimately incorporated into nodules. Several reactions may be involved here, although the most important is likely the release of metals from organic matter during oxic diagenesis (as described in the previous section). Other possible reactions include equilibrium adsorption/desorption interactions between nodules and sediments, or the formation of nontronite; this latter reaction between iron oxides and dissolved silica fixes iron and silica into this authigenic clay and releases manganese and other trace metals for uptake into nodules. The third mode of accretion involves *suboxic diagenesis*, i.e., reductive dissolution of sedimentary manganese oxides and transfer of this manganese and associated metals to nodules.

As shown in table 13.3, these modes of accretion result in material with differing elemental composition and mineralogy, and they also lead to different nodule growth rates. Dymond et al. (1984) were also able to show that these different modes of accretion can explain regional differences in nodule composition at three contrasting MANOP sites. At MANOP site R (the most "oxic" of the sites they studied, with the smallest carbon rain rate to the sediments) hydrogenous accretion and oxic diagenesis are both important, while at MANOP site S higher biogenic (organic matter) fluxes lead to a more important role for oxic diagenesis in nodule formation. Nodule accretion as a result of suboxic diagenesis appears to be important only at MANOP site H, a hemipelagic site that underlies productive waters in the eastern tropical Pacific.

Although nodule composition data at site H suggest the occurrence of the suboxic diagenesis accretionary process in the formation of nodules at this site, pore-water data are somewhat inconsistent with the possibility. In these sediments, the major zone of manganese reduction, based on pore-water Mn^{2+} profiles, occurs at a sediment depth of ~10 cm (Klinkhammer, 1980; Bender and Heggie, 1984), and because this is not sufficiently close to the sediment-water interface, diffusion or bioturbation (at deep-sea rates) is likely too slow to transport this solubilized manganese to the sediment surface before it is reoxidized. As a result, manganese and associated trace metals remobilized in this deep region of the sediments at this site are likely retained by the upper, oxic sediments, and therefore not available for nodule accretion. Boudreau and Scott (1978) came to similar conclusions regarding the source of manganese to nodules in general.

To reconcile these contrasting observations Dymond et al. (1984) suggest that nodule accretion as a result of suboxic diagenesis is not a steady-state process. As they discuss, short-term (seasonal) variations in surface water productivity may lead to pulses of organic matter deposition to the sediments that then lead to transient (micro ?) "zones" of manganese reduction in oxic surficial sediments. Episodic deposition of phytodetritus, i.e., fresh phytoplankton debris, on the deep-sea floor (generally following surface-water phytoplankton blooms) may also represent another mechanism for this organic matter input (e.g., see Smith et al., 1996, and references therein). Measurements of the sediment manganese oxidation state (Kalhorn and Emerson, 1984) and interfacial pore-water manganese profiles (Heggie et al., 1986) provide additional evidence in support of the occurrence of some types of manganese reduction in surficial sediments such as those at site H, and of the possible role that this might play in remobilizing metals for nodule formation by this suboxic accretion mechanism (also see Burdige, 1993, for a more detailed discussion of these results). Finally, consistent with its presumed nonsteady state (or episodic) nature, nodules at MANOP site H appear to grow by this suboxic diagenesis mechanism only ~7% of the time that growth by oxic diagenesis occurs (Dymond et al., 1984).

The discussion above suggests that the growth of manganese nodules may occur by three possible mechanisms; however, it is also important to examine the relative significance of these three processes in terms of overall manganese nodule formation in the deep sea. First, as noted above, nodule growth by suboxic diagenesis appears to be an episodic process, and not associated with suboxic diagenesis within the sediments. Thus while certain aspects of the growth histories of some nodules may be explained by this process, suboxic diagenesis is likely not the primary mechanism by which most manganese nodules grow in the deep sea.

This then suggests that hydrogenous precipitation and/or oxic diagenesis represent the dominant accretionary processes for deep-sea manganese nodules. This suggestion is also consistent with the general prevalence of nodules in highly oxidizing deep-sea sediments with low sediment accumulation rates and low sediment TOC contents. However, an attempt to further differentiate between the importance of these two modes of accretion is somewhat difficult.

For example, model calculations by Boudreau and Scott (1978) suggest that the growth of Mn nodules is controlled largely by mass transfer (diffusion) of Mn through the diffusive boundary layer above the nodule surface (e.g., see section 12.2). Based on typical current velocities in the deep-sea and seawater manganese concentrations, they then calculate that the uptake rate of manganese by nodules should range from 0.02 to 1.7 g Mn cm$^{-2}\cdot 10^6$ yr^{-1}, a range consistent with bulk nodule Mn accumulation rates (e.g., Bender et al., 1971; Moore et al., 1981). Furthermore, based on the results in table 13.3, Mn accumulation rates by hydrogenous precipitation are ~0.06–0.11 g Mn cm$^{-2}\cdot 10^6$ yr^{-1} and ~0.8–4 g Mn cm$^{-2}\cdot 10^6$ yr^{-1} for nodule formation by oxic diagenesis. As can be seen here, both estimates are roughly consistent with Boudreau and Scott's (1978) diffusion-controlled uptake rates.

At the same time, depending on whether the mode of oxic diagenesis responsible for nodule formation occurs in the sediments or at the sediment-water interface, the ability to differentiate between hydrogenous precipitation or nodule formation by oxic diagenesis may be difficult (e.g., the uptake by nodules of metals remobilized by oxic diagenesis on the sediment surface in the benthic boundary layer could still be mass-transport controlled). Therefore, on the one hand the results from Dymond et al. (1984) may indeed indicate that trace elements in nodules have multiple sources, i.e., hydrogenous versus diagenetic. However, on the other hand, the calculations presented by Boudreau and Scott (1978) may similarly suggest that the overall growth rate of nodules (and perhaps ultimately the source of the major components of nodules, e.g., Mn and Fe) is largely controlled by mass transport from the water column to nodules through the diffusive boundary layer.

The discussion above focuses, relatively speaking, on the bulk accretion of nodule material. However, it does point out that top/bottom differences in nodule chemistry may be explained by the occurrence of these different accretionary processes, i.e., nodule tops accumulating by hydrogenous accumulation and nodule bottoms accumulating by some combination of oxic and suboxic diagenesis. At the same time, though, numerous studies have shown that a common feature of many nodules, in both marine and freshwater environments, is their laminated structure, with concentric internal bands often alternatingly rich in iron and manganese (Sorem and Fewkes, 1977; Dean and Ghosh, 1978; Moore, 1981; Piper and Williamson, 1981; Han et al.,

2003; and others). Visual observations of these bands indicate that their thicknesses can be quite variable, i.e., some may be of near equal thickness while others may vary in thickness.

Early models for the occurrence of this layering (see Glasby, 1977) generally proposed mechanisms in which changes in the environment at the nodule surface alternately favored the accretion of manganese or iron-rich layers. Piper and Williamson (1981) also suggested that individual layers result from differences in the relative contributions of a seawater source versus a sediment interstitial water source to the nodule. Therefore, when re-examined in the context of the discussions above, changes in the relative contributions of hydrogenous accretion versus diagenetic accretion (either oxic or suboxic) could, under some conditions, be a more plausible explanation for this layering. For example, visual observations of nodule fields (Dymond, 1984; Gardner et al., 1984) suggest that cycles of nodule burial by sediment followed by exhumation by benthic organisms are common. Such cycles could then lead to the required changes in nodule accretion discussed above. Similarly, if nodules are periodically turned over (as discussed in Moore et al., 1981; Mangini et al., 1990), this could also lead to changes in the relative roles of different nodule accretion processes to a given portion of a nodule.

Studies of freshwater nodules in Oneida Lake, New York (USA), provide additional information on a slightly related mechanism for alternating manganese- and iron-rich layers in nodules. Here Moore (1981) has shown that temporary burial of nodules by organic-rich sediment, e.g., sedimentation of phytoplankton debris after the crash of a bloom in the surface waters, leads to the development of anoxic conditions at the nodule surface. With this, the selective remobilization of manganese in the surface layer of the nodule then occurs, since manganese reduction is thermodynamically favored over iron reduction (see section 7.3.3). This then transforms the nodule surface, which initially accretes in Oneida Lake nodules as a manganese-rich layer, into one that is now enriched in iron, because of selective removal of manganese from the outer layer of the nodule. The thickness of this "new" iron-rich layer relative to the now deeper manganese-rich layer depends on the time of nodule accretion under oxic conditions versus the time of nodule burial under reducing conditions. Removal of the anoxic sediment cover, e.g., by a storm event, leads to the resumption of fresh manganese-rich material accreting at the nodule surface, and the continued repetition of this process should then lead to the observed banding in these nodules. Consistent with this model is the observation that Oneida Lake nodules undergo periods of rapid growth (>1 mm·10^2 yr^{-1}) separated by periods of either no net growth or net chemical erosion (Moore et al., 1980).

This mechanism could also play a role in the growth history of deep-sea nodules, although these nodules have not been examined in the context of such processes. However, in light of the discussions above the occurrence of such processes with deep-sea nodules does not appear to be unreasonable. A growth hiatus in deep-sea nodules, which must accompany this mechanism, is difficult to detect in light of the extremely slow growth rates of deep-sea nodules as compared to freshwater nodules. However, evidence for this does exist in high-resolution radiochemical profiles of some Pacific nodules (Han et al., 2003; also see related discussions in Mangini et al., 1990).

As a final possible explanation for the occurrence of this layering, Han et al. (2003) used spectral analysis of these layers in a nodule collected in the central Pacific to suggest that the frequencies of layering strongly resemble that seen in Milankovitch orbital cycles. The mechanism by which this rhythmic growth of nodules is linked to changes in Earth's orbital parameters is not well understood or constrained, if indeed there is such a linkage; however, these authors speculate that changes in the strength and intensity of Antarctic bottom water flow may play a role (see related discussions in Mangini et al., 1990).

As noted above, one of the intriguing questions associated with manganese nodules is how objects with such extremely slow growth rates (relative to local rates of sediment accumulation) remain at the sediment surface. This problem may be answered with relatively simple models that suggest a steady state likely exists between nodule accretion and growth at the sediment-water interface, and nodule burial in the sediments (Bender et al., 1966; Heath, 1978). For example, if the average age of nodules in a given region is ~ 1 million years and the average sediment accumulation rate at this site is 4 m·10^6 yr^{-1} ($= 0.4$ cm·10^3 yr^{-1}), then the number of nodules buried in the upper 4 m of sediment (per m^2 of seafloor) should roughly be the same as the number found on the sediment surface (see similar arguments in Broecker and Peng, 1982). Data presented in Bender et al. (1966) and Heath (1978) are roughly consistent with this model, within a factor of ~ 2–5.

Related to this observation is that when one examines manganese accumulation rates in both nodules and underlying sediments it is seen that sediments actually accumulate more manganese than do nodules (by a factor of ~2 or more; Bender et al., 1970). Thus while nodules *appear* to be richer in manganese, their much slower growth rates leads to nodules actually accumulating *less* manganese (again per m^2 of seafloor) than do the sediments on which they rest (also see Broecker and Peng, 1982).

Taken together, these two observations imply that controls on the growth of manganese nodules are related to processes that lead to nodule formation (e.g., the availability of a suitable nucleating agent) and factors that then keep nodules at the sediment-water interface (e.g., the probability of burial as discussed above). Identifying the processes that keep nodules at the sediment surface has been, and continues to be, a challenging problem. Over the years, a wide range of explanations have been proposed, although present consensus is that the activity of benthic macrofauna may play the dominant role (Heath, 1978; Broecker and Peng, 1982; Gardner et al., 1984; Sanderson, 1985). While the precise details of how this occurs are not known, they may include nodule excavation (or upward nudging) during bioturbation, or feeding of sediment on nodule tops and redeposition of sediment under the nodule (upward wedging). The required frequency of these events needed to maintain nodules at the sea floor does not appear to be unreasonable, based on nodule dynamics and sediment accumulation rates in the deep sea. However, given the slow growth rates of nodules, direct observation (and therefore "proof") of the occurrence of these processes has proven to be difficult to obtain on human time scales (Gardner et al., 1984).

13.4 Diagenesis of Opaline Silica

Biogenic silica, or opal, is an important component of many marine sediments, particularly those underlying upwelling areas (see section 2.2 and fig. 2.1). However, only a small fraction of the opal produced in surface waters escapes dissolution in either the water column or the sediments. Nevertheless, understanding the factors controlling opal diagenesis in sediments, i.e., opal burial in sediments, is important for several reasons. The first is that biogenic silica deposition in sediments has the potential to be useful as an indicator of paleoproductivity and past carbon export, assuming that the relationship between opal production (and primary productivity) in surface waters

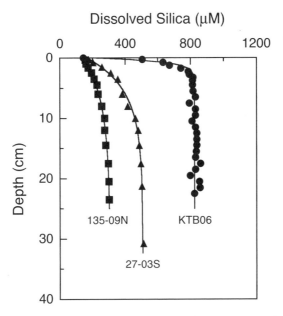

Figure 13.5 Pore-water silica profiles from sites along 140°W in the equatorial Pacific (135-09N and 27-03S) and from a site in the Indian sector of the Southern Ocean (KTB06). Data sources: McManus et al., 1995 (equatorial Pacific); Rabouille et al., 1997 (Southern Ocean).

and opal burial in sediments can be well constrained (McManus et al., 1995; Nelson et al., 1995; Ragueneau et al., 2000). The second is that work in recent years has lead to a re-evaluation of the role of silica diagenesis in sediments in the low-temperature formation of authigenic clay minerals through reverse-weathering type reactions (e.g., Mackin, 1987; Michalopoulos and Aller, 1995; Dixit et al., 2001). Additionally, this recent work suggests that authigenic mineral formation may have important implications for silica burial in sediments and the oceanic silica budget (Michalopoulos and Aller, 2004).

Pore-water silica profiles have proven to be an important tool for studying silica diagenesis in marine sediments (see recent summaries in Emerson and Hedges, 2003; Martin and Sayles, 2003). In most deep-sea sediments, silica profiles increase in an asymptotic fashion, approaching a constant value at depth (fig. 13.5). A number of different diagenetic models, of varying degrees of complexity, have been developed to explain these profiles (see Boudreau, 2000, for a summary). Regardless of the model, all require (predict) that the rate of biogenic silica dissolution goes to zero with sediment depth. This is necessary to produce profiles with the shape of those observed (see analogous discussions in section 6.3.4), and is also consistent with measured rates of opal dissolution in sediments (Van Cappellen and Qiu, 1997a).

The equation used in most of these models for the rate of silica dissolution is based on early laboratory experiments in which the rate of biogenic silica dissolution was observed to be a linear function of the degree of solution undersaturation (see Van Cappellen and Qiu, 1997a, for a summary of this work). This rate equation (R) is expressed as

$$R = k_s B(C_\infty - C) \qquad (13.2.)$$

where k_s is the rate constant for dissolution, B is the concentration of biogenic silica in the sediments, and C_∞ is the asymptotic pore-water concentration observed at depth (its relationship to the solubility of biogenic silica will be discussed below).

While diagenetic models that use eqn. (13.2) provide an adequate fit to the observed data, there are several conceptual problems that limit the interpretation of these model results. The first is that values of k_s determined with these models are generally substantially lower than analogous laboratory-derived values (see Dixit et al., 2001, for a summary of these observations). Recent studies using flow-through reactors (versus batch reactors) also suggest that the rate of biogenic silica dissolution is a nonlinear function of the degree of undersaturation, particularly at low degrees of undersaturation (Van Cappellen and Qiu, 1997a). Such results are, however, consistent with some models that assume k_s decreases exponentially with depth (e.g., McManus et al., 1995).

Most importantly, though, these results also imply that C_∞ values should equal C_{sat}, the solubility of biogenic silica. However, in sediments where biogenic opal is preserved C_∞ values are highly variable geographically. These C_∞ values range from ~100 to 900 μM (see recent summaries in Ragueneau et al., 2000; Emerson and Hedges, 2003)—typically smaller than the reported solubility of "pure" biogenic silica, which is generally > 1,000 μM (see Dixit et al., 2001, for a summary). At the same time, studies using stirred flow-through reactors to determine the "apparent" solubility of biogenic silica in natural sediments, containing ~4–80% biogenic silica, show apparent biogenic silica solubilities (C_{sat}) that have a range comparable to that observed for pore-water C_∞ values (Van Cappellen and Qiu, 1997a,b;

Gallinari et al., 2002). However, in all cases the asymptotic pore-water silica concentrations observed in sediment cores from the same sites were lower than these apparent solubility concentrations (i.e., at any given site $C_\infty < C_{sat}$).

At least three models have been put forth to explain these observations. All three predict that rates of opal dissolution decrease with sediment depth, which is required (as discussed above) to explain the observed pore-water profiles. However, the three models differ in terms of the factors ultimately controlling opal dissolution in sediments. They also have differing implications in terms of the paleoceanographic interpretation of sediment opal profiles.

The equilibrium model assumes that these asymptotic values of C_∞ represent the solubility of opal in a given sediment and that opal solubility varies geographically (see discussions in Archer et al., 1993). While the discussion above indicates that the apparent opal solubility does vary geographically (for reasons that will be further discussed below), it has also been shown that C_∞ values are less than measured C_{sat} values in the same sediment. Thus this model likely does not explain the field observations.

The reactive opal model assumes that there are at least two fractions of biogenic opal: at least one fraction that is reactive and completely dissolves before being buried, and one that is preserved. In this model the interplay among the flux of reactive opal to the sediments, opal dissolution kinetics, and bioturbation controls the magnitude of pore-water C_∞ values, which then never actually reach the solubility concentration of this reactive phase. While quantitative diagenetic models can be developed based on this conceptual model that fit observed pore-water profiles (Schink et al., 1975), other evidence from a variety of sources does not fully support this model (McManus et al., 1995; Dixit et al., 2001; Emerson and Hedges, 2003).

The third model, the surface coatings model, assumes that new silica-containing surface coatings, e.g., authigenic clays, form on biogenic opal particles during diagenesis/dissolution. The formation of such surface coatings, which are often enriched in Fe and Al, is well documented (see Michalopoulos and Aller, 2004, for a summary), and these coatings then lower the opal surface area available for dissolution. Because silica dissolution is surface controlled (Van Cappellen et al., 2004), these coatings will also eventually lower the opal dissolution rate. As a result, here the value of C_∞ will be controlled by a balance between opal dissolution and silica uptake into these newly

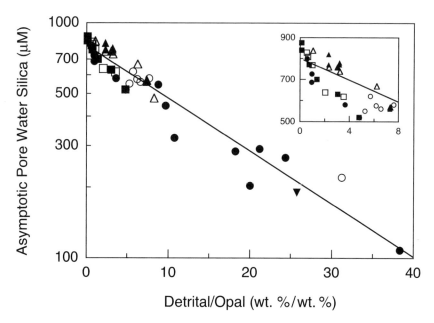

Figure 13.6 Asymptotic pore-water silica concentrations at depth (e.g., see fig. 13.5) as a function of the ratio of detrital material to opaline silica in the core. Also shown as an insert are the results from the upper left corner of this plot on an expanded scale. Data sources: ● = Boetius et al. (2000) for sites in the Scotia Sea, Norwegian Sea, Peru Basin, and Northeast Atlantic and along the Juan de Fuca ridge; ▼ = Gallinari et al. (2002) for a site on the Porcupine Abyssal Plain; ■ = Rabouille et al. (1997) for sites in the Indian sector of the Southern Ocean; □ = King et al. (2000) for sites in the Antarctic South Atlantic; ▲ and △ = Sayles et al. (2001) for sites in the Southern Ocean (also see Martin and Sayles, 2003); ○ = Koning et al. (1997) for sites in the Northwest Indian Ocean.

forming phases (Mackin, 1987; McManus et al., 1995; Rabouille et al., 1997; Dixit et al., 2001).

Recent evidence suggests that this third model is likely the best explanation for the controls on pore-water silica concentrations and opal preservation in sediments. Most relevant to this suggestion is the observation that values of C_∞ in sediments show a systematic decrease as the ratio of detrital material to biogenic silica in the sediments increases (fig. 13.6). Originally observed by Van Cappellen and Qiu (1997b) in studies of Southern Ocean sediments (siliceous oozes with >50% biogenic silica), recent work has extended this general trend, albeit with a bit more scatter, over a broader range of sediments with lower biogenic silica contents. Similar trends are also observed in lab studies in which different types of biogenic silica, i.e., siliceous ooze sediments or freeze-dried diatom cultures, are mixed with varying amounts of model detrital components (e.g., kaolinite or fresh basalt) and then allowed to come to steady state with respect to apparent opal solubility (Dixit et al., 2001; Van Cappellen et al., 2004).

In these studies the controls on opal solubility have been examined as a function of the amounts of detrital material present; however, the dissolved Al derived from this detrital material actually appears to be the main controlling factor. Uptake of Al into biogenic opal, i.e., direct incorporation of Al^{3+} into the silica crystal lattice, is known to decrease silica solubility (see Dixit et al., 2001, and references therein). While some amount of Al can be directly incorporated into siliceous frustules when they initially form in the surface ocean, it appears that the more important process is Al uptake by siliceous debris very quickly after the material is deposited in sediments (from Al release into sediment pore waters by the dissolution of detrital aluminosilicates; Van Cappellen and Qiu, 1997b; Dixit et al., 2001).

While these equilibrium effects depress biogenic opal solubility after its deposition, they only partially explain pore-water silica profiles, because even when the effects of Al uptake by opal are taken into account, sediment pore waters never achieve solubility with respect to opal dissolution (Van Cappellen et al., 2004). Therefore a second possible mechanism is the inhibition of silica dissolution by Al adsorption onto reactive surface sites of biogenic silica particles (Dixit et al., 2001; Dixit and Van Cappellen, 2002). This process is likely part of more general changes in the surface chemistry of biogenic silica particles as they age, both in the water column and in sediments, that decrease the overall reactivity of the particles toward dissolution (also see Van Cappellen and Qiu, 1997a).

Results in Dixit et al. (2001) also suggest that as pore-water Al concentrations build up above some threshold level, they induce the formation of a new authigenic aluminosilicate phase. As discussed above, if this material then coats opal particles it will also decrease the surface area available for biogenic silica dissolution. This reduction in reactive surface area then likely leads to a situation in which the balance between opal dissolution and silica uptake by these newly forming phases controls the value of C_∞ in the pore waters. The composition—Si/Al ratio, for example—of the dissolving detrital phases likely affects the composition of the newly forming aluminosilicates and, hence, the values of C_∞ in pore waters (Mackin, 1987). This explanation may then explain some of the observed scatter in fig. 13.6, i.e., different C_∞ values for similar detrital/opal ratios from different regions, since detrital components in these different regions will vary in terms of their specific mineral content, weathering history, and therefore reactivity toward dissolution.

The discussion here has focused on the implication of these results to silica diagenesis in sediments. However, from a broader standpoint, these results also provide evidence for the importance of authigenic mineral formation in the oceans, and the role of reverse weathering processes in balancing the marine geochemical budgets for major elements brought to the oceans by weathering reactions on land (Mackenzie and Garrel, 1966). A more detailed discussion of these processes will be presented in chapter 18.

As noted in the beginning of this section, biogenic silica deposition in sediments has the potential to be useful as an indicator of paleoproductivity and past carbon export. However, this use requires that the relationship between opal productivity and primary production in surface waters and opal burial in sediments be well constrained. An initial examination of this problem indicates that opal burial efficiency (OBE[4]) in surface sediments shows large variability, both across as well as within different geographic areas. In a recent summary, Martin and Sayles (2003) report OBE values that range from less than a few percent in opal-poor sediments to >30% in Southern Ocean siliceous ooze sediments. Understanding the factors controlling OBE values is important in this paleoceanographic context because changes in biogenic silica in longer cores, e.g., on glacial-interglacial time scales, may then reflect either changing productivity, i.e., silica rain rate, or preservation, i.e., burial efficiency (see discussions in Archer et al., 1993; McManus et al., 1995).

In attempts to determine at least empirical relationships that begin to separate these effects, two recently proposed nonlinear equations relating OBE and sedimentation rate appear to show good fits (within ~25%) to the available data (Ragueneau et al., 2000; Sayles et al., 2001). Since work discussed above suggests the importance of detrital matter input along with opal rain in controlling opal dissolution and preservation, it is interesting to speculate that these empirical rela-

---

[4] See eqn. (8.19) and associated discussions for a definition of burial efficiency.

tionships between OBE and sedimentation rate somehow incorporate these effects; however, this suggestion will need to be further examined and tested.

The next important step in the use of opal burial as a paleoproxy involves the ability to link opal rain to the sediments to opal production in surface waters, and then ultimately to surface water primary production. Recent summaries have examined some of the problems associated with making this linkage (Nelson et al., 1995; Ragueneau et al., 2000); these studies suggest that while progress is being made, work still remains to make opal a robust indicator of paleoproductivity.

13.5 Diagenesis of Calcium Carbonate

Calcite is the most abundant biogenic component in pelagic sediments, with around half of all surface sediments in the deep sea and on the continental slope containing >50% $CaCO_3$ and some sediments containing >80% calcite (Kennett, 1982). General factors related to carbonate production in surface ocean waters, its dissolution in the water column, and its deposition in sediments are discussed in sections 2.2.1 and 2.2.3. After deposition in sediments, some of this calcite undergoes further dissolution on early diagenetic time scales, while some can also undergo recrystallization on slightly longer time scales, eventually becoming limestone rock (see Morse and Mackenzie, 1990, and discussions therein).

Carbonate preservation in sediments is important for several reasons. First, calcite burial plays an important role in controlling the alkalinity and ΣCO_2 content of seawater, and thus ultimately impacts atmospheric CO_2 concentrations (see discussions in Sundquist, 1990; Archer et al., 2000). Calcium carbonate in marine sediments also represents the largest eventual buffer for neutralizing anthropogenic CO_2 produced by fossil fuel combustion (Walker and Kasting, 1992; Archer et al., 1998); therefore, understanding the occurrence of carbonate dissolution in sediments is important for determining the time scales over which this neutralization will occur. Finally, because patterns of calcite preservation in sediments appear to correspond to glacial cycles, understanding the controls on these preservation cycles, i.e., preservation versus productivity effects, should provide important information on the past history of the oceans (e.g., Arrhenius, 1952; Archer et al., 2000; Mekik et al., 2002).

Although surface waters are supersaturated with respect to both calcite and aragonite, the solubility of both minerals increases with decreasing temperature and increasing pressure. To describe these changes, the calcite saturation state of seawater is often examined in terms of the saturation index Ω:

$$\Omega = [Ca^{2+}][CO_3^{2-}]/K'_{sp} \qquad (13.3)$$

where K'_{sp}, the apparent solubility product constant for calcite, is related to K_{sp}, the thermodynamic constant at a given T and pressure, by the calcium and carbonate activity coefficients. An examination of eqn. (13.3) indicates that Ω is greater than 1 in supersaturated waters, equals 1 for saturated waters and is less than 1 in undersaturated waters. Since calcium is nearly constant in seawater, Ω can also be described simply in terms of the saturation carbonate ion concentration ($[CO_3^{2-}]_{sat'n.}$) as

$$\Omega = \frac{[CO_3^{2-}]_{in\ situ}}{[CO_3^{2-}]_{sat'n.}} \qquad (13.4.)$$

where $[CO_3^{2-}]_{in\ situ}$ is the carbonate ion concentration at a given water depth. This concentration is generally calculated with pCO_2, alkalinity, ΣCO_2, or pH measurements (only two of the four factors are needed) and the two carbonic acid dissociation constants (e.g., Millero, 1996; Lewis and Wallace, 1998). For reasons outlined above, $[CO_3^{2-}]_{sat'n}$ ranges from ~40 µmol kg^{-1} in surface waters to ~120 µmol kg^{-1} in deep ocean waters (e.g., Broecker and Peng, 1982; Emerson and Archer, 1990). Calcite solubility is also defined in terms of the carbonate ion concentration difference,

$$\Delta[CO_3^{2-}] = [CO_3^{2-}]_{in\ situ} - [CO_3^{2-}]_{sat'n.} \qquad (13.5)$$

Here $\Delta[CO_3^{2-}]$ is greater than zero for saturated waters and decreases as the waters become less supersaturated; in saturated waters it equals zero, and it is less than zero in undersaturated waters.

Changes in calcite saturation in the water column are also affected by organic matter remineralization, since this process produces aqueous CO_2 which can react with the carbonate ion through the reaction,

$$H_2O + CO_2(aq) + CO_3^{2-} \rightarrow 2HCO_3^- \qquad (13.6)$$

By decreasing the carbonate ion content of seawater, rxn. (13.6) further contributes to a decrease in the degree of saturation of ocean waters with respect to both calcite and aragonite. As a result of decreasing water column carbonate ion concentrations via rxn. (13.6) and increasing $[CO_3^{2-}]_{\text{sat'n}}$ with increasing pressure and decreasing temperature, calcite and aragonite saturation indexes, either Ω or $\Delta[CO_3^{2-}]$, there fore decrease with water column depth. The depth at which $\Omega = 1$ or $\Delta[CO_3^{2-}] = 0$ then defines what is referred to as the *calcite saturation horizon* (see fig. 13.7).

The transition from super- to undersaturated waters leads, in part, to the onset of calcite dissolution in sinking particles and/or in sediments.[5] However, the occurrence of calcite dissolution is not controlled simply by these thermodynamic considerations, but also by kinetic factors. Furthermore, the interplay between these thermodynamic considerations and deep-water ocean (thermohaline) circulation leads to a situation in which bottom water carbonate ion concentrations also continually decrease along the deep-water flow path from the North Atlantic to the South Atlantic and then up into the Indian or Pacific Oceans. As a result, the depth of the calcite saturation horizon shallows along this flow path (table 13.4). Calcite is therefore more prevalent in Atlantic Ocean sediments than in Pacific Ocean sediments and, in general, calcite is more prevalent in marine sediments on oceanic topographic highs.

The rate of calcite dissolution (R) is usually written as equations of the form

$$R = k(1 - \Omega)^n \qquad (13.7)$$

$$R = k^* \cdot ([CO_3^{2-}]_{\text{sat'n}} - [CO_3^{2-}]_{\text{in situ}})^n = k^* \cdot (-\Delta[CO_3^{2-}])^n \qquad (13.8)$$

(e.g., Morse and Arvidson, 2002; Emerson and Hedges, 2003) where $k^* = k([Ca^{2+}]/K'_{sp})^n$. Because the rate of carbonate dissolution is not instantaneous, it requires some degree of undersaturation for the reaction to occur. Modeling carbonate diagenesis in sediments, e.g.,

---

[5] As noted earlier, the lower solubility of aragonite leads to virtually complete aragonite dissolution in the water column at depths much shallower than that for calcite. Although the discussion here focuses on calcite dissolution, other workers have discussed the significance of aragonite dissolution in sediment carbonate diagenesis (Morse and Mackenzie, 1990; Hales et al., 1994).

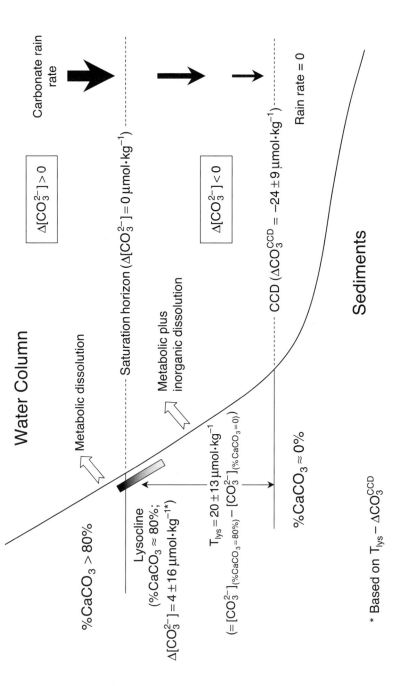

Figure 13.7 A schematic representation of the geochemistry of calcium carbonate dissolution in deep-sea sediments. See the text for the definitions of the terms used here, as well as the sources for their values.

TABLE 13.4
Water Column Depths of the Lysocline and CCD Along
the Bottom-water Thermohaline Flow Path*

Ocean Basin	Lysocline	CCD
North Atlantic	4.5–5 km	>5 km
South Atlantic	3.5–4 km	4–5 km
South Pacific	3.5–4 km	4–5 km
North Pacific	2.5–3 km	3–4 km

* Taken from several sources (Berger, 1976; Broecker and Peng, 1982; Seibold and Berger, 1996).

pore-water alkalinity or pH profiles, using eqn. (13.7) requires estimates of k and n. The value of n that is often used (=4.5) is based on lab studies examining the dissolution of synthetic and biogenic calcites (Keir, 1980). In contrast, values of k are generally derived from fitting pore-water profiles and range from <0.1% d^{-1} to ~100–200% d^{-1} (e.g., see Martin and Sayles, 1996, for a summary). These model-derived k values are significantly lower than laboratory-determined k values.

Hales and Emerson (1997b) re-examined Keir's results using better-defined K'_{sp} values, and suggest that lower-order dissolution kinetics may be more appropriate ($n = 1-2$). They used this lower-order dissolution kinetics in a reanalysis of their pore water data from deep-sea sites on the Ontong-Java Plateau and the Ceara Rise. They observed that modeled k values (which differed by nearly three orders of magnitude with $n = 4.5$) now differed by only roughly one order of magnitude with $n = 1$. Furthermore, much of the difference in these $n = 1$ k values may be attributable to grain size differences in the carbonates at the two sites.[6] In a similar fashion, Green and Aller (2001) suggested, based on results from nearshore Long Island Sound sediments, that n values of ~2 (±1) are more appropriate. At the same time, though, there still appear to be substantial differences between k values determined in the lab and those obtained from modeling deep-sea pore-water profiles (e.g., Hales and Emerson, 1997b), even

---

[6] Although not explicitly shown here, the value of k in rxn. (13.7) is actually a function of the available carbonate surface area (Morse and Arvidson, 2002). Thus, with all other things being equal, fine-grained, high surface area carbonate particles will have larger k values (i.e., show faster rates of dissolution).

taking into account possible lower-order dissolution kinetics. As several authors have noted (Morse and Arvidson, 2002; Emerson and Hedges, 2003), resolving difference such as these will be important to further our understanding of the controls on deep-sea carbonate dissolution and burial.

In addition to describing calcite dissolution in marine systems using chemical data (i.e., the saturation horizon), one can also describe the process in terms of the *lysocline*. The lysocline represents the water column depth at which one first sees evidence of calcite dissolution in the sediments, i.e., the boundary between well-preserved and poorly preserved calcareous shell assemblages (Berger, 1976; Peterson and Prell, 1985; Seibold and Berger, 1996; also see discussions in Morse, 2003). One can further define the calcite compensation depth (CCD) as the depth at which the downward flux of calcite in the water column exactly balances calcite dissolution. Below the CCD there is no calcite accumulation in marine sediments (fig. 13.7).

A number of factors make it difficult to estimate the depths of either the lysocline, the saturation horizon, or the CCD to within roughly a few hundred meters (see discussions in Emerson and Hedges, 2003; Morse, 2003). In some modeling studies (Emerson and Archer, 1990; Archer, 1991b) the lysocline is therefore defined as the depth at which $\Delta[CO_3^{2-}] = 0$ and/or %$CaCO_3$ in sediments drops below 80% (Archer, 1996). At the same time, using more general sediment properties such as %$CaCO_3$ to estimate the onset of carbonate dissolution in sediments is equally difficult. This occurs primarily because the $CaCO_3$ content of sediments is very insensitive to dissolution effects until ~50% or more of the calcite rain to the sediment surface undergoes dissolution (fig. 13.8). Recently, Mekik et al. (2002) have proposed a new method for estimating percent calcite dissolution in sediments using a calibrated foram fragmentation index. Results with this index support general observations about the controls on sediment carbonate dissolution (see below), and may also allow one to less ambiguously separate productivity versus preservation effects in sediment carbonate profiles.

Within the uncertainties mentioned above, global maps of the lysocline and CCD indicate that both get shallower as one moves along the bottom water flow path (table 13.4), consistent with the discussion above regarding analogous changes in bottom water carbonate ion concentration. However, to better understand the controls on carbonate dissolution in the marine environment, and to ease com-

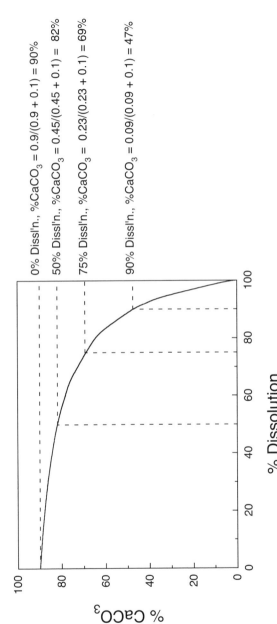

Figure 13.8 The %$CaCO_3$ in the sediments as a function of the % carbonate dissolution in the water column for sinking particulate material that is initially 90% $CaCO_3$ and 10% noncarbonate. Because %$CaCO_3$ in the sediments is determined by flux balance, it is insensitive to the amount of carbonate dissolution until ~50% or more of the carbonate has dissolved (see similar discussions in Broecker and Peng, 1982, and Emerson and Hedges, 2003).

parisons of data from different ocean basins, workers in recent years have begun to use $\Delta[CO_3^{2-}]$ "depth" units (see eqn. 13.5; e.g., Archer, 1991b, 1996). Using this approach, Archer (1996) has generated a map of bottom water $\Delta[CO_3^{2-}]$ that can be compared to maps of %CaCO$_3$ to begin to understand the controls on the transition from carbonate-rich to carbonate-poor sediments. Based on the results of this model Archer defines a parameter T_{lys}, which is the carbonate ion content at the water column depth of the lysocline minus that observed at the depth of the CCD. Globally this value averages 20 ± 13 μmol kg^{-1}. The fact that this transition from carbonate-rich to carbonate-depleted sediments occurs over a wide degree of undersaturation ($\Delta[CO_3^{2-}]$ between roughly −7 to −33 μmol kg^{-1}) strongly argues for kinetic, rather than thermodynamic, controls on calcite dissolution in the marine environment (also see discussions in Emerson and Hedges, 2003).

Archer (1996) also defines a term ΔCO_3^{CCD} which is essentially the value of $\Delta[CO_3^{2-}]$ at the CCD. Globally this value averages −24 ± 9 μmol kg^{-1}. A comparison of T_{lys} and ΔCO_3^{CCD} then indicates that $\Delta[CO_3^{2-}]$ at the lysocline should equal 4 ± 16 μmol kg^{-1}. One interpretation of this result is that this value is indistinguishable from zero, and thus the lysocline and saturation horizon occur at the same water column depth (also see Broecker and Peng, 1982, who reached the same conclusions using a slightly different approach). Given the uncertainties in defining the depths of these carbonate horizons, this conclusion does not appear to be unreasonable.

At the same time, though, an alternate explanation of this observation is that the lysocline occurs over a range of $\Delta[CO_3^{2-}]$ values that span supersaturated to undersaturated bottom water conditions (approx. +21 to −13 μmol kg^{-1}). The negative end of this range is not difficult to understand in terms of the kinetics of carbonate dissolution, since calcite in sediments underlying undersaturated bottom waters will undergo "inorganic" dissolution in the sediments as corrosive bottom water is transported into the sediments by such processes as diffusion, bioturbation, or bioirrigation. However, the positive end would appear at first glance to be counterintuitive, i.e., sediment carbonate dissolution occurring in sediments underlying supersaturated bottom waters.

Nevertheless, there is evidence for the occurrence of sediment carbonate dissolution above the lysocline/saturation horizon by a process often described as metabolic carbonate dissolution (e.g., see

Emerson and Bender, 1981, and references cited below). Because aerobic respiration dominates organic matter remineralization in pelagic sediments, this process adds metabolic CO_2 to the pore waters, as it also does in the water column. Rxn. (13.6) then drives supersaturated pore waters toward undersaturation and once the sediments become sufficiently undersaturated, carbonate dissolution occurs. Metabolic dissolution is also important for sediments below the lysocline/saturation horizon, although here it also competes with inorganic dissolution by understaurated bottom waters.

In the most simplistic sense the coupling of aerobic respiration and carbonate dissolution can be described as,

$$CH_2O + O_2 + CaCO_3 \rightarrow Ca^{2+} + 2HCO_3^- \quad (13.9)$$

which then predicts 1:1 relationships between alkalinity, Ca^{2+} or ΣCO_2 production, or a 1:2 relationship between oxygen consumption and the production of either alkalinity or ΣCO_2.[7] Evidence from a variety of sources indicates the occurrence of metabolic carbonate dissolution in deep-sea sediments (although also see discussions below for evidence against its occurrence). This includes pore-water stoichiometric models (Sayles, 1981), sediment diagenesis models (Gehlen et al., 1999), and high resolution in situ pore-water profiles of pH, O_2,

---

[7] This equation is, however, somewhat of an oversimplification, because the remineralization of organic nitrogen in marine organic matter coupled to nitrification (e.g., see eqns. 7.5 or 11.7) also "produces" nitric acid, i.e., a proton plus the nitrate ion. Furthermore, we can think of phosphorus regeneration as "producing" phosphoric acid, which at pore-water pH values partially dissociates and produces two protons. Since these protons are also capable of dissolving carbonate, during metabolic carbonate dissolution this will lead to the production of slightly more alkalinity per mole of ΣCO_2 produced. Assuming a Redfield C/N ratio of 106/16, a C/P ratio of 106/1 and a carbon oxidation state of 0, the oxidation of 106 moles of organic carbon produces 2·106 moles of alkalinity from rxn. (13.8) and 18/106 moles of alkalinity from the protons produced by nitrification and phosphorus regeneration ($CaCO_3 + H^+ \rightarrow Ca^{2+} + HCO_3^-$). As a result, the $\Delta Alk/\Delta \Sigma CO_2$ ratio will equal $(2 \cdot 106 + 18)/(2 \cdot 106) = 1.08$. Similarly, stoichiometric relationships between the amount of O_2 consumed and CO_2 produced (and therefore ultimately the amount of carbonate dissolved) require that one also take into account the oxygen consumed by nitrification and the fact that organic matter undergoing remineralization may not have an oxidation state of ~0 (i.e., it is not "CH_2O"; see section 11.2.1 for further details and also see discussions in Hales et al., 1994). For example, based on the numbers above, the remineralization of this organic matter will consume 138 moles of O_2 and as a result $\Delta O_2/\Delta Alk$ will equal $138/(2 \cdot 106 + 18) = 0.6 = 1/1.67$.

CO_2(aq), and/or Ca^{2+}, often in conjunction with in situ benthic lander studies (Archer et al., 1989b; Berelson et al., 1990, 1994; Cai et al., 1995; Martin and Sayles, 1996; Hales and Emerson, 1997a; Jahnke et al., 1997; and others). In situ pore-water pH profiles are also discussed in section 16.1.

Early studies of metabolic carbonate dissolution demonstrated the importance of the relative rain rates of organic matter and calcite to the sediments in affecting the occurrence of metabolic dissolution (Emerson and Bender, 1981). More recent studies have also shown the importance of the depth distribution in the sediments of the rates of calcite dissolution and organic matter remineralization (e.g., Boudreau and Canfield, 1993). This then suggests that the reactivity of both calcium carbonate and organic matter in sediments is important in controlling the occurrence of metabolic carbonate dissolution.

Although we can write a coupled reaction such as rxn. (13-9), not all metabolic CO_2 necessarily dissolves calcium carbonate. This then leads to the concept of *metabolic dissolution efficiency* (MDE), which can be thought of as the likelihood that metabolic CO_2 dissolves calcium carbonate versus, for example, exchange with bottom waters and neutralization by bottom water $[CO_3^{2-}]$. Quantitatively we can define MDE as the depth-integrated rate of carbonate dissolution divided by the depth-integrated rate of organic matter remineralization, i.e., metabolic CO_2 production. Based on this discussion, the efficiency of metabolic dissolution appears to be related to the relative rates and depth distributions of sediment organic matter remineralization and calcite dissolution as compared to the rates of transport across the sediment-water interface (e.g., see discussions in Jahnke and Jahnke, 2004).

At one extreme, calcite dissolution is rapid as compared to organic matter remineralization and benthic exchange rates. Under these conditions, metabolic carbonate dissolution will have a high efficiency; i.e., metabolic CO_2 that is produced will have a high likelihood of dissolving carbonate. A high metabolic dissolution efficiency is also expected if bioturbation effectively mixes organic matter into the sediments, since CO_2 production will occur deeper in the sediments and therefore have a greater likelihood of dissolving calcite in the sediments (versus being transported out of the sediments by diffusion).

Calculations presented by Martin and Sayles (2003) demonstrate this effect. In their calculations they fixed the total (depth-integrated) sediment organic matter remineralization rate and increased the e-folding depth[8] of aerobic respiration from 0.3 cm to 3.5 cm, values that are typical for many deep-sea settings. This increase partitions metabolic CO_2 production deeper into the sediments, and MDE values increased from 20% to 60% for supersaturated bottom waters ($\Delta[CO_3^{2-}] = +20$ μmol kg^{-1}). Similarly, MDE values increased from 55% to 85% for undersaturated bottom waters ($\Delta[CO_3^{2-}] = -10$ μmol kg^{-1}).

At the other extreme, if carbonate dissolution is slow compared to organic matter remineralization, organic matter remineralization will be localized close to the sediment-water interface and benthic exchange rates of dissolved CO_2 species will be relatively fast. In this case, regardless of the bottom water saturation state, much of the metabolic CO_2 that is produced will simply be transported out of the sediments before reacting with sediment calcite (Hales and Emerson, 1996). In addition, for sediments underlying supersaturated waters, much of this metabolic CO_2 will be neutralized by the influx of CO_3^{2-} into the sediments from the bottom waters, rather than by dissolving sediment carbonate (Emerson and Bender, 1981; Jahnke et al., 1994). As a result, metabolic dissolution efficiency will be low.

For sediments underlying undersaturated bottom waters, regardless of their %CaCO$_3$ content, benthic flux studies and in situ pore-water profiles indicate that MDE is generally >60% and sometimes >90% (Berelson et al., 1990, 1994; Hales et al., 1994; Hales and Emerson, 1996; Martin and Sayles, 1996; Jahnke and Jahnke, 2004). Note that for these sediments it is possible to obtain MDE values that are greater than 100% if all metabolic CO_2 dissolves carbonate *and* there is also some inorganic carbonate dissolution from corrosive bottom water influx into the sediments. Looked at from another perspective, such high MDE values imply that a large percentage of the total sediment carbonate dissolution occurs by metabolic versus inorganic dissolution, e.g., ~60–100%, based on the calculations in Martin and Sayles (1996). Thus the interplay between sediment processes and sediment-water exchange leads to a situation in which metabolic dissolution is apparently more important than inorganic dissolution, in

---

[8] This is defined as the depth over which the rate of respiration decreases to 1/e ($\approx$37%) of its value at the sediment surface.

spite of undersaturated bottom water conditions. Finally, as might be expected these sediments are very efficient at dissolving the rain of calcite to the sediments, with carbonate burial efficiencies (CBE) that are generally <50%. For comparison, in Milliman's (1993) oceanic calcium carbonate budget, he assumes an average CBE of 40% for sediments below the lysocline.

For sediments underlying supersaturated waters, similar comparisons yield results that appear to be a function of the $CaCO_3$ content of the sediments. For low %$CaCO_3$ sediments, metabolic dissolution efficiency is generally >60%, based on both benthic lander studies and in situ pore-water profiles (Hales et al., 1994; Jahnke and Jahnke, 2004). Because carbonate dissolution must occur by metabolic dissolution in these sediments, MDE values here cannot exceed 100%. Carbonate burial efficiencies in these sediments are in the range of ~40–60% (Hales et al., 1994).

At the same time, there appears to be a discrepancy among results obtained for sediments rich in $CaCO_3$ underlying supersaturated bottom waters. Evidence obtained from in situ pore-water profiles in such sediments indicates the occurrence of metabolic carbonate dissolution, with MDE values of ~30–50% (Martin and Sayles, 1996, 2003; Hales and Emerson, 1997a). In contrast, benthic lander studies at such sites, including some of the same sites where in situ pore-water profiles were also obtained, show no apparent evidence of metabolic carbonate dissolution (Jahnke et al., 1994, 1997; Jahnke and Jahnke, 2004). The exact causes of the discrepancy between benthic lander and pore-water results are not well understood at the present time. However, reactions occurring quite close to the sediment-water interface may be important; for example, reprecipitation of fresh $CaCO_3$ or surface exchange reactions between protons and carbonate mineral phases (Broecker and Clark, 2003; Jahnke and Jahnke, 2004). There are potential problems with such explanations (e.g., see Hales and Emerson, 1996; Emerson and Hedges, 2003; Martin and Sayles, 2003) that clearly require further study and examination.

This discrepancy also impacts the quantification of carbonate burial efficiency in sediments above the lysocline. The Jahnke and Jahnke (2004) lander results predict a CBE value of ~100% in high carbonate sediments, while estimates based on pore-water studies predict CBE values of ~50–80% (Sayles, 1981; Martin and Sayles, 1996, 2003). In his global sediment carbonate budget Milliman (1993) assumes a CBE value of 80% for sediments above the lysocline, based in part on faunal assemblage results (e.g., Peterson and Prell, 1985) that suggest less dissolution above the lysocline than that predicted by the pore-water geochemical data discussed above.

Globally, ~20–30% of the carbonate production in surface waters appears to escape dissolution in either the water column or the sediments, to become buried in marine sediments (Archer, 1996; Milliman et al., 1999). However, the extent to which this 70–80% calcite loss occurs in sediments, both below and above the lysocline, or in the water column is uncertain. Historically it has been assumed that much of this dissolution occurs in sediments (see discussions in Milliman, 1993; Archer, 1996). However, in a recent oceanic carbonate budget, Milliman et al. (1999) suggest that ~60% of the carbonate production in open-ocean surface waters actually dissolves above the lysocline in the upper 1,000 m of the water column. e.g., through biological processes associated with particle sinking. The remaining ~40% of this production is about equally divided between dissolution in deep-sea sediments and burial in sediments. In a related fashion Chen (2002) has suggested that alkalinity sources from suboxic and anoxic remineralization processes occurring in continental margin sediments (see section 16.1) may be larger than previously thought. Since this shelf alkalinity production is independent of carbonate dissolution, it suggests that oceanic alkalinity budgets, which are often used to develop carbonate budgets such as those discussed here, may need to be re-examined.

Finally, an implicit assumption in all of these budgets and calculations is that oceanic, or sediment, calcium carbonate budgets are in steady state. However, studies in equatorial Pacific sediments (Berelson et al., 1997; Stephens and Kadko, 1997) suggest that the Holocene sediment carbonate budget is not in steady state, with the sum of dissolution and burial exceeding the rain of carbonate to the sediments. Looked at from the standpoint of the terminology in chapter 8, this implies that $J_{in} < (J_{out} + J_{bur})$. Possible explanations for this are either a recent increase in the rate of sediment carbonate dissolution, or a decrease in the calcium carbonate rain rate to the sediment. Based on the available data, both groups suggest that the former is the more likely explanation. Berelson et al. (1997) further suggest that the carbonate ion content of Pacific Ocean bottom waters has decreased by ~10–15 μmol kg^{-1} in the late Holocene, i.e., within the last ~3,000 yr. This then implies that this enhanced dissolution has occurred because the present-day ocean, with its lower carbonate ion concentration, is more corrosive to calcite than it was in the early Holocene. Consistent with these observations, several other studies have suggested that at least in the Pacific, glacial bottom waters contained perhaps ~5–20 μmol kg^{-1} more dissolved carbonate ion than they do today (Emerson and Archer, 1990; Anderson and Archer, 2002).

≈ **CHAPTER FOURTEEN** ≈

Nonsteady-State Processes in Marine Sediments

IN ALMOST ALL OF THE MODELS and budget calculations discussed previously we have either implicitly or explicitly assumed that steady-state conditions exist. The steady-state assumption is one that is often made, even if generally not explicitly discussed. Many times, this assumption is based simply on practicality: there is only one set of observations taken at a single point in time, with no additional information to critically examine whether or not the system shows any temporal variability. Nevertheless, evidence increasingly suggests that true steady-state conditions may be far less common in many marine sediments than are nonsteady-state (or time-dependent) conditions. However, when the question of nonsteady-state diagenesis is explicitly examined, it can also be shown that a steady-state assumption may still be reasonably valid, at least over certain time scales.

In this chapter we will discuss the relationship between steady-state and nonsteady-state conditions in marine sediments (also see related discussions in section 5.2), how we can differentiate between the two, and the impact of nonsteady-state diagenesis on biogeochemical processes in marine sediments.

14.1 GENERAL CONSIDERATIONS

Many sediment systems show some kind of time-varying behavior. For example, environmental changes on time scales that range from seasonal cycles to glacial-interglacial periods, or longer, can lead to the occurrence of nonsteady-state sediment diagenesis. Time-varying behavior may also occur as the result of a single event, such as the deposition of a relatively organic-rich turbidite in a slowly accumulating, organic-poor pelagic sediment.

In muddy deltaic sediments, such as those on the Amazon continental shelf, physical reworking on daily to annual time scales leads to nonsteady-state behavior in these sediments (Mackin et al., 1988).

The occurrence of these processes significantly affects sediment carbon remineralization and preservation and may also play a role in the oceanic cycles of other elements (see discussions in sections 5.5, 7.6.3, and 11.2.3 and chapter 18).

Developing a simple rule of thumb or scaling calculations to examine the importance of nonsteady-state processes is difficult because of the variety of ways that these processes may occur. Often, however, scaling calculations such as those discussed in section 5.4 can be used to provide some constraint on the occurrence of nonsteady-state sediment processes (e.g., Froelich et al., 1979; Gobeil et al., 1997). This approach will be discussed in greater detail later in this chapter.

For a single impulse input of nonreactive material such as the deposition of a volcanic ash layer, sedimentation will eventually bury the signal produced by this event. However, if simple bioturbation/biodiffusion also occurs and this process is sufficiently rapid (see below), it will lead to a "smearing" of the initial signal (fig. 14.1). Continued long-term burial plus bioturbation will then produce a broad, more diffuse peak whose maximum concentration at any point in time will be less than that at the time of deposition. However, the depth-integrated amount of material in this peak remains constant with time because of the conservative nature of the material.

For a single impulse input of reactive material, the same general physical processes will operate. However, both the maximum concentration of this material at any depth and its depth-integrated amount will decrease with time. If the rate of reaction of this material also decreases with increasing sediment burial, then the potential exists for the long-term preservation of at least some of this reactive material. As we will see below, this appears to be the case after the deposition of an organic-rich turbidite in deep-sea sediments.

14.2 Periodic Input Processes

For sediments affected by periodic processes, the extent to which this periodicity impacts sediment properties depends in part on the frequency and magnitude of these changes in comparison to the rates of transport and reaction. In an early study of this problem Lasaga and Holland (1974) developed a model with a time-varying (sinusoidal) input of reactive organic matter to the sediment surface; this then allowed them to examine whether this periodicity was observable in

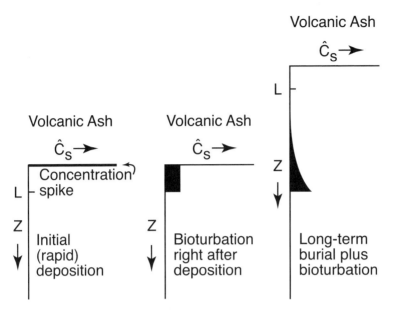

Figure 14.1 Schematic representation of the effects of bioturbation on the distribution of an "instantaneous" spike of volcanic ash material at the surface of a marine sediment (redrawn after Ruddiman and Glover, 1972, and Berner, 1980). Immediately after deposition this ash spike is assumed to be homogenized throughout the sediment mixed layer. Subsequently, sediment in the mixed layer is buried below the depth $z = L$ as deposition of fresh sediment (which does not contain volcanic ash) occurs at the sediment surface. The mixing of this fresh sediment with the remaining sediment in the mixed layer then leads to a situation in which the volcanic ash in the mixed layer is slowly diluted out. Over time this leads to a record of this event being preserved as an upward exponentially decreasing profile of ash concentration versus sediment depth.

pore water profiles of remineralization end-products. They showed that molecular diffusion effectively damps out the effects of high frequency (short period) events in pore-water profiles, and that periodicity will be seen in these profiles only when

$$\alpha < \frac{\omega^2}{D_s}\frac{\pi}{4} \ (= \alpha_{\max}) \tag{14.1}$$

where α, the event frequency, is $2\pi/\tau$ when τ is the event period. For a typical diffusion coefficient of 5×10^{-6} cm^2 sec^{-1} and a sedimentation rate of 1 cm yr^{-1}, e.g., a coastal sediment, this implies that α_{max} is 0.005 yr^{-1}. Similarly τ_{min}, the minimum periodicity that can be preserved in pore water profiles ($= 2\pi/\alpha_{max}$) is 1,260 yr. Decreasing ω to 1 cm 10^3 yr^{-1}, e.g., a deep-sea sediment, decreases α_{max} to 5×10^{-9} yr^{-1}, and increases τ_{min} to 1.26×10^9 yr. The depth scales over which this periodicity occurs is also of the order of $\sim \pi\omega/\alpha$, i.e., meter-scale, for the coastal sediment parameters discussed above.

Other types of periodicity observed in sediment pore-water profiles result from temporal changes in the concentration of the overlying seawater, i.e., a time-dependent upper boundary condition. This has been examined, for example, in terms of modeling changes in bottom water salinity or δ^{18}O over glacial-interglacial time scales (McDuff, 1985; Schrag and DePaolo, 1993; Schrag et al., 2002). In such systems the preservation of this signal in pore waters is a function of time (sediment burial), since over long time scales diffusion will attenuate the downward propagation (or preservation) of this signal. Intuitively one might expect that the long-term preservation of a pore-water peak at a depth l should be a function of the "age" of the sediment horizon (roughly equal to l/ω) and the magnitude and periodicity of the changing upper boundary condition. However, the details of this calculation remain to be worked out.

For the periodic input of material to a bioturbated sediment, subject only to biodiffusion, the ability of the sediment to record this nonsteady-state input depends on the relationship between the periodicity of this input and the time scales of sedimentation and bioturbation. To examine this problem, it is tempting to simply modify eqn. (14.1) by substituting D_b for D_s. However, unlike diffusion, bioturbation acts only upon a finite region of the sediments, i.e., the sediment mixed layer; therefore this approach would appear to be incomplete.

In their work modeling bioturbation, Guinnasso and Schink (1975) define a dimensionless parameter G ($= D_b/L\omega$) and showed that for values of G greater than $\sim$3, a pulse input to the sediments is homogenized in the mixed layer. Similarly, when G is less than $\sim$0.3 this input will simply pass through the sediments as a slightly broadened peak. While this approach provides some constraints on the mixing of such an instantaneous pulse input, it is difficult to relate G to the factors controlling the preservation of periodic sediment inputs.

PERIODIC INPUT PROCESSES

An alternate approach to this problem begins with the assumption that if, roughly speaking, half the period of the input function is greater than the age of the sediment mixed layer (M_a), then bioturbation will not filter out (or dampen) this periodicity. Mathematically, this implies that if

$$0.5\tau > M_a \quad (14.2)$$

then this periodicity will be preserved in the sediment.

One way to approximate the "age" of the mixed layer ($\tau_{\text{mixed layer}}$) is with the equation

$$M_a \approx L/\omega \quad (14.3)$$

where L is the mixed layer depth. In deep-sea sediments M_a can also be estimated using ^{14}C sediment profiles and the equation,

$$M_a = \frac{1}{\lambda}\ln\left(1 + \frac{\lambda L}{\omega}\right) \quad (14.4)$$

where λ is the rate constant for ^{14}C decay (Berner, 1980), and the implicit assumption is that bioturbation is sufficiently rapid to destroy all ^{14}C gradients in the sediment mixed layer. For small values of $\lambda L/\omega$ these two equations yield nearly identical results since $\ln(1 + x) \approx x$ for small values of x. Re-expressing eqn. (14.2) in terms of the input frequency then implies that if

$$\alpha < \pi/M_a \quad (14.5)$$

then the periodicity in the input signal will be preserved in sediment profiles; that is, the signals will not be completely damped out as a result of passage through the sediment mixed layer.

This approach is somewhat incomplete in the sense that it does not explicitly account for the effect of the magnitude of D_b on the preservation of periodic inputs. Thus, for example, it seems possible that eqn. (14.5) might not be valid for sediments in which bioturbation rates (D_b or G values) are low. However, results obtained using a time-dependent sediment diagenesis model (Dhakar, 1995) for model sediments in which $G \approx 10$ appear to broadly support eqn. (14.5).

14.3 SEASONALITY IN SEDIMENT PROCESSES

In many marine sediments changes on seasonal time scales play an important role in sediment diagenesis. In estuarine and coastal sediments, seasonality in sediment pore-water depth profiles and in rates of sediment biogeochemical processes are often observed in response to annual changes in temperature, the composition and reactivity of carbon deposited in the sediments, the rates of benthic macrofuanal activity, or bottom water conditions (Klump and Martens, 1989; Sun et al., 1991; Aller, 1994b; Gerino et al., 1998). Similar temporal changes may also occur in deep-sea sediments under some circumstances, presumably in response to seasonal, or pulsed, inputs of organic matter, since temperatures in such environments are essentially invariant (see Sayles et al., 1994; Soetaert et al., 1996b; and references therein).

In deep-sea sediments, model studies by Martin and Bender (1988) show that for seasonally varying organic matter deposition to express itself as seasonality in sediment remineralization processes, the mean lifetime of the organic matter undergoing remineralization must be shorter than a seasonal period (1 yr). Since the mean lifetime for material undergoing first-order decomposition is $1/k$, this then implies that when

$$k > \sim 1 \text{ yr}^{-1} \tag{14.6}$$

seasonal behavior in deep-sea sediment remineralization processes should be observed. For more refractory organic matter, the rate of decomposition is too slow compared to the time scale of sediment mixing processes. Therefore, mixing damps out seasonal signals in surface organic carbon inventories, and hence any seasonal variation in sediment organic matter remineralization rates.

Data from two deep-sea sites of comparable water depths in the North Atlantic (NA) and the North Pacific (NP) can be used to test this observation.[1] At both sites there are seasonal variations in the rain rate of organic matter to the sediments (Smith et al., 1992; Sayles et al., 1994), although the absolute values of fluxes to the NP site are ~3–5

---

[1] The NA site is at a water depth of 4,400 m near Bermuda, while the NP site is at a water depth of 4,100 m near the base of the Monterey Submarine Fan off the coast of California.

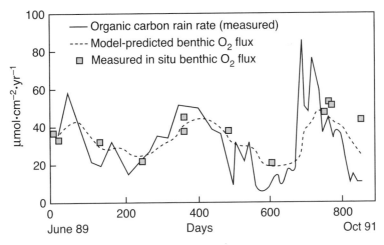

Figure 14.2 Application of time-dependent sediment diagenesis model to data from a deep-sea site in the Pacific Ocean (modified after Soetaert et al., 1996b). Shown as a solid line is the organic carbon rain rate to the sediments (Smith et al., 1992), which is used as a forcing function for the model. Shown as a dashed line is the sediment O_2 flux predicted by the model, and shown as solid squares are measured in situ sediment O_2 fluxes (Jahnke et al., 1990; Reimers et al., 1992; Smith et al., 1992). The slight offset between the O_2 flux curve and carbon rain rate curve is caused by the effective reaction kinetics of sediment organic matter remineralization. As discussed in section 7.4.3, the relatively good agreement between carbon fluxes to the sediments and measured (or calculated) sediment oxygen fluxes provides evidence for the fact that sediment oxygen uptake is often a good indicator of integrated (overall) sediment organic carbon remineralization rates.

times larger than they are at the NA site (see fig. 14.2 for carbon rain rate and benthic flux data from the NP site). At the NP site benthic O_2 fluxes are also similarly larger than they are at the NA site (Jahnke et al., 1990; Reimers et al., 1992; Sayles et al., 1994), as are model-derived k values for organic matter degradation, i.e., $\sim$5–10 yr^{-1} at the NP site versus <1 yr^{-1} at the NA site (Sayles et al., 1994; Soetaert et al., 1996b).

More importantly, however, these differences impart seasonality in the rates of sediment organic matter remineralization at the NP site (fig. 14.2) and not at the NA site (Sayles et al., 1994), consistent with eqn. (14.6). While it is difficult (and perhaps dangerous) to generalize based on such results from just two sites, it is interesting to note that these differences in sediment organic matter reactivity at sites of roughly the same water depths in the Atlantic versus the Pacific are consistent with results in fig. 7.9, which suggest that for the same water column depths the sediment oxygen penetration depth is shallower in Pacific sediments than it is in Atlantic sediments. The reasons for these differences, if indeed they are real, are not well understood.

In many estuarine and coastal sediments seasonality in both amount and reactivity of organic matter deposited in the sediments (Sun et al., 1991; Canuel and Martens, 1993; Gerino et al., 1998) can lead to temporal variability in sediment organic matter remineralization rates. In temperate sediments an additional source of such temporal variability in the rates of sediment processes results from annual temperature changes of up to ~20°C between winter and summer. Given the observed temperature dependence of sediment remineralization rates (see section 7.6.2), this implies that remineralization rates in temperate sediments may vary by up to a factor of ~8 over an annual cycle.

In estuarine and coastal sediments, seasonality in remineralization rates expresses itself in time-varying sediment pore-water profiles of remineralization end-products, e.g., ammonium or phosphate). However, because diffusion damps out seasonal periodicity in these profiles (see eqn. 14.1), what is generally observed are smooth, asymptotic profiles whose concentrations simply grow in and out over the annual seasonal cycle (e.g., Aller, 1980a; Klump and Martens, 1989). At the same time, though, such seasonality can also lead to significant (centimeter-scale) vertical migration in the depth zonation of some sediment remineralization processes (fig. 14.3).

Despite the nonsteady-state seasonality in these sediments, under some circumstances it is possible to assume that pore-water profiles are in quasi steady state over short time periods. In the absence of significant advection this occurs because the characteristic time scales of diffusion are sufficiently short that diffusive fluxes rapidly reach steady state with respect to the current (or "instantaneous") rates of reaction, i.e.,

$$D_s \frac{\partial^2 C}{\partial z^2}\bigg|_t \approx -\sum R_i \big|_t \qquad (14.7)$$

(also see the discussion of eqn. 6.30 for further details). This then allows one to apply steady-state diagenetic models to individual, time-varying

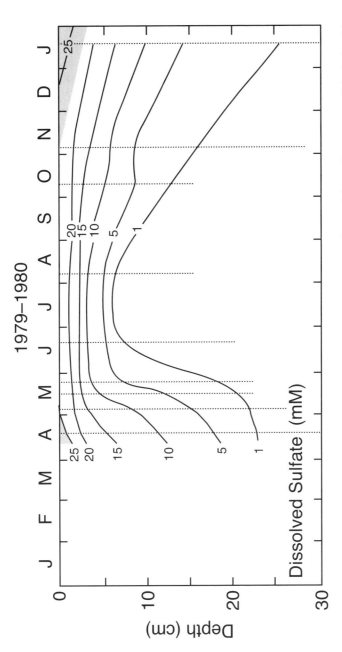

Figure 14.3 Seasonal variations in pore-water sulfate concentrations in Cape Lookout Bight sediments (modified after Martens and Klump, 1984). The grey area near the sediment surface indicates the observed depth distribution of light brown sediment, presumably as a result of seasonal in-growth of iron oxides near the sediment surface and their subsequent reduction (see discussions in section 16.3 and fig. 16.9).

pore-water profiles to estimate instantaneous sediment remineralization rates at the same times as pore-water collection (e.g., Martens and Klump, 1984; Soetaert et al., 1996b). Because of the way these rates are calculated they are essentially snapshots of time-varying rates at a particular point in time; therefore, they are not necessarily good estimates of the rates of sediment processes on more appropriate seasonal or annual time scales. To determine seasonal- or annual-averaged rates one must then average these instantaneous rate estimates over a seasonal cycle (e.g., Burdige and Zheng, 1998). Alternately, one can fit a group of individual pore-water profiles and, if available, profiles of rate measurements (both collected over time), with a time-dependent sediment diagenesis model (Klump and Martens, 1989; Soetaert et al., 1996b). In the sediments of Cape Lookout Bight these two approaches yield comparable seasonally averaged rates of sediment processes such as sulfate reduction (Klump and Martens, 1989). Such agreement reinforces the fact that on annual time scales Cape Lookout Bight sediments appear to be in steady state with respect to organic matter deposition and remineralization, despite significant seasonal variation in sediment rates and pore-water profiles (Haddad and Martens, 1987; Martens et al., 1992).

14.4 Diagenetic Processes in Deep-Sea Turbidites

Nonsteady-state processes occurring on longer, i.e., greater than annual, time scales also impact sediment geochemistry, particularly in the deep sea. The relevant processes here include turbidite deposition, climate change on glacial-interglacial time scales, and the formation of organic-rich sapropels (such as in eastern Mediterranean sediments). Discussions in the next two sections will focus on sediments affected by the first two of these processes; however, an examination of the sapropel literature indicates that studies of these other sediment processes can be applied in some degree to the understanding of diagenetic processes in sapropels (e.g., Passier et al., 1996; Jung et al., 1997).

As described in section 5.2, turbidites are gravity-driven flows of continental margin sediment to the deep sea. The thickness of an individual turbidite sequence can range from several centimeters to several meters, and in many settings multiple turbidites are interbedded with pelagic sediment (de Lange, 1986; Buckley and Cranston,

1988). From a geochemical perspective the study of turbidites has proven extremely important for understanding metal redox cycling in sediments and controls on sediment carbon preservation (Middelburg, 1993).

The conceptual model of turbidite geochemistry used here (fig. 14.4) first assumes that an organic-rich turbidite of sufficient thickness is instantaneously deposited over a carbon-limited pelagic sediment (see figs. 7.8A or 13.1A). For now, we ignore processes occurring at the lower contact of the turbidite with the previous pelagic sediment-water interface. This then creates a situation in which there is organic-rich turbidite sediment at the sediment-water interface that upon emplacement (or shortly thereafter) consumes all available oxygen and nitrate in the turbidite pore waters. Oxygen and nitrate then begin to diffuse into the turbidite sediment and oxidize the reactive organic matter in the turbidite down to its background refractory level (Colley et al., 1984; Wilson et al., 1985). As this occurs, oxygen and nitrate are able to diffuse deeper into the turbidite, which then leads to a downwardly progressing redox "front" within the turbidite. This process has been described as one in which these oxidants "burn down" through the turbidite sediment, hence this redox front is also described as a burn-down front.

At any given time, the burn-down front serves as a locus for subsurface aerobic respiration, denitrification, and iron and manganese oxidation; this can be seen in pore-water, solid phase, and denitrification rate depth profiles (also see Sørensen et al., 1984). As a result of turbidite burn-down, a growing zone of altered (or oxidized) turbidite sediment is produced through which oxygen and nitrate continually diffuse downward to the burn-down front, which roughly represents the sediment TOC discontinuity and the level of zero oxygen or nitrate in the pore waters (see fig. 14.5). Although the majority of the oxygen consumption at the burn-down front occurs in the remineralization of organic carbon, some is also consumed by the oxidation of the upwardly diffusing Mn^{2+} and Fe^{2+} from the deeper sediments (Wilson et al., 1986). Similar processes are also responsible for nitrate consumption at the burn-down front (e.g., see table 7.2).

Returning to fig. 14.4, we note that pelagic sedimentation also resumes after turbidite deposition. Because this sediment contains low levels of reactive TOC, as compared to the turbidite, what then develops is a situation in which some oxygen consumption and nitrate production occurs initially in the upper portions of the pelagic

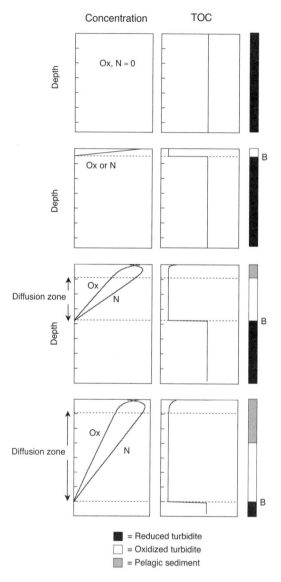

Figure 14.4 The evolution of geochemical properties in a pelagic sediment after the deposition of a more organic-rich turbidite. Note that time increases as one moves down the panels, and the tick marks on the y-axes indicate equivalent depth intervals. (Top panels) Either before or just after turbidite deposition this sediment is (or quickly becomes) devoid of pore-water oxygen (Ox) and nitrate (N). (Second panels) Oxygen and nitrate begin to diffuse into the turbidite sediment and oxidize the organic carbon in the turbidite

sediment. This is then followed by O_2 and nitrate diffusion through the lower pelagic sediment and oxidized turbidite to the burn-down front, where O_2 and nitrate are both consumed.

While the oxidized and reduced portions of the turbidite constitute a single unit in terms of their sedimentology and other compositional parameters, the burn-down process also leads to several distinct differences between the upper, oxidized and lower, reduced (or unaltered) portions of the turbidite. The first is that there is often a distinct color change at this redox boundary related to the brown-green sediment color change associated with sediment iron reduction (see section 7.3.3.2). Furthermore, in addition to the organic carbon discontinuity at the redox boundary, the cycling of redox sensitive elements here leads to their diagenetic redistribution during the burn-down process.

As a combined result of continued pelagic sedimentation and the burn-down process, the linear gradients in oxygen and nitrate continually decrease (or shallow) over time (fig. 14.4). This occurs because there is a continuous increase in the length of the linear portions of these gradients for approximately similar concentration changes

---

down to its background (refractory) level at the turbidite burn-down front (B). (Third panels) With the resumption of pelagic sedimentation after turbidite deposition, organic matter remineralization also resumes in the surface sediments (driven by the carbon rain from above). Because the pelagic sediment contains low levels of reactive TOC (as compared to the turbidite) a situation develops in which some oxygen consumption and nitrate production occur in the upper portions of the pelagic sediment, i.e., above the upper dotted lines in these panels and the panels below. This is then followed by O_2 and nitrate diffusion through the lower pelagic sediment and oxidized portion of the turbidite (the diffusion zone) to the burn-down front, where these oxidants are consumed. Although the oxidized turbidite and the deeper reduced turbidite constitute a single sedimentological unit, the burn-down process leads to distinct geochemical gradients across this boundary. (Bottom panels) As a combined result of pelagic sedimentation and turbidite burn-down, the linear gradients of oxygen and nitrate in the diffusion zone continually decrease with time; this slows the downward diffusive transport of these solutes. Since the nitrate and oxygen diffusive fluxes to the burn-down front control the rate of organic carbon oxidation at the burn-down front, this results in a negative feedback that similarly slows the rate of turbidite burn-down over time.

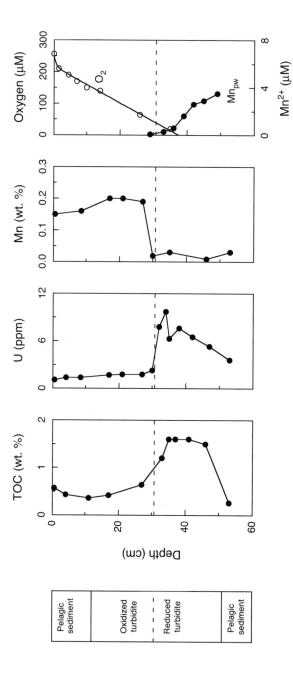

Figure 14.5 Sediment profile of TOC, uranium, and manganese, and pore-water profiles of Mn^{2+} and O_2 in sediments at a site (sta. 10554) on the Madeira Abyssal Plain in the Northeast Atlantic (redrawn using results in Wilson et al., 1985). Note the distinct concentration changes across the current boundary between the reduced and oxidized turbidite for all of these sediment constituents.

between the upper pelagic sediment, where surficial organic matter diagenesis occurs, and the burn-down front. Since the diffusive fluxes of nitrate and oxygen control the rate of burn-down, this results in a negative feedback that slows down the rate of turbidite burn-down over time. Early in the burn-down process (on the order of $\sim 10^2$ to 10^3 yr after turbidite emplacement), the burn-down rate is much faster than typical pelagic sedimentation rates, implying that significant burn-down occurs in the turbidite before significant amounts of overlying pelagic sediment are deposited (Wilson et al., 1986; Thomson et al., 1987). However, after several thousand years the rate of turbidite burn-down decreases exponentially, to rates that are comparable to local sedimentation rates (see Colley et al., 1984, and Wilson et al., 1985, for more details).

Because of these diffusion limitations, the relative contribution of subsurface carbon remineralization to that occurring in the surficial pelagic sediments varies with time. After several thousand years, deep metabolism at the burn-down front represents only a small proportion of the total sediment organic matter remineralization (Wilson et al., 1985; also see related discussions in Jahnke et al., 1989). At the same time, though, the occasional deposition of organic rich turbidites in otherwise oxic deep-sea sediments can have a significant effect on the occurrence of certain redox processes (see below), because of the resulting nonsteady-state conditions. Furthermore, this occurs beyond that which one might think should occur based simply on the amounts of organic carbon added to the sediments by turbidite deposition.

Both time-dependent and steady-state diagenetic models of turbidite burn-down have been developed (Wilson et al., 1995, 1996), and the results from one such model are shown in fig. 14.6. The ability to use steady-state models to examine some aspects of turbidite diagenesis is based on the fact that pore water solutes respond rapidly to changes in environmental conditions, as has been discussed earlier. Therefore, when solid-phase profiles change slowly, that is, when the rate of turbidite burn-down is comparable to the rate of pelagic sedimentation, the observed pore-water profiles will appear to be in "instantaneous" steady state with the existing solid-phase profiles.

The nonsteady-state processes that occur because of this downwardly propagating burn-down front also have a major impact on the sediment cycling of redox sensitive elements. For an element such as Mn, which is more soluble under reducing versus oxidizing conditions,

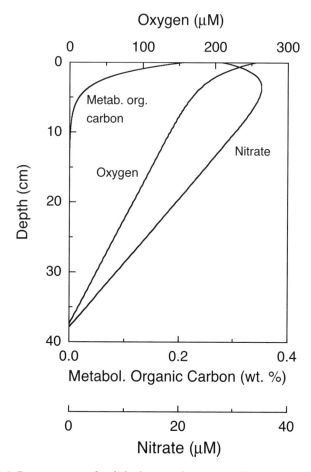

Figure 14.6 Pore-water and solid-phase sediment profiles of oxygen, nitrate, and metabolizable organic carbon predicted by the steady-state model equations of Wilson et al. (1985) for turbidite sediments. In this calculation metabolizable organic carbon in the surface pelagic sediments is consumed by ~9 cm, while the depth of the burn-down front (i.e., the boundary between oxidized and reduced turbidite sediment) is at 37.5 cm. The parameters used in these calculations are listed in table 2 of Wilson et al. (1985) for sta. 10554, and provide an excellent fit to the data.

its upward flux from the reduced turbidite sediments leads to precipitation (oxidation) at or just above the redox boundary (fig. 14.5). At the same time, however, the downward movement of the redox boundary also leaves behind manganese that has been precipitated. Therefore manganese accumulating in the oxidized turbidite sediment, and perhaps the pelagic sediment, can range from fairly broad peaks, localized around the redox boundary, to simply a manganese-enriched surface sediment layer (Wilson et al., 1985; Thomson et al., 1987, 1993). The factors that control this are complex, and likely include the time scales of redox burn-down, including both rate and duration, the magnitude of the upward Mn^{2+} flux, and the amount of bioturbation in the surface sediments (see discussions in sections 13.2 and 14.5 and in Thomson et al., 1984b, 1996; Dhakar, 1995; Dhakar and Burdige, 1996).

For an element such as uranium, which is more soluble under oxidizing versus reducing conditions, fig. 14.5 shows that U forms a distinct peak just below the redox boundary that gradually decreases with depth (Colley et al., 1984; Colley and Thomson, 1985; Thomson et al., 1987, 1993). Two processes appear to be responsible for these profiles. The first is that U associated with deposited organic matter is released into the pore waters as soluble U^{6+} (uranyl ion) during turbidite burn-down, and then immobilized in the reduced turbidite sediments as insoluble U^{4+} upon downward diffusion. In addition, the downward diffusion of uranyl ion from bottom waters to the redox front, where it is reduced and precipitates out, is also an important source of uranium to these peaks. Regardless of the relative roles of these two sources, the important point is that turbidite burn-down continuously recycles and concentrates this U peak in conjunction with downward movement of the burn-down front.

Studies have also shown that the same general processes that affect U and Mn during turbidite burn-down control the cycling of other redox sensitive elements that undergo similar phase changes in association with oxidation state changes (see Jarvis and Higgs, 1987; Thomson et al., 1993, 1996; Crusius and Thomson, 2003; and references therein). The remobilization and redistribution of all of these metals, as a result of turbidite burn-down, appears to be largely controlled by the redox potential of the relevant half-reactions and any associated changes in solubility. Thus an element such as V, which has a soluble oxidized form and insoluble reduced form, will be concentrated below the redox boundary, as is uranium); in contrast, Fe (like

Mn) will be concentrated at or above the redox boundary. Similarly, elements that can strongly adsorb to Fe and Mn oxides such as Ni, Co, Zn, and P, as phosphate, often show diagenetic behavior similar to that seen for their carrier phases.

The discussion above describes the time evolution of a system with a single turbidite deposited in an otherwise continuously accumulating pelagic sediment. Depending on the thickness of the turbidite, the burn-down process may completely oxidize the turbidite. At the other extreme, the turbidite may be sufficiently thick such that after long time periods the oxygen and nitrate gradients become essentially zero, i.e., there is effectively no concentration change in the diffusion zone (see fig. 14.4). This then leads to a situation in which the downward diffusion of oxygen or nitrate essentially ceases, at which point turbidite burn-down also ceases and the remaining turbidite carbon is preserved at this great depth. While in both cases the resulting porewater profiles will now look like steady-state carbon-limited profiles (fig.7.8A or 13.1A), solid-phase sediment profiles will still contain some evidence of the turbidite burn-down process (e.g., Colley et al., 1984).

More commonly, though, what occurs in many turbidite sediments is that the partial oxidation of a turbidite sequence is interrupted by the deposition of a new turbidite. For example, on the Madeira Abyssal Plain there are 19 turbidites in the upper 35 m of sediment that are each separated by pelagic sediment units 5–20 cm thick. These turbidites range in thickness from ~40 cm to ~3–4 m, and 12 appear to have undergone partial oxidation through the burn-down process described above (Buckley and Cranston, 1988).

The deposition of a new turbidite covers up the pelagic sediment along with the original partially oxidized and unaltered, reduced turbidite, and therefore isolates the reduced turbidite from its oxygen and nitrate source in the overlying waters. As a result, the burn-down process and associated redox reactions cease at the now relict (buried) burn-down front. Since any oxygen or nitrate in the pore waters of the relict pelagic sediment and oxidized turbidite will be consumed quickly after this isolation process, these sediments will now be exposed to some degree of suboxic and perhaps anoxic conditions. This isolation process then leads to the occurrence of secondary reactions that partially modify the primary diagenetic signatures that developed during the original burn-down processes (de Lange, 1986; Jarvis and Higgs, 1987; Buckley and Cranston, 1988; Thomson et al., 1989, 1996, 1998;

Van Os et al., 1993). Associated with these reactions can be distinct sediment color changes that occur in and around the relict burn-down front (e.g., dark purple bands extremely enriched in Cu, possibly as CuS_2)

Although these secondary reactions are not fully understood, several factors appear to be important. One is whether the sediment enrichment (or peak) that formed during turbidite burn-down occurred above the redox boundary by oxidative processes (e.g., as is the case for Mn), or below the redox boundary by reductive processes (e.g., as is the case for U and V). Peaks formed by reductive processes may be more stable after deposition of a new turbidite (Thomson et al., 1998), although at least for Mn, discussions in the next section suggest that the long-term preservation of Mn peaks as carbonate, rather than oxide, phases may complicate this explanation.

A second factor of importance appears to be the occurrence of suboxic conditions in these sediments, such that the primary Mn oxides are reduced, perhaps coupled to iron oxidation (rxn. 9 in table 7.2). This may then lead to the solubilization of some metals associated with Mn oxides (e.g., Co, Ni, Zn, and Cu), followed by their reuptake into either primary or secondary iron oxides in the relict turbidite sediments (Van Os et al., 1993; Thomson et al., 1996, 1998). Finally, in sediments where anoxic conditions occur (i.e., some degree of sulfate reduction), some metals that have been solubilized (e.g., Cu, Cd, Ni, Co, and Zn) may reprecipitate as insoluble sulfides (Van Os et al., 1993; Thomson et al., 1998; also see related discussions in section 17.3).

Understanding the occurrence of these secondary reactions is of some importance for deciphering the history of interbedded pelagic-turbidite sediments such as those on the Madeira Abyssal Plain. This also has implications for the use of these redox sensitive elements as paleoenvironmental indicators of changes in sediment redox conditions associated with glacial-interglacial transitions (e.g., see Crusius and Thomson, 2003, and references therein).

14.4.1 Organic Geochemical Studies of Turbidite Diagenesis

Studies of turbidite sediments have also proven useful in furthering our understanding of the factors controlling organic matter diagenesis in sediments. Much of this work has involved studies of a relatively shallow relict oxidation front associated with the f-turbidite on the Madeira Abyssal Plain (Keil et al., 1994a; Cowie et al., 1995; Prahl et al.,

1997, 2003; Hoefs et al., 2002; and others). This organic-rich turbidite (~1% TOC) was deposited ~140 kyr BP as a homogeneous deposit 4–4.5 m thick (de Lange et al., 1987), and subsequently underwent burn-down to a depth of ~0.5 m into the turbidite, before the process was halted by the deposition of another turbidite. The exact length of this burn-down event is not well constrained, but is likely on the order of tens of thousands of years (compare discussions in Buckley and Cranston, 1988; Cowie et al., 1995; Thomson et al., 1998).

Organic geochemical studies of the MAP f-turbidite have focused generally on comparisons between the composition of the upper, oxidized turbidite sediment and that of the lower, reduced turbidite sediment. Because the entire unit material was deposited as a homogeneous unit, this comparison then allows for discussions of the redox controls on organic matter remineralization in these sediments.

Perhaps the most important general observation from these studies is that the unaltered organic matter in the lower, reduced portion of the turbidite has remained essentially undegraded in the presence of pore-water sulfate for ~140 kyr, yet during the much shorter time of oxic, and perhaps suboxic, burn-down, ~80% of the TOC and ~60% of the TN in the turbidite was remineralized (Cowie et al., 1995). The final concentrations of TOC in the oxidized section of the turbidite are extremely low and comparable to those generally observed in deep-sea sediments, i.e., ~0.2% versus ~1% in the lower, reduced portion of the turbidite. Furthermore, the oxidized turbidite contains high levels (>30 mole%) of the nonprotein amino acids β-ala and γ-aba, a parameter shown to be indicative of highly degraded organic matter, i.e., of high diagenetic maturity (Cowie and Hedges, 1994; also see figs. 9.1 and 9.5A)

Together, these observations provide strong evidence for an oxygen effect associated with the remineralization of at least some fraction of the organic matter found in marine sediments. Some of the possible reasons for this effect have been discussed previously (e.g., sections 7.6 and 11.2.4). A hypothetical model illustrating this is shown in fig. 14.7. An important point of this model is that it assumes this oxygen effect is associated only with less reactive organic matter fractions, such as those possibly associated with the turbidite sediments. As a result, then, this oxygen effect will not be apparent during early stages of diagenetic maturity, i.e., in coastal sediments. This observation is consistent with results from studies of organic matter diagenesis in

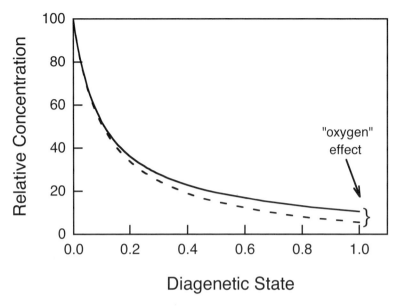

Figure 14.7 Hypothetical curves for the degradation of two mixtures of five equal amounts of organic matter (modified after Cowie et al., 1995). In each mixture, the four most reactive types of organic matter degrade by first-order kinetics with rate constants that successively decrease by a factor of 2 (e.g, $k = 8, 4, 2$, and 1). The dashed line then illustrates results in which the fifth, and least reactive, component in the mixture undergoes "oxic" degradation, and is twice as reactive as this last component in the mixture that undergoes "anoxic" degradation (solid line; i.e., here this last k in the "oxic" experiment is 0.5 while in the "anoxic" experiment it is 0.25). As discussed in the text, if this oxygen effect is associated only with less reactive organic matter fractions (or late stage diagenetic processes), then the effect will not be seen during early stages of diagenesis.

coastal sediments (e.g., Cowie and Hedges, 1992b; Emerson and Hedges, 2003), and will be further discussed in chapter 15.

Studies of organic matter degradation during turbidite burn-down have also allowed for an examination of the selectivity of remineralization. Although the loss of measured major biochemicals such as amino acids, aldoses, and lignins was largely nonselective (Cowie et al., 1995), these analytically determined components comprise only ~10%

of the total organic matter in these sediments, and therefore roughly the same percentage of the total organic matter that was oxidized during turbidite burn-down (see similar discussions in chapters 9 and 11). In contrast, stable isotope and lipid biomarker studies suggest that some selectivity does occur during organic matter oxidation in these turbidites, although results to date are equivocal on the factors controlling this selectivity (Hoefs et al., 1998, 2002; Prahl et al., 2003).

Focusing on the relative reactivity of marine versus terrestrial organic matter, Prahl et al. (1997) used a binary mixing model with data on the $\delta^{13}C$ of the sediment TOC and plantwax n-alkanes to examine this problem. This model indicates that while ~15% of the original TOC in the (reduced) turbidite was of terrestrial origin, it represented only ~7% of the total organic matter that was remineralized during turbidite oxidation. This suggests the strong preservation of terrestrial organic matter in the f-turbidite during oxidation; based on these calculations ~40–60% carbon that is preserved after oxic burn-down is of terrestrial origin.

In contrast, other biomarker studies are not necessarily consistent with this explanation of preferential preservation of terrestrial organic matter in the f-turbidite during oxic burn-down (Hoefs et al., 1998). Furthermore, studies of older MAP relict turbidites (i.e., back to the late Miocene, versus the Pleistocene-age f-turbidite) suggest that the interpretation of biomarker records, in terms of broader questions of marine versus terrestrial organic matter preservation, may be more complex than that discussed by Prahl et al. (1997), because of the possible uncoupling of biomarker remineralization versus that of its source material (compare discussions in Prahl et al., 2003, and Hoefs et al., 2002; also see general discussions of problems such as this in Hedges and Prahl, 1993). At the same time, these biomarker studies do provide evidence for other possible mechanisms of carbon preservation, including the preservation of membrane-bound bacterial lipids (Prahl et al., 2003; sections 9.7 and 11.1.6) and aliphatic refractory biomacromolecules (Hoefs et al., 1998; section 9.8.4).

One factor that may partially contribute to some of these differences is that on extremely long time scales, i.e., tens of millions of years, at least some fraction of the organic matter in unaltered (reduced) relict turbidite sections is susceptible to degradation by sulfate reduction (Meyers et al., 1996; Thomson et al., 1998). This degree of reactivity is, however, several orders of magnitude lower than that in the presence of oxygen, and perhaps nitrate, during the primary burn-down process. Because of this difference, this anoxic degradation likely has little impact on any oxygen effects associated with carbon burial on time scales that are less than $\sim 10^5$–10^6 yr. However, the effect of this slow, long-term anoxic remineralization on lipid biomarkers has not been examined.

14.5 Multiple Mn Peaks in Sediments: Nonsteady-State Diagenetic Processes Associated with Paleoceanographic Change

A common observation in many pelagic and hemipelagic sediments is the occurrence of multiple solid-phase manganese peaks (see fig. 14.12). Such manganese peaks occur over a range of depth and time scales, in some cases occurring as millimeter-scale laminae, rich in both manganese and iron (see discussions in Wilson et al., 1986). Multiple manganese-rich layers of centimeter-scale thickness are found in the upper one to several meters of other sediments, on time scales that go back not only to the most recent glacial-interglacial transition, i.e., the Glacial-Holocene boundary, but also through much of the late Quaternary (Burdige, 1993).

Interest in understanding the formation of these peaks stems, in part, from a desire to determine the paleoceanographic information contained in these peaks. However, because these Mn peaks are postdepositional features, extracting such paleoceanographic information, e.g., the time scales over which they form or their relationship to other environmental parameters, requires the use of flux calculations (Froelich et al., 1979; Mangini et al., 2001) or time-dependent models of sediment diagenesis (Dhakar, 1995). Similarly, because these manganese peaks are postdepositional diagenetic features, an a priori relationship between the age of the peak, or the time of its formation, and the age of the host sediment may not exist (Thomson et al., 1996; Mangini et al., 2001). At best, the age of the host sediment provides only an upper limit for the age of the peak.

In this discussion these manganese peaks are taken to be indicators of either the present-day location or the previous location(s) of the manganese redox boundary.[2] This assumption is based in part on the

---

[2] Recall that the manganese redox boundary was defined here as the depth in the sediments where pore-water oxygen goes to zero (see section 13.2).

assumption that the dominant form of manganese in these peaks is an oxide phase (see section 13.2). However, there is also some evidence in the literature that some of these peaks may actually be manganese carbonates. The extent to which this affects the discussion below depends in part on how such Mn carbonate peaks form, and their possible relationship to precursor Mn oxide peaks; these points will also be discussed below.

The diagenetic models discussed in section 13.2 predict that under some steady-state conditions a single manganese peak should be observed in sediments just above the sediment redox boundary (e.g., scc fig. 13.3). Therefore, the existence of multiple manganese peaks has been attributed to the occurrence of nonsteady-state manganese redox cycling. To further understand the history of the formation of these peaks, manganese pore-water profiles can be useful in distinguishing which of the manganese peaks is the active diagenetic feature associated with the current redox boundary; this is possible because pore-water manganese profiles adjust relatively rapidly to changes in the depth of the manganese redox boundary and begin precipitating a new solid-phase manganese peak at the new redox boundary (e.g., Froelich et al., 1979; Pedersen et al., 1986).

In the simplest sense, the nonsteady-state conditions affecting manganese diagenesis result from the net migration, relative to the sediment-water interface, of the manganese redox boundary, since under steady-state conditions the redox boundary moves upward at a rate equal to the sedimentation rate and therefore remains at a fixed depth relative to the sediment-water interface. A complex interplay of changes in such factors as bottom water oxygen concentrations, carbon rain rate to the sediments (both the absolute value and the reactivity of the material), and sedimentation rate control this movement (Thomson et al., 1984b; Wilson et al., 1986; Burdige, 1993; Dhakar, 1995; Mangini et al., 2001). While the details of this process are not fully understood, key aspects of the problem can be illustrated using several simple examples.

With this approach, the depth of the manganese redox boundary will increase with time if, for example, bottom water oxygen concentrations increase, the flux of organic carbon to the sediments decreases, and/or the rate constant for sediment organic matter oxidation decreases (as a result of a decrease in the reactivity of this material). Similarly, the depth of the redox boundary will shallow if these parameters change in the opposite direction. General evidence in support of

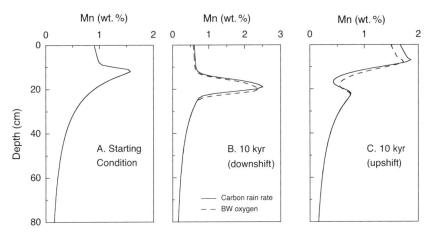

Figure 14.8 Results from a nonlinear, time-dependent model of sediment diagenesis in which there is an instantaneous change in either the bottom-water oxygen concentration (dashed line) or the organic carbon rain rate to the sediments (solid line). A. The initial steady-state manganese profile. B. The manganese profile at 10 kyr after a redox downshift caused by either a decrease in the carbon rain rate (from 46 to 35 mmol m^{-2} yr^{-1}) or an increase in the bottom-water concentration (from 167 to 218 μM). C. The manganese profile at 10 kyr after a redox upshift caused by either an increase in the carbon rain rate (from 46 to 66 mmol m^{-2} yr^{-1}) or a decrease in the bottom water concentration (from 167 to 120 μM). Additional details about the model used here, along with other specific information about these calculations can be found in Dhakar (1995).

such relationships between the depth of the manganese redox boundary and these parameters comes in part from experimental and modeling results (Emerson, 1985; Emerson et al., 1985; Dhakar, 1995; Mangini et al., 2001; also see discussion in section 7.5.2). In addition, other studies have observed that the depth of the present-day manganese redox boundary, as inferred from manganese pore-water profiles, is inversely correlated with the flux of organic carbon to the sediments (Lynn and Bonatti, 1965; Swinbanks and Shirayama, 1984; Shimmield and Price, 1986; Finney et al., 1988).

A deepening redox boundary (redox downshift) leads to a situation in which the downward progressing redox front produces a broad Mn oxide peak that essentially "scavenges" manganese during the downshift (fig. 14.8). In contrast, a shallowing redox boundary (redox upshift) leads to a situation in which the Mn peak becomes trapped below the upwardly moving redox boundary and begins to dissolve. This leads to the formation of a transient second peak just above the relict peak, although with time the newly precipitated manganese in the transient peak generally becomes redistributed by bioturbation into a Mn-rich surface sediment layer. An earlier conceptual model of such upward and downward movement of the sediment redox boundary (Burdige, 1993) suggested that these simple processes could lead to the formation of multiple Mn peaks in the sediments. However, this model neglected to consider the impact of bioturbation in controlling the formation of Mn peaks in sediments (Dhakar and Burdige, 1996). It also implicitly assumed that the migration of the sediment redox boundary is faster than was indicated by later modeling studies of Dhakar (1995).[3]

Dhakar's (1995) model results suggest that for deep-sea sediment conditions, redox upshifts or downshifts occur on time scales of $\sim 10^3$ -10^4 yr, with the rate of movement being initially relatively rapid and then decreasing exponentially with time. Note that similar results were also obtained in models of turbidite burn-down discussed in the previous section (Colley et al., 1984; Wilson et al., 1986). In their time-dependent model, Wilson et al. (1986) also discuss how a redox downshift that slows exponentially over time leads to a situation in which the redox front eventually hovers near its maximum depth of penetration (or balance point) for "a considerable period." These general observations appear broadly consistent with Dhakar's (1995) model results, which also show that the sediment redox boundary reaches a new steady state after ~ 10 kyr.

---

[3] An interesting contrast to these model results for deep-sea sediments is presented by Mucci et al. (2003). Here, in a coastal fjord 10–50 cm of relatively organic-rich sediment was deposited in the fjord sediments as a single depositional event after widespread flash flooding in the surrounding watershed (also see discussions in Deflaundre et al., 2002). Under these circumstances, this does result in the near-instantaneous upward shift in the sediment redox boundary, and leads to the trapping at depth of the manganese peak formerly at the sediment surface, in the now reducing sediments. Furthermore, over the next four years of study, a transient pair of double Mn peaks was observed because of the reductive dissolution of the lower peak (at the relict sediment surface) and its reprecipitation just below the new sediment-water interface. In contrast, though, to the behavior of Mn following this depositional event, Fe and As that were remobilized in association with this flood event were largely retained at depth (i.e., close to the former sediment surface), through precipitation of authigenic sulfides (e.g., AVS; see section 17.3).

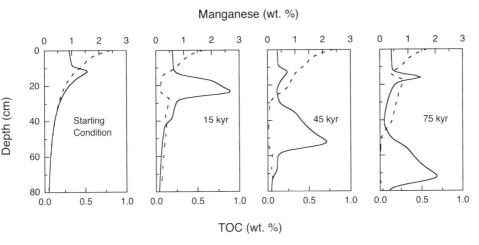

Figure 14.9 Results from a nonlinear, time-dependent model of sediment diagenesis in which there is a temporal change in the carbon rain rate (from 46 to 35 mmol m^{-2} yr^{-1}) with a periodicity of 60 kyr. The solid line is the sediment Mn profile; the dashed line is the TOC profile. Additional details about the model used here, along with other specific information about these calculations can be found in Dhakar (1995). During the redox downshift that occurs in the first half of the period, the original Mn-peak near the current redox boundary grows in, as is seen in the 15 kyr results. When the redox boundary next moves upward during the second half of the period (15 → 45 kyr), a new Mn-peak begins to form above this relict Mn-peak, in a manner that was discussed in the text for the case of a singular redox upshift (also see fig. 14.8c). During the next period (45 → 75 kyr), the redox boundary again moves downward and this small upper Mn-peak now continues to grow in (as in the 15 kyr results). At the same time, the relict Mn-peak (at ~70 cm) is also buried below the redox boundary. However, because of the decreased amount of organic carbon near this relict peak, the rate of manganese reduction at this sediment horizon is significantly reduced. This then leads to preservation (or "permanent" burial) of this peak.

At the same time, Dhakar's (1995) model results do show that multiple Mn peaks can form in sediments in response to periodic, versus singular, changes in either the bottom water oxygen concentration or the carbon rain to the sediments (e.g., see fig. 14.9). However, not all changes lead to the formation of multiple Mn peaks (e.g., high frequency changes), and the relationship between the periodicity of these changes and the formation of multiple Mn peaks is similar to that discussed in section 14.2 (also see Dhakar, 1995, for further details). Nevertheless, for appropriate periodicities, changes in the carbon rain rate to the sediments or the bottom water oxygen concentration can lead to the formation of a new Mn peak during each cycle. Consistent with these model results are those obtained with the slightly less complex time-dependent model of Wilson et al. (1986). Their results show how overlapping peaks in Mn or Fe can form in a system close to the "balance point" in response to small (10%) changes in the O_2 flux. In addition to linking the formation of these metal-rich layers to nonsteady-state phenomena, the Wilson et al. (1986) model results also provide a possible explanation for the formation of laminated sediment structures below the zone of active bioturbation.

14.5.1 Multiple Mn Peaks and the Glacial-Holocene Transition

The TOC content of Glacial sediments is, in general, higher than that in Holocene sediments (Müller and Suess, 1979; Pedersen, 1983; Coppedge and Balsam, 1992). Such differences could be related either to an increased organic carbon rain rate to Glacial sediments because of higher surface ocean productivity, or to more efficient carbon preservation in Glacial sediments. More efficient carbon preservation, at least in pelagic sediments, could be caused by lower bottom water oxygen concentrations that result from decreased ventilation of the deep ocean during Glacial times (see discussions of these contrasting explanations in Emerson, 1985; Lyle et al., 1988; Pedersen et al., 1988; Verardo and McIntyre, 1994; Yu et al., 1996; and references therein). At the same time, Glacial sedimentation rates are also higher than Holocene rates (Coppedge and Balsam, 1992; Thomson et al., 1996). Based on discussions above, any (or all) of these differences would have led to changing sediment conditions that should then have caused a redox downshift in the sediments starting around the Glacial-Holocene transition.

Numerous studies present evidence for the in-growth of sediment Mn and Fe peaks near the depth of the Glacial-Holocene boundary (Thomson et al., 1984b, 1996; Wilson et al., 1986; Wallace et al., 1988; Mangini et al., 2001). Furthermore, consistent with past discussions of how such peaks form, distinct sediment layers of other redox-sensitive elements, such as U, V, and Mo, are also observed in and around this boundary (e.g., Thomson et al., 1996). However, because

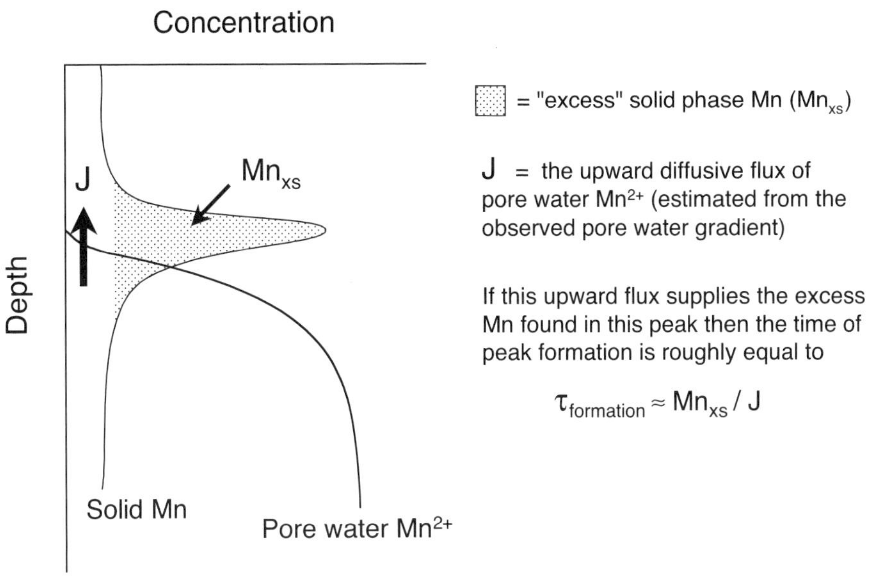

Figure 14.10 Schematic illustration of a method that can be used to estimate the time required to form a manganese peak in the sediments. In this approach upward diffusion supplies the manganese to this peak, and it is assumed that the pore-water gradient has remained constant over time. Under these conditions, then, the time of formation is approximately equal to the ratio of the "excess" Mn in the peak divided by the upward Mn^{2+} flux into the peak.

Diagram labels: Concentration; Depth; J; Mn_{xs}; Solid Mn; Pore water Mn^{2+}; = "excess" solid phase Mn (Mn_{xs}); J = the upward diffusive flux of pore water Mn^{2+} (estimated from the observed pore water gradient). If this upward flux supplies the excess Mn found in this peak then the time of peak formation is roughly equal to

$$\tau_{formation} \approx Mn_{xs} / J$$

these peaks are diagenetic features, the fact that they are found near the Glacial-Holocene boundary does not necessarily imply that they formed as a result of the Glacial-Holocene transition.

One approach to examining the timing of the formation of these peaks is illustrated in fig. 14.10. Here, pore-water manganese gradients and the inventory of manganese in a solid phase peak are used to obtain a rough estimate of time of formation of this peak. Unfortunately, estimates from this type of calculation are equivocal. Results presented by Mangini et al. (2001) for some Atlantic cores suggest that the time scale of peak formation is ~11–26 kyr, roughly consistent, given the uncertainties in such calculations, with the timing of the Glacial-Holocene transition (11 kyr). However, similar calculations for other sediment cores (Wallace et al., 1988; Thomson et al., 1996) predict significantly longer formation times, >50 kyr, that are themselves greater than the age of the host sediments. This strongly suggests that the upward diffusive fluxes of Mn^{2+} observed today are lower than those in the past, leading to an overestimate of the peak formation time. This observation does not appear to be unreasonable, since the upward diffusion of Mn^{2+} from Glacial sediments might be expected to decrease over time without replenishment by burial and reductive dissolution of Mn oxides.

In another approach to this problem, Wilson et al. (1986) used their time-dependent model to simulate the formation of Fe-rich layers in sediments that are thought to form by the same redox downshift mechanism discussed above. Using input parameters for western equatorial Atlantic sediments, their model demonstrates that peaks similar to those observed in these sediments can be produced in ~11 kyr, consistent with a redox downshift associated with the Glacial-Holocene transition.

While the data discussed above are consistent with a redox downshift associated with the Glacial-Holocene transition, such a redox downshift may be caused by either an increase in the bottom-water oxygen concentration or a decrease in the organic carbon rain rate to sediments. Thus interest in modeling the nonsteady-state diagenesis of Mn in sediments was motivated, in part, by a desire to examine this problem, potentially allowing for the use of Mn sediment profiles to determine the causes of the sediment redox downshift associated with the Glacial-Holocene transition. More importantly, in a general sense, such results would then have provided additional insight into the validity of different models for the operation of the Glacial ocean (see the discussion at the beginning of this section).

Unfortunately, model results presented by Dhakar (1995) show that opposing changes in organic carbon rain rate or bottom water oxygen concentration lead to very similar time-dependent Mn sediment profiles (fig. 14.8B). As such, this modeling approach is unable to distinguish between the two different forcing functions that could have caused a sediment redox downshift in association with the Glacial-Holocene transition. Mangini et al. (2001) came to a similar conclusion based on their examination of Mn profiles in North Atlantic sediment cores.

14.5.2 Multiple Mn Peaks and Pleistocene Climate Cycles

In gravity cores collected in the hemipelagic sediments at MANOP site H, peaks in both solid phase manganese and organic carbon

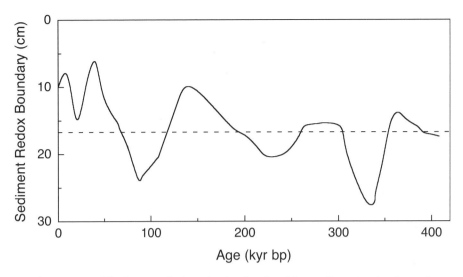

Figure 14.11 Hindcast variations in the depth of the sediment redox boundary at MANOP site H determined using sediment manganese and TOC profiles. Shown as a dashed line is the mean value of this redox boundary over this time period (redrawn from results presented in Finney et al., 1988).

concentrations occur over the past ~400,000 yr, and Finney et al. (1988) have suggested that they are created by periodic changes in sediment redox conditions, similar to those discussed above for the most recent glacial-interglacial transition. To further examine these changes they developed a simple model that uses the organic carbon and manganese profiles to hindcast the depth of the manganese redox boundary over the age of these cores (fig. 14.11). The results of this model predict a periodicity of ~100,000 yr in the depth of the manganese redox boundary at this site, consistent with other information on paleoproductivity changes during the Quaternary at this site that should affect the depth of the redox boundary (Finney et al., 1988; Lyle et al., 1988).

Finney et al. (1988) also suggested that a relict manganese peak trapped in the manganese reduction zone by a redox upshift can become permanently buried in the sediments, because they assumed that little of this peak is remobilized once it is buried below the redox boundary. While this suggestion is consistent with their data from this site and model results in fig. 14.9, data from other sites do not always show similar preservation of such relict Mn peaks (Dean et al., 1989; Thomson et al., 1998).

The causes of these preservation differences are not well understood (see discussions in de Lange et al., 1994). One possible explanation for the preservation of such relict Mn peaks is that under some circumstances they are actually Mn carbonate, and not oxide, peaks (Gingele and Kasten, 1994), since the reduced form of Mn in a carbonate phase is likely to be more stable under the suboxic or (anoxic) conditions in which these relict peaks are buried. Work by Thomson et al. (1986) suggests that high sediment carbonate content may play an important role here. In such sediments, Mn^{2+} produced by the reductive dissolution of Mn oxides may be efficiently adsorbed to calcite phases, because of the high Mn^{2+} partition coefficient, before it can diffuse any significant distance away from the site of reductive dissolution. Over time, this adsorbed manganese may then become incorporated into a mixed $(Mn,Ca)CO_3$ phase, and a resulting (though slightly broader) carbonate "peak" will then be found close to the depth of the original oxide peak (see analogous discussions in Thomson et al., 1998, of related dissolution/reuptake processes in turbidite sediments). In the absence of such an efficient uptake mechanism, Mn^{2+} produced in this manner is likely to diffuse in both directions away from its source, i.e., the relict Mn oxide peak. Upward diffusion should lead to reoxidation of the Mn^{2+} at the current redox boundary, thus destroying any record of the relict oxide peak (Thomson et al., 1996). Downward diffusion may either simply dissipate this Mn^{2+} into the deeper pore waters, or, if pore-water alkalinity levels are sufficiently high, induce the precipitation of a mixed carbonate phase at some depth below the relict oxide peak (see section 13.2 for further details). Although one might envision that this could lead to the formation of a new Mn carbonate peak, the fact that this carbonate peak would likely form some distance from the relict oxide peak suggests that it would be very difficult to interpret this carbonate peak in a paleoceanographic context.

14.5.3 Multiple Mn Peaks in Holocene Sediments

In cores collected in both Atlantic and Pacific equatorial upwelling systems, pairs of Mn peaks are often observed in the upper ~40–50 cm of sediment (Froelich et al., 1979; Berger et al., 1983; Price, 1988; fig. 14.12). Furthermore, pore-water Mn^{2+} data suggest that the upper peak likely defines the region where manganese is currently being precipitated, while the lower peak is apparently a relict peak. Although

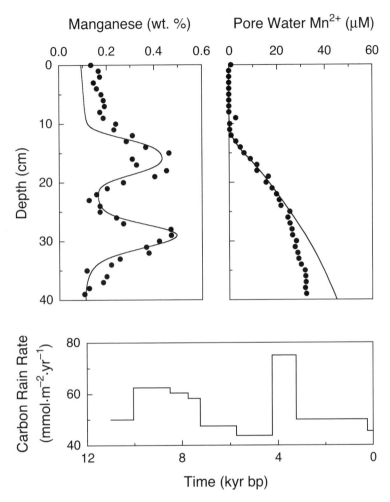

Figure 14.12 Solid-phase and pore-water manganese profiles at a site in the equatorial Pacific (RC 52) fit with a nonlinear, time-dependent sediment diagenesis (data from P. N. Froelich, pers. comm.). The temporal changes in the organic carbon rain rate to these sediments (used as a forcing function to generate this fit to the data) are also shown here. Additional details about the model used here, along with other specific information about these calculations can be found in Dhakar (1995).

interest in understanding the formation of these peaks came initially from a desire to use this information to examine the Glacial-Holocene transition (Berger et al., 1983), it appears, in actuality, that the processes leading to the formation of these peaks have occurred solely within the Holocene (Burdige, 1993).

Initial evidence for this comes from calculations similar to those described in fig. 14.11 that suggested that the upper Mn peak in a core collected in the equatorial Atlantic could have formed in ~0.7–5 kyr (Froelich et al., 1979; Price, 1988). Subsequent modeling studies by Dhakar (1995) of this core and other cores in the equatorial Atlantic and Pacific (references cited above) provide best fits to these data consistent with millennial-scale changes in organic carbon rain rate to these sediments during the Holocene (fig. 14.12). The causes of these changes are not well understood (e.g., see earlier discussions in Price, 1988). However, it seems possible that they could be related to other, more general millennial-scale changes in climate and/or ocean processes that appear to have occurred during the Holocene (Bond et al., 1997; Cane and Clement, 1999; Indermühle et al., 1999).

Evidence for the more recent occurrence of such processes can be seen in a core collected near the East Pacific Rise (Pedersen et al., 1986). In this core solid-phase Mn decreases exponentially with sediment depth, while the pore-water Mn^{2+} profile suggests that active manganese oxidation is now occurring at a sediment depth of ~10 cm, where much lower sediment Mn levels occur (fig. 14.13). Pedersen et al. (1986) suggest that these profiles are caused by a redox downshift that resulted from a decrease in the carbon rain rate to these sediments within the past 1,000 yr. Modeling of the data from this core by Dhakar (1995) provides support for this suggestion.

In this discussion of changes in the depth of the redox boundary in pelagic and hemipelagic sediments we have seen that these fluctuations generally occur on time scales greater than ~10^2–10^3 yr. Although pore-water profiles can respond more rapidly to environmental changes, other sediment processes generally limit the overall response time of these sediment systems. However, in studies of continental margin sediments, metals that show opposing responses to changing redox conditions, i.e., Mn and Cd (Gobeil et al., 1997) and Mn and Re (Gobeil et al., 2001), appear to provide evidence for the occurrence of relatively large (centimeter-scale) migration of the sediment redox boundary on decadal or perhaps annual time scales. Analogous seasonal changes in sediment redox conditions can be seen in coastal

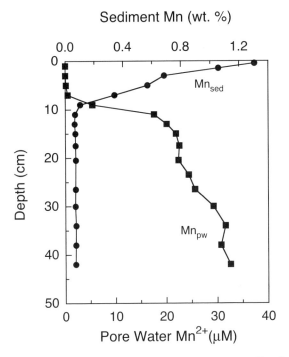

Figure 14.13 Pore-water and solid-phase manganese profiles from a core collected in hemipelagic sediments near the East Pacific Rise (core HUD 22, water depth = 2,768 m). Redrawn using data presented in Pedersen et al. (1986).

sediments (e.g., see section 14.3 and fig. 14.3), provided seasonal changes in organic matter remineralization rates are sufficiently large. In contrast, though, evidence for this type of rapid change in margin sediments has not been described. One explanation for the profiles observed by Gobeil et al. (1997, 2001) is that they occur as a result of seasonal (?) pulses of reactive organic matter to these sediments. Scaling calculations that use reasonable, though perhaps upper limit, estimates for the reactivity of this organic matter are consistent with such an explanation (Gobeil et al., 1997, 2001). However, time-dependent sediment diagenesis models have not yet been applied to the data from these sediments to further examine this explanation.

≈ **CHAPTER FIFTEEN** ≈

The Controls on Organic Carbon Preservation in Marine Sediments

THROUGHOUT EARLIER CHAPTERS OF THIS BOOK we have considered different aspects of organic carbon preservation and burial in marine sediments. Understanding the controls on carbon preservation is important for a number of reasons. In the simplest sense the burial of organic carbon in marine sediments represents the major link between "active" surface pools of carbon in the oceans, in the atmosphere, on land, and in marine sediment, and carbon pools that cycle on much longer, geological time scales, i.e., carbon in sedimentary rock, coal, and petroleum deposits. Thus, organic carbon burial in marine sediments plays an important role in controlling atmospheric CO_2 and O_2 on these long time scales (see section 8.3 and, for example, Berner, 1989; Hedges, 2002).

In this chapter we will discuss recent results that begin to provide mechanistic explanations for the controls on organic matter preservation in marine sediments. As we will see, a variety of factors appear to play a role in carbon preservation, including organic matter–mineral interactions, organic matter composition and reactivity, and the time of sediment oxygen "exposure."

The process of carbon preservation in marine sediments is often thought of in an equivalent sense to carbon remineralization. However, this view may be somewhat misleading, because <0.5% of the gross production/photosynthesis on the Earth escapes remineralization; that is, for every 100 units of organic matter produced on land or in the oceans, >99.5 units are remineralized, and <0.5 units are buried in marine sediments. Looked at somewhat differently, the burial efficiency (BE[1]) of carbon in marine sediments with respect to production on land and in the surface ocean is <0.5%. This fact might, therefore, lead one to conclude that information on the controls of carbon remineralization will not be particularly useful in understanding the controls on carbon burial and preservation. At the same time, when

---

[1] Recall that BE is defined in eqn. (8.19).

we examine preservation and remineralization in marine sediments we see that this "mismatch" between these processes may not be as severe.

Relative to that found in surface water organisms, the organic matter deposited in marine sediments has decreased in absolute amount (on a weight % basis), and undergone some amount of fractionation, i.e., diagenetic maturity, prior to deposition (fig. 9.1); as a result its reactivity has also decreased substantially (e.g., see section 8.1 and fig. 8.1). Because of these changes, organic carbon burial efficiencies with respect to carbon rain rate to the sediments are generally ~10–20%, or more, particularly in sediments that represent the major sites of carbon burial in the oceans (Henrichs and Reeburgh, 1987; Aller, 1998; also see fig. 15.1). Thus, based on these observations, it seems more likely that the factors controlling sediment preservation and remineralization could be more linked.

As was discussed in chapter 9, preservation and remineralization are related somehow, if only because preservation is the absence of remineralization, and visa versa. However, as the discussions in chapter 9 began to demonstrate, preservation does not necessarily result simply from the absence of remineralization per se. Rather, specific factors may also directly enhance preservation. At the same time, another important question involves determining what factors, if any, may inhibit remineralization and thus enhance preservation.

Teasing out the factors that control carbon preservation versus remineralization is difficult for a number of reasons. One problem here is the fact that these effects appear to be most apparent only for less reactive types of organic matter. This was illustrated, for example, in the model results in fig. 14.7. Thus, the ability to examine controls on preservation versus remineralization are strongly affected by the time scales of the experimental study or field observation (also see section 8.1 for further details).

Controls on carbon preservation in sediments are often examined in terms of organic carbon burial efficiency (BE; see fig. 15.1). Examining carbon preservation in terms of BE has certain advantages over, for example, examining preservation using TOC content or carbon burial rates at depth (see discussions in Henrichs and Reeburgh, 1987; Hedges and Keil, 1995). Burial efficiency avoids problems associated with dilution effects such as tailing off of sediment TOC values at very high sedimentation rates, because of dilution of TOC with organic-poor inorganic material; it also avoids mathematical interdependencies

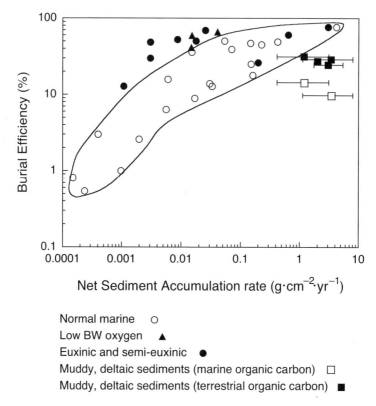

Figure 15.1 Burial efficiency of sediment organic carbon versus sedimentation rate for a range of sedimentary environments (modified after Aller, 1998, using data cited therein). The envelope shown here defines the commonly observed pattern in normal marine sediments of burial efficiency increasing with increasing sedimentation rate (Henrichs and Reeburgh, 1987; Canfield, 1994). This figure also illustrates three other important points: sediments underlying low, or zero, bottom-water oxygen concentrations do not show uniformly enhanced carbon preservation (high BE values) as compared to normal marine sediments; muddy, deltaic sediments generally show lower burial efficiencies than normal marine sediments at the same sedimentation rate; marine organic carbon is more efficiently remineralized than terrestrial organic carbon in muddy deltaic sediments. The data shown here for muddy, deltaic sediments are for the Amazon and Mississippi deltas (marine and terrestrial organic carbon) and Fly and Chiangjiang deltas (terrestrial organic carbon only).

between carbon burial rates and sedimentation rates, because the former is a function of the latter.

A further examination of fig. 15.1 in the context of past discussions of organic carbon remineralization in sediments (Henrichs and Reeburgh, 1987; Canfield, 1994; McKee et al., 2004) indicates the following. Low BE sites are low sedimentation-rate sites, which predominantly occur in pelagic or abyssal regions. Aerobic respiration dominates in these sediments (section 13.1), and the organic matter being preserved here is of low reactivity (e.g., see fig. 8.1). With increasing sedimentation rate burial efficiency generally increases, as does the importance of suboxic and eventually anoxic remineralization (sections 7.5.2 and 11.2.4), along with the preservation of more reactive organic matter. However, as will be discussed in later sections, nearshore, muddy deltaic sediments tend to be exceptions to these general trends.

Based on observations such as those in fig. 15.1, it can be seen that in many marine sediments, virtually complete remineralization of sediment organic matter can occur, as indicated by the extremely low TOC content of deep-sea sediments, ~0.1–0.2% (e.g., Premuzic et al., 1982), and their low burial efficiencies, on the order of a few percent. In contrast, the TOC content of ancient shales and modern fine-grained continental margin (deltaic, shelf, and upper slope) sediments (their presumed precursors)[2] are both on the order of ~1–2% (Berner, 1982; Premuzic et al., 1982; Hunt, 1996), with correspondingly higher burial efficiencies in these sediments. Therefore, specific aspects of the biogeochemical processes occurring in continental margin sediments may play an important role in the overall controls on sediment carbon preservation (e.g., Hedges and Keil, 1995; Aller, 1998; Hedges et al., 1999a).

Previous studies of carbon preservation in sediments have often involved attempts to relate the controls on sediment carbon preservation to specific environmental characteristics. Two parameters commonly considered are sediment accumulation rate and bottom water oxygen concentration (see, for example, discussions in Henrichs and Reeburgh, 1987; Calvert and Pedersen, 1992; Hedges and Keil, 1995;

---

[2] This inferred relationship between ancient shales and modern continental margin sediments stems from the fact that >80% of the carbon burial and preservation in marine sediments and remineralization as well, occurs in continental margin sediments (see table 8.1).

Hedges et al., 1999a). The role of bottom water oxygen, and oxygen in general, on sediment carbon preservation will be discussed in sections 15.2 and 15.4.

In a general sense, sediment accumulation rate may exert control on carbon preservation by moving organic matter rapidly away from the surface sediments, where rates of remineralization are generally highest (see discussions in Hedges and Keil, 1995). If oxidant availability limits organic carbon degradation in any fashion, then rapid deposition also increases the path length that soluble oxidants such as O_2, nitrate, and perhaps sulfate must travel from the bottom waters in order to reach reactive organic matter at depth (Henrichs, 1982; Aller and Mackin, 1984; Emerson, 1985). In both cases, this decreased degradation could lead to an increase in preservation.

Consistent with this suggestion, the diagenetic model of Tromp et al. (1995) does predict a general increase in carbon burial efficiency with increasing sediment accumulation rate. Since this is also seen in the envelope of results shown in fig. 15.1, these two observations might be taken as evidence for at least some type of causative relationship between sediment accumulation and carbon preservation. At the same time, Hedges and Keil (1995) discuss some of the difficulties in relating sediment accumulation rate to sediment carbon preservation. Furthermore, as Aller (1998) has also observed, many nearshore, muddy deltaic sediments show (for similar sedimentation rates) lower burial efficiencies than that predicted by the overall pattern shown in fig. 15.1. This then implies that at least in these sediments, organic matter preservation and remineralization may be decoupled from sediment accumulation. This point will be explored below in further detail.

15.1 Organic Matter–Mineral Interactions

In general, organic matter is preferentially found in fine-grained sediment fractions (Mayer, 1994b; Hedges and Keil, 1995). For example, Premuzik et al. (1982) observed a positive correlation between sediment TOC concentrations and the clay content (sediment particles <2 μm dia.) for sediments in water depths <2,000 m. Such observations may be explained two ways. The first is that there is preferential attachment, e.g., adsorption, of organic matter to fine-grained, high surface area sediment particles. The second is that fine-grained inorganic sediment particles and low-density organic particles are hydrodynamically equivalent, which then leads to these two physically separate materials ending up in the same size class during conventional grain-size separation and analysis. Studies to date, however, have not supported this latter explanation, and the vast majority of the organic matter in sediments, up to ~90%, appears to be intimately associated with mineral particles (Keil et al., 1994b; Hedges and Keil, 1995). Such observations strongly suggest that organic matter–mineral interactions are important in terms of understanding organic matter dynamics in sediments.

In further examining the covariance of grain size and TOC content Mayer (1994b) suggested that sediment surface area might be more important controlling parameter than grain size (also see earlier studies by Suess, 1973). Evidence in support of this comes from observations showing that sediment surface area and sediment TOC content are indeed positively correlated, at both the bulk sediment level (e.g., Mayer, 1994b) and in discrete grain-size separates, e.g., sand, silt, and clay fractions (Keil et al., 1994b). Furthermore, in many continental margin sediments this covariance leads to what appears to be a diagenetically stable surface-area normalized TOC concentration[3] of ~1 mg OC m^{-2}. Interestingly, this concentration is equivalent to the expected concentration of a monolayer of moderately sized organic molecules evenly adsorbed over all mineral surfaces (Mayer, 1994b). In some sediments this diagenetically stable, monolayer equivalent (ME) concentration is seen throughout the zone of sediment diagenesis. However, in other cases, sediments appear to have "excess" organic matter near the sediment-water interface and approach this ME value at depth during diagenesis, where sediment TOC levels also reach asymptotic concentrations (fig. 15.2).

Further analysis of riverine suspended matter and sediments across a broad spectrum of depositional environments has shown that there actually appear to be five broad types of carbon loading on sediment particles (Mayer, 1994a,b; Hedges and Keil, 1995; Keil et al., 1994b, 1997):

1. Continental shelf and upper slope sediments with loadings of ~0.5–1.1 mg OC m^{-2}, i.e., the so-called ME level of carbon loading;

---

[3] This surface-area normalized TOC concentration is often referred to as the OC:SA ratio or loading.

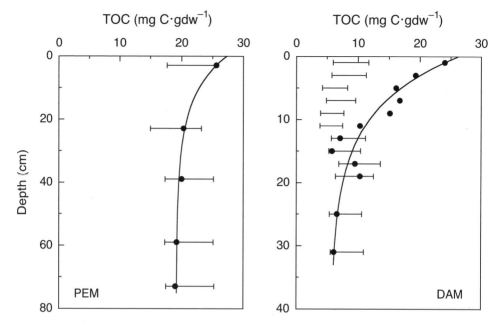

Figure 15.2 Total organic carbon concentration versus sediment depth in cores from the Damariscotta estuary, Maine (USA) [*right*: DAM; water depth = 33 m] and the central Gulf of Maine [*left*: PEM; water depth = 89 m]. Filled circles are measured TOC concentrations; horizontal error bars represent the TOC concentrations predicted by the measured sediment surface area at each sediment depth horizon and the typical ME carbon loading of ~0.5–1.1 mg OC m^{-2} observed in many continental margin sediments (see Mayer, 1994b, for further details). The best-fit curves through the measured TOC concentrations were obtained by fitting the data to eqn. (8.4). Redrawn using data presented in Mayer (1994b).

2. Sediments accumulating in either high-productivity regions or under low-oxygen bottom waters, with loadings >2 mg OC m^{-2}, i.e., greater than ME levels. These sediments also show no apparent approach to an ME level of stabilization with sediment depth;

3. Sediments associated with major river deltas that have loadings of ~0.3 mg OC m^{-2}, i.e., less than ME levels. Interestingly, because of the high sedimentation rates in deltaic environments, these sediments still account for ~40% of the carbon burial in all marine sediments (see table 8.3), which is roughly equivalent to the amount buried in other continental margin sediments with higher carbon loadings (see 1 above);

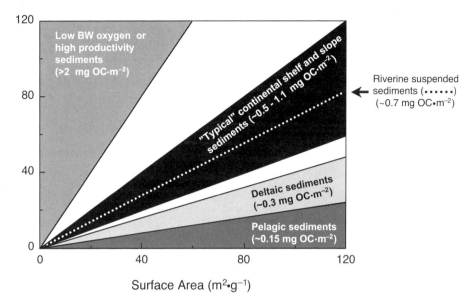

Figure 15.3 Sediment organic carbon content versus surface area for sediments from a range of depositional settings. Modified after Hedges (2002) based on results presented elsewhere (Mayer, 1994a,b; Hedges and Keil, 1995; Keil et al., 1997).

4. Pelagic sediments that have very low carbon loadings (~0.15 mg OC m^{-2});

5. Riverine suspended sediments that show a range of carbon loadings averaging ~0.7 mg OC m^{-2}.

These observations are also illustrated in fig. 15.3, and their significance will be discussed throughout the rest of this chapter.

Although organic matter loading on sediment particles was initially described in terms of monolayer coverage, Mayer (1994a,b) also noted that most minerals in nature have rough surfaces, with mesopores (width ≈ 2–50 nm) that contribute to much of the total mineral surface area. As a result, Mayer initially suggested that most of this organic matter is likely adsorbed to sediments in these mesopores and not uniformly spread across mineral surfaces.

Extending these observations to the examination of the controls on sediment carbon preservation further led to what is sometimes referred to as the *surface adsorption/mesopore protection hypothesis* (Mayer, 1994b). In this model, organic matter adsorbed to mineral surfaces inside of mesopores may be protected from bacterial exoenzymes in at least two ways. First, these pores are likely too small to allow entry of these enzymes and thus prevent the direct contact needed by most enzymes to carry out their activity. Second, steric constraints within the pores may further reduce the effectiveness of an enzyme even if it is able to enter a pore.

At the same time, because reactive oxygen species such as peroxide or the superoxide radical should be able to enter these mesopores, this hypothesis may provide a mechanistic explanation for the possible role of anoxia in enhancing carbon preservation or, conversely, for the enhanced remineralization of organic matter when exposed to oxygen (Aller, 1994a; Hedges and Keil, 1995; Hartnett et al., 1998; Hedges et al., 1999a; also see discussions in section 7.6.3 and 15.2). As we will also see below, low carbon loadings in deltaic and pelagic sediments (see above) may similarly be explained by this oxygen effect.

Subsequent work (Ransom et al., 1997; Mayer, 1999; Bock and Mayer, 2000; Arnarson and Keil, 2001) has further demonstrated that organic matter is indeed unevenly distributed (patchy) on sediment surfaces. As a result much of the mineral surface is exposed to the surrounding aqueous solutions. High-resolution microscopy further suggests that organic matter is not contained inside of *inter*granular mesopores (on mineral surfaces) but is found in *intra*granular pores (or slits) between mineral grains. Again the small size of these slitlike spaces between particles still provides the physical protection required in the original mesopore protection hypothesis. These recent observations imply that sediment-organic matter interactions may be less like adsorption and more akin to glue that forms organosediment aggregates (also see related discussions in section 9.8.5). Furthermore, the nature of these interactions also suggests that particulate organic matter does not necessarily have to undergo partial degradation/solubilization to be protected by interactions with sediment particles and surfaces (compare discussions in Hedges and Keil, 1995, and Hedges, 2002).

Finally, regardless of the mechanism(s) of sediment-organic matter interaction, another important observation to come from these studies is that there appears to be some type of active partitioning of this particle-associated organic matter in seawater and perhaps in pore waters. In particular, particles transported from land to the ocean, and eventually to marine sediments, exchange (or lose) much of their terrestrial organic matter loading and take up marine organic matter

(Hedges et al., 1997; Keil et al., 1997; Mayer et al., 1998). The significance of this exchange/loss process, particularly in terms of the remineralization of terrestrial organic matter in the oceans, will be discussed in later sections of this chapter.

15.2 The Role of Oxygen in Sediment Carbon Remineralization and Preservation

The role of oxygen in carbon preservation has long been discussed from a variety of standpoints. Historically this problem has been examined from the perspective of the role of bottom water oxygen concentration in controlling the TOC content of marine sediments, in part because petroleum geochemists have often considered organic-rich sediments underlying anoxic bottom waters (e.g., the Black Sea) as modern counterparts of petroleum source rocks (see discussions in Demaison and Moore, 1980; Pedersen and Calvert, 1990; Hunt, 1996). However, studies of recent marine sediments reveal no clear-cut relationships between bottom water O_2 concentration and either sediment TOC content, organic carbon BE, or the preservation of type II "oil-prone" kerogen (Betts and Holland, 1991; Calvert and Pedersen, 1992; Ganeshram et al., 1999; also see fig. 15.1).

At the same time, these observations do not imply that oxygen per se has no effect on carbon preservation or remineralization; for example, see discussions on organic matter diagenesis in turbidites in section 14.4.1. Rather, it suggests that bottom water O_2 concentrations may not necessarily be the proper "metric" with which to examine these effects. Moreover, results that will be discussed in section 15.4 show that the average time that organic matter in surface sediments is exposed to oxic conditions does appear to play at least some role in controlling the extent of organic matter remineralization. By definition this process then *indirectly* enhances preservation, although section 15.6 will also discuss how factors that inhibit remineralization may also *directly* enhance preservation.

In examining the role of oxygen in controlling organic matter remineralization, several broad conclusions can be reached. The first is that, in general, the remineralization of relatively fresh, i.e., labile, organic matter occurs at roughly the same rate under oxic versus anoxic conditions (Westrich and Berner, 1984; Cowie and Hedges, 1992b; Lee, 1992; Kristensen and Holmer, 2001; also see Calvert and Pedersen,

1992, for a review of the earlier literature). A second broad conclusion is that aged or refractory organic matter is degraded more slowly, if at all, under anoxic versus oxic conditions (e.g., Cowie et al., 1995; Kristensen et al., 1995; Hulthe et al., 1998). An exception to these apparent trends, though, are results presented by Harvey et al. (1995) in which the loss of TOC, TN, protein, and carbohydrate during the microbial decay of fresh phytodetritus (from diatom and coccolith cultures) was faster under oxic versus anoxic conditions. However, an examination of the decay rate constants in this study, for both oxic and anoxic conditions, indicates that they are in the upper end of the rate constants shown in fig. 8.1. This latter observation suggests that anoxic remineralization of this very fresh organic matter is still significant.

In examining these results, some caution must be applied to subjective terms such as "labile," "refractory," or "aged" (see, for example, discussions in section 8.1). However, the works cited above include both short-term laboratory experiments and field observations that cover longer time scales. Thus, it appears reasonable to conclude that oxygen-sensitive organic matter is selectively concentrated with increasing organic matter diagenetic maturity (e.g., Hedges, 2002; also see fig. 14.7).

One possible explanation for this observation involves controls on the remineralization process by the initial depolymerization of sedimentary organic matter. As discussed in section 7.6.2, certain complex materials, such as proteins or carbohydrates, can undergo hydrolysis in the presence or absence of oxygen. In contrast other types of organic matter may be effectively degraded only by O_2-requiring enzymes or by strong oxidants derived from O_2, such as the hydroxyl radical (see section 7.6.1). Thus, the preferential hydrolysis of reactive proteins and carbohydrates during early stages of diagenetic maturity, under both oxic and anoxic conditions, may then concentrate hydrolysis-resistant materials or carbon-rich substrates, e.g., lignin or some lipids, that can be effectively degraded only when O_2 is present (also see discussions of this latter point in section 9.7).

Consistent with this discussion are, for example, general suggestions (Burdige, 1991; Aller et al., 1996; Prahl et al., 1997; Aller and Blair, 2004) that marine organic matter, e.g., proteins and carbohydrates, can be preferentially degraded over terrestrial organic matter, e.g., lignins, or perhaps soil organic microbial residues (see section 11.1.2). Results in fig. 15.1 also show that terrestrial organic matter has a higher burial efficiency in muddy deltaic sediments than does marine organic matter.

Other factors may also play a role in the oxygen sensitivity of some types of sediment organic matter toward degradation. Physical protection of reactive (non-O_2 sensitive ?) materials by O_2-sensitive compounds could, for example, shield reactive material from degradation; this was discussed in section 11.2.4 in terms of the "remineralization constancy" observed across a broad range of marine sediments. Discussions in section 15.6 will also describe ways in which the occurrence of other types of abiotic processes, e.g., geopolymerization, could lead to analogous oxygen effects associated with the decomposition of some types of sedimentary organic matter.

Another possible explanation of this apparent oxygen effect on organic matter remineralization comes from results presented by Dauwe et al. (2001). In this work, they observed that enhanced rates of aerobic, versus anaerobic, remineralization were a function of the overall remineralization rate—i.e., low rates favored aerobic versus anaerobic remineralization—and that this was independent of the inherent reactivity of the material being degraded. Using a simple model for organic matter decomposition, Dauwe et al. (2001) showed that such an oxygen effect at low remineralization rates could be explained by assuming that anaerobic bacteria have a lower growth efficiency and/or a higher maintenance efficiency than aerobic bacteria (also see discussions in Fenchel et al., 1998). These results also suggest that a critical amount of metabolizable organic matter is needed to sustain anaerobic metabolism in sediments. Thus as organic matter concentration and reactivity both decrease with diagenetic maturity, so does the overall remineralization rate; hence the oxygen effect observed by Dauwe et al. (2001) would become increasingly important.

15.3 THE ROLE OF BENTHIC MACROFAUNAL PROCESSES IN SEDIMENT CARBON REMINERALIZATION AND PRESERVATION

The role of benthic macrofaunal processes in affecting sediment carbon preservation is often regarded in terms of their role in adding oxygen to marine sediments. While in many sediments this may indeed be important, the role of benthic macrofauna in affecting sediment carbon preservation is actually more complex, as is summarized in table 15.1. In general, it appears that macrofaunal processes impact rates of sediment organic matter degradation as well as overall organic matter oxidation efficiency in sediments (as defined in section

TABLE 15.1
Macrofaunal Effects on Organic Matter Decomposition and Remineralization in Sediments (Modified from Aller et al., 2001, and References Cited Therein)

Macrofaunal Process[a]	Effect	Decomposition Rate and/or Extent[b]
Particle manipulation (7.6.3, 15.3)	Substrate exposure, increase of surface area via fragmentation	+
Grazing, digestion (7.5.1, 7.6.3, 11.1.5)	Microbe consumption, bacterial growth stimulation, surfactant production, enzyme bath, redox oscillation	+
Excretion, secretion (12.5, 12.5.1)	Mucus substrate, nutrient release, bacterial growth stimulation	+
Construction, secretion (9.8.4, 11.1.5)	Synthesis of refractory or inhibitory structural products (tube linings, halophenols, body structural products)	–
Irrigation (7.6.3, 12.5, 12.5.1, 15.3)	Soluble reactants supplied, metabolite build-up lowered, increased reoxidation, redox oscillation	+
Particle transport (7.6.3, 12.4, 15.3)	Transfer between major redox zones, increased reoxidation, redox oscillation, substrate priming	+

a. Listed in parentheses below each macrofaunal process are the sections in the book where these are discussed.

b. + = stimulation, – = inhibition

8.3.2; see Aller et al., 2001). Therefore, the enhancement of carbon remineralization by benthic macrofaunal processes translates into a decrease in carbon burial efficiency.

Because bioirrigation increases the diffusive openness of sediments (see section 12.5.1), it plays a role in the enhanced input of oxygen to sediments (Aller et al., 1996; Kristensen and Mikkelsen, 2003). At the same time, diffusive openness in general appears to enhance remineralization rates in sediments, beyond effects simply related to enhanced O_2 input into bioirrigated sediments (Aller et al., 1996; Aller and Aller, 1998).

Bioturbation, i.e., particle reworking, also appears to enhance carbon remineralization, although in somewhat different ways. In general these processes do not strongly enhance oxygen input into sediments in the way that bioirrigation does (see, for example, recent discussions in Kristensen and Mikkelsen, 2003). In part, this occurs because D_B values are generally smaller than D_s values (see discussions in sections 12.4 and 12.5). However, oxygen input into sediments by bioturbation may impact other aspects of the redox conditions in bioturbated sediments, such as iron redox cycling (Ingalls et al., 2000), which may then affect carbon remineralization (see section 7.6.3). Although the effects of bioturbation on solute transport may not always be significant, its impact on particle transport can be important. As is discussed in section 12.4.3 bioturbation, and in particular nonlocal sediment mixing, can transport reactive organic matter from the sediment surface to greater sediment depths, and may, therefore, play a role in priming the decomposition of refractory organic matter through cometabolism or co-oxidation (as defined in section 7.6.3). The passage of refractory organic matter through macrofaunal guts also appears to enhance the subsequent microbial degradation of this material (e.g., Kristensen and Mikkelsen, 2003) by mechanical, chemical, and/or enzymatic digestion.

15.4 Oxygen Exposure Time as a Determinant of Organic Carbon Preservation in Sediments

Previous discussions strongly suggest a role for oxygen, in a very broad sense, in controlling carbon preservation in sediments. However, rather than examining these effects in terms of bottom water O_2 concentration, results from a number of studies suggest that the more

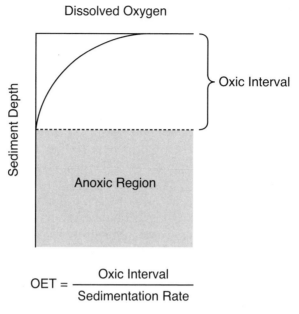

Figure 15.4 An illustration of how oxygen exposure time (OET) is calculated in the work of Hartnett et al. (1998) and Hedges et al. (1999a). As discussed in the text (see section 15.4.1), this approach does not consider the effects of transport processes such as bioturbation, bioirrigation, or physical resuspension on estimates of OET (also see related discussions in Keil et al., 2004). This figure is modified after Hedges (2002).

important variable may be the average time that organic matter in sediments is exposed to oxic conditions (Aller, 1998; Hartnett et al., 1998; Hedges et al., 1999a; Hedges, 2002).

In the simplest sense this *oxygen exposure time* (OET) may be viewed as the average time organic matter is exposed to oxic conditions in the sediments before being permanently buried in oxygen-deficient sediments at depth. The approach shown in fig. 15.4 illustrates how OET may be determined from pore-water O_2 profiles and rates of sediment accumulation; estimates of OET may also be obtained using benthic O_2 flux measurements and simple diagenetic models (see Hartnett et al., 1998, for further details).

To begin to quantitatively examine the role of oxygen exposure on carbon preservation, Hartnett et al. (1998) examined the relationship between sediment OET and organic carbon burial efficiency for sediments along the western Pacific continental margin. These sediments were found in water depths ranging from ~100 m to 3,500 m and had bottom water oxygen concentrations ranging from essentially zero to greater than ~100 μM. Results from this study indicated that organic carbon BE varies inversely with log(OET), over more than 3 orders of magnitude of OET; specifically, values of OET varied from essentially zero for sediments deposited under the anoxic bottom waters of the intense oxygen minimum zone on the Mexican continental slope, to values of ~1,000 yr for sediments on the lower slope of the California and Washington margins. Organic carbon BE values at these sites ranged from ~50–60% to <5%, i.e., within the broad range of BE values seen in fig. 15.1.

Follow-up studies of these Washington margin sediments (along an offshore transect ranging from ~100 m water depth on the mid-shelf, down the continental slope and into the Cascadia Basin at ~2,500–3,000 m water depths) allowed Hedges et al. (1999a) to further examine the role of sediment oxygen exposure on organic matter preservation. Because these sediments are uniform in terms of elemental and isotopic composition, the effects of oxygen exposure on sediment organic matter remineralization could be examined without some of the ambiguities that can arise when comparing results from differing sedimentary systems.

In this work, Hedges et al. (1999a) observed that OET increases offshore along this transect as a result of the combined effects of increasing oxygen penetration depth and decreasing sedimentation rate. The OC:SA ratio also decreases as one moves offshore and, as a result, varies inversely with OET (fig. 15.5) from typical continental margin to deep-sea values (~1 mg OC m^{-2} to ~0.3 mg OC m^{-2}; see fig. 15.3). Organic geochemical indicators of organic matter "freshness," such as mole% nonprotein amino acids (fig. 9.1) and 100/mole% glucose (section 9.5), further demonstrate that the organic matter in these sediments becomes increasingly degraded with increasing oxygen exposure time. Pollen analysis offers a third tracer of oxic effects on organic matter diagenesis since pollen is degraded only under oxic conditions (Keil et al., 1994a). In these Washington margin sediments the percentage of corroded pollen increases with increasing OET (Hedges et al., 1999a).

In summarizing these results Hedges (2002) suggests that most of the sediment organic matter along the Washington margin is degraded only when exposed to oxygen and undergoes little, if any

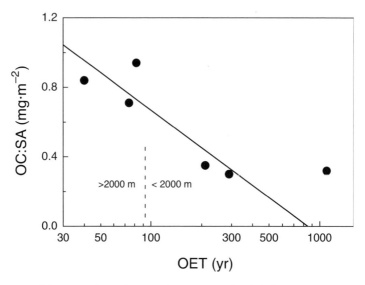

Figure 15.5 The organic carbon:surface area ratio (OC:SA) versus oxygen exposure time for sediments along an offshore transect on the Washington (USA) continental margin (redrawn from data presented in Hedges et al., 1999a). Assuming sediment surface area is conserved along this transect, the best-fit line through these data (excluding the data point with the highest OET value) predicts that this organic matter has a half-life with respect to remineralization of ~100 yr.

degradation, in the absence of O_2. Thus oxygen exposure clearly controls carbon remineralization and, hence, preservation in these sediments, consistent with results from the MAP-f turbidite discussed in section 14.4.1. Of equal importance, this comparison suggests that this oxic effect associated with organic matter–mineral association operates over a fairly broad range of time scales, i.e., for organic matter with a large range in reactivity, with these time scales ranging from ~10^2 yr on the Washington shelf to ~10^4 yr in the MAP-f turbidite.

Conversely, these observations do not imply that anoxic remineralization is not important in other sediments, or on time scales longer than those of early diagenesis, since other discussions throughout this book have clearly shown otherwise. Rather, these observations do seem to imply that oxidant availability plays a role in sediment organic carbon burial and preservation for some types of organic matter as one moves offshore across the continental margin (see additional discussions in Hedges and Keil, 1995; Aller et al., 2001). Thus, while the early rates of organic matter remineralization may be largely controlled by the quality and quantity of available organic matter (Westrich and Berner, 1984; Henrichs, 1992), the extent of preservation, particularly after some degree of organic matter diagenetic maturity, appears to involve a set of processes associated somehow with exposure of sediment organic matter to oxygen and/or oxygen input to the sediments.

15.4.1 What Exactly Does Sediment Oxygen Exposure "Mean"?

The previous discussion has been presented in terms of the specific role of molecular oxygen exposure on carbon preservation/remineralization. However, other discussions (e.g., section 7.6.3) have also noted that the specific mechanisms by which this occurs may not necessarily *directly* involve O_2 itself. Rather, it is equally likely that oxygen exposure actually leads to other processes or conditions that enhance remineralization over that which occurs under more strict anoxic conditions (also see related discussions in Aller, 1994a; Hedges and Keil, 1999; Hedges, 2002). Past discussions have indicated several possibilities for how this might occur, including effects associated with redox oscillations (section 7.6.3), benthic macrofaunal processes (section 15.3), the diffusive openness of bioirrigated sediments (section 12.5.1), and the role of Fe and Mn redox cycling and metal oxide reduction in promoting organic matter remineralization (section 7.6.3). Oxygen exposure time may therefore, under some circumstances, simply be a reasonable proxy, in some way, for these effects.

A specific example of how these effects might work can be seen in results from muddy tropical deltaic sediments such as those on the Amazon continental shelf and in the Gulf of Papua (also see previous discussions in sections 5.5, 7.5.2, and 7.6.3). As seen in fig. 15.1, these sediments have lower burial efficiencies of both terrestrial and marine organic matter than other more "normal" marine sediments of comparable sediment accumulation rates. These normal marine sediments also tend to be sulfidic sediments in which anoxic remineralization is the predominant mode of organic matter remineralization. In contrast, though, the physical reworking and continual reoxidation of these deltaic sediments leads to nonsulfidic sediments in which iron redox cycling is extremely important and iron reduction dominates sediment organic matter remineralization. Although these processes require repeated oxygen exposure of the sediments, this occurs in a nonsteady-state fashion that cannot be characterized in terms of the OET calculations shown in fig. 15.4. These results therefore also point out that OET:BE relationships such as that presented in Hartnett et al. (1998) would not be appropriate for examining carbon preservation in these muddy deltaic sediments.

It has also been argued that bioturbation has a minimal impact on sediment organic matter OET because particles will be randomly mixed both into and out of the oxic region of the sediments at roughly the same rate (Hartnett et al., 1998). However, this steady-state argument does not take into account the role of asymmetric redox oscillations in enhancing the overall degradation of sediment organic matter (see sections 7.5.1 and 7.6.3, and discussions in Hedges et al., 1999a; Aller et al., 2001; Sun et al., 2002).

Finally, in some sediments organic matter may be extensively degraded prior to deposition in a sediment (e.g., as a result of lateral versus vertical transport; see section 5.2 and related discussions in Martens et al., 1992; Hedges and Keil, 1995). In such settings, organic matter reactivity, and hence burial efficiency, is likely not simply related to OET values calculated in the manner of fig. 15.4. In part, this may occur because OET as calculated in this manner underestimates the true oxygen exposure of sediment organic matter because of this lateral transport (e.g., Keil et al., 2004).

In spite of such uncertainties in the relationship between organic carbon BE and OET, these observations are not meant, however, to minimize the importance of the observations of Hartnett et al. (1998) and Hedges et al. (1999a). Their results, along with the work of Aller and coworkers in muddy deltaic sediments, provide important mechanistic information on the possible controls of carbon preservation in sediments. At the same time, though, all of these results appear to suggest that sediment organic matter oxygen *exposure* in a broader sense (and perhaps other co-related processes or phenomena) rather than oxygen exposure *time* (specifically as calculated in fig. 15.4) is the more important general controlling parameter.

As discussed previously, the broad range of OC:SA ratios observed in different sedimentary environments (<0.15 to >2 mg OC m^{-2}; see fig. 15.3) indicates that surface interactions alone do not control carbon burial or preservation. In re-examining these observations here, we first note that the OC:SA ratios for riverine sediments or typical continental shelf and slope sediments (~0.5–1.1 mg OC m^{-2}) may be viewed here as a baseline loading, although the reasons (or explanation) for this are not well understood at the present time. Nevertheless, in spite of this uncertainty, OC:SA ratios that are higher or lower than this level of carbon loading may now possibly be explained in terms of the oxygen exposure controls on sediment carbon remineralization discussed here.

Based on discussions above, pelagic sediments and shelf-deltaic sediments are both expected to have the greatest amount of oxygen exposure; as can be seen in fig. 15.3, both also have OC:SA values that fall below this baseline range. This trend is also broadly consistent with the results on the Washington shelf (fig. 15.5), which show an exponential-like decrease in OC:SA loading with increasing oxygen exposure. Thus, long-term exposure to oxic conditions can lead to sediments that are depleted in organic matter, as compared to these baseline loadings. This then suggests that such long-term oxygen exposure (and any associated processes) can apparently overcome the controls on these baseline carbon loadings.

At the opposite end of this spectrum, sediments from low oxygen bottom water environments have relatively high OC:SA values, and show no apparent approach to this baseline OC:SA range with sediment depth/diagenesis. This suggests that the high carbon input to these sediments, and their relatively brief exposure to oxygen, results in a high degree of organic matter preservation. While these observations are mechanistically important, in a larger sense they have little impact on the global rate of sediment carbon burial and preservation. In spite of this high carbon preservation at low bottom water oxygen concentrations, table 8.3 demonstrates that these low oxygen environments are minor sites of carbon burial.

As noted above (section 15.1) such oxygen effects are broadly consistent with the mesopore protection hypothesis (Mayer, 1994b), particularly if reactive oxygen species such as peroxide or the superoxide radical are required for the degradation of certain types of refractory sediment organic matter (see sections 7.6.1 and 15.2). Similarly, other types of oxic effects (e.g., see discussions in sections 7.6.3, 11.2.4, 15.2, and 15.3) could also possibly explain the results in fig. 15.3. Examining these issues will be important in the further development of a more complete mechanistic understanding of carbon preservation in marine sediments, and in developing more robust predictors (as compared to OET as defined in fig. 15.4) of sediment carbon preservation.

15.4.2 Organic Carbon Burial and Controls on Atmospheric O_2

Over the Phanerozoic (0–600 my BP) atmospheric O_2 levels have apparently been maintained within a relatively constant range (roughly ±50% as compared to present oxygen levels; Berner, 1989). Given that the atmospheric residence time of oxygen is ~4 my, this argues for some degree of negative feedback control on atmospheric O_2. Evidence for such control also comes from the fact that there are relatively large gross fluxes of O_2 into and out of the atmosphere, driven by photosynthesis and respiration, with significantly smaller net input or removal fluxes, i.e., carbon burial in sediments and oxidation of kerogen and reduced inorganic minerals on land. Because these net fluxes are ~0.1% of gross fluxes (see, for example, a global oxygen budget in Hedges, 2002), marine photosynthesis in the absence of respiration or any nutrient limitations has the potential to double atmospheric O_2 concentrations in only several thousand years. This observation strongly argues for the need for negative feedback control of atmospheric O_2. At the same time, evidence from a number of studies (e.g., Garrels et al., 1976) suggests that this control operates on the oxygen source, rather than sink, namely with the net oxygen production associated with carbon burial in marine sediments.

In a recent paper, Hedges (2002) addresses this problem in terms of the discussions above regarding mineral–organic matter associations and the role of oxygen in sediment carbon burial and preservation. In this work he hypothesizes that the control of atmospheric O_2 over the Phanerozoic has been maintained by a "mineral conveyor belt" (fig. 15.6), modulated by negative feedback controls associated with carbon burial in continental margin sediments (also see related discussions of this hypothesis in Hedges et al., 1999b). One key assumption in this model is that mineral transport from weathered continental rocks to continental margin sediments occurs with little net change in the amount, but not the composition, of organic matter loading. A second assumption is that mineral surface area is roughly conserved during the cycle of weathering, sediment burial, and continental uplift. This second assumption will not be discussed here; the supporting arguments are presented in Hedges (2002).

Evidence for the former assumption comes in part from the similarity of the OC:SA loading of riverine suspended particles and most continental margin sediments (see fig. 15.3). Discussions at the beginning of this chapter also indicate the similarity of the TOC content of ancient shales and sediments found at depth on modern continental margins, i.e., below the zone of early diagenesis. Both of these observations appear to suggest that transport of mineral material (sediment) from weathered rocks to riverine suspended matter to marine sediments occurs without large changes in its organic matter loading. At the same time, the composition of this organic matter does change dramatically, from ancient kerogen, to terrestrial organic matter (e.g., slowly cycling soil humus), to more recently formed marine organic matter. In concert with the active partitioning of these different types of organic matter during mineral transport is the oxidation (remineralization) of the organic matter that is being removed. This occurs either on the particles as a part of the removal process, or in solution after the organic matter is removed from the particles (see discussions in section in 15.5 and in Keil et al., 1997; Aller, 1998).

One portion of the negative feedback in this model for the control of atmospheric oxygen occurs through the *tectonic cycle* on the left side of fig. 15.6. As illustrated here, increased weathering will increase the flux of mineral surface area to the oceans and, in a proportional sense, presumably also increase the kerogen oxidation rate and thus atmospheric oxygen consumption. Increasing the flux of mineral surface area to the oceans then allows for the increased uptake and burial of marine organic matter, and an increased amount of oxygen is then added back to the atmosphere to compensate for the O_2 taken up by kerogen weathering.

This tectonic cycle is intriguing partly because it provides a physical link between distant sources and sinks of oxygen. However, evidence from a number of modeling studies (summarized in Hedges, 2002) also suggests that additional negative feedback control is required to fine-tune the global oxygen cycle and obtain the high degree of stability in atmospheric O_2 levels observed over the Phanerozoic. In the model shown here this additional negative feedback control comes from the *weathering cycle* shown on the right-hand side of fig. 15.6, which builds on the observation of sharp offshore gradients in sediment TOC concentration along the lower (>2,000 m) continental margin (see summaries in Hedges and Keil, 1995; Hedges, 2002). Furthermore, at least on the Washington shelf this also corresponds to the sediment region where increasing oxygen exposure leads to large decreases in the OC:SA ratio (fig. 15.5).

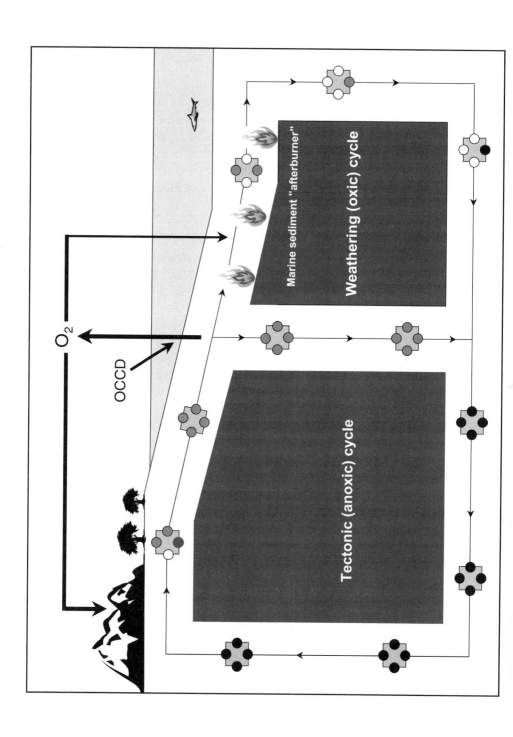

Figure 15.6 A conceptual model illustrating the proposed mineral conveyor belt model for the control on atmospheric O_2 (modified after Hedges, 2002). The two cycles shown here represent the conservative transport of sediment particles/mineral surface area (squares with circles) with differing organic matter loadings (shaded or black circles). The outer (weathering) cycle represents deposition of very low TOC-content sediments below the marine organic carbon compensation depth (OCCD), where long term exposure to oxic conditions is sufficient to remineralize almost all sediment organic matter. As discussed in the text, the response of the weathering cycle to changing environmental conditions, e.g., imbalances in the tectonic cycle, leads to the movement of the OCCD in response to such perturbations. The negative feedback associated with the weathering cycle, in conjunction with the negative feedback associated with the tectonic cycle, is proposed by Hedges (2002) to lead to the observed long-term stability of atmospheric O_2 during the Phanerozoic.

By analogy to the calcite compensation depth (see section 13.5), Hedges (2002) refers to this transition depth as the *organic carbon compensation depth* (OCCD). At this depth, oxic controls on the degradation of less reactive fractions of organic matter in sinking particles and sediments result in extremely small amounts of TOC burial in sediments below the OCCD, relative to that which occurs in continental margin sediments above the OCCD (i.e, in table 8.3 compare carbon burial rates in continental margin sediments to those in open ocean sediments). The basis of this control mechanism assumes that an increase in atmospheric O_2 will eventually lead to an increase in the oxygenation of bottom waters; this should then increase oxygen penetration into sediments, and in a broader sense increase oxygen exposure of sedimentary organic matter. This enhanced degradation of sediment TOC should lead to the shoreward movement (or shallowing) of the OCCD to a new water column depth that results in less organic carbon burial in sediments and thereby decreases the rate of oxygen input to the atmosphere. Hedges (2002) refers to this as a marine sedimentary "afterburner" and suggests that it should be capable of operating on fast enough time scales, i.e., millions of years, to compensate for any perturbations in tectonic controls on atmospheric O_2.

From a qualitative standpoint the mineral conveyor belt model in fig. 15.6 appears to provide a reasonable explanation for the long-term controls on atmospheric O_2 that builds on these recent studies of the controls on carbon burial in marine sediments. At the same time, the model may also aid in resolving difficulties of past quantitative models of atmospheric O_2 models in generating reasonable simulations of atmospheric O_2 over the Phanerozoic (also see discussions in section 17.1.2). However, the next step will clearly be to incorporate the concepts of the mineral conveyor belt model into such quantitative models of atmospheric O_2 (e.g., Berner and Canfield, 1989; Petsch and Berner, 1998; Paytan and Arrigo, 2000; also see a summary in Berner, 1999).

15.5 The Composition of Organic Matter Preserved in Marine Sediments and the Fate of Terrestrial Organic Matter in Marine Sediments

For a number of reasons, there may be a relationship between the fate of terrestrial organic matter (TOM) in marine sediments and the controls on sediment carbon burial and preservation. In part, this may stem from the observation that in a broad sense marine organic matter is more reactive than terrestrial organic matter (see discussion above in section 15.2 and in sections 11.2.3 and 14.4.1). As a result, one might suppose that much of the marine organic matter deposited in sediments should be preferentially remineralized, and thus subject to less efficient burial. Similarly, terrestrial organic matter deposited in marine sediments might then be expected to undergo less efficient remineralization and therefore be preferentially buried. However, an examination of carbon budgets for the oceans suggests that roughly two to three times more organic carbon is transported to the oceans from land as riverine POC and DOC than is buried in marine sediments (Berner, 1989; Hedges and Keil, 1995). Therefore, depending on the amount of TOM that does escape remineralization in the oceans to be buried in marine sediments, the burial efficiency of TOM in marine sediments must be less than ~33–50%. In fact, as will be discussed below, TOM burial efficiency is indeed relatively low, or conversely, its remineralization efficiency is relatively high (also see discussions in Hedges et al., 1997). This then implies that the oceans are fairly efficient at the remineralization of refractory terrestrial organic matter in the oceans, yet also bury, or preserve, marine organic matter that is presumably more reactive. As Hedges et al. (1997) note, the resolution of this geochemical "conundrum" has important implications from a number of standpoints for understanding the ocean carbon cycle.

At the same time, only a small fraction (~20% or less) of riverine suspended matter is deposited in deep-sea sediments (Berner and Berner, 1996; McKee et al., 2004), and much of the particulate TOM transported by rivers to the oceans is therefore deposited in continental margin sediments (also see discussions in section 11.1.3). Since continental margin sediments are also the major sites of sediment carbon burial and remineralization (table 8.1), this further suggests that there should be a linkage between TOM burial and remineralization in sediments and sediment carbon preservation in general. Finally, the fate of TOM in marine sediments is of importance since it is generally assumed that most terrestrial (riverine) DOM is remineralized in the water column (Hedges, 1992). However, even if this were not the case, in the absence of long-term accumulation of terrestrial DOM in the water column, terrestrial DOM would have to be transported into sediments in some fashion for long-term preservation.

TABLE 15.2
The Fate of Terrestrial Organic Matter (TOM)
in Continental Margin Sediments

Sediment Type	TOM Burial Efficiency	TOM as a % of the Total Organic Matter Buried at Depth (TOM/ΣOM_{bur})
Deltaic sediments[a]	22 ± 5%	67 ± 24%
	~25–30%[b]	
Non-deltaic, continental margin sediments[a]	16 ± 4%	17 ± 4%

a. Based on estimates of the marine versus terrestrial contribution to the sediment organic matter pool that use stable isotope mass balance calculations along with OC:SA measurements of riverine suspended particles, deltaic sediments, and nondeltaic continental margin sediments. Given the approach of these calculations, burial efficiency is calculated here with respect to riverine input of terrestrial POM (rather than deposition in the sediments). From Aller (1998) and references therein (e.g., Showers and Angle, 1986; Keil et al., 1994b, 1997, Mayer, 1994a,b). Also see Burdige (2005) for further details about these calculations.

b. Based on sediment carbon budgets that use pore-water data and direct measurements of remineralization rates (Aller et al., 1996; Aller, 1998). Given this approach, burial efficiency is calculated here with respect to deposition of terrestrial POM to the sediments (rather than river input). In these calculations the use of stable isotope measurements of pore-water ΣCO_2, sediment TOC, and authigenic carbonates in the sediments also demonstrates that the loss of sediment TOM results from remineralization, as opposed to, e.g., desorption or other partitioning mechanisms of mineral-bound TOM into the water column.

Table 15.2 synthesizes the results of recent studies that allow us to examine this problem, by comparing TOM burial in deltaic versus nondeltaic continental margin sediments (the former representing continental margin sediments associated with large river systems). The primary reason for making this distinction come from results indicating distinct differences in the patterns of organic matter preservation and remineralization in deltaic versus nondeltaic, continental margin sediments (e.g., Aller, 1998; McKee et al., 2004). Some of these observations have been discussed in the beginning of this chapter and can be seen, for example, in fig. 15.1 (also see discussions in section 7.5.2). The first observation that comes from this table is that the burial efficiency of TOM in both sediment types is quite low, and in fact the value in the nondeltaic sediments may be lower than that in deltaic sediments. Given the small size of the dataset used in these calculations, some care must be taken in the interpretation of this observation, although these differences may indeed be real (see Burdige, 2005, for further details).

The factors that contribute to the low TOM burial efficiency in deltaic sediments have been discussed earlier in this chapter, and in sections 7.6.3 and 11.2.3, and are largely related to the physical reworking of these sediments and the associated sediment oxygen exposure. In contrast, a different set of factors contribute to the low TOM burial efficiency in nondeltaic continental margin sediments. First, TOM associated with riverine particles that are eventually deposited in nondeltaic continental margin sediments is more likely, as compared to that deposited in deltaic sediments, to be either remineralized in the marine water column or simply desorbed from the suspended particles prior to its deposition (see section 15.1). In addition, some amount of the material deposited in deltaic sediments is subsequently resuspended and redeposited in nondeltaic continental shelves and slopes (see discussions in Hedges and Keil, 1995; McKee et al., 2004). As a result, TOM in such sediments will have undergone at least one remineralization cycle in deltaic sediments prior to redeposition in nondeltaic, continental margin sediments. Therefore, because burial efficiencies have been calculated here with respect to riverine input, these predepositional processes lead to this low TOM burial efficiency in nondeltaic continental margin sediments. Interestingly, though, a comparison of the results in table 15.1 suggests that estimates of TOM burial efficiency in deltaic sediments are similar regardless of whether BE is calculated with respect to riverine input of TOM or TOM deposition in the sediments. Apparently because of their close proximity to river mouths, terrigenous particles are deposited in deltaic sediments before they lose much of their TOM loading.

Table 15.2 also indicates that there is a strong difference in the relative TOM content of the organic matter buried at depth ($TOM/\Sigma OM_{bur}$) in these two different types of continental margin sediment. For deltaic sediments, the large value of $TOM/\Sigma OM_{bur}$ is likely due to the significant input (deposition) of TOM to such sediments (e.g., Aller and Blair, 1996), and the fact that while these sediments are efficient at remineralizing terrestrial organic matter, they are even more efficient at remineralizing marine organic matter (MOM; see fig. 15.1 and discussions in Aller et al., 1996, 2004). Furthermore, because of the dynamics of deltaic sediments, the loss of TOM during sediment diagenesis is apparently not compensated for by the uptake of MOM, as appears to be the case for most riverine particles entering the oceans, including those eventually deposited in nondeltaic margin sediments. Thus, along with the high relative concentration of TOM buried in these sediments, this also leads to a net decrease in the OC:SA loading of deltaic sediments (~0.3 mg OC m^{-2}) as compared to riverine suspended particles (average of ~0.7 mg OC m^{-2}; also see fig. 15.3).

In contrast, while nondeltaic continental margin sediments lose much of their TOM by processes discussed above, they also apparently replace it with MOM (see section 15.1). In this way, they then maintain more constant OC:SA loadings as compared to riverine particles (see fig. 15.3). When coupled with the relatively high BE values for all organic matter in nondeltaic continental margin sediments (see fig. 15.1) this then leads to a low percentage of TOM being buried at depth in these sediments.

As a cross-check on these estimates of $TOM/\Sigma OM_{bur}$ in continental margin sediments, we can compare the estimates made here with similar estimates based on lipid biomarkers, lignin oxidation products, and compound-specific stable isotope measurements used in conjunction with bulk tracers such TOC $\delta^{13}C$ (e.g., see section 11.1.3). Such a comparison (see Burdige, 2005, for details) suggests that the observations in Table 15.2 are not unreasonable. However, it is also clear that more work will be needed to further verify these estimates (also see analogous discussions in McKee et al., 2004).

Combining the results in table 15.2 with those in table 8.1, the estimated rate of TOM burial in continental margin sediments is

$$(0.67 \pm 0.24)(70 \text{ Tg C yr}^{-1}) + (0.16 \pm 0.04)(68 \text{ Tg C yr}^{-1})$$
$$= 58 \pm 17 \text{ Tg C yr}^{-1} \tag{15.1}$$

Assuming that no significant TOM burial occurs in other sediment regimes listed in table 8.3, e.g., deep-sea sediments (see discussions below), this observation implies that about one-third of the organic matter buried in marine sediments is of terrestrial origin. For comparison, Schlünz and Schneider (2000) also determined a very similar TOM burial rate in continental margin sediments (43 Tg C yr^{-1}) using a slightly different approach.

Assuming a riverine organic matter input (both POC and DOC) of ~400–430 Tg C yr^{-1} (Hedges et al., 1997; Schlünz and Sneider, 2000) the results of eqn. (15.1) also imply that the overall burial efficiency of TOM in the oceans is ~9–17%, again assuming that all terrestrial DOM is remineralized in the water column. Although much of the TOM transported by rivers to the oceans is presumed to be relatively refractory (see discussions in Hedges et al., 1997) this calculation indicates that TOM has what appears to be a relatively high remineralization efficiency in the oceans. The effective remineralization of terrestrial organic matter in the oceans (and its magnitude in comparison to marine carbon burial rates in sediments) presumably contributes to the oceans being a net heterotrophic system (Smith and Hollibaugh, 1993).

When compared to the rate of total carbon burial in marine sediments (~160 Tg C yr^{-1}), the results of this calculation also indicate that the majority of the organic matter buried in marine sediments (~100 Tg C yr^{-1}) is of marine origin, and in an absolute sense the burial of MOM is roughly twice that of TOM. However, based on earlier discussions, MOM is (in general) presumed to be more reactive than TOM (see sections 11.2.3 and 14.4.1). This apparently counterintuitive observation makes more sense, though, when looked at in terms of the overall burial efficiency of MOM with regards to its ultimate oceanic source, primary production in the water column.

Shown in table 15.3 are estimates of MOM burial efficiency with respect to surface water primary productivity, either for all marine sediments globally or solely for continental margin sediments. As can be seen here, these estimates of MOM burial efficiency (0.25% to <1.3%) are less than that for TOM (9–17%), by at least an order of magnitude. Thus, from this perspective, it is clear that marine organic matter is much more efficiently remineralized than terrestrial organic matter in the oceans, in spite of the observed trends in the composition of the organic carbon that is buried in marine sediments. Expressed another way, because of the shear magnitude of marine productivity versus riverine input of TOM, more marine organic matter is buried in marine sediments in spite of these differences in MOM versus TOM burial efficiency. Even when burial efficiencies are calculated with respect to carbon deposition (rain rate) to the sediments as in fig. 15.1, MOM still has a lower burial efficiency than TOM, at least in deltaic sediments. However, in these estimates there is now a difference of only ~2 between TOM and MOM burial efficiencies.

Based on past studies of the fate of TOM in the oceans, Hedges et al. (1997) discuss the "geochemical conundrum" of the efficient reminer-

TABLE 15.3
Estimates of the Burial Efficiency of Marine Organic Carbon
in Marine Sediments with Respect to Primary Productivity
in the Surface Waters

Oceanic Regime	Primary Production $(Gt\ C\ yr^{-1})$[a]	Burial Efficiency
Global ocean[b]	40	0.25%
Coastal ocean[c,d]	7.8	<1.3%

a. All productivity estimates are from Liu et al. (2000).

b. Based on an assumed marine organic carbon burial rate in the global ocean of 100 Tg C yr^{-1} (= 0.1 Gt Cyr^{-1}).

c. In these productivity estimates the coastal ocean is defined as the oceanic region in which water depths are less than 200 m. Since this region is clearly smaller than that of deltaic plus continental shelf and upper slope sediments, these productivity estimates are a lower limit for the region of the oceans in which the majority of the organic carbon is buried. As a result these estimates of burial efficiency are upper limits.

d. Based on an assumed marine organic carbon burial rate of 86 Tg Cyr^{-1} in deltaic plus continental shelf and upper slope sediments. This is calculated as 100 Tg Cyr^{-1} minus that accumulating in biogenous sediments (high-productivity zones; = 10 Tg Cyr^{-1}), pelagic sediments (low-productivity zones; 5 Tg Cyr^{-1}) and anoxic basins (1 Tg Cyr^{-1}). See table 8.3 for further details.

alization of refractory terrestrial organic matter in the oceans and the burial of (presumably) more reactive marine organic matter. However, when this problem is re-examined in the context of the results presented here this conundrum may not be as severe as once thought. At the same time, though, a number of key problem areas still remain and will require further examination to verify these calculations and therefore better understand TOM burial in marine sediments (see Burdige, 2005, for further details).

As table 8.3 indicates organic carbon burial in deep-sea sediments is a minor component of the total amount of organic carbon buried in marine sediments. Nevertheless, examining TOM burial in deep-sea sediments is of some interest, in part because its possible occurrence has implications in terms of using sediment TOC depth profiles as paleoproxies of global ocean productivity (also see discussions in section 11.1.3). Evidence-to-date suggests that the majority of the organic matter preserved in deep-sea sediments is of marine origin, based on its elemental, isotopic, and molecular (e.g., lignin) composition (Degens, 1969; Emerson et al., 1987; Emerson and Hedges, 1988;

Hedges, 1992; Gough et al., 1993). In part, this is likely related to differences in the relative inputs of marine versus terrestrial organic matter to these sediments (see section 11.1.3), which are themselves controlled by the input of terrigenous materials, in general, to deep-sea sediments (see sections 2.1 and 2.5).

However, if we also look at the processes affecting organic matter deposition to deep-sea sediments in the context of the discussions in this chapter, we can also see why the input of TOM to these sediments is likely to be low, and therefore why it should also be a minor component (at best) of the organic matter buried in these sediments. Since deep-sea sediments are far from terrestrial sources, terrigenous particles deposited in these sediments should have ample time, prior to deposition, to lose their TOM loadings. This may occur by oxidative degradation in the water column or in continental margin sediments, for materials brought to deep-sea sediments by lateral transport of resuspended margin sediments, as well as by partitioning of TOM off terrigenous particles into the water column, e.g., desorption. In the absence of *substantive* fractionation of terrestrial versus marine organic matter during oxic remineralization in deep-sea sediments, which one might presume to be unlikely, based on the comparison above of burial efficiencies of TOM and MOM in deltaic sediments, small inputs of TOM to typical deep-sea sediments are not likely to be amplified into large relative amounts of TOM being preserved in these sediments.

15.6 THE RELATIONSHIP BETWEEN PHYSICAL PROTECTION, OXYGEN EXPOSURE, AND POSSIBLE ABIOTIC CONDENSATION REACTIONS IN SEDIMENT CARBON PRESERVATION

Past interest in abiotic condensation reactions such as humification, or geopolymerization, has stemmed from the possible role that these processes play both in the formation of molecularly uncharacterized organic matter (MU-OM; section 9.8) and in controlling carbon preservation in marine sediments. Although the classical mechanisms of humification (fig. 9.11) do not appear to adequately explain the formation of MU-OM in sediments, related abiotic reactions, perhaps in association with inorganic or organic matrix protection (see section 9.8.5), may play some role in forming MU-OM, affording the material some degree of protection from biological degradation, and therefore enhancing its preservation in sediments.

For example, Collins et al. (1995) suggested that organic matter adsorption and protection in sediment particle mesopores could promote the occurrence of geopolymerization reactions by either steric- or concentration-related phenomena (also see similar discussions in Mayer, 1994b, and Hedges and Keil, 1995). If these or other types of condensation reactions occur in association with adsorption, and also operate in concert, then this could lead to a positive feedback in which adsorption promotes condensation, which might then enhance the strength of adsorption of the resulting condensate. This mechanism also requires that increased condensation leads to an increase in the number of adsorption binding sites within the condensate. Calculations presented by Henrichs (1995) support this suggestion and show that the strength of adsorption, i.e., the adsorption coefficient, increases with the number of attachment points between the molecule and a particle surface.

Studies of adsorption of monomeric amino acids, small peptides, and proteins to sediments (Henrichs and Sugai, 1993; Ding and Henrichs, 2002) have shown that some of this adsorption is not readily reversible and that adsorbed amino acids may undergo what appear to be melanoidin-type condensation reactions (also see section 10.5). At the same time, protein degradation in seawater appears to occur more slowly when proteins are associated with either submicron particles or bacterial membranes (Nagata and Kirchman, 1996; Nagata et al., 1998; Borch and Kirchman, 1999). Such decreased rates of protein degradation may then enhance the occurrence of other condensation reactions that ultimately lead to protein preservation (also see similar discussions in Nguyen and Harvey, 1998).

In summary, these observations (and others discussed in earlier chapters) suggest the following scenario. During early stages of diagenetic maturity selective utilization of reactive organic matter occurs, leading to an enrichment of more refractory, and O_2-sensitive (?), material. Associated with these trends, processes such as physical protection or adsorption also decrease the rate of biological degradation of some types of sedimentary organic matter. This may then increase the probability that this organic matter becomes involved in abiotic reactions such as geopolymerization. If such processes occur, they have the potential to operate in a positive feedback mode, leading to organic mater that becomes increasingly more refractory as well as increasingly physically protected from degradation, e.g., irreversibly adsorbed. The net result is that this material becomes (either operationally or in actuality) MU-OM, and may be preserved in sediments (see Mayer, 2004, for a discussion of many of these same concepts).

The occurrence of abiotic condensation reactions in concert with physical protection could also play some role in explaining the oxygen effects associated with the remineralization of some diagenetically mature fractions of sediment organic matter (see section 15.4). As discussed in sections 7.6.1 and 7.6.3, O_2-requiring enzymes or strong chemical oxidants such as H_2O_2 are very effective at cleaving nonhydrolyzable bonds in materials such as lignins, hydrocarbons, and other more refractory organic compounds in the first steps of their depolymerization. Furthermore, because many of these enzymes and oxidants are also nonspecific, they are useful in degrading materials that are randomly polymerized. In the presence of oxygen, then, such enzymes and oxidants could likely play a role in decomposing not only refractory biopolymers (i.e., those that are nonhydrolyzable and/or randomly polymerized), but also abiotic condensates that have an analogous random structure. This could then possibly help explain why aged organic matter is degraded more slowly, if at all, in the absence of oxygen.

Discussions above (section 15.4) have further suggested that other factors, e.g., Mn and Fe redox cycling, may actually be responsible, to some degree, for these oxic effects on organic matter degradation. Specifically, discussions in section 7.6.3 indicate that Mn and Fe redox cycling in mixed redox sediments could play a role similar to that of oxygen and associated enzymes in depolymerization of refractory and nonhydrolyzable organic matter. Again this recalcitrant organic matter may comprise natural biopolymers as well as abiotic condensates or geopolymers.

≈ **CHAPTER SIXTEEN** ≈

Biogeochemical Processes in Continental Margin Sediments. I. The CO_2 System and Nitrogen and Phosphorus Cycling

AS COMPARED TO PELAGIC SEDIMENTS, continental margin sediments, which include continental shelf and upper slope sediments, as well as nearshore estuarine and coastal sediments, generally have larger carbon rain rates and higher sedimentation rates. As a result, suboxic and anoxic remineralization processes play more important roles in sediment early diagenesis. Continental margin sediments are also important sites for nitrogen, phosphorus, and sulfur cycling and represent the major sites of carbon burial and preservation in the oceans (as discussed in section 8.3 and chapter 15). In this chapter, I discuss early diagenetic processes in continental margins sediments, focusing on the controls on the CO_2 system (alkalinity) in sediment pore waters and the cycling of nitrogen and phosphorus. Sulfur, methane, and trace metal cycling in these sediments will be discussed in the next chapter.

16.1 Pore-Water pH and Carbonate Chemistry under Suboxic and Anoxic Conditions

Many processes occurring in marine sediments are either implicitly or explicitly acid-base reactions. This includes not only organic matter remineralization reactions, but also Mn^{2+} and Fe^{2+} oxidation as well as calcium carbonate dissolution and precipitation. Consequently, understanding the factors that control pore-water pH values is of great interest and importance in sediment geochemistry. The controls on pore-water pH values are strongly connected to sediment alkalinity production, and, as a result, impact carbonate mineral saturation and precipitation/dissolution.

In deep-sea sediments aerobic respiration plays an important role in controlling the occurrence of sediment carbonate dissolution (see section 13.5). Since aerobic respiration leads to a decrease in pore-water pH values (as a result of carbonic acid production) and calcite dissolution causes pH values to rise, their interplay (e.g., rates and depth distribution) controls pore-water pH profiles in deep-sea sediments. In sediments where one process dominates the other, *in situ* pH studies often show relatively smooth, exponential-like pH profiles that either increase or decrease on the order of ±0.1–0.2 pH units over the upper ~5–10 cm of sediment (e.g., Archer et al., 1989b; Hales and Emerson, 1997a). However, at some sites, the interplay between these processes leads to a pH minima in the upper ~1–2 cm of sediment (e.g., Hales and Emerson, 1996).

In continental margin sediments suboxic and anoxic remineralization similarly play important roles in alkalinity production and in the controls on pH and carbonate mineral saturation. For the sake of clarity in this discussion sediment organic matter will be assumed to have a Redfield C:N:P stoichiometry and a carbon oxidation state of zero. Interested readers, if they so desire, can modify the equations presented here for other C:N:P ratios and/or carbon oxidation states. More general forms of the equations presented here that use arbitrary C:N:P stoichiometry can also be found in the literature (e.g., Van Cappellen and Wang, 1996).

With this approach, the process of denitrification can be written as,

$$94.4 NO_3^- + (CH_2O)_{106}(NH_3)_{16}(H_3PO_4) \rightarrow 13.6 CO_2 \\ + 92.4 HCO_3^- + 84.8 H_2O + 55.2 N_2 + HPO_4^{2-} \quad (16.1)$$

where the implicit assumption is that all of the nitrogen from sediment organic matter is subject to ammonification (i.e., ammonium production in either fig. 16.2 or 16.6) and subsequent oxidation by nitrate in the anammox (anaerobic ammonium oxidation) reaction (see table 7.2). Other workers have written analogous equations in which this remineralized ammonium is not oxidized (e.g., Froelich et al., 1979; Aller, 1980a), which then results in a slightly different stoichiometry for denitrification. This point will be discussed in the next section in further detail.

The consumption of nitrate, a strong acid anion, results in alkalinity production, although the imbalance between nitrate consumption and organic carbon oxidation also results in a slight "excess" of CO_2 production. Thus in spite of this alkalinity production, denitrification also has the potential to dissolve small amounts of calcium carbonate

(Froelich et al., 1979; Emerson et al., 1982). However, because the production of both CO_2 and HCO_3^- affects the overall equilibrium of the carbonate system,

$$H_2O + CO_2(aq) + CO_3^{2-} \rightleftharpoons 2HCO_3^- \quad (16.2)$$

not all of the CO_2 produced by rxn. (16.1) will lead to carbonate dissolution (Emerson et al., 1982).

Manganese and iron reduction differ from denitrification in that these reactions lead to a net production of alkalinity without CO_2 production. These reactions can be written in terms of either H^+ consumption or bicarbonate production (= CO_2 consumption), since protons are equivalent to "negative" alkalinity (e.g., Millero, 1996), as can be seen in

$$H^+ + HCO_3^- \rightleftharpoons H_2CO_3 (\rightleftharpoons H_2O + CO_2) \quad (16.3)$$

which consumes protons and bicarbonate and produces CO_2, but also conserves alkalinity. Expressing manganese and iron reduction using the latter formalism, these reactions can be written as

$$236MnO_2 + (CH_2O)_{106}(NH_3)_{16}(H_3PO_4) + 364CO_2$$
$$+ 104H_2O \rightarrow 470HCO_3^- + 8N_2 + 236Mn^{2+} + HPO_4^{2-} \quad (16.4)$$

$$424Fe(OH)_3 + (CH_2O)_{106}(NH_3)_{16}(H_3PO_4) + 756CO_2 \rightarrow$$
$$862HCO_3^- + 16NH_4^+ + 424Fe^{2+} + HPO_4^{2-} + 304H_2O \quad (16.5)$$

Again note that nitrogen regeneration during manganese reduction is written assuming that the organic nitrogen is oxidized to N_2 (rxn. 6 in table 7.2).

Two important points need to be made here. The first is that because the pK_a for NH_4^+ is 9.5, any NH_3 produced during iron reduction will be protonated according to the reaction

$$NH_3 + CO_2 + H_2O \rightarrow HCO_3^- + NH_4^+ \quad (16.6)$$

thus representing another source of alkalinity (e.g., Berner et al., 1970). More importantly, though, CO_2 consumption or alkalinity production drive rxn. (16.2) to the left, producing carbonate ion. This alkalinity increase can then make sediment pore waters supersaturated with respect to carbonate minerals such as calcite. The extent to which authigenic carbonate mineral precipitation occurs as a result of these processes will be discussed below.

While these suboxic processes produce alkalinity, their ultimate impact on carbonate mineral saturation and sediment pH values is a function of their relative importance in sediment organic mater remineralization (e.g, see discussions in Boudreau, 1987). For example, given the fact that denitrification generally accounts for <20% of organic matter remineralization in most sediments, it is generally assumed that it does not significantly impact sediment carbonate chemistry (Canfield and Raiswell, 1991a; Boudreau and Marinelli, 1994).

In contrast, sulfate reduction plays a major role in controlling sediment pH values and carbonate mineral saturation in anoxic sediments. This reaction is commonly written as

$$53SO_4^{2-} + (CH_2O)_{106}(NH_3)_{16}(H_3PO_4) \rightarrow 106HCO_3^- + 53H_2S \\ + 16NH_3 + H_3PO_4 \quad (16.7)$$

although if we take into account alkalinity production via ammonia production and alkalinity consumption via phosphoric acid dissociation, the reaction can be rewritten as

$$53SO_4^{2-} + (CH_2O)_{106}(NH_3)_{16}(H_3PO_4) + 14CO_2 \\ + 14H_2O \rightarrow 120HCO_3^- + 53H_2S + 16NH_4^+ + HPO_4^{2-} \quad (16.8)$$

An examination of this reaction indicates that sulfate reduction not only produces alkalinity, but also produces the weak acid H_2S. The effect of these two opposing processes on pH values and carbonate mineral saturation in closed system calculations is shown in fig. 16.1. During the very early stages of sulfate reduction carbonate mineral saturation initially decreases; this occurs because the impact of the lowering of pH, which can be thought of as driving rxn. (16.2) to the right, is initially more important than the impact of the increase in alkalinity, which drives rxn. (16.2) to the left. However, with increasing sulfate reduction the pH of the system reaches a steady state of ~6.6, and the increase in alkalinity then drives this closed system back to supersaturated conditions. The net result of this is that during the

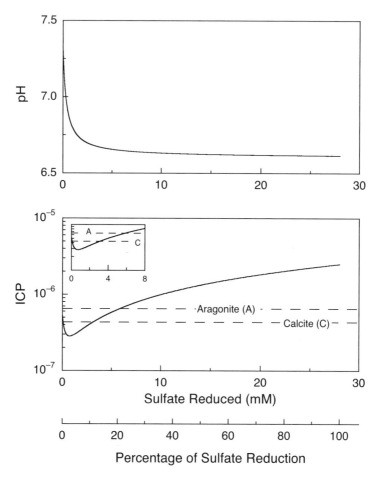

Figure 16.1 The effect of sulfate reduction on the degree of calcite and aragonite saturation and pH in closed-system calculations (also see similar plots in Morse and Mackenzie, 1990; Canfield and Raiswell, 1991a; Boudreau and Canfield, 1993). Solubility is expressed as ICP ($=[Ca^{2+}][CO_3^{2-}]$) and the dashed lines represent the K'_{sp} values for calcite and aragonite. Aragonite is more soluble than calcite and thus, has a higher K'_{sp}. In this format, values above each of these lines represents supersaturated conditions with respect to a given carbonate phase.

early stages of sulfate reduction (up to ~3–6 mM sulfate reduced), pore waters are actually undersaturated with respect to both calcite and aragonite. Thus carbonate dissolution may occur in response to small amounts of sulfate reduction, a fact that is not always appreciated in the literature. It can also be shown that by increasing the C/N ratio of the organic matter undergoing remineralization the general shape of curves such as that in fig. 16.1 essentially remain the same, but are shifted downward (Morse and Mackenzie, 1990). This occurs because of the decrease in ammonium, and hence alkalinity, production with increasing C/N ratio.

Closed-system calculations such as these do a good job of qualitatively predicting the general trends observed in pore-water pH depth profiles in anoxic marine sediments. However, because these calculations do not take into account transport processes such as diffusion or bioturbation they also have limitations in terms of examining many of the details of the controls on sediment pore-water pH values and ultimately carbonate mineral saturation in sediments. To overcome some of these difficulties, open system calculations that are essentially advection/diffusion/reaction models for pH or H^+, have been presented (Boudreau and Canfield, 1993; Luff et al., 2001; Meysman et al., 2003b). Interestingly, while closed-system calculations indicate that carbonate mineral understauration can occur during the early stages of sulfate reduction, open-system calculations suggest that this may be highly unlikely in most marine sediments (Boudreau and Canfield, 1993). At the same time, though, the success of open-system models to examine such problems is driven, in part, by the extent to which they include all of the relevant processes affecting pH in sediments (see below). Furthermore, transport processes such as bioirrigation may allow for carbonate mineral understauration during early stages of sulfate reduction; more work is needed to examine this possibility (see Boudreau and Canfield, 1993, and discussions below for further details).

An important consideration in the interpretation of these closed- and open-system calculations is the fate of the H_2S produced during sulfate reduction (Ben Yaakov, 1973; Boudreau and Canfield, 1988). In iron-poor sediments where sulfide precipitation does not occur, pH trends can be similar to those seen in closed-system calculations, although differences may occur depending on the occurrence of carbonate mineral dissolution, precipitation, or chemical equilibrium (see discussion below and in Thorstenson and Mackenzie, 1974; Tribble et al., 1990; Walter et al., 1993).

In terrigenous sediments with high levels of iron, dissolved sulfide may precipitate out as an iron sulfide, the dominant sulfide mineral phase that forms in marine sediment (see section 17.1 for further details). There are several possible sources of iron for this reaction; these can be broken down essentially into two classes based on their impact on pore-water pH values and carbonate mineral saturation. One class is iron sources that do not involve proton consumption. This includes Fe^{2+} exchange with other cations from ion exchange/adsorption sites on mineral surfaces. Assuming that sulfide precipitates as FeS, this then implies that this process liberates the two protons from the H_2S molecule, and pH values drop even lower than that observed in the absence of sulfide precipitation (Boudreau and Canfield, 1988).

In contrast, iron may be made available for sulfide precipitation by redox reactions such as

$$2Fe(OH)_3 + 3H_2S \rightarrow 2FeS + 6H_2O + S^o \qquad (16.9)$$

or, less likely, through the dissolution (chemical weathering) of iron-containing minerals such as biotite or chlorite. These reactions then either directly or indirectly neutralize the acid H_2S that is produced during sulfate reduction. This not only results in much higher pH values, but sediment pore waters also become supersaturated with respect to carbonate mineral phases at very low levels of sulfate reduction (>0.2–0.3 mM sulfate reduced; e.g., Boudreau and Canfield, 1988; Canfield and Raiswell, 1991a).

Finally, an important consideration is the extent to which carbonate precipitation actually occurs under supersaturated or saturated conditions in sediments. In seawater, calcium carbonate precipitation is inhibited, most likely by the magnesium ion (Morse and Mackenzie, 1990), although evidence from a number of sedimentary systems indicates that carbonate precipitation does occur when high levels of supersaturation in the pore waters are reached (Sholkovitz, 1973; Boudreau and Canfield, 1988; Aller et al., 1996; also see discussions below).

Model calculations that include suboxic and anoxic remineralization along with carbonate dissolution or precipitation and FeS precipitation generally predict smooth exponential-like changes in pH with sediment depth over the region of diagenesis (Boudreau and Canfield, 1988). In contrast, high-resolution sediment pH profiles in some continental margin and coastal sediments show well-defined pH minima, on the order of ~0.2–2 pH units, in the upper 10–20 mm of sediment (Jørgensen and Revsbech, 1983; Cai and Reimers, 1993; Reimers et al., 1996; Komada et al., 1998). In general, these minima occur in close proximity to the zero oxygen level in the sediments and have been proposed to result from the oxidation of upwardly diffusing Mn^{2+}, Fe^{2+}, and/or HS^- by O_2, according to

$$4Mn^{2+} + 2O_2 + 4H_2O \rightarrow 4MnO_2 + 8H^+ \qquad (16.10)$$

$$4Fe^{2+} + O_2 + 10H_2O \rightarrow 4Fe(OH)_3 + 8H^+ \qquad (16.11)$$

$$HS^- + 2O_2 \rightarrow SO_4^{2-} + H^+ \qquad (16.12)$$

Manganese and sulfide oxidation coupled to nitrate reduction (see table 7.2) are also proton-producing reactions; therefore the downward diffusion of nitrate could also contribute to the occurrence of these pH minima. Diagenetic models that include rxns. 16.10–16.12 (Boudreau, 1996; Van Cappellen and Wang, 1996) predict pH profiles with minima similar to those observed in some sediments.

Processes such as bioturbation and bioirrigation also impact surficial pH values and carbonate mineral saturation. The effects of bioirrigation appear to be complex and may include acid production due to enhanced O_2 input into sediments and stimulation of aerobic respiration, and/or the oxidation rxns. 16.10–16.12 (e.g., Aller, 1982a). However, other modeling and experimental studies suggest that bioirrigation may buffer pH changes in irrigated sediments by enhancing either the input of overlying water (and carbonate ion) into the sediments (Marinelli and Boudreau, 1996) or the transport of bicarbonate out of the sediments (Boudreau, 1996).

Another important aspect of the effect of bioturbation on sediment pH values and carbonate mineral saturation results from sediment mixing that may lead to anoxic FeS-containing sediments being mixed with surface sediments containing Mn and Fe oxides (Aller and Rude, 1988). This mixing may allow for the occurrence of redox reactions between these solids such as those listed in table 7.1. An examination of these reactions, and other similar proposed reactions (also see Hulth et al., 1997; Green and Aller, 1998) suggests that both consumption and production of protons are possible. The effect of these reactions on pore-water pH, alkalinity, and carbonate mineral saturation may therefore be complex. Furthermore, the occurrence of such reactions is likely to be highly variable in space and time as a result of differences in particle reworking rates and reactant concentrations.

In developing some general conclusions about all of these processes, two end-member situations may be envisioned. In one case, redox reactions involving Fe, Mn, and S are coupled with suboxic/anoxic organic matter remineralization (e.g., figs. 7.3 and 7.4) or may use O_2 as the oxidant; for example,

$$4FeS + 9O_2 + 10H_2O \rightarrow 4Fe(OH)_3 + 4SO_4^{2-} + 8H^+ \quad (16.13)$$

Reactions such as this result in oxidized end-products and net acid production and can, therefore, promote carbonate dissolution. In the other case, these processes involve the net production of reduced end-products such as Mn^{2+}, for example,

$$4.5MnO_2 + FeS + 7CO_2 + 5H_2O \rightarrow 4.5Mn^{2+}$$
$$+ Fe(OH)_3 + SO_4^{2-} + 7HCO_3^- \quad (16.14)$$

(Aller and Rude, 1988). Since these reactions produce alkalinity/consume protons (also see rxns. 5–9 in table 7.2), they may then lead to the formation of mixed phase Ca-Mn carbonates (as is discussed in section 13.2). Calculations presented by Aller and Rude (1988) suggest that reactions such as rxn. (16.14) could lead to authigenic carbonate mineral formation in nearshore, bioturbated sediments.

Discussions above (and in section 13.2) describe the factors that may lead to the formation of Mn carbonates or, more likely, mixed phase Ca-Mn carbonates, in organic-rich, temperate nearshore sediments. Similar factors might also be expected to lead to the formation of iron carbonates like siderite in these same sediments. However, the high sulfide levels found in many of these settings, and the insolubility of these sulfides (see section 17.1), generally minimizes the importance of iron carbonate formation. In contrast, in muddy tropical sediments such as the Amazon shelf or the Gulf of Papua, the lack of significant net sulfate reduction, coupled with extensive iron reduction, leads to situations in which iron carbonate formation may be of much greater importance (Aller and Blair, 1996; Aller et al., 2004). At the same time, though, calculations presented by Aller et al. (2004) for Gulf of Papua sediments suggest that the vast majority (>70%) of the Fe(II) in these sediments occurs in forms other than authigenic sulfides or carbonates, e.g., authigenic clays (see chapter 18).

In studies specifically examining the relationship between sediment biogeochemical processes and carbonate chemistry in nearshore Long

Island Sound (LIS) sediments, Green and Aller (1998, 2001) have observed that the calcite and aragonite saturation state of the sediments follows a regular seasonal pattern. Beginning in spring, rising temperatures and the deposition of reactive organic matter from the spring bloom increase rates of anoxic sediment metabolism. This then leads to shallower sediment oxygen penetration along with FeS formation and seasonal storage. Together, these effects drive the sediments to carbonate mineral supersaturation by late spring–early summer. During fall/winter, however, decreasing rates of anoxic sediment metabolism and greater oxygen penetration into the sediments promote acid production through the oxidation of reduced metabolites, such as pore-water Mn^{2+} or ammonium, or the oxidation of seasonally stored solid phase FeS (also see Sampou and Oviatt, 1991). This acid production then drives these supersaturated sediments back to undersaturation.

While seasonal changes in sediment carbonate saturation state promote $CaCO_3$ dissolution and precipitation in these LIS sediments, on an annual basis these processes are almost completely balanced, and ~90% of the carbonate precipitated in the sediments is subsequently dissolved. There is also a strong correlation between these changes in sediment chemistry and benthic foraminifera inventories, although in the spring there is a slight decoupling between foram recruitment in response to the spring bloom and the transition of the sediments from undersaturated to supersaturated conditions. These results and those of other studies all suggest that carbonate dissolution may be a more important aspect of sediment diagenesis in shallow water environments than previously thought. More importantly, though, this may be the case not only in carbonate-rich sediments (Walter and Burton, 1990; Rude and Aller, 1991; Burdige and Zimmerman, 2002), but also in carbonate-poor, muddy (siliciclastic) coastal estuarine sediments.

Finally, it is important to note that this discussion implicitly assumes the reactions presented here and associated equilibria with carbonate minerals control sediment pH values. In particular, reactions such as H^+ adsorption/desorption on mineral surfaces (e.g., Meysman et al., 2003b) or other types of clay mineral equilibrium reactions (e.g., reverse weathering type reactions; Mackin and Aller, 1984b; Aller and Rude, 1988) are assumed to be of minor significance. However, the extent to which this assumption is reasonable will require more careful examination.

16.2 Sediment Nitrogen Cycling

In oxic sediments, nitrogen cycling (in a net sense) is dominated by aerobic respiration coupled to nitrification (see section 13.1). Although ammonium can be detected in these sediment pore waters (e.g., Berelson et al., 1990), the levels are generally low ($< \sim 5 \, \mu M$). Ammonium pore-water profiles are also generally structureless and show no obvious evidence of net sediment production.[1] Benthic ammonium fluxes from such sediments are also generally indistinguishable from zero (Hammond et al., 1996). In contrast, pore-water nitrate concentrations often increase with depth in an exponential-like fashion (fig. 13.1A), indicating net sediment production. These pore-water profiles and benthic nitrate flux measurements, which indicate a net flux of nitrate out of these sediments, are both consistent with nitrate being the dominant end-product of organic matter remineralization in such oxic sediments (also see Martin et al., 1991).

As carbon fluxes increase and sediments become suboxic, denitrification (see table 7.1 or rxn. 16.1) becomes increasingly important (fig. 16.2; also see discussions in section 7.5.2). The nitrate source for sediment denitrification can be the overlying water column; this is often termed direct denitrification, although much of the nitrate used in sediment denitrification is actually produced in situ by nitrification—that is, coupled nitrification-denitrification (Jenkins and Kemp, 1984; Middelburg et al., 1996).

Another related aspect of nitrogen cycling under suboxic conditions is the fate of particulate organic nitrogen during suboxic organic matter remineralization. Implicit in rxns. (16.1), (16.4), and (16.5) is that regenerated ammonium is oxidized to N_2 during denitrification and Mn reduction, but not during iron reduction. This dichotomy assumes that other processes must be coupled with these organic matter remineralization processes, including ammonium oxidation by nitrate (the anammox reaction; rxn. 1 in table 7.2) or manganese oxides (rxn. 6 in table 7.2).

As discussed in section 7.4.2, experimental studies provide evidence for the possible occurrence of these redox reactions in marine

---

[1] The "exception" to this observation is an apparent artifact associated with the shipboard extraction of pore waters from deep-sea sediments that causes elevated ammonium and nitrate concentrations in near-surface pore waters (see Hammond et al., 1996; Martin and Sayles, 2003; and references therein).

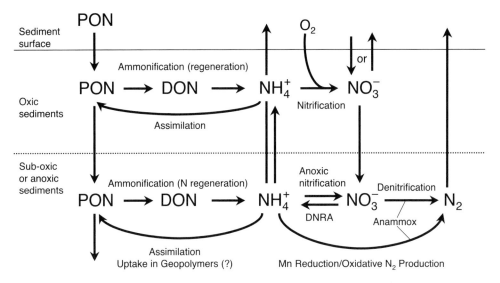

Figure 16.2 A conceptual model illustrating the processes associated with nitrogen cycling in marine sediments (based on information from several sources).

sediments. Pore-water nitrate and ammonium profiles (fig. 16.3) also generally indicate that significant ammonium accumulation does not occur until nitrate is depleted (Canfield et al., 1993; Hyacinthe et al., 2001; Hartnett and Devol, 2003), consistent with the occurrence of the anammox reaction. In contrast, pore-water Mn^{2+} profiles are not always entirely consistent with the occurrence of coupled manganese reduction and ammonium oxidation to N_2. However, as discussed in the previous section and in chapter 7, the occurrence of Mn/N redox reactions such as this one is likely mediated by bioturbation or bioirrigation, and this fact may complicate such a "simple" interpretation of pore water Mn^{2+} and ammonium profiles.

To begin to examine how these processes affect sediment nitrogen cycling in continental margin sediments, model-derived benthic nitrogen fluxes from continental slope and continental shelf sediments are shown in fig. 16.4 (Middelburg et al., 1996). Because this model is parameterized using field data, the trends shown here are broadly consistent with those observed in field studies (Bender et al., 1989; Devol and Christensen, 1993; Berelson et al., 1996; Devol et al., 1997; Jahnke and Jahnke, 2000; and others).

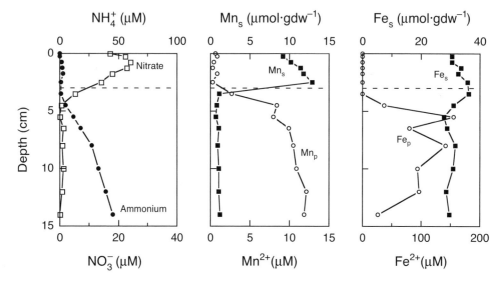

Figure 16.3 Pore-water profiles of ammonium, nitrate, Mn^{2+} (Mn_p), and Fe^{2+} (Fe_p), and solid-phase profiles of reactive (HCl-extractable) manganese (Mn_s) and iron (Fe_s), all in continental margin sediments at 1,000 m water depth in the Bay of Biscay, France (redrawn using results in Hyacinthe et al., 2001). The horizontal dashed line represents the oxygen-penetration depth, based on oxygen microelectrode profiles. As discussed in the text, ammonium does not accumulate in pore waters until nitrate is depleted. Also note the surface enrichment in solid-phase Mn, and the evidence of a slight Mn peak, just above the oxygen penetration depth (e.g., see discussions in section 13.2).

As noted above, most of the nitrate involved in sediment denitrification is produced in the sediments by nitrification. In some pelagic or hemipelagic sediments this occurs through nitrification coupled to aerobic respiration. This nitrate diffuses both downward to be denitrified, and upward, where it escapes the sediment as a benthic flux (see figs. 13.1B and 15.2). While these sediments are still a source of nitrate to the overlying waters (fig. 16.4), denitrification attenuates the net amount of nitrate returned to the overlying waters. In other sediments, e.g., continental margin sediments, ammonium, which is produced at depth, diffuses upward, where it can be nitrified.

With increasing carbon rain rate, i.e., as one moves into shallower water depths, denitrification becomes an increasingly important process in sediments (see figs. 7.7 and 15.4), despite the fact that, under some circumstances, these sediments may still be a source of nitrate

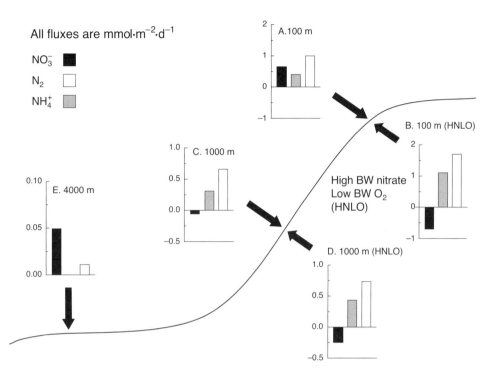

Figure 16.4 Model-derived benthic nitrogen fluxes from sediments with "normal" bottom waters at 4,000 m (E), 1,000 m (C), and 100 m (A), and with high nitrate (40 μM), low oxygen (80 μM) bottom waters at 1,000 m (D) and 100 m (F). Redrawn after Middelburg et al. (1996).

to the overlying waters (e.g, Devol et al., 1997). However, in many continental margin settings, the sediments are actually a net *sink* for bottom-water nitrate; that is, there is now a benthic flux of nitrate into sediments that fuels direct denitrification, as well as denitrification coupled to sediment nitrification (see fig. 16.4, Middelburg et al., 1996, and references therein). In such sediments, pore-water nitrate concentrations often simply decrease exponentially with depth, much like pore-water oxygen profiles (e.g., Hartnett and Devol, 2003).

In examining this problem it is important to note that the environmental controls on direct denitrification are often different than those for coupled nitrification-denitrification. Thus understanding the controls on total denitrification in sediments generally requires that these distinctions be considered (Cornwell et al., 1999). With increasing carbon flux to the sediments, or decreasing bottom-water oxygen concentrations, the sediment oxygen penetration depth decreases (fig. 7.9), which may limit coupled nitrification-denitrification by decreasing the depth interval in the sediments where nitrification can occur (Rysgaard et al., 1994; Cornwell et al., 1999). At the same time decreasing sediment oxygen penetration also increases direct sediment denitrification by decreasing the diffusive path length between the sediment surface and the zone of denitrification (Berelson et al., 1996). Evidence for these complementary effects can also be seen in the model results in fig. 16.4, i.e., compare panels A vs. B and C vs. D.

Overall, bottom water oxygen and nitrate concentrations and carbon rain rate to the sediments are the important controlling parameter for sediment denitrification (Middelburg et al., 1996; Cornwell et al., 1999). In many coastal settings, though, high carbon loading and low bottom-water oxygen often coincide, thus making it difficult to independently examine their effects on sediment denitrification (Cornwell et al., 1999). However, results from Monterey Bay, Calif. sediments (100 m water depth) allow us to begin to examine these effects separately. These sediments underlie relatively high oxygen ($\sim$100–180 μM) and low nitrate ($\sim$15–30 μM) bottom waters, and seasonal studies at this site demonstrate that both the sign and magnitude of the benthic nitrate flux are a function of the carbon rain rate to the sediments (fig. 16.5). However, because these sediments are extensively bioirrigated, seasonality in bioirrigation rates may also play a role in explaining these results (see section 12.5.3 and the discussion below).

Nitrogen redox cycling associated with sediment bioirrigation can play an important role in mediating the occurrence of denitrification in continental margin sediments (see section 12.5). Figure 16.6 illustrates the tight coupling between ammonium production (ammonification), nitrification, and denitrification that occurs in irrigated sediments. Ammonification occurs in the anoxic sediments, leading to ammonium diffusion into the inner, oxic sediment ring around a burrow, where it is then oxidized to nitrate. Subsequent radial diffusion of nitrate in both directions may lead to some nitrate loss to the overlying waters (see below), as well as nitrate diffusion back into the anoxic sediments, where it is denitrified.

Model calculations demonstrate that the geometry of burrowed sediments has a significant impact on both coupled nitrification-denitrification as well as direct denitrification in bioirrigated sediments (Aller, 2001; Gilbert et al., 2003). The exact reasons this occurs

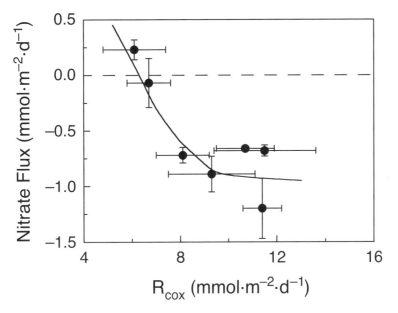

Figure 16.5 Benthic nitrate flux versus R_{cox} (the depth-integrated rate of organic carbon oxidation) for bioirrigated sediments at 95 m water depth in Monterey Bay (CA). Redrawn using data in Berelson et al. (2003).

are complex, although they appear to be related to the interplay between the diffusive and reaction length scales of the processes shown in fig. 16.6. Of particular importance in these calculations (and in related experimental studies cited in Aller et al., 2001) is that this interplay can lead to irrigated sediments being either a net source or sink for bottom-water nitrate (fig. 16.7). For example, in bioirrigated sediments where denitrification rates are large relative to nitrification rates, or where bottom-water nitrate levels are sufficiently high, the sediments will be a net sink for reactive nitrogen.

In some continental margin sediments the coupling between nitrification and denitrification is sufficiently tight that there is no net flux of ammonium out of the sediments. Although not illustrated in the model calculations shown in fig. 16.4, this can be seen in data from California Borderland sediments (Bender et al., 1989; Berelson et al., 1996) and from some eastern North Pacific continental margin sediments (Devol and Christensen, 1993). In other margin sediments, ammonium production (ammonification) exceeds the ability of the sediment to nitrify this material, and not all of this ammonium is oxidized in the

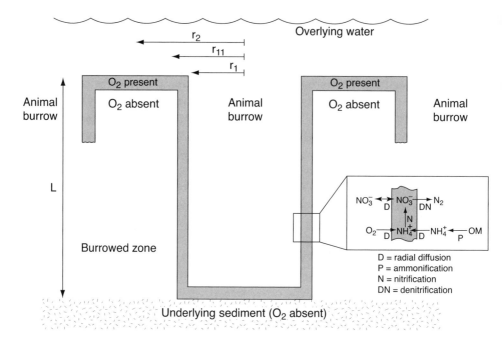

Figure 16.6 A conceptual model illustrating the processes affecting nitrogen cycling in a bioirrigated sediment. Based on Aller's (1988) model for the cylindrical redox zonation around a burrow tube (see fig. 12.11) and discussions in Gilbert et al. (2003).

sediments. Benthic ammonium fluxes therefore become an increasingly important component of nitrogen cycling in these sediments.

In a general sense, this balance between sediment nitrate uptake (direct nitrification) and net ammonium release (ammonification minus coupled nitrification-denitrification) can lead to the situation in which sediment nitrate uptake actually exceeds ammonium release (e.g, Berelson et al, 1996, 2003; Hartnett and Devol, 2003). Continental margin sediments in which this occurs are consequently not only net sinks for water column nitrate, but are also sinks for fixed (reduced) nitrogen in general. On local scales, this may provide a negative feedback to productivity cycles (Berelson et al., 2003); on a more global scale it amplifies the importance of sediment denitrification as a sink for reactive nitrogen in the oceans.

To a first order, the marine nitrogen cycle is dominated by nitrogen fixation, i.e, production of fixed or reactive nitrogen from atmospheric N_2, and denitrification, the reduction of nitrate to N_2 (see budget

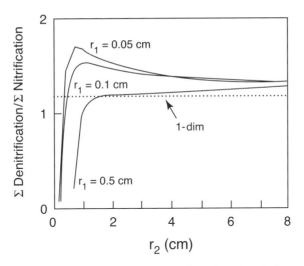

Figure 16.7 Model calculations illustrating the changing balance between denitrification and nitrification in a bioirrigated sediment as a function of the burrow geometry parameter r_2, the half-spacing between burrows (see fig. 12.11). The three curves here are for r_1, the burrow radius, equal to 0.05, 0.1 cm, and 0.5 cm. In these calculations the burrow wall oxic zone thickness was kept constant at 0.1 cm, implying that in fig. 12.11 $r_{11} = 0.1 + r_1$. All nitrification and denitrification rate parameters were also kept constant. Also shown as a horizontal dotted line is this reaction rate ratio based on a 1-dimensional, two-layer model with a surface oxidized zone 0.1 cm thick overlying anoxic sediments having the same nitrification and denitrification rate parameters as in the other calculations. Note that when the ratio on the y-axis is >1 the sediments are a nitrate sink, and a nitrate source when this ratio is <1. Redrawn after Aller (1988, 2001).

calculations most recently in Codispoti et al., 2001). Furthermore, sediment denitrification dominates total marine denitrification, being ~2–4 times that of water column denitrification. The global magnitude of sediment denitrification has been estimated using two very different approaches. One approach (Middelburg et al., 1996) is based on results from a sediment diagenesis model (e.g., fig. 16.4) that are scaled up using empirical relationships between rates of sediment remineralization processes and water column depth. The other approach uses a nitrogen stable isotope and mass balance calculation (Brandes and Devol, 2002). Both approaches, however, predict very similar global sediment denitrification rates (~250–300 Tg N yr^{-1}).

The results of Middelburg et al. (1996) also indicate that the integrated rate of denitrification in continental slope and deep-sea sediments exceeds that on the continental shelf, the latter defined in this study as sediments in water depths <150 m. As discussed here and in Codispoti et al. (2002) this observation runs counter to conventional thinking that shallow-water shelf sediments represent the major sites of marine sediment denitrification. To explain these observations Middleburg et al. (1996) note that in the deep sea, as carbon fluxes increase, denitrification also increases at the expense of aerobic respiration. However, as carbon fluxes continue to increase up the continental slope and onto the shelf, sulfate reduction now becomes increasingly more important at the expense of both aerobic respiration and denitrification (see fig. 7.7 and analogous discussions in section 7.5.2). This interplay therefore constrains the importance of denitrification to sediment regions between the deep sea and the continental shelf.

In the development of the Brandes and Devol (2002) nitrogen stable isotope budget discussed above, an interesting point that emerged from this work and their earlier work (Brandes and Devol, 1997) was that the relatively large fractionation of nitrate during denitrification ($\varepsilon \approx$ 20–40‰; see section 3.6) appears to occur only in lab studies and in open ocean (water column) settings. In contrast, there appears to be much less fractionation of nitrate during sedimentary denitrification ($\varepsilon \approx$ 0–3‰; also see results in Lehmann et al., 2004, that support these observations). To explain this apparent decrease in ε, Brandes and Devol (1997) suggest that steep nitrate concentration gradients in diffusive layers around reactive microzones in sediments may reduce the effective fractionation factor for denitrification in these microzones. The occurrence of nitrate consumption (and N_2 production) in reactions other than respiratory denitrification (i.e., those discussed above and in section 7.4.2) may also play a role in explaining these observations, if these other reactions have intrinsically small fractionation factors (again also see Lehmann et al., 2004).

As noted above, in coastal and estuarine sediments increasing carbon fluxes lead to the increasing importance of anoxic remineralization (sulfate reduction), with ammonium production and benthic fluxes becoming the major component of sediment nitrogen cycling (see, for example, pore-water profiles such as those in fig. 6.4). In such sediments nitrate fluxes, either into or out of the sediments, are also generally insignificant (<10%) in comparison to benthic ammonium fluxes. However, nitrogen loss by denitrification can still account for a significant fraction of the reactive nitrogen deposited in these sediments, again indicating the importance of coupled nitrification-denitrification (table 16.1). The factors controlling the relative importance of denitrification in coastal sediments are therefore similar in some ways to those discussed above for continental margin sediments.

In the "extreme" case of organic-rich, sulfidic coastal sediments such as those of Cape Lookout Bight, other factors may also limit the importance of denitrification relative to benthic ammonium fluxes. The presence of sulfide has been shown to inhibit nitrification (Joye and Hollibaugh, 1995) and may also enhance dissimilatory nitrate reduction to ammonium (see section 7.3.2 and discussions in Tobias et al., 2001; An and Gardner, 2002). At the same time, oxygen uptake by these sediments can also occur by other processes in addition to nitrification. These include not only aerobic respiration and the oxidation of solid phase FeS, but also the oxidation of other soluble reduced species that are transported upward to the sediment-water interface, e.g., sulfide, and perhaps reduced iron and manganese. Competition among these different reactions may then limit the importance of nitrification, and therefore denitrification, in sulfidic, organic-rich sediments.

In some coastal systems, seasonal anoxia or hypoxia in the bottom waters limits sediment nitrification and denitrification, because of the importance of coupled nitrification-denitrification in these sediments (Kemp et al., 1990). A decrease in the importance of sediment denitrification as a nitrogen sink may enhance primary production in the overlying waters (see below) and subsequently enhance water column respiration. This then acts as a positive feedback that further exacerbates low dissolved oxygen conditions in the bottom waters (also see discussions in Kemp and Boynton, 1992).

As discussed in section 14.3 rates of sediment remineralization in many coastal and estuarine sediments often vary seasonally. The resulting seasonality in benthic fluxes is often closely coupled with seasonality in water column productivity (e.g., Kemp and Boynton, 1984, 1992). This coupling stems from the fact that sediment nutrient regeneration and benthic fluxes often provide a significant fraction, sometimes >50%, of the nitrogen as well as phosphorus required by primary producers in the water column (also see Nixon, 1981; Klump and Martens, 1983; Cowan and Boynton, 1996).

TABLE 16.1
Nitrogen Cycling Processes in Estuarine and Coastal Sediments as Compared to Similar Values for Model Sediments

Location	Benthic Ammonium Flux[a]	Benthic Ammonium Flux[b]	Denitrification[b]
Ochlockonee Bay FL (USA)[c]	0.63	22%	63%
Long Island Sound[d]	0.63–1.9		<25%
Tomales Bay CA (USA)[e]	2.0	69%	43%
Narragansett Bay RI (USA)[c]	2.2	57%	37%
Southern Chesapeake Bay[f]	1.6–3.6		nd
Mesohaline Chesapeake Bay[f]	4.0–5.4	~83%	~12–37%
Cape Lookout Bight[g]	10.1 ± 3.6	>94%	<6%
Jamaica Bay, NY[h]	~13	~80%	~20%
Model sediments (Middelburg et al., 1996)[i]			
Shelf sediments (100 m)—NBW	0.4	20%	49%
Shelf sediments (100 m)—HNLO	1.1	51%	81%
Slope sediments (1,000 m)—NBW	0.3	34%	73%
Shelf sediments (1,000 m)—HNLO	0.4	47%	80%
Deep-sea sediments (4,000 m)—NBW	0	0%	18%

a. Units of mmol m^{-2} d^{-1}.

b. As a percentage of reactive N input. Reactive N input is defined here as the input of particulate organic nitrogen remineralized to inorganic nutrients and DON (where measured), plus (or minus) the benthic nitrate flux (depending on whether the sediments are a source or sink for water column nitrate).

c. From a compilation in Seitzinger (1988). These are annual averages based on seasonal studies at these sites.

d. These are annual averages of benthic ammonium fluxes for three sites, calculated with results in Aller (1980a). The denitrification estimate is based on discussions in Aller and Benninger (1981).

e. From Dollar et al. (1991). These are annual averages based on seasonal studies at this site.

f. From Cowan and Boynton (1996) and Burdige and Zheng (1998). These are annual averages based on seasonal studies at these sites.

g. From Klump and Martens (1987). This benthic ammonium flux is an annual average based on seasonal studies at this site. The denitrification estimate is based on the assumption that denitrification accounts for less than 1% of the total carbon remineralized in these sediments (Klump and Martens, 1987), and is converted to a denitrification rate using eqn. (16.1). Studies of dissolved N_2 in the pore waters of these sediments also provide little evidence for the occurrence of denitrification in these sediments (Kipput and Martens, 1982).

h. From Cornwell et al. (1999). Based on a single flux measurement.

i. NBW = normal bottom-water conditions; HNLO = high nitrate (40 μM), low oxygen (80 μM) bottom waters. See the text and the caption to fig. 16.4 for further details.

16.2.1 Benthic DON Fluxes

The discussion so far has focused largely on inorganic nitrogen cycling and benthic fluxes. However, inorganic nitrogen cycling is intimately tied to the cycling of organic nitrogen compounds. General aspects of the diagenetic processes affecting organic nitrogen have been discussed in section 9.9, and as can be seen in figs. 10.2 and 16.2, pore-water dissolved organic nitrogen (DON) compounds are produced and consumed as intermediates in organic matter remineralization. Since discussions in chapter 10 address the processes affecting DON cycling in sediment pore waters, the discussion here will focus on benthic DON fluxes and their role in the marine nitrogen cycle.

Interest in this topic is similar to that discussed in section 10.4 for benthic DOC fluxes and the marine carbon cycle. However, because nitrogen can be a limiting nutrient in marine ecosystems and because marine phytoplankton can use DON as their nitrogen source (Bronk, 2002), there is additional interest in understanding the role of sediments as a source of DON to the water column.

A number of studies have examined benthic DON fluxes from coastal and estuarine sediments, as well as from high-latitude continental margin sediments (see references in Burdige, 2002). These fluxes exhibit a tremendous range both in absolute magnitude and in direction, i.e., into and out of the sediments. However, at estuarine or coastal sites where repeated, or seasonal, studies have been carried out, mean or annual averages generally suggest that DON fluxes are small, usually out of the sediments, and only a small percentage of the benthic DIN flux[2] (see Burdige and Zheng, 1998, for more details).

Historically, interest in benthic DON fluxes also came from a desire to better understand the relative importance of these fluxes as compared to sediment denitrification rates in sediment nitrogen budgets (Nixon, 1981; Bender et al., 1989; Kemp et al., 1990; Devol and Christensen, 1993; Burdige, 2002). Until recently, sediment denitrification rates were often determined indirectly because of a number of inherent difficulties in determining these rates (Devol, 1991; Cornwell et al., 1999). As a result, in many early studies denitrification was used to balance sediment nitrogen budgets by implicitly assuming that benthic DON fluxes were of minor significance.

---

[2] DIN = dissolved inorganic nitrogen, i.e., ammonium plus nitrate + nitrite.

At a site in the mesohaline Chesapeake Bay, Burdige and Zheng (1998) were able to develop a nitrogen budget for these sediments using data on sediment nitrogen deposition and burial, DIN and DON benthic fluxes, and independent estimates of sediment denitrification rates. This approach led to the conclusion that here benthic DON fluxes were <20% of sediment denitrification and only ~3–4% of benthic DIN fluxes. Results from irrigated Washington State (USA) continental margin sediments further suggest that benthic DON fluxes are not an important component of nitrogen cycling in these sediments (Devol and Christensen, 1993). These workers were able to balance sediment nitrogen budgets at these sites using only DIN benthic fluxes and measured denitrification rates, determined with benthic lander N_2 measurements. In contrast, in Arctic sediments (water depths 170–2,600 m; Blackburn et al., 1996) and continental margin sediments of the Skagerrak (Landén-Hillmeyr, 1998), benthic DON fluxes were observed to be much more comparable in magnitude to benthic DIN fluxes and/or denitrification rates. The reasons for these differences are not well understood (see further discussions in Burdige, 2002).

16.3 Sediment Phosphorus Cycling

Phosphorus is an important nutrient in marine ecosystems that on long (geologic) time scales may play a role in controlling oceanic primary productivity (see reviews in Föllmi, 1996; Delaney, 1998; Ruttenberg, 2003). Unlike nitrogen, phosphorus cycling does not have a significant atmospheric component; as a result, burial in marine sediments represents the only significant mechanism by which it is removed from the oceans.

The major source of P to the oceans is riverine input of dissolved and particulate phosphorus, ultimately derived from continental weathering. In seawater, phosphorus is found as both dissolved inorganic phosphate (largely HPO_4^{2-} at typical seawater and pore-water pH values) and dissolved organic phosphorus (DOP). In sediments, the dominant solid forms of P are organic phosphorus, phosphorus associated with iron oxyhydroxides (Fe-bound P), and phosphate minerals such carbonate fluoroapatite.

Studies of organic phosphorus in DOM, sinking particles, and marine sediments (Ingall and Van Capellen, 1990; Kolowith et al., 2001;

Paytan et al., 2003) have shown that phosphate esters predominate. Phosphate esters also dominate the organic phosphorus found in marine organic matter sources such as phytoplankton, where they occur in nucleic acids and membrane phospholipids. Phosphonates, i.e., compounds containing a C-P bond, are found in differing amounts in nonliving organic matter, up to 25% in HMW DOM and <6% in sinking marine particles. Phosphonates are also only a few percent or less of the organic phosphorus in living organic matter. Because of analytical limitations (e.g., see discussions throughout chapter 9) little more is known about the composition of organic phosphorus in marine sediments and the controls on its preservation in sediments (see Ruttenberg, 2003, for further discussions).

Iron oxides, and possibly manganese oxides, have a strong affinity for binding/adsorbing phosphate, and phosphorus cycling in coastal and continental margin sediments is strongly tied to iron redox cycling (Krom and Berner, 1981; Klump and Martens, 1987; Sundby et al., 1992; Jensen et al., 1995; Slomp et al., 1996; McManus et al., 1997). Freshly precipitated iron oxides that form in the water column in association with midocean ridge/hydrothermal vent processes, and which are ultimately deposited in deep-sea sediments, also scavenge phosphate from seawater (see Ruttenberg, 2003, for a summary).

Phosphate adsorption to iron oxides is a function of factors such as the mineralogy, crystallinity, and surface area of the oxide phase (Ingall and Jahnke, 1997; Colman and Holland, 2000; Haese, 2000). The ability of iron oxides to participate in sediment phosphate cycling depends on the reactivity of these different phases toward reductive dissolution. In surface marine sediments Fe-bound P may be a significant component of the total sediment P (Ruttenberg and Berner, 1993). At the same time, Fe-bound P does not likely represent an important phase in "permanent" sediment P burial, because of diagenetic remobilization (reductive dissolution) of these oxides in many sediments (see discussions below).

Mineral-bound phosphate can take a number of different forms. The ability to examine these phases, and better define their role in sediment P cycling and burial, was for many years hampered by the fact that the different forms were analytically difficult to separate, detect, and/or quantify in sediments. The development of selective leaching techniques, which allow one to chemically distinguish these different phases, has proven invaluable in such studies (Ruttenberg, 1992; Anderson and Delaney, 2000; Schenau and de Lange, 2000).

Biogenic hydroxyapatite [$Ca_{10}(PO_4)_6(OH)_2$] is a major component of the hard parts of marine fish such as scales, bones, and teeth (Suess, 1981; Lowenstam and Weiner, 1989; Schenau and de Lange, 2000). Because seawater is undersaturated with respect to this material, it is subject to dissolution in both the water column and sediments. Detrital fluoroapatite [$Ca_5(PO_4)_3F$] is of igneous as well as metamorphic origin and transported to the oceans by rivers. This latter form of fluoroapatite is generally only a small percentage of the total P in most sediments and is likely to be fairly refractory (Ruttenberg, 1993; Ruttenberg and Berner, 1993).

Perhaps the most important form of mineral phosphate in marine sediments is authigenic carbonate fluoroapatite (CFA), also know as the mineral francolite (Ruttenberg and Berner, 1993; Delaney, 1998). The formation of authigenic CFA is often referred to as phosphogenesis. The chemical formula of CFA can be expressed as $Ca_{10}(PO_4,CO_3)_6F_2$ or $Ca_{10}[(PO_4)_{6-x}(CO_3)_x]F_{2+x}$ ($x \leq 1.5$), although some amount of Na and Mg may also be substituted for Ca (see discussions in Froelich et al., 1988; Föllmi, 1996). In sediments, particularly those underlying coastal upwelling areas such as, for example, the Peru margin, authigenic CFA occurs in several different forms, including phosphorite nodules (found both at the sediment-water interface and buried in sediments), discrete centimeter-scale francolite-rich layers in the upper ~1 m of sediment, and more diffuse (disperse) precipitates, which form in the upper ~5–10 cm of sediment (Froelich et al., 1988; Reimers et al., 1996; Schuffert et al., 1998; Schenau et al., 2000; Ruttenberg, 2003). Recent work has also shown that authigenic CFA can form in nonupwelling continental margin sediments, as well as in deep-sea sediments (Ruttenberg and Berner, 1993; Filippelli and Delaney, 1996). As will be discussed below these observations have led to a re-evaluation of the role of CFA burial in the oceanic phosphorus cycle.

Phosphorites are defined as marine sedimentary deposits containing ~5–40 wt% P_2O_5; in contrast, most marine sediments contain <0.3 wt% P_2O_5. Many phosphorite deposits on modern continental margins appear to be relict features and are generally thought to be postdepositional (secondary) deposits, in which more dispersed (primary, authigenic) sedimentary CFA is concentrated by physical processes such as winnowing and sediment reworking (see discussions in Föllmi, 1996; Ruttenberg, 2003). However, there are also a limited number of settings—the Peru margin, for example—where modern

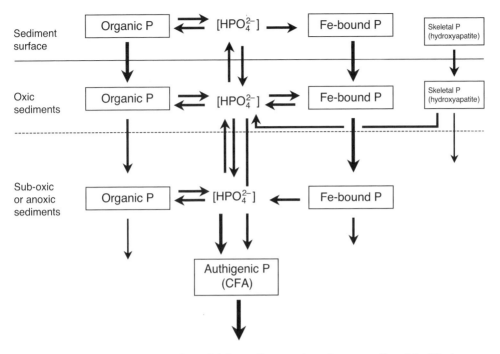

Figure 16.8 A conceptual model for sediment phosphorus cycling. Modified after Slomp et al. (1996).

phosphorite formation apparently occurs (Burnett and al, 1982; Froelich et al., 1983; Schuffert et al., 1998; Schenau et al., 2000).

To examine the cycling of reactive phosphorus[3] in sediments we will use the conceptual model shown in fig. 16.8. Historically it has been assumed that organic P represents the dominant carrier phase of reactive P to sediments (e.g., Froelich et al., 1982), although recent work suggests that ~30–50% of the P in sinking marine particles is found in an inorganic form (Paytan et al., 2003). Next in this model, organic matter remineralization releases phosphate into sediment pore waters. As a result, pore-water phosphate concentrations increase with depth, and under some circumstances have asymptotic profiles (e.g., see section 6.3.4 and Berner, 1980; Klump and Martens,

---

[3] "Reactive" phosphorus consists of forms of P that have the potential to undergo alteration or transformation on early diagenetic time scales. Thus, for example, detrital apatite is generally not considered reactive, since it does not appear to undergo alteration once deposited in a sediment.

1987). However, the shape(s) of these profiles can also be modified by reversible adsorption of phosphate onto iron oxides (Sundby et al., 1992) or the precipitation of authigenic CFA (Van Cappellen and Berner, 1988; Ruttenberg and Berner, 1993). As will be discussed below, authigenic CFA forms at the expense of both phosphate regenerated during organic matter remineralization (e.g., Ruttenberg and Berner, 1993) and phosphate liberated by the reductive dissolution of iron oxides (e.g., Slomp et al., 1996).

While the increase in pore-water phosphate concentrations generally leads to a benthic flux of phosphate out of sediments, the magnitude of the flux can be strongly attenuated by the presence of iron oxides at the sediment-water interface. Furthermore, under some circumstances, surficial iron oxides not only trap all of the upwardly diffusing phosphate, but can also lead to the net uptake of phosphate by the sediments from the water column (e.g., Klump and Martens, 1987; Cowan and Boynton, 1996; McManus et al., 1997).

As the model in fig. 16.8 illustrates, Fe-bound P can be input directly to sediments, and iron oxides at the sediment surface may also scavenge phosphate from bottom waters prior to their burial. The significance of Fe-bound P as a source of phosphate to the sediments is that its input is uncoupled from organic carbon input and remineralization. At the same time, iron oxides that form in situ in sediments during internal iron redox cycling also play an important role as a scavenging phase for phosphate that is regenerated in sediments.

In some situations, these oxides form as a thin surface layer as a result of the oxidation of upwardly diffusing Fe^{2+} or sediment FeS or pyrite by bottom-water oxygen or perhaps nitrate that is input into the sediments. In many coastal settings the formation of this surface oxide layer is a seasonal phenomenon (see Klump and Martens, 1981; Jensen et al., 1995; Colman and Holland, 2000; and references therein; also see figs. 14.3 and 16.9). As a result, there is an additional component of seasonality in benthic phosphate fluxes, based on iron redox cycling, above that predicted by, e.g., changing rates of sediment organic matter remineralization (also see section 14.3).

Evidence for the importance of these processes can be seen not only in comparisons of measured versus calculated benthic phosphate fluxes (fig. 16.9), but also in comparisons of measured benthic phosphate fluxes versus predicted fluxes based on measured rates of sediment carbon oxidation and the C/P remineralization ratio, i.e., this elemental ratio in the sediment organic matter undergoing rem-

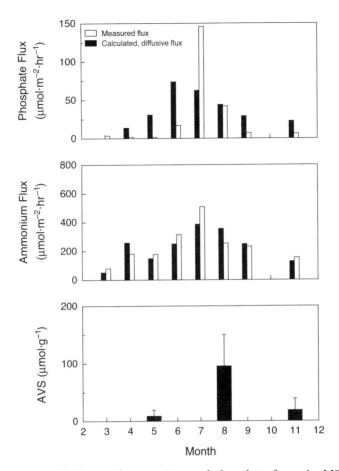

Figure 16.9 Benthic fluxes of ammonium and phosphate from site M3 in the mesohaline Chesapeake Bay (upper two panels). Directly measured fluxes (open bars) are from Cowan and Boynton (1996), while calculated diffusive fluxes (solid bars) were determined with pore-water profiles using procedures discussed in chapter 12 (J. Cornwell, unpub. data). Shown in the lower panel is the concentration of acid volatile sulfide (AVS) in the surface sediments (0–0.5 cm) of sites in an east-west transect across the Bay near site M3 (Cornwell and Morse, 1987).

The similarity in measured and calculated ammonium fluxes indicates that molecular diffusion dominates sediment-water exchange at this site; that is, bioturbation and bioirrigation are negligible (also see Burdige, 2001). In contrast, there are distinct seasonal differences between measured and calculated phosphate fluxes. During periods when rates of sediment organic matter ineralization (fig. 16.10). In coastal sediments when these benthic fluxes are examined over seasonal time scales (see discussions in section 14.3 and in Colman and Holland, 2000), this transient uncoupling of sediment carbon remineralization and benthic phosphate fluxes is often averaged out, such that integrated annual average remineralization ratios do not necessarily suggest enhanced sediment retention of phosphorus relative to carbon (e.g., compare Klump and Martens, 1987, Dollar et al., 1991, and Jensen et al., 1995).

In continental margin and some deep-sea sediments similar iron-phosphate interactions occur, and also impact benthic phosphate fluxes. Here, though, bottom-water oxygen concentrations can be an important controlling factor, since many of these sediments, particularly on continental margins, lie beneath the oxygen minimum zone.

At low rates of sediment carbon oxidation, the C/P remineralization ratio is generally similar to that predicted by the Redfield ratio (fig. 16.10), and roughly independent of bottom-water oxygen concentration. In this case there is presumably either no iron redox cycling occurring in the sediments, i.e., the sediments are "completely" oxic, or iron redox cycling occurs sufficiently deep in the sediments and does not impact benthic phosphate fluxes (McManus et al., 1997).

In contrast, as carbon oxidation rates increase (above ~2–3 mmol m^{-2} d^{-1} in the data shown in fig. 16.10) the behavior of benthic phosphate fluxes appears depend on bottom-water oxygen levels. At high bottom-water oxygen concentrations, the data generally tend to fall

←――――――――――――――――――――――――――――――――――

remineralization rates are low and oxygen penetration into the sediments presumably increases, e.g., winter and fall months, iron oxides seasonally form near the sediment surface (also see fig. 14.3) and temporarily trap and store much of the phosphate produced at depth that diffuses upward toward the sediment-water interface. These oxides are seasonally reduced because of higher rates of (anoxic) sediment organic matter remineralization that lead to the shallowing of the oxygen penetration depth, and/or to seasonal anoxic bottom waters; this is seen here here by. seasonal production of AVS (see discussions in section 17.1 for further details, and also compare with fig. 14.3). Phosphate "stored" on these oxides is then initially released, leading to a transient enhancement of phosphate fluxes from the sediments. Later in the summer, when iron sulfides still predominate in the surface sediments, measured phosphate fluxes are essentially equal to calculated diffusive fluxes.

SEDIMENT PHOSPHORUS CYCLING

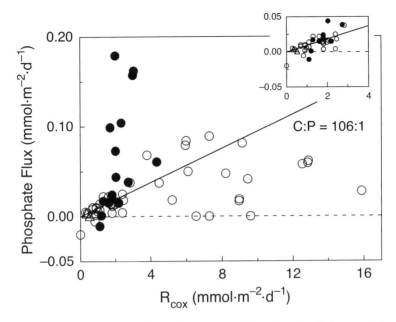

Figure 16.10 Benthic phosphate fluxes versus R_{cox} (the depth-integrated rate of organic carbon oxidation) for sediments along the California and Washington (USA) continental margin and in the equatorial Pacific (3,300–4,500 m water depth). Shown as an insert are the data in the lower left portion of this figure on an expanded scale. Open symbols are for sediments overlying bottom waters with >20 μM O_2 (Δ = EqPac sites), while filled circles are low oxygen sites ($[O_2] < 20$ μM). Data from several sources (Smith et al., 1979; Berelson et al., 1987b; Bender et al., 1989; Jahnke, 1990; Reimers et al., 1992; Devol and Christensen, 1993; Hammond et al., 1996; Ingall and Jahnke, 1997; McManus et al., 1997).

on or below the Redfield ratio line. This suggests that in many of these sediments less phosphate is being returned to the overlying waters than that predicted by the observed rates of sediment organic matter remineralization. The most likely explanation for these observations is that iron redox cycling occurs sufficiently close to the sediment-water interface, and acts to trap some of the phosphate regenerated in the sediments (Ingall and Jahnke, 1997; McManus et al., 1997). In contrast, though, to the seasonal occurrence of this phenomenon in coastal sediments (fig. 16.9), this process may lead to a situation in which much of this Fe-bound P is eventually transported, by either sedimentation or bioturbation, deeper into the sediments, where it can be transformed into authigenic CFA (see below).

At low bottom-water oxygen concentrations we see again that there is some scatter around the marine organic matter line, although a large number of these observations fall above the Redfield ratio trend line. This implies that there is a return flux of phosphate to the bottom waters in excess of that predicted by organic matter remineralization. There may be several possible explanations for these observations. The first is that oxygen penetration depths are sufficiently shallow in this case, and the diagenetic formation of a surficial iron oxide layer is essentially suppressed. Phosphate release from the reductive dissolution of deposited iron oxides then adds to the benthic phosphate flux that results from organic matter remineralization (Ingall and Jahnke, 1997; McManus et al., 1997). Calculations presented by Ingall and Jahnke (1997) and Colman and Holland (2000) provide some evidence in support of the possibility that this external deposition of Fe-bound P could provide the additional phosphate flux seen in fig. 16.10, although the uncertainties in these estimates are large.

Consistent with this suggestion is that shallow oxygen penetration depths, i.e., less than ~0.3 cm, result in both enhanced benthic phosphate and iron (Fe^{2+}) fluxes from some California continental margin sediments. Note that these observations represent a subset of the results in fig. 16.10 (McManus et al., 1997). However, because the oxygen-penetration depth in sediments is a function of both bottom-water oxygen content and sediment carbon oxidation rates (e.g., see discussions in sections 7.5.2 and 14.5), bottom-water oxygen is not necessarily the sole parameter that may lead to such excess phosphate fluxes (see discussions in McManus et al., 1997, for further details).

Colman and Holland (2000) also suggest that this apparent uncoupling of sediment carbon oxidation and benthic phosphate fluxes in low bottom-water oxygen settings could be the result of temporal variability in carbon rain rates to continental margin sediments (see section 14.3), in a manner analogous to that which occurs in some coastal sediments (also see similar discussions in Ingall and Jahnke, 1997; McManus et al., 1997). In this case, such snapshots capture only that portion of a time-varying P cycle in these margin sediments where measured fluxes exceed predicted fluxes; analogous, for example, to sampling the Chesapeake Bay site described in fig. 16.9 only in month 7. Based in part on arguments presented at the end of section 14.3, seasonal benthic flux studies and/or seasonal studies of sediment P distributions would help in further examining this suggestion.

Another possible explanation for the observations in fig. 16.10 involves redox-driven bacterial sequestration and release of phosphate (see Ingall and Jahnke, 1997, and references therein). This mechanism postulates that under oxic conditions bacteria accumulate and store phosphate as the end-product of oxidative phosphorylation (see section 7.1). Upon depletion of oxygen, bacteria then utilize this stored P for energy and release phosphate back into the surrounding waters (also see discussions in Aller, 1994a). Under nonsteady-state conditions similar to those discussed above, this process also has the potential to lead to a short-term uncoupling of sediment carbon oxidation rates and benthic phosphate fluxes (Ingall and Jahnke, 1997). At the same time, however, this bacterial redox cycling of phosphate cannot lead to a steady-state enhancement of benthic phosphate fluxes unless bacteria actually acquire this excess phosphate from the bottom waters.

These observations therefore point to a more general conclusion that under steady-state conditions, over the proper time scales, benthic phosphate fluxes can exceed those predicted by rates of sediment organic matter remineralization only if there is an additional external source of phosphate to the sediments that is uncoupled from organic matter deposition and remineralization. Deposition of phosphate associated with sedimenting iron oxides, or scavenged by these oxides from bottom waters, represents one such way for this to occur, as does deposition and dissolution of fish debris, or bacterial uptake of phosphate from the water column.

In a similar fashion, benthic phosphate fluxes lower than those predicted by rates of sediment organic matter remineralization require the transfer of this regenerated phosphate to a stable burial phase. Except in highly oxic (deep-sea) sediments, Fe-bound P is likely not to be this stable phase (e.g., Jensen et al., 1995). For example, in deep-sea calcareous sediments these oxides appear to form as coatings on sediment particles (Sherwood et al., 1987), and in early sediment P budgets (Froelich et al., 1982) it was incorrectly inferred that biogenic calcite, rather than the associated iron oxides, was the actual carrier phase for P burial in deep-sea sediments. However, except in shallow-water tropical carbonate sediments where phosphate adsorption is an important component of phosphorus biogeochemistry (e.g., Jensen et al., 1998), calcium carbonate is a minor burial phase for P in most sediments (Ruttenberg, 1993; Delaney, 1998).

16.3.1 Formation of Authigenic CFA and Phosphorus Burial in Sediments

While the reductive dissolution of iron oxides limits the importance of Fe-bound P as a permanent phosphorus burial phase, in some sediments this process may play an important role in the formation of authigenic CFA (see fig. 16.8 and Slomp et al., 1996). This then suggests that the temporary storage of phosphate as Fe-bound P in sediments provides a mechanism by which phosphate is transferred into a stable burial phase.

Several lines of evidence support the possible role of Fe-bound P in the formation of CFA. First, iron oxides are, as noted above, a trap that can concentrate phosphate liberated during organic matter remineralization. Downward mixing, by bioturbation, or burial, by sedimentation, of this Fe-bound P into regions in the sediment where the oxides undergo reductive dissolution can subsequently release this phosphate back into the pore waters. This suggestion is consistent with some observations in the literature that pore-water phosphate profiles in sediments where phosphorites form, or francolite precipitates, appear to be uncoupled from organic matter remineralization and require an "extra" source of phosphate to explain the profiles (Froelich et al., 1988; Schuffert et al., 1994). Diagenetic models of such pore-water nutrient profiles that include the appropriate transport processes and nutrient production by organic matter remineralization can reproduce ammonium, alkalinity, and/or ΣCO_2 pore-water profiles but not phosphate profiles. At the same time, though, other processes may also contribute to the excess phosphate needed to induce authigenic CFA formation. These include the dissolution of fish debris (Suess, 1981) as well as bacterial redox cycling of phosphate (Ingall and Jahnke, 1997).

A second piece of evidence in support of the role of Fe-bound P as an intermediate in the formation of CFA involves examining the source(s) of the fluoride incorporated into authigenic CFA. In some systems, diffusion from overlying seawater appears to be the major F^- source (Froelich et al., 1983); however, fluoride also adsorbs to iron oxides (Ruttenberg and Berner, 1993) and, along with phosphate, may be concentrated in sediments by iron redox cycling. Dissolution of these oxides may then also provide the necessary fluoride for CFA precipitation (Ruttenberg, 2003). Finally, a quantitative diagenetic model for the dissolved and solid P phases described in fig. 16.8 supports the important role that Fe-bound P may play in authigenic CFA formation (Slomp et al., 1996).

The precipitation of CFA, actually francolite, generally occurs from supersaturated pore waters, although the kinetics and mechanism of precipitation vary depending on the degree of supersaturation. Francolite precipitation can also occur through the dissolution/precipitation replacement of existing minerals such as calcium carbonate. In either case, once francolite nuclei form, they tend to serve as a template for further precipitation until environmental conditions limit further precipitation (see Föllmi, 1996, for a summary).

Historically, in the classical sites of phosphogenesis and phosphorite formation, for example, productive coastal upwelling regions such as the Peru margin, authigenic CFA formation has been assumed to occur in the near-surface sediments, i.e., upper ~5–10 cm (Froelich et al., 1988; Glenn et al., 1988; Reimers et al., 1996; Schuffert et al., 1998; Schenau et al., 2000). This apparently is controlled by the alkalinity increase that occurs in such organic-rich sediments as a result of anoxic organic matter remineralization, primarily sulfate reduction. More specifically, increasing carbonate substitution into CFA decreases CFA solubility; this therefore results in decreasing CFA solubility with increasing pore-water carbonate ion concentration (Jahnke, 1984). Increasing carbonate substitution into CFA also limits the extent of CFA crystallite growth (Glenn et al., 1988) and leads to a dramatic increase in the CFA equilibrium phosphate concentration (Jahnke, 1984). Thus assuming there is a source of excess phosphate in near-surface sediments (as described above), increasing alkalinity with sediment depth will likely limit CFA precipitation to this upper region of the sediments.

In contrast, it has also been shown in recent years that authigenic CFA can form in nonupwelling continental margin sediments, as well as in deep-sea sediments (Ruttenberg, 1993; Ruttenberg and Berner, 1993; Filippelli and Delaney, 1996; Slomp et al., 1996; Delaney, 1998). More importantly, in these sediments the region of authigenic CFA formation is not limited to the near-surface sediments, but in some cases can extend down several hundred meters. The mechanism(s) of CFA precipitation and the needed sources of phosphate and fluoride appear to be similar to those discussed above. Yet because such sediments are generally less organic rich than sediments in upwelling regions, alkalinity accumulation in these pore waters is smaller. This then avoids the inhibition of CFA precipitation by alkalinity discussed above. Iron redox cycling also likely occurs at greater sediment depths, implying that this source of phosphate and fluoride releases these solutes into the pore waters at greater sediment depths. Conversely, in many of these sediments phosphate liberated by organic matter remineralization may directly contribute to CFA precipitation (fig. 16.8), without the need for Fe-bound P as an intermediate (Ruttenberg and Berner, 1993; Slomp et al., 1996).

The recognition that authigenic CFA formation is more widespread than previously thought has led to the re-evaluation of several aspects of the oceanic and sediment P cycles. The first is that estimates of the sediment P burial flux have increased, with a concomitant decrease in the oceanic residence time of phosphate, from previous estimates of ~80 kyr to values as low as ~10–20 kyr (Ruttenberg, 1993, 2003; Filippelli and Delaney, 1996; Colman and Holland, 2000). Discussions in these papers also examine the broader significance of this downward revision of the phosphate residence time in terms of oceanic phosphate cycling and its possible relationship to glacial-interglacial changes in atmospheric CO_2 concentrations.

Although there is still large uncertainty in phosphorus burial estimates, authigenic CFA appears to represent roughly half of the P buried in marine sediments, with continental margin sediments constituting the major site of this burial (Ruttenberg, 1993). Organic P and Fe-bound P comprise the remaining half of the P buried in marine sediments, although there is some degree of uncertainty in the magnitude of the organic P burial flux (compare discussions in Ruttenberg, 1993, Ruttenberg and Berner, 1993, and Delaney, 1998). Another important point to come out of these observations is the concept of sink-switching during sediment P diagenesis (Ruttenberg and Berner, 1993). This implies that reactive P is deposited in sediments in relatively labile forms, i.e., organic P, Fe-bound P, and perhaps biogenic mineral P, and that which is not returned to the overlying waters as a benthic phosphate flux is largely transformed in the sediments and buried as authigenic CFA.

The majority of the reactive P deposited in sediments is input as organic P, and examining the factors controlling organic phosphorus burial and preservation in sediments leads to questions and concerns similar to those that have been discussed previously for carbon preservation (see discussions in Ingall and Van Capellen, 1990; Ruttenburg and Goñi, 1997; Delaney, 1998; Ruttenberg, 2003). However, because of sink-switching, phosphorus burial in sediments is not controlled simply by the occurrence of organic P preservation versus remineralization. Rather, extensive remineralization of organic phosphorus does not necessarily preclude significant P burial if the formation of authigenic CFA also occurs.

Redox-associated bacterial cycling of phosphate also apparently leads to the production of refractory organic P compounds as a part of the newly produced bacterial biomass that forms during this process (Gächer et al., 1988). Based on the way this cycling occurs, this production of refractory organic P bacterial necromass (see section 11.1.5) should be most important in mixed redox sediments subject to bioturbation, biorrigation, or physical reworking (Aller, 1994a; 1998). At the same time, diffusive openness of sediments also appears to enhance the microbial production, and retention, of organic P in sediments (Aller and Aller, 1998). Conversely, it has also been suggested that refractory organic matter derived from bacterial necromass may be responsible for the relatively low C/P (and C/N) ratios observed in very organic-poor, oxic deep-sea sediments (see section 11.1.3). As noted earlier (also see chapter 9) a better characterization of organic phosphorus compounds in marine sediments will be important in further understanding the controls on the production of this apparent refractory material, and its relationship to sediment P cycling and burial.

Finally, as discussed above, phosphate benthic fluxes appear to be enhanced under anoxic conditions i.e., low bottom-water oxygen concentrations or shallow sediment oxygen penetration depths, because of the couplings between sediment Fe and P cycling (see figs. 16.9 and 16.10 and Ingall and Jahnke, 1997; Colman and Holland, 2000). As a result, the burial of authigenic CFA, and perhaps bacterially derived organic P, may be enhanced by oxic, suboxic, or mixed redox conditions. This inverse relationship between sediment and/or bottom-water redox conditions and sediment P burial may exert a negative feedback on atmospheric oxygen during the Phanerozoic, perhaps through long-term controls on oceanic primary productivity by phosphate availability (Van Cappellen and Ingall, 1996). However, the details of how this may occur, if indeed it does occur, are not worked out (compare discussions in Colman et al., 1997; Colman and Holland, 2000; Kump et al., 2003; Ruttenberg, 2003).

≈ **CHAPTER SEVENTEEN** ≈

Biogeochemical Processes in Continental Margin Sediments. II. Sulfur, Methane, and Trace Metal Cycling

IN THIS CHAPTER we will continue the discussion of biogeochemical processes in continental margin sediments, focusing on sulfur, methane, and trace metal cycling.

17.1 Sediment Sulfur Cycling

In earlier sections of this book we examined aspects of sediment sulfur cycling, discussing sulfur isotope geochemistry (section 3.7), the role of bacterial sulfate reduction in sediment organic matter remineralization (sections 7.3.4 and 7.5.2), sulfide oxidation in sediments (section 7.4), and the occurrence of organic sulfur compounds in sediments (sections 9.7 and 9.8.3). In this section we will re-examine these and other topics in a more thorough discussion of sediment sulfur cycling.

The dominant source of sulfur to the oceans is sulfate in rivers derived from weathering of continental rocks. Burial in marine sediments is the main sink for seawater sulfate. However, the sediment geochemistry of sulfur is centered around bacterial sulfate reduction since most sulfur is buried as pyrite (FeS_2) or as diagenetically produced organic sulfur (as discussed in section 9.7). These latter two processes appear to represent the dominant sulfur removal processes, since large-scale precipitation of calcium sulfate, either in association with low-temperature marine evaporites or high-temperature hydrothermal processes, is relatively insignificant in the modern ocean (Vairavamurthy et al., 1995; Berner and Berner, 1996; Goldhaber, 2003).

In addition to sulfate, there are several other forms of dissolved sulfur that will be considered here. Dissolved sulfide, generally HS^- at pore-water pH values,[1] is perhaps the other most important dissolved species, although other forms of dissolved sulfur with oxidation states intermediate between sulfide and sulfate also occur, generally at much lower levels (see summaries in Kasten and Jørgensen, 2000; Goldhaber, 2003). The most important of these other forms include thiosulfate ($S_2O_3^{2-}$), sulfite (SO_3^{2-}), and zero-valent sulfur, either as $S°_{aq}$ or as polysulfides, S_x^{2-}. These compounds can be important intermediates in different aspects of sedimentary sulfur cycling, as illustrated in fig. 17.1.

In anoxic or mixed redox sediments where sulfate reduction occurs, pyrite (FeS_2) is the dominant form of solid-phase sulfur (see fig. 17.2). Pyrite has a range of crystal morphologies, although in marine sediments only a limited number are known to occur, with single euhedra and framboids generally dominating (Canfield and Raiswell, 1991b; Wang and Morse, 1996; Goldhaber, 2003). Additional details about the morphology and crystal structure of pyrite in marine sediments will be presented in discussions below.

Other sulfide minerals found in marine sediments include amorphous-FeS and mackinawite (FeS_{1-x}, $X \leq \sim 0.1$). Both of these phases are here termed 'FeS' as is common in the literature. Another metastable sulfide phase is greigite (Fe_3S_4), the sulfur analog to magnetite (Fe_3O_4). All of these phases are generally referred to as *acid volatile sulfide* (AVS) because of their solubility in nonoxidizing mineral acids such as HCl. Pyrite is not soluble in HCl and requires more stringent, i.e., oxidizing or reducing, conditions to be solubilized. These differences in solution chemistry have allowed for the development of selective leaching techniques that, at least operationally, allow one to quantify AVS and pyrite in marine sediments (Canfield et al., 1986; Cornwell and Morse, 1987; Morse et al., 1987).

The black color common in many organic-rich coastal sediments is due to the presence of sulfide phases. Acid volatile sulfides are generally considered metastable phases that will eventually transform into pyrite (fig. 17.2), the thermodynamically stable sulfide phase under early diagenetic conditions (e.g., Berner, 1984). Another form of sulfur found in marine sediments is solid elemental S°, which generally forms during sulfide oxidation and can also play an important role in pyrite formation.

---

[1] In many works dissolved sulfide is also referred to as ΣH_2S, by analogy to ΣCO_2, and is similarly defined as the sum $[H_2S]_{aq} + [HS^-] + [S^{2-}]$.

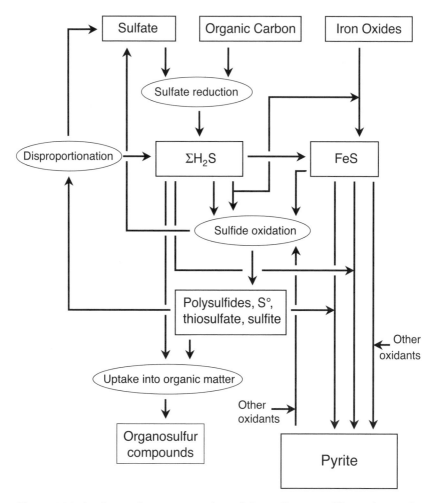

Figure 17.1 A schematic representation of the sediment sulfur cycle. Modified after Cornwell and Sampou (1995) and Vairavamurthy et al. (1995).

The conceptual model in fig. 17.1 used to describe sulfur cycling in recent marine sediments includes many processes discussed previously, including bacterial sulfate reduction and sulfide oxidation. One set of processes not yet discussed are sulfur disproportionation reactions involving elemental sulfur, thiosulfate, and sulfite. These reactions are the inorganic equivalent of organic fermentation reactions (see section 7.1) and result in the production of reduced sulfide and oxidized sulfate from these intermediate oxidation state sulfur species

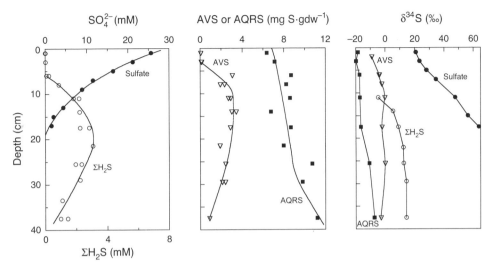

Figure 17.2 The depth distribution in Cape Lookout Bight sediments of: (Left panel) pore water sulfate and sulfide; (Center panel) solid phase AVS (acid volatile sulfide ≈ FeS or iron monosulfides) and AQRS (aqua regia soluble sulfide ≈ FeS_2 or pyrite); (Right panel) $\delta^{34}S$ of these various sulfur pools (redrawn using data for a core collected on 3/9/82, presented in Chanton, 1985; Chanton et al., 1987b). Note that with depth all sulfate is reduced and essentially buried as pyrite. In a completely closed system, this should then result in pyrite being buried with a $\delta^{34}S$ value identical to that of the starting seawater (+20‰). However, because this sediment is open to sulfate via diffusion (which preferentially adds light sulfate to the sediments) the pyrite that is buried is distinctly lighter than the starting seawater (approx. −10‰; see discussions in the text and in Goldhaber, 2003).

(Jørgensen and Bak, 1991). For example, thiosulfate disproportionation can be written as

$$S_2O_3^{2-} + H_2O \rightarrow H_2S + SO_4^{2-} \quad (17.1)$$

and analogous reactions can be written for S° and sulfite disproportionation. An important aspect of sulfur disproportionation is that it results in significant sulfur isotope fractionation. In particular, the sulfide produced by disproportionation can be depleted in ^{34}S by ~3–28‰ relative to the initial intermediate oxidation state sulfur compound (Canfield and Thamdrup, 1994; Habicht et al., 1998; Habicht and Canfield, 2001).

In examining fig. 17.1 we start with bacterial sulfate reduction, which produces dissolved sulfide as one of its end-products. In nearshore, coastal sediments with TOC contents of ~1–3%, sulfate reduction occurs over the upper tens of centimeters to several meters of sediment. In some offshore continental margin sediments, sulfate reduction occurs much deeper in the sediments and sulfur cycling in these sediments can be quite different from that which occurs in these nearshore settings. This problem will also be discussed in more detail in section 17.2.2.

Dissolved sulfide produced by bacterial sulfate reduction has several possible fates: reaction with iron oxides or Fe^{2+} to form FeS (AVS), incorporation into sediment organic matter (formation of organosulfur compounds), reaction with FeS to form pyrite, or oxidation to form $S°$ or other increasingly oxidized sulfur species up to, and including, sulfate.

The reaction of sulfide with iron oxides is an important process in sediments and plays a key role in explaining why sulfide does not accumulate in pore waters until some depth below that where sulfate reduction, i.e., sulfide production, begins to occur (fig. 17.2). In such regions of the sediment, reactive iron oxides (section 7.3.3.1), if present, are rapidly reduced by dissolved sulfide, effectively removing sulfide from the pore waters. Once these reactive oxides are exhausted, more refractory forms of oxidized iron (e.g., more crystalline oxides, magnetite, and iron in silicates) are less efficient at reacting with sulfide. As a result, sulfide then begins to accumulate in the pore waters (Canfield et al., 1992; Raiswell and Canfield, 1998).

Because FeS is a metastable mineral phase, once it is produced it is either reoxidized (see below) or transformed into pyrite. The oxidation of FeS in sediments, along with dissolved sulfide and pyrite, can occur either chemically or microbially, using oxidants such as O_2, nitrate, or metal oxides (see section 7.4). In mixed redox sediments these processes are also mediated by bioturbation and/or bioirrigation (e.g., see section 7.4.3), while in nonbioturbated or nonbioirrigated sediments these processes may occur as a result of the upward diffusion of sulfide and/or the downward diffusion of oxygen or nitrate. The occurrence of these reactions plays a role in the extent to which sulfides produced by bacterial sulfate reduction are retained in the sediments, i.e., buried as reduced sulfur. At the same time, though, because the disulfide in pyrite (FeS_2) is more oxidized than its initial source, e.g., sulfide in either dissolved sulfide or FeS, these oxidative processes may also play a role in pyrite formation.

Evidence in support of these oxidative processes comes in part from observations that the sulfur isotope fractionation seen in sedimentary sulfides generally exceeds that observed during bacterial sulfate reduction (see Habicht and Canfield, 2001; Goldhaber, 2003; and references therein). In both field studies and lab studies with pure cultures of sulfate reducing bacteria, the sulfide produced by bacterial sulfate reduction is generally depleted in ^{34}S by 5–46‰ relative to the starting sulfate, i.e., average value of ~21‰ (Habicht et al., 1998). On the other hand, sedimentary sulfides can show depletions in ^{34}S of ~45–70‰ relative to starting seawater (also see discussions in Canfield and Thamdrup, 1994).

To understand the role that sulfide oxidation processes may play in pyrite formation, and in controlling its δ^{34}S value, I first note that much of the pyrite that forms in marine sediments does so close to the sediment-water interface, where rates of sulfate reduction are usually the highest and the availability of reactive iron likely to be the greatest (e.g., Goldhaber, 2003). More importantly, sulfide reoxidation, through either mixed redox conditions or upward sulfide diffusion, also occurs in this portion of the sediments. This reoxidation results in fairly intense sulfide redox cycling, in which sulfide oxidation produces intermediate sulfur compounds (S°, sulfite, and thiosulfate) that can disproportionate back to sulfide and sulfate. Given the associated fractionation factors associated with sulfur disproportionation (Habicht et al., 1998), repeated cycling of sulfide in this manner generates sulfide that becomes increasingly depleted in ^{34}S relative to the sulfide initially produced during sulfate reduction (Canfield and Thamdrup, 1994; Habicht and Canfield, 2001). Thus in surface sediments, these processes can produce pyrite precursors, and eventually pyrite, that are extremely depleted in ^{34}S.

Results to date examining this type of sulfur cycling and pyrite formation have generally been obtained either in pure culture studies or in closed system manipulative sediment studies (Canfield and Thamdrup, 1994; Habicht et al., 1998; Habicht and Canfield, 2001). Attempts to examine these processes in surface marine sediments under true in situ conditions are difficult, in part because of the low concentrations and rapid cycling of these intermediate sulfur compounds during sulfur redox cycling and disproportionation (see discussions in Canfield and Thamdrup, 1994; Goldhaber, 2003). However, based on a summary of results across a range of mainly coastal sediments, Canfield and Thamdrup (1994) have observed that $\Delta_{\text{sulfate–sed. sulfides}}$, the difference in $\delta^{34}S$ between the starting sulfate and observed sedimentary sulfides, becomes increasingly more negative as $\%S_{\text{oxidized}}$ increases. Note that $\%S_{\text{oxidized}}$ is defined here as the percentage of sulfide produced by bacterial sulfate reduction that is not buried, and therefore lost in a net sense by oxidation. This observation is consistent with the model that increasing sulfide redox cycling produces increasingly lighter dissolved sulfide, and consequently light sedimentary sulfides.

Pyrite formation also occurs deeper in the sediments, i.e., below the surface mixed or bioirrigated layer, where this type of sulfur redox cycling is unlikely to be important. Here, extremely depleted sulfide minerals may result from the fact that bacterial sulfate reduction in sediments results in not only bulk concentration gradients for sulfate and dissolved sulfide, but also differential isotope concentration gradients for the ^{32}S- and ^{34}S-containing forms of these solutes (Jørgensen, 1979; Chanton et al., 1987b). Thus in these deeper sediments, where diffusion dominates pore-water transport, this situation leads to enhanced downward diffusion of ^{32}S-enriched sulfate and enhanced upward diffusion of ^{34}S-enriched dissolved sulfide. As a result, differential transport of sulfur isotopes leads to the net burial of sulfur, as pyrite, that is isotopically lighter than that based solely on the instantaneous isotopic fractionation associated with sulfate reduction (see fig. 17.2).

Experimental studies examining pyrite formation have led to suggestions of three possible mechanisms for pyrite formation in marine sediments (Goldhaber, 2003, presents an excellent summary of the recent literature on this topic; also see Morse et al., 1987, and Rickard et al., 1995, for earlier summaries). In all three cases, FeS is the initial precursor for pyrite formation, because of the ease with which it forms by the reaction of dissolved sulfide with either Fe^{2+} or iron oxides and also because direct precipitation of pyrite from solution appears to be kinetically inhibited (Schoonen and Barnes, 1991).

One mechanism for pyrite formation involves the addition of zero-valent sulfur to FeS by polysulfides through a dissolution/reprecipitation reaction (Luther et al., 1991; Rickard et al., 1995)

$$FeS(s) + S_x^{2-} \rightarrow FeS_2(s) + S_{x-1}^{2-} \qquad (17.2)$$

A second mechanism can be thought of as forming pyrite through iron loss (coupled to FeS oxidation), as opposed to sulfur addition

(Wilkin and Barnes, 1996). Here pyrite forms from FeS via a greigite intermediate:

$$4FeS \rightarrow Fe_3S_4 \rightarrow 2FeS_2 \tag{17.3}$$

(also see Sweeney and Kaplan, 1973; Schoonen and Barnes, 1991; Morse and Wang, 1997; Benning et al., 2000). Because of the nature of this process, a number of possible oxidizing agents, in addition to zero-valent forms of sulfur, may play a role in this type of pyrite formation. These include not only other sulfur species with intermediate oxidation states, such as thiosulfate, but also O_2 itself (Wilkin and Barnes, 1996; Benning et al., 2000; Goldhaber, 2003). The factors controlling the occurrence of these two possible modes of pyrite formation are not well understood, although pyritization through a greigite intermediate (eqn. 17.3) may be favored under slightly more oxidizing conditions (Schoonen, 1991). Pyrite formation via a greigite intermediate also appears to be faster than that occurring directly from mackinawite (Schoonen and Barnes, 1991; Wang and Morse, 1996), and a greigite intermediate may be required for the formation of framboidal pyrite (see Goldhaber, 2003, and references therein).

Given the potential oxidants required in these two modes of pyrite formation, their occurrence is consistent with discussions above regarding pyrite formation near the sediment surface, for example, in association, with mixed redox conditions (Chanton et al., 1987a; Canfield et al., 1992; Goldhaber, 2003). However, as is also noted above, pyrite also forms deeper in sediments (see fig. 17.2) where dissolved sulfide levels are high and the availability of such oxidants is likely to be low to nonexistent. Here a third mechanism of pyrite formation may be

$$FeS(s) + 2H_2S \rightarrow FeS_2(s) + 2H_2(g) \tag{17.4}$$

where H_2S now serves as the oxidizing agent of FeS, and reduced hydrogen gas is produced as an end-product (Rickard, 1997).

Reaction (17.4) has a negative $\Delta G°$ (Wilkin and Barnes, 1996) and some experimental evidence supporting its occurrence has been presented (see Rickard et al., 1995; Rickard, 1997; and references therein). However, other experimental results (Wilkin and Barnes, 1996; Benning et al., 2000) suggest that pyrite formation does not occur with H_2S as an oxidant, and that the process requires an oxidizing agent more oxidized than sulfide. The occurrence of reaction (17.4) is appealing as an explanation of pyrite formation in deeper, highly anoxic regions of sediments, explaining observations such as those in fig. 17.2, in which pyrite increases slightly with sediment depth as sulfide in the pore waters and solid phase FeS both decrease. However, these equivocal experimental results must also be reconciled. At the same time, it is interesting to recall that H_2 is an excellent substrate for both sulfate reducers and methanogens, and it is generally found at nM levels in the regions of the sediment that these organisms inhabit (see section 7.6.2). Since this region of the sediments is also where rxn. (17.4) is likely to be most important, the low hydrogen levels observed there could provide a strong kinetic driving force for the occurrence of this reaction (as well as provide a ΔG for the reaction that is significantly more negative than its $\Delta G°$).

17.1.1 Sulfur Burial Efficiency

Much of the pyrite found in marine sediments appears to form as a result of the interaction between oxidative, or mixed redox, processes and bacterial sulfate reduction. The interactions of these same processes also leads to only a small fraction of the total sulfide produced by sulfate reduction eventually being buried as pyrite. Expressed another way, sulfur ($\approx$ pyrite) burial at depth is generally much smaller than the integrated gross rate of sulfate reduction; the ratio of the two is thus defined as the *sulfur burial efficiency*. Historically, oxygen has been assumed to be the predominant oxidant of pyrite (Morse, 1991). However, recent studies have shown that other substances, such as nitrate, manganese, and iron oxides, may also oxidize pyrite anaerobically (Luther and Church, 1992; Luther et al., 1992; Schippers and Jørgensen, 2002). As is also the case for anaerobic FeS oxidation (see section 7.4.2), evidence to date regarding chemical versus microbial mediation of anaerobic pyrite oxidation is equivocal.

In most sediments sulfur burial efficiency is generally $< \sim 10–20\%$ (Jørgensen, 1982; Berner and Westrich, 1985; Chanton et al., 1987a; Canfield, 1989a; Swider and Mackin, 1989), although there are examples where burial efficiencies are much higher; in Cape Lookout Bight sediments, for example, sulfur burial efficiency is 77%. Sediments with such high sulfur burial efficiencies are often sites with high sedimentation rates and high rates of sulfate reduction, and are also generally not extensively bioturbated or bioirrigated (also see

Canfield, 1994; Goldhaber, 2003). In further examining the controls on sulfur burial efficiency, the availability of sulfate, reactive iron, and metabolizable organic matter are likely the most important parameters; of these, the availability of the last two appears to be most critical (e.g., Goldhaber, 2003).

The effect of iron availability on pyrite formation is a complex function of the relationship between the production of sulfide by sulfate reduction and its reaction with iron minerals in sediments. This is often examined in the context of the parameter DOP (degree of pyritization):

$$\text{DOP} = \frac{\text{pyrite-Fe}}{\text{pyrite-Fe} + \text{'reactive'-Fe}} \quad (17.5)$$

'Reactive'-Fe is defined here as Fe that is soluble in HCl (as noted above pyrite is not soluble in HCl), and is likely to include iron in phases whose reactivity toward sulfide ranges from hours to perhaps $\sim 10^4-10^5$ yr (Raiswell and Canfield, 1998; Goldhaber, 2003). Studies to date suggest that it is not unreasonable to assume that only the most highly reactive iron in sediments, mainly amorphous oxides (see section 7.3.3.1), is available for pyritization on early diagenetic time scales (e.g., Canfield et al., 1992). Because this highly reactive iron is a subset of the total reactive (or HCl soluble) iron, an upper limit for DOP is estimated to be ~ 0.5 (± 0.1) for normal marine sediments (Raiswell and Canfield, 1998).[2,3] Actual values of DOP in such "normal" marine sediments span roughly the same range (Berner, 1970; Canfield et al., 1992; Raiswell and Canfield, 1998), which suggests that at this upper limit, the availability of highly reactive iron could limit pyritization.

However, in general, reactive organic matter (rather than reactive iron) appears to limit pyritization. Furthermore, in some sediments the inputs of reactive iron and organic carbon may be directly linked, if

---

[2] Normal marine sediments are defined here as fine-grained, siliciclastic sediments underlying well-oxygenated seawater (>80 μM) of typical oceanic salinity. This definition excludes carbonate- and opal-rich sediments (because of their low detrital iron content), sediment underlying low bottom-water oxygen concentrations (i.e., dysaerobic settings with bottom-water [O_2] between ~ 5 and 80 μM), and anoxic/euxinic settings with bottom water [O_2] <5 μM and/or with dissolved sulfide present. In these latter two settings the occurrence of iron and sulfur redox chemistry in the water column adds an additional level of complexity to the occurrence of sedimentary pyrite formation.

[3] Note that Raiswell and Canfield (1992) refer to this upper limit as the potential DOP.

both are associated with the surfaces of sediment particles in a constant proportion (Berner, 1984; Hedges and Keil, 1995; also see discussions in Goldhaber, 2003). The association of these materials with sediment surfaces appears to be a reasonable assumption for both reactive iron oxides (e.g., see section 7.3.3) and organic matter (section 15.1).

The assumption that organic matter abundance is the primary control on sediment pyrite formation comes largely from the observation that there is a strong linear relationship between organic carbon and total sulfur burial in normal marine sediments, for sediment TOC levels of ~5-6% and total sulfur levels up to ~2% (Berner, 1982; Berner and Raiswell, 1983; Morse and Berner, 1995). In such sediments, the C/S ratio (wt/wt) below the zone of early diagenesis is 2.8 ± 0.8, implying that this ratio varies by only a factor of ~2. In contrast, this C/S ratio varies more widely in other depositional environments (e.g., normal marine versus nonmarine, freshwater or marine, euxcinic), suggesting the possible use of the C/S ratio as a paleoenvironmental indicator (see Berner and Raiswell, 1983, Morse and Berner, 1995, and discussion in note 2 in this section).

The factors controlling the C/S ratio in normal marine sediments appear to be: the fraction of total sediment organic matter that is metabolized, the fraction of metabolized sediment organic matter that is decomposed by sulfate reduction, and the fraction of the total sulfide produced by sulfate reduction that is buried as pyrite (Morse and Berner, 1995, and references therein). Morse and Berner (1995) have also developed a simple model for the C/S ratio as a function of these three parameters. An interesting outcome of this model is that the three parameters apparently scale with each other in a way that leads to a relatively constant C/S ratio, although the cause of this constancy is unclear.

This near-constant sediment C/S ratio across a broad range of marine sediments implies that sulfur burial is strongly tied to carbon burial, in terms of both the magnitude of burial and its distribution among marine sediments. In fact, global estimates of sulfur burial have been obtained by dividing the global carbon burial rate by this C/S ratio (Berner, 1982). Implicit in this approach is the assumption that organic sulfur burial is not significant. However, this may not necessarily be an appropriate assumption, and more data are needed to better quantify the role of organic sulfur burial in terms of overall sulfur removal from the oceans (e.g., Vairavamurthy et al., 1995; Werne et al., 2003).

17.1.2 Long-Term Changes in the Sedimentary Sulfur Cycle

The residence time of sulfate in seawater is $\sim 10^7$ yrs (Pilson, 1998). Because this value is long compared to the mixing time of the oceans, the concentration of seawater sulfate (and its sulfur isotope composition) can be considered constant over short geologic time periods. As discussed in the beginning of this section, the sulfate concentration in seawater is roughly controlled by sulfate input from rivers, which is itself derived largely from pyrite weathering on land, and pyrite burial in sediments. Therefore, isotope fractionation associated with bacterial sulfate reduction and pyrite oxidation (section 3.7) impacts the sulfur and oxygen isotopic composition of seawater sulfate, and an examination of the sulfur isotope age-curve in the evaporite sulfate mineral gypsum has the potential to provide information on the relative importance of these two processes over geologic time scales greater than ~ 10 million years (across major geological time periods; see discussions in Hoefs, 1997). For example, during periods when sulfate reduction predominates, the sulfur isotopes in seawater sulfate should become progressively heavier (i.e., enriched in $\delta^{34}S$). Conversely, during periods when the oxidation of isotopically light pyrite in terrestrial shale is of greater importance, seawater sulfate should become progressively lighter. Turchyn and Schrag (2004) also discuss the use of the oxygen isotope record of seawater sulfate (as recorded in marine barite) to further examine sulfur cycling on slightly shorter time scales, i.e., over the past 10 million years.

Long-term changes in the distribution of sulfur between its major oxidized and reduced reservoirs (i.e., sulfate and sulfide) also impact the distribution of carbon in its reduced and oxidized reservoirs (i.e., organic matter and CO_2 or carbonate/bicarbonate) and ultimately play a role in controlling atmospheric oxygen level on long time scales (>100 my; see Berner, 1999, for a summary). One equation commonly written to express the network of reactions responsible for long-term global redox cycling and balance is

$$4FeS_2 + CaCO_3 + 7CaMg(CO_3)_2 + 7SiO_2 + 15H_2O \rightleftharpoons$$
$$15CH_2O + 8CaSO_4 + 2Fe_2O_3 + 7MgSiO_3 \quad (17.6)$$

The apparent exclusion of atmospheric O_2 and CO_2 from this equation recognizes that these gases can be viewed as intermediates in this long-term redox cycling associated with rock weathering (for details see Berner, 1989; Hedges, 2002; and references therein).

Given the large sizes of the reservoirs of these redox-sensitive minerals in this cycle, any imbalances in this reaction have the potential to rapidly change the much smaller amounts of O_2 or CO_2 in the atmosphere (see discussions in section 15.4.2 and Hedges, 2002). Thus for O_2, for example, the negative feedback controls described by the mineral conveyor belt model in fig. 15.6 potentially play an important role in avoiding such large-scale fluctuations. The individual reactions coupling the reduced and oxidized forms of carbon and sulfur in eqn. (17.6) can also be thought of roughly in terms of the following oxygen producing and consuming reactions:

$$CO_2 + H_2O \rightleftharpoons CH_2O + O_2 \quad (17.7)$$

$$4FeS_2 + 15O_2 + 16CaCO_3 + 8H_2O \rightleftharpoons 2Fe_2O_3 + \\ 16Ca^{2+} + 16HCO_3^- + 8SO_4^{2-} \quad (17.8)$$

Thus any long term imbalance in the net flux between the oxidized and reduced carbon reservoirs may also be balanced by opposing fluxes between reduced and oxidized sulfur reservoirs, to similarly avoid large-scale fluctuations in atmospheric O_2 (Berner, 1999).

Linkages between the mineral conveyor belt model and oxygen controls described by eqns. (17.7) and (17.8) remain to be quantitatively explored. They are, however, not necessarily incompatible with each other, given some of the difficulties in past modeling efforts to generate reasonable simulations of atmospheric oxygen levels over the Phanerozoic (see most recently Paytan and Arrigo, 2000, as well as general discussions in section 15.4.1 and in Hedges, 2002).

17.2 Methanogenesis and Anaerobic Methane Oxidation

17.2.1 Shallow (Coastal) Sediments

In nearshore, coastal sediments, sulfate reduction occurs over the upper tens of centimeters to several meters of sediment. Depending on whether sulfate or metabolizable organic carbon is limiting (see section 7.3.5), sulfate concentrations either decrease to zero in an exponential-like fashion (see figs. 6.4, 8.5, or 16.2), or reach some asymptotic nonzero concentration at depth, in a manner analogous to that seen in fig. 7.8A for pore-water oxygen profiles in carbon-limiting deep-sea sediments. Upon complete sulfate depletion, methanogenesis occurs below the zone of sulfate reduction, consistent with the concepts of biogeochemical zonation discussed in chapter 7 (also see figs. 8.5 and 17.2). For reasons that will become apparent later, this type of methanogenesis will be referred to as "normal" methanogenesis.

Given the high concentration of sulfate in seawater, the occurrence of methanogenesis in shallow marine sediments requires a relatively large flux of reactive organic matter to the sediments (see sections 7.2 and 7.3.5). As a result, methanogenesis is generally responsible for a small amount of sediment carbon remineralization. Globally, organic carbon remineralization by methanogenesis in sediments is ~12% that of organic carbon remineralization by sulfate reduction (Henrichs and Reeburgh, 1987), although in some sediments methanogenesis can be more important. For example, in Cape Lookout Bight sediments, organic carbon remineralization by methanogenesis is ~43% that of remineralization by sulfate reduction[4] (Martens and Klump, 1984).

Another important aspect of the geochemistry of methanogenic sediments is the occurrence of anaerobic methane oxidation (AMO) in the transition zone between sulfate reduction and methanogenesis (e.g., Martens and Berner, 1977; Reeburgh, 1982; Alperin and Reeburgh, 1985; Iversen and Jørgensen, 1985). This reaction can be written as

$$CH_4 + SO_4^{2-} \rightarrow HCO_3^- + HS^- + H_2O \qquad (17.9)$$

although as discussed in section 7.3.5.1, evidence to date suggests that the process is not mediated solely by sulfate reducing bacteria. Regardless of the exact mechanism by which AMO occurs, it often leads to characteristic concave-up methane pore-water profiles (fig. 17.3). In marine sediments where methane gas ebullition (section 12.6) does not occur, anaerobic methane oxidation serves as a quantitative sink for virtually all of the methane produced in the sediments

---

[4] In these sediments, sulfate reduction accounts for ~70% of total sediment organic matter remineralization and methanogenesis accounts for ~30%. Hence, 0.3/0.7 = 0.43.

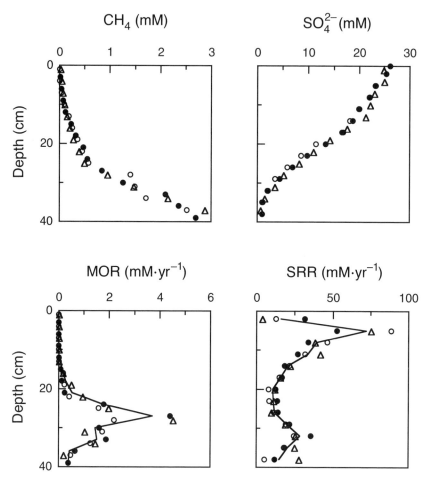

Figure 17.3 The depth distribution in Skan Bay sediments of: (A) methane concentration, (B) sulfate concentration, (C) methane oxidation rate, (D) sulfate reduction rate. The different symbols represent replicate subcores collected from the same box core on which concentration and rate measurements were both performed. The solid lines in panels C and D are curves through the mean values at a given depth (redrawn using data presented in Alperin and Reeburgh, 1985).

(Henrichs and Reeburgh, 1987). This process therefore significantly limits the global importance of the oceans as a source of methane to the atmosphere.

In and around the methane-sulfate boundary, i.e., the transition zone between the regions of sulfate reduction and methanogenesis, AMO can account for varying amounts of the observed sulfate reduction. In Skan Bay sediments, measured rates of sulfate reduction in the region of anaerobic methane oxidation are ~10 times larger than methane oxidation rates (fig. 17.3; Alperin and Reeburgh, 1985). At the other extreme, in Danish continental margin sediments AMO accounts for ~60–90% of the observed sulfate reduction in this transition zone (Iverson and Jørgensen, 1985). However, because the global rate of methanogenesis is only ~12% of the sulfate reduction rate (in similar carbon-equivalent units), this sets the upper limit for the overall contribution of AMO to sulfate reduction. In a given sediment, the relative contribution of anaerobic methane oxidation to sulfate reduction may be inversely related to the reactivity of the sediment organic matter (Devol et al., 1984; Blair and Aller, 1995), although this suggestion has not been well tested.

Under some circumstances, AMO can account for a significant amount of the observed sulfate reduction in a specific region of the sediments, i.e., the transition zone between sulfate reduction and methanogenesis. In general, though, anaerobic methane oxidation does not account for a major fraction of the total sulfate reduction that occurs in shallow sediments. Rather, most sulfate reduction in these surficial sediments is fueled by the remineralization of particulate sediment organic matter.

17.2.2 Continental Margin Sediments

As one moves offshore from coastal to continental margin sediments, the importance of sulfate reduction in sediment organic matter remineralization decreases (e.g., see section 7.5.2 and Fig. 7.7). As a result, pore-water sulfate gradients decrease, since the sediment uptake of sulfate is essentially equal to the depth-integrated rate of sulfate reduction, and methanogenesis occurs at increasingly deeper sediment depths, if indeed it occurs at all (Borowski et al., 1999). In an analysis of data from the first 182 legs of the Deep Sea Drilling Project and the Ocean Drilling Program, D'Hondt et al. (2002) observed that complete sulfate reduction does not occur along the ocean margins until, in general, a few tens of meters below the sediment surface, versus tens of centimeters or a few meters in coastal sediments. In pelagic sediments, sulfate concentrations remain high down to sediment depths of several hundred meters. Methanogensis is therefore significantly inhibited in these sediments and, as a result, methane concentrations are generally at background levels.

The patterns and occurrence of sulfate reduction and anaerobic methane oxidation (AMO) also differ in continental margin versus coastal sediments. In many outer continental margin sediments sulfate reduction is largely coupled to anaerobic methane oxidation, rather than to the remineralization of particulate sediment organic matter, as is the far more common mode of sulfate reduction in coastal sediments. At the same time, in some continental margin settings, AMO takes on a greater importance in sediment remineralization processes as a result of nonsteady-state conditions in the sediments that alter the occurrence of normal methanogenesis and its linkage to sulfate reduction and anaerobic methane oxidation (Blair and Aller, 1995; Burns, 1998).

A common observation in many continental margin sediments are linear sulfate profiles such as those shown in fig. 17.4 (Niewöhner et al., 1998; Borowski et al., 1999; Fossing et al., 2000; Dickens, 2001; Hensen et al., 2003). Such profiles occur on depth scales of ~4 m to >20 m, i.e., the depth of the sulfate-methane boundary or z_{sm} in fig. 17.4. These linear profiles suggest that sulfate diffuses through the upper sediments with little or no apparent sulfate reduction until the methane-sulfate boundary is reached, where sulfate is quantitatively consumed in the oxidation of methane. Consistent with this suggestion, diffusive flux calculations in some settings, i.e., where high resolution methane data can be used to accurately estimate the upward methane gradient, indicate that the downward sulfate flux at z_{sm} balances the upward methane flux, as is required by rxn. (16.9) (Niewöhner et al., 1998; Fossing et al., 2000).

The methane found at depth in these sediments is of biogenic origin, based on its $\delta^{13}C$ and δD content (Borowski et al., 1997; Hoehler et al., 2000). However, this methane does not appear to be produced by normal methanogenesis, since the sediment organic matter being buried in these sediments appears to be degraded not by sulfate-reducing bacteria, but solely by methanogens. At first glance, such an explanation clearly seems at odds with the concepts of biogeochemical zonation discussed earlier (e.g., section 7.2).

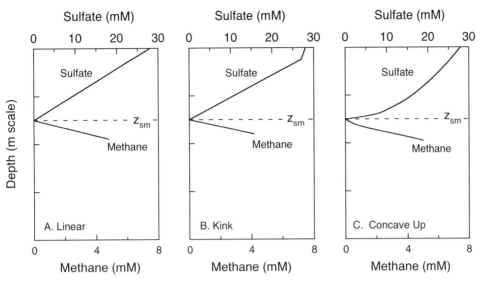

Figure 17.4 Typical sulfate and methane profiles in continental margin sediments containing a deep zone of anaerobic methane oxidation at depth z_{sm}. Linear profiles (A) apparently represent a steady-state situation, while kinked profiles (B) may be nonsteady-state profiles resulting from sediment slides that overthrust sediments downslope, assuming that the original pore-water signals are preserved in the upper sediments (see Hensen et al., 2003, for further details). Kinked profiles may also be steady-state profiles that occur when there are significant amounts of sulfate reduction above depth z_{sm} (as compared to that which occurs at z_{sm}) as well as deep bioirrigation that resupplies the deep sediments with sulfate down to approximately the depth of the slope break in the sulfate profile (Fossing et al., 2000). Concave-up profiles (C) may also be nonsteady-state profiles that result from an increase with time in the upward flux of methane in the sediment. Calculations presented in Hensen et al. (2003) for continental margin sediments in the Argentine Basin suggest that such profiles can develop in $\sim 10^2 - 10^3$ yr after reasonable changes in the methane flux, and can result in an upward shift of z_{sm} on the order of 5 m over this time period. In spite of the curvature of these profiles, these are still conservative (but by nonsteady-state) profiles in the sense that there is no sulfate reduction in the sediments except by anaerobic methane oxidation at depth z_{sm}.

A possible explanation for these observations is shown in fig. 17.5. This model is based on results from the Blake Ridge on the southeast US continental margin (water depths ~2,800 m), although analogous processes are likely to occur at other continental margin sites where linear sulfate gradients are observed. A key aspect of this model is the increase in sediment temperature during deep burial, i.e., from deep-sea values of ~2–3°C to ~20–30°C at sediment depths of several hundred meters. This temperature increase leads to a consequent increase in the rate of organic matter degradation at depth. The reason(s) this occurs is not well understood, although it seems possible that the temperature increase with sediment burial decreases the activation energy of a key step in the overall remineralization process (see section 7.6.2). This then allows for the decomposition of organic matter that would otherwise be refractory toward decomposition at the lower temperatures found in shallower sediments.

All of these changes occur in concert with a general increase in microbial activity at depth, in what is often referred to as the "deep marine biosphere." As a result, pore-water acetate concentrations increase with sediment depth below ~300 m on the Blake Ridge, with concentrations as high as 15 mM at 700 m (Wellsbury et al., 1997). Methanogenesis from both acetate disproportionation (rxn. 7.8) and CO_2 reduction (rxn. 7.9) is also stimulated at these depths; this then produces the methane that diffuses upward to be oxidized by sulfate reduction much higher in the sediment column (on the Blake Ridge at sediment depths of ~20 m; Dickens, 2001).

The occurrence of this methane production at great depth also leads to the formation of methane gas hydrates; the methane produced at depth must therefore pass through the gas hydrate zone to be oxidized (Dickens, 2001). Gas hydrates, or methane clathrates, are solid, crystalline substances containing water and gas, mainly methane (Kvenvolden, 1993). They are stable at high pressure, low temperature, and high gas concentration. Because of this, increasing temperature and pressure with increasing sediment depth defines a fixed depth interval in sediments where gas hydrates are stable[5] (e.g., Rempel and Buffett, 1997; Xu and Ruppel, 1999). Marine gas hydrates

---

[5] The lower boundary of marine gas hydrates is generally defined by the appearance of an anomalous bottom simulating reflector (BSR) in marine seismic reflection profiles (Kvenvolden, 1993; Borowski et al., 1999). This inference is based on the fact that the depth of this reflection coincides with the depth predicted from thermodynamic phase diagrams as the base of the gas hydrate stability zone.

METHANOGENESIS AND ANAEROBIC METHANE OXIDATION

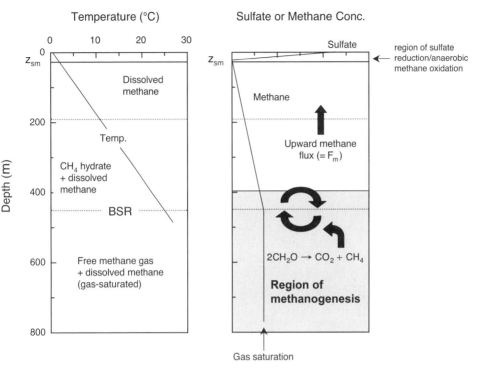

Figure 17.5 A schematic representation of the processes affecting methane in deeply buried marine sediments (based on results presented in Dickens, 2001, and Wellsbury et al., 1997, from the Blake Ridge off the southeast coast of the US; also see a related conceptual model in Xu and Ruppell, 1999). (Left) Temperature gradient and key regions in these sediments, including: the surface zone (0 to ~20 m [= z_{sm}]) where sulfate gradients are linear; the region between 190 m and the bottom simulating reflector (BSR) at 450 m, where gas hydrates are stable; the region below the BSR in which free methane gas (bubbles) and methane-saturated pore waters occur. (Right) Sulfate and methane profiles in these sediments, including the region below ~400 m where microbial activity is stimulated and methanogenesis occurs (see the text for further details). Also indicated is the methane cycling that occurs around the BSR as a result of changes in hydrate stability with depth (primarily associated with increasing temperature during burial). Finally, under steady-state conditions deep methane production and its upward flux into the gas hydrate region should be balanced by the methane flux out of the upper portion of the gas hydrate (= F_m). This methane then continues to diffuse upward to be oxidized at depth z_{sm} by downwardly diffusing sulfate.

appear to be common along the outer continental margin; consistent with this model, the methane in these hydrates appears to be of biogenic origin. Gas hydrates are also found on land in high-latitude permafrost regions.

With increased burial, gas hydrate–containing sediments are brought to sediment depths below the bottom simulating reflector (BSR), where the gas hydrate is no longer stable. The hydrate then dissociates to form gas-saturated pore waters and free gas (Dickens, 2001), which can then diffuse upward to be reincorporated into gas hydrate; the result is methane cycling around the BSR (Egeberg and Dickens, 1999) that has some similarity to the Mn redox cycling that occurs around the sediment redox boundary in much shallower sediments (see section 13.2 and figs. 13.2–13.3). At the same time, if gas hydrate formation is at steady state, then deep methane production and the resulting upward flux of methane into the gas hydrate should be balanced by a flux of dissolved methane out of the upper portion of the gas hydrate. This soluble methane at the top of the hydrate diffuses upward to be oxidized by downwardly diffusing sulfate at the methane-sulfate boundary z_{sm}.

Taken together, these results suggest a strong causative link between the deep marine biosphere, the presence of methane gas hydrates, upward methane fluxes, and linear sulfate gradients. Furthermore, it has been suggested (Borowski et al., 1999) that the shape and magnitude of these sulfate gradients may be useful in the initial identification of potential settings for underlying gas hydrates. For example, under steady-state conditions the stoichiometry of rxn. (16.9) predicts that the downward flux of sulfate should be related to the upward methane flux, and therefore that z_{sm} should be inversely proportional to F_m, the methane flux from the top of the gas hydrate (see Dickens, 2001, and references therein).[6]

At the same time, many of the settings where gas hydrates exist appear to be in nonsteady state on time scales of tens to hundreds of thousands of years (Dickens, 2001; Hensen et al., 2003). This has the potential to complicate the interpretation of pore-water sulfate gradients (Dickens, 2001), and, for example, lead to situations in which

---

[6] To see this, note that we can express the downward sulfate flux in a profile such as that in fig. 16.4A as $D_s C_o / z_{sm}$. This flux also equals F_m in Fig. 16.5, if AMO is the predominant process by which sulfate reduction occurs. Since C_o is fixed in this flux equation (and assuming a constant D_s), we see the inverse relationship between F_m and z_{sm}. Therefore if, for example, we double F_m, then z_{sm} must be halved.

there is curvature in conservative, but time-dependent, sulfate profiles (Hensen et al., 2003; also see fig. 17.4C). Another complication in the interpretation of these profiles occurs in some sediments that show kinked sulfate profiles (fig. 17.4). Here deep irrigation of sediments (down to ~2–3 m) may occur in conjunction with sulfate reduction driven by the oxidation of particulate sediment organic matter in the sediment column above z_{sm} (Fossing et al., 2000). While this does not imply that sulfate reduction coupled to AMO at z_{sm} is not significant in these sediments, it does imply that the depth of z_{sm} may be controlled by more than just the upward methane flux.

Interest in such linear sulfate profiles and their relationship to underlying methane gas hydrates stems from three factors. The first is that the amount of methane in gas hydrates appears to be quite large, perhaps twice that present in all known fossil fuel deposits (Kvenvolden, 1993). Although this number is highly uncertain (also see recent estimates in Kvenvolden, 1999), the potential size of these gas hydrate deposits has made them highly attractive (at least to some) as a potential energy resource, technological obstacles of recovery notwithstanding.

Second, under some circumstances marine gas hydrates can dissociate and be destabilized in ways that lead to sediment failure, i.e., submarine landslides and slumps. In his recent examination of marine gas hydrates, Kvenvolden (1999) suggests that gas hydrate failure as a geohazard may have the greatest impact on human welfare as the peoples of the Earth move to exploit the oceans at ever increasing depths.

Finally, methane release from destabilized gas hydrates is of potential concern also because methane is a greenhouse gas, and the amount of methane trapped in gas hydrates, both on land and in marine sediments, could be ~3,000 times that in the atmosphere. Thus, the release of large amounts of methane from marine sediment gas hydrates into the water column and possibly the atmosphere could have a large impact on the radiative properties of the atmosphere and therefore on global climate.

One way that marine gas hydrates may become destabilized is through increased bottom-water temperatures and/or decreased hydrostatic pressure, i.e., falling sea level. Several workers have therefore suggested that Pleistocene climate cycles could have led to significant amounts of methane being released into the atmosphere, when associated environmental changes destabilized marine gas hydrates (see Kennett et al., 2000, and references therein). In contrast, Kvenvolden (1999) suggests that in many of these cases, little methane may actually be released to the atmosphere as a result of marine gas hydrate destabilization, thus possibly minimizing this link between gas hydrate destabilization and past climate change.

Moving to the present and future, rising levels of atmospheric CO_2 and possible global warming may similarly affect the stability of marine, and continental, gas hydrates. However, many aspects of this still remain highly uncertain. This includes not only the potential magnitude of the effect of this methane release on global or regional climate (if indeed it occurs), but also whether this effect (if significant) will act as a strong or weak positive feedback to global warming (e.g., see Kvenvolden 1999; Buffett and Archer, 2004; and references therein).

17.3 Trace Metal Cycling

In deep-sea sediments trace metal diagenesis is strongly affected by manganese and iron redox cycling (see section 13.2). However, in addition to metal oxides, sulfide minerals (metal sulfides in general and iron monosulfides and pyrite in particular) are also important carrier phases for many trace metals. Therefore, as one moves into continental margin and coastal sediments, the increasing importance of sulfate reduction and sulfide production implies that uptake of metals into AVS (acid volatile sulfides; i.e., iron monosulfides) and pyrite can play an important role in the sediment cycling of trace metals. In addition, trace metal cycling in coastal and continental margin sediments can be affected by nonsteady-state conditions in these sediments. This includes not only regular seasonal changes such as those discussed in section 14.3 (also see discussions below and at the end of section 14.5.3) but also episodic events associated with, for example, storm or flooding events (see note 3 in section 14.5).

Examining the uptake of an operationally defined "reactive" metal fraction into pyrite has been done by examining what is termed the degree of trace metal pyritization, or DTMP (Huerta-Diaz and Morse, 1992), a parameter defined in the same manner as the degree of pyritization is for iron (eqn. 17.5). Comparisons of DTMP and DOP in a large number of sediments has proven useful for understanding the relative degree to which various metals are incorporated into pyrite

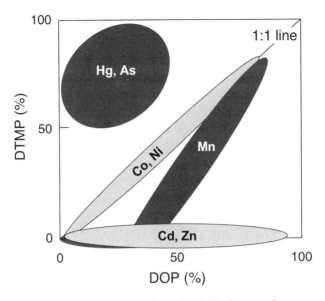

Figure 17.6 A schematic representation of DTMP (degree of trace metal pyritization) versus DOP (degree of pyritization) for selected trace metal (see the text for further details). Based on results presented in Huerta-Diaz and Morse (1992) and Morse and Luther (1999).

relative to the formation of pyrite itself (fig. 17.6). Such studies have shown the following general pattern of DTMP values for these metals:

$$Hg > As = Mo > Cu = Fe > Co > Ni$$
$$\gg Mn > Zn > Cr = Pb > Cd \qquad (17.10)$$

While sulfidic sediments can be an important sink for all these trace metals (see discussion below), an examination of DTMP values determines whether this occurs in metal (Me) monosulfide phases (MeS) versus pyrite phases, or phases that are operationally extracted with pyrite. This has some impact on the stability of these metals in anoxic or mixed redox sediments (see discussions at the end of this section for further details).

Morse and Luther (1999) have examined the thermodynamic and kinetic controls on the observed trends in eqn. (17.10). In their approach they use the rate of water exchange of the soluble metal, i.e., the extent to which water molecules in an aqueous complex $Me(H_2O)_6^{2+}$ exchange with the surrounding water molecules in solution, as a first-order proxy for the relative reactivity of a metal to form a MeS phase. Using this approach, they show that elements such as Zn, Cd, and Pb have faster water kinetics than Fe^{2+} and therefore form extremely insoluble MeS phases prior to the formation of FeS. Since FeS is generally the precursor to pyrite formation (see section 17.1) this leads to these elements having low DTMP values even at high DOP values (fig.17.6). Elements such Co and Ni have slower water kinetics than Fe^{2+} and therefore tend to coprecipitate with or adsorb to iron sulfides; consequently, they show a regular increase in DTMP with DOP. For Mn, water exchange kinetics predict that Mn should behave like Zn or Cd in terms of uptake into pyrite. At low DOP values this appears to be the case, although this may also be more of a function of Mn uptake into carbonate mineral phases. At high DOP values, Mn does appear to be incorporated into pyrite and this may be a function of Mn^{2+} adsorption to FeS and its subsequent pyritization (Morse and Luther, 1999).

This general approach does not, however, explain all of the observations in eqn. (17.10) and in fig. 17.6. For elements such as As and Mo, which show high DTMP values that are almost independent of DOP, they are first reduced by sulfide, i.e., As^{5+} to As^{3+} and Mo^{6+} to Mo^{4+}, before forming sulfide phases. They then undergo a more complex set of redox reactions than the simple substitution reactions discussed above, and these reactions appear to strongly favor the coprecipitation of these elements with pyrite. High DTMP values for Hg may be a function of several poorly understood factors, including more complicated Hg solution chemistry that should favor Hg incorporation into pyrite. At the same time, because HgS is extremely insoluble in HCl (Cooper and Morse, 1998b), it will also operationally be extracted with the pyrite fraction.

A conceptual model of trace metal cycling in continental margin sediments is shown in fig. 17.7. The figure is presented as a generic model for trace metal, and trace element, cycling; thus, different aspects of the processes shown here may be more or less important for specific trace metals. The redox cycling of Mn and Fe oxide and Fe sulfide phases plays an important role in controlling not only trace metal geochemistry in coastal and continental margin sediments, but also their benthic flux from the sediments. Furthermore, in many cases the reactions of interest are actually redox reactions directly between Mn and Fe oxides and iron sulfides, which can be mediated by macrofaunal activity (see sections 7.4.2 and 16.1.1, and discussions below).

TRACE METAL CYCLING

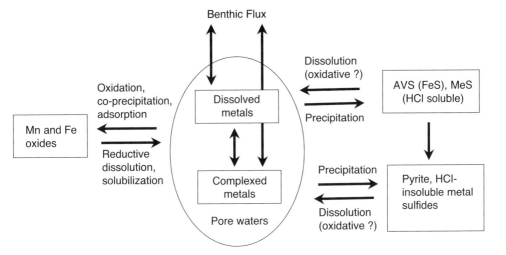

Figure 17.7 A conceptual model for trace metal cycling in coastal and nearshore sediments, presented in terms of the major sediment carrier phases of these metals (e.g., oxides and sulfides). Although not explicitly shown here, some of the relevant redox reactions these sulfides and oxides undergo involve both as reactants (see rxn. 16.14 or rxn. 7 in table 7.2).

Since most of the reactions in this figure are dissolution/reprecipitation processes, the cycling processes generally go through a dissolved phase. Furthermore, as trace metals move from oxide phases to sulfide phases, and visa versa, during sediment redox cycling, release from one phase may sometimes be faster than uptake by the other phase. This imbalance can, under these conditions, allow metals to accumulate in sediment pore waters. The depth zonation of such processes may then play an important role in regulating the magnitude and direction of benthic fluxes of trace metals from sediments. Another important point to consider is that organic complexation of these metals also likely occurs in the pore waters (Elderfield, 1981; Skrabal et al., 2000). This complexation might be of greatest importance in anoxic sediments where sulfate reduction occurs because of the high DOC levels generally seen in such sediments (see chapter 10 and Burdige, 2002). The potential importance of this metal complexation in pore waters has been discussed in terms of its impact on benthic fluxes (also see Skrabal et al., 1997), although it also appears to play a role in the solubilization and precipitation of metals in sediments (Luther et al., 1992).

In the situation outlined above, the release of dissolved metals from sediments is generally observed, and this can play an important role in affecting dissolved metal concentrations in shallow-water coastal ecosystems (e.g., Warnken et al., 2001, and references therein). Bottom-water oxygen concentration, carbon rain rate, and benthic macrofaunal activity all play important roles in regulating these benthic trace metal fluxes, in part because of their impact on iron, manganese, and sulfur redox cycling in sediments (Sundby et al., 1986; Shaw et al., 1990; Aller, 1994b; Riedel et al., 1997; and references discussed below). At the high anthropogenic metal levels that can often be found in some nearshore sediments, benthic metal fluxes may consequently play some role in controlling the toxicity and bioavailability of these metals to water column organisms (see Riedel et al., 1997; Shine et al., 1998; and references therein). This observation is in contrast to that observed in open ocean settings, where the availability of certain trace metals can limit phytoplankton growth (e.g., Donat and Bruland, 1995).

The relative balance of the diagentic processes discussed above, and also shown in fig. 17.7, can also, under some circumstances, lead to sediments taking up certain metals from the water column. For example, Colbert et al. (2001) demonstrated with benthic flux measurements that coastal and continental margin sediments go from being a source of Cd to the water column to being a Cd sink with increasing rates of sediment sulfate reduction. The sediment uptake of Cd at high sulfate reduction rates appears to be related to the precipitation of CdS in the sediments. At low rates of sulfate reduction, i.e., sulfide production, other oxidative processes (discussed below) apparently contribute to the net release of dissolved Cd from the sediments. Earlier studies by Elderfield et al. (1981a,b) similarly suggested that anoxic sediments could be sinks for both Cd and Ni as a result of formation of highly insoluble sulfides, based here on both pore-water profiles and measured benthic fluxes.

Except for sediment underlying dysoxic, or anoxic, bottom waters, metals (including iron and manganese) deposited in continental margin sediments with organic matter carrier phases will be regenerated by aerobic respiration quite close to the sediment-water interface because of the shallow sediment oxygen penetration depths in many of these sediments. In the absence of bioturbation or bioirrigation this may lead to a benthic flux of these metals and/or uptake into Mn and Fe oxide phases (see analogous discussions in section 13.2).

In some suboxic sediments, these manganese and iron oxides and their associated trace metals will undergo further diagenetic redox cycling, as is also discussed in section 13.2 and illustrated in fig. 13.2. The relationship between the time scale of transport though surface sediments, versus the time scale of either oxidative removal (for Mn and Fe) or uptake onto freshly formed oxides (for the other trace metals) will then determine the extent to which metals regenerated in the sediments by suboxic processes are retained in the sediments, or released as a benthic flux. Both bottom-water oxygen concentrations and carbon rain rate to the sediments play important roles here, in part because of their impact on the sediment oxygen penetration depth (also see section 16.3 and the discussion of the relationship between sediment iron redox cycling and sediment phosphorus cycling). Because Fe oxidation kinetics are much faster than those of Mn (section 7.3.3.1) this implies that such processes are more likely to lead to Mn versus Fe benthic fluxes (Aller, 1980b; Balzer, 1982).

In sediments where organic matter remineralization ceases with suboxic processes, i.e., there is no sulfate reduction, trace metal cycling is largely dominated by processes associated with Mn and Fe redox cycling (as is discussed in section 13.2). However, because sulfate reduction is an important process in most continental margin and coastal sediments, suboxic remobilization of metals also leads to the *downward* diffusion of regenerated metals into the deeper, anoxic sediments. Here uptake of some of these metals into sulfide phases (as outlined above) will be important. In the absence of bioturbation these sulfides will not be reoxidized and will simply be buried in the sediments.

In contrast, the occurrence of bioturbation and/or bioirrigation and the resulting mixed redox conditions have a more complex effect on sediment trace metal cycling and benthic fluxes. Metals that are sequestered as sulfides can be mixed upward into surficial portions of the sediments where they can be reoxidized by a number of oxidants such as dissolved O_2 and nitrate, and solid-phase manganese and iron oxides. This sulfide oxidation liberates metals into the pore waters, so their potential flux out of the sediments may be enhanced. The diffusive openness of mixed redox bioirrigated sediments (section 12.5.1) also provides an additional means of transporting dissolved oxidants into the sediments and, as a result, allows metals that are remobilized by oxidative processes at depth a greater likelihood of escaping the sediments as a benthic flux (e.g., Emerson et al., 1984).

Consistent with these suggestions, in coastal sediments the magnitude of benthic fluxes of metals such as Cd, Cu, Zn, and Ni (and in some cases also the direction of this flux) is strongly affected by the occurrence of these oxidative processes (Westerlund et al., 1986; Riedel et al., 1997; Skrabal et al., 1997; Shine et al., 1998; Colbert et al., 2001). Of equal importance is also the fact that metal sulfide production at depth does not necessarily imply permanent burial of metals.

Metals that are remobilized and solubilized at depth in association with the reduction of manganese and iron oxides show somewhat differing behaviors under mixed redox conditions. For Mn, for example, it has been observed that while mixed redox conditions promote greater internal redox cycling of manganese, benthic Mn fluxes do not appear to be greatly enhanced (Sundby et al., 1986; Aller, 1990, 1994b; Aller and Aller, 1998; also see related discussions in section 7.4.3).

Both experimental and field studies suggest that low dissolved oxygen in bottom waters and/or increased carbon fluxes to the sediments will enhance benthic Mn fluxes along with the fluxes of other metals like, for example, Co, associated with both Mn and Fe oxides (Sundby et al., 1986; Aller, 1994b; Riedel et al., 1997). Seasonal studies of benthic Mn fluxes also show that these fluxes increase with increasing temperature (Aller and Benninger, 1981; Elderfield et al., 1981a; Hunt, 1983). However, identifying the exact causes of such seasonal trends is usually hampered by the fact that temperature, bottom-water oxygen, changes in carbon input, and sediment oxygen penetration depth often covary, although not always by direct causality (see discussion in Aller, 1994b). In a related fashion, studies of iron fluxes from California continental margin sediments suggest that these fluxes are significant only for sediments with shallow oxygen penetration depths, i.e., <20 mm, and/or those underlying low bottom-water oxygen concentrations of <20 μM (McManus et al., 1997).

The discussion above illustrates several ways that benthic macrofauna affect sediment geochemical processes and impact sediment metal cycling. Deposit-feeding organisms are also capable of solubilizing metals such as Cu, Cd, and Pb from contaminated sediments during their passage through macrofaunal digestive systems (Mayer et al., 1996; Chen and Mayer, 1998). This process appears to be mediated by soluble proteins in the animal's digestive fluids, and thus a large fraction of these solubilized metals are also complexed. This process not only increases the bioavailability of the metals to the deposit feeder organism itself, and can therefore lead to metal accumulation in the organism, but may also lead to the release of these metals into the pore waters.

Seasonal variations of redox conditions in coastal sediments affect the sediment cycling and benthic fluxes of many other trace metals in addition to Mn (e.g., Hines et al., 1984; Aller, 1994b; Shine et al., 1998). The results of Cooper and Morse (1998a) suggest further that for metals that form highly insoluble sulfides, e.g., Cu and Zn in their study, both the amounts and seasonal changes in the sediment AVS content play an important role in regulating seasonal sediment metal cycling. Based on their comparative studies of two sites in Chesapeake Bay, they suggest that if sediment Fe in AVS is <~20% of HCl-soluble Fe, then seasonal changes in sediment AVS content can affect the sediment cycling of these other metals. In contrast, when Fe in AVS is >~20% of this reactive Fe, then there appears to be enough sulfide available in the sediments to effectively trap these metals, thereby buffering temporal changes in metal dynamics.

Such factors may also play a role in explaining metal toxicity among organisms dwelling in coastal sediments. For metals that show a low degree of trace metal pyritization and form highly insoluble monosulfides (e.g., Cd, Ni, Zn, Pb, and Cu) the molar ratio of the acid (HCl)-extractable metal to acid-volatile sulfide is strongly correlated with sediment metal toxicity (see DiToro, 2001, and references therein). When this ratio is <1 the excess AVS in the sediment apparently titrates all of the available metal, such that no free metal is available to the sediment organisms. In contrast, as this ratio increases, sulfide availability limits metal uptake as a sulfide precipitate and thus makes metals more available to organisms.

Although pyrite is the ultimate burial phase for most sulfur in sediment, the low sulfur burial efficiency discussed in section 16.1.1 implies that both pyrite and AVS undergo extensive cycling in surface sediment. While AVS appears to be more reactive than pyrite (Morse, 1991), pyrite oxidation does occur on early diagenetic time scales. As has been discussed previously this process occurs both chemically and biologically, using O_2 and other oxidants such as nitrate and manganese oxides (Luther and Church, 1992; Luther et al., 1992; Schippers and Jørgensen, 2002). Laboratory studies by Morse (1994) have also shown that pyrite oxidation releases significant amounts of metals such as As, Cu, Hg, and Mo, all of which generally show high values of DTMP. More importantly, the percentage loss of these metals from the pyrite fraction is similar to or greater than the percentage loss of Fe during pyrite oxidation. Thus, pyrite oxidation, driven by sediment resuspension or in situ sediment redox processes, also has the potential to increase the bioavailability of pyritized metals.

≈ **CHAPTER EIGHTEEN** ≈

Linking Sediment Processes to Global Elemental Cycles: Authigenic Clay Mineral Formation and Reverse Weathering

MARINE SEDIMENTS can be thought of as representing the link between surficial processes that generally operate on relatively short time scales and longer-term geological cycles. This linkage was discussed in chapter 15 in terms of the controls on organic carbon burial in marine sediments. In this last chapter these linkages will be discussed in terms of the occurrence of authigenic clay mineral formation, i.e., reverse weathering, in marine sediments, and the role that this process may play in controlling the chemical composition of seawater.

The possible low-temperature formation of authigenic clay minerals was first proposed by Sillén (1967) as the primary removal mechanism that balances the riverine input of major ions and bicarbonate to the oceans. Because the ultimate source of these ions is weathering reactions such as rxn. (2.1), this process has come to be known as *reverse weathering*. In a general sense this process can be written as,

$$\begin{array}{l}\text{clay minerals (i.e., cation-}\\ \text{"poor" degraded Al-silicates)}\\ + SiO_2(aq) + \text{dissolved}\\ \text{cations} + HCO_3^-\end{array} \rightarrow \begin{array}{l}\text{authigenic clays (i.e., cation-}\\ \text{"rich" Al-silicates)} + CO_2 + H_2O\end{array} \quad (18.1)$$

By acting as a sink for major elements brought to the oceans by rivers, reverse weathering would achieve chemical mass balance for these elements in the oceans. Furthermore, by consuming bicarbonate, i.e., producing protons, the reaction also returns CO_2 consumed by weathering to the atmosphere (see section 2.1). Taken together, these two aspects of reverse weathering would prevent the oceans from having the chemical composition of an alkali soda lake (also see discussions in Mackenzie and Garrel, 1966).

The occurrence of these reactions and their impact on elemental cycles has remained controversial since their introduction in the 1960s (see, e.g., Mackenzie et al., 1981; Mackin and Aller, 1989). In part this has occurred because studies beginning in the 1970s (including those discussed in section 3.3.2) that looked for evidence of the occurrence of these authigenic clays often yielded inconclusive results (also see Mackenzie and Kump, 1995, and Michalopoulos and Aller, 2004, for this historical perspective). During this time there were also other studies, in some cases not fully appreciated, that were consistent with the occurrence of low-temperature authigenic clay mineral formation (see discussions below). In this context, and viewed more as an exception than as a general occurrence, were studies of some Pacific sediments where authigenic Fe-rich smectites (nontronite) were shown to form through the reaction between hydrothermally derived iron oxides (see section 13.2) and biogenic silica (Heath and Dymond, 1977; Hein et al., 1979; Cole, 1985).

As interest in the study and importance of authigenic clay mineral formation waned in the 1970s, early studies of the chemistry of hydrothermal vents at mid-ocean ridge spreading centers suggested that processes occurring there might be responsible for the required chemical mass balance attributed to reactions such as rxn. (18.1) (Edmond et al., 1979a,b). However, subsequent work (see Elderfield and Schultz, 1996, for a summary) has shown that hydrothermal vent systems are more complex than previously thought; as a result their impact on chemical mass balance in the oceans is now much more uncertain.

In light of these observations, a re-examination of early studies of authigenic clay mineral formation suggests several factors that could have precluded the observation of a definitive reverse weathering "signal." The first is that the calculations of Mackenzie and Garrels (1966) indicate that only ~7% of the mass of sediments accumulating in the marine environment need to undergo reverse weathering to achieve mass balance between rivers and the oceans. Thus, many early studies may have been unable to resolve such small changes in the composition of marine sediments. In deep-sea sediments, this is likely to have been exacerbated by the slow rates of these presumed reactions, and the fact that only a small fraction (~20% or less) of riverine suspended matter, i.e., one of the proposed reactants in the reverse weathering process, is deposited in deep-sea sediments (Berner and Berner, 1996; McKee et al., 2004).

These arguments suggest that the formation of authigenic clay minerals is likely to be difficult to detect with measurements of the sediments themselves. However, pore-water profiles of major solutes may be more sensitive indicators of the occurrence of these reactions, and this can be seen, for example, in major ion pore-water studies of Sayles (1979, 1981); these results will be discussed below in greater detail (also see general discussions in chapter 1 about the sensitivity of pore-water measurements in elucidating the occurrence of sediment geochemical processes).

In more recent years, there have been also two broad sets of results that have led to a significant re-evaluation of past assumptions regarding the significance of authigenic clay mineral formation in marine sediments on rapid, i.e., early diagenetic, time scales. The first of these was discussed in section 13.4, where it was suggested that pore-water silica profiles, and perhaps silica preservation, in deep-sea marine sediments are controlled by the formation of authigenic aluminosilicate phases that coat the dissolving biogenic silica particles. As was also noted here, the availability of dissolved Al played a key role in the occurrence of these processes.

At the same time, similar relationships between sediment silica and Al cycling and authigenic mineral formation have been observed in rapidly accumulating deltaic and organic-rich estuarine sediments (Mackin, 1987; Mackin and Aller, 1989; and references therein). These studies were then followed up by more detailed studies of authigenic mineral formation in Amazon delta sediments (Rude and Aller, 1989, 1994; Michalopoulos and Aller, 1995, 2004; Michalopoulos et al., 2000). This latter work has involved pore-water and solid phase sediment studies (including wet chemical analyses, X-ray diffraction studies, and scanning and transmission electron microscopy), sediment incubation studies (to quantify reaction rates of these processes), and laboratory experiments (in which natural substrates such as iron oxides or glass beads are incubated in natural sediments to more directly examine the formation of new clay minerals).

Based on the results of these studies authigenic clay mineral formation is proposed to occur somewhat differently than that described earlier (e.g., rxn. 18.1), and this process appears to be intimately associated with sediment iron redox cycling and organic matter remineralization. A general form of this reaction can be written as,

$$\begin{array}{c}\text{cation-poor clay minerals} \\ \text{(i.e., degraded Al-silicates)} + SiO_2 + \\ \text{iron oxides} + \text{organic carbon} + \\ \text{dissolved cations}\end{array} \rightarrow \begin{array}{c}\text{authigenic clays} + \\ CO_2 + H_2O\end{array} \quad (18.2)$$

(also see Michalopoulos and Aller, 1995, for more specific examples of this proposed reaction). In actuality, this net reaction is likely a combined dissolution-reprecipitation process, in which the solid reactants on the left side are first solubilized and then recombine as new clay minerals (see the references cited above as well as Mackenzie et al., 1981; Aplin, 1993). Also note that the reprecipitation half of this process is often referred to as *clay mineral neoformation.*

An examination of this reaction suggests that the formation of iron-rich authigenic clays involves the availability of three reactants. The first is reactive silica, which in many cases comes from biogenic silica deposited in sediments. This has important implications for understanding the controls on silica burial in sediments, as has been discussed in section 13.4 and will be discussed below in further detail. The source of iron in these clays is likely reactive iron oxides, requiring iron redox cycling to remobilize the iron and make it available for uptake in new clays.

Finally, this reaction requires a source of reactive Al. In rxn. (18.2) this is presumed to come from the dissolution of highly weathered aluminosilicates. However, in settings such as the Amazon delta sediments the dissolution of amorphous Al oxides (gibbsite; see discussions below), or the release of adsorbed Al during iron reduction are likely to be equally or even more important. In contrast, though, to silica and iron diagenesis in sediments, Al remobilization generally does not lead to the build-up of high levels of Al in most sediment pore waters. Representative profiles in Mackin and Aller (1989) indicate that pore-water Al concentrations are $<1\,\mu M$ and either show minimal depth variation across the sediment-water interface, or actually decrease with sediment depth. The apparent immobility of Al that might be inferred from such profiles simply requires a tight coupling between Al remobilization (dissolution) and its uptake in clay mineral neoformation.

Studies in Amazon sediments suggest that a variety of iron-rich authigenic silicates form there, including 7Å Fe-Mg rich clays, 10Å Fe-K rich clays, and 7Å Fe-rich K-poor clays (Rude and Aller, 1989; Michalopoulos and Aller, 1995, 2004).[1] The factors that control the formation of specific clays appear to depend in part on the type of reactants and sites of reaction, i.e., biogenic silica or iron oxide coated

---

[1] See the caption to fig. 2.2 for a discussion of this terminology for clay identification, and the relationship between these basal spacings and specific clay mineral structures.

quartz grains. Analogous Fe-rich authigenic clays (including clays such as glauconite or odinite) are also proposed to form in other nearshore (Mackenzie et al., 1981) and continental margin (Odin, 1988; Aplin, 1993) settings. However, clays in these latter environments are inferred to form on time scales of >10^3 years. In contrast, the formation of new clay minerals in Amazon sediment occurs much more rapidly, on time scales of months to years (Michalopoulos and Aller, 2004).

In comparing results in Amazon sediments with those from deep-sea, primarily siliceous, sediments (section 13.4) it appears that Amazon sediments represent an end-members in a spectrum in which either biogenic silica or other reactants, e.g., reactive Al and Fe, limit authigenic clay mineral formation (Michalopoulos and Aller, 2004). In deep-sea sediments, sources of Al and Fe likely limit clay mineral formation. Such an explanation appears consistent with results in fig. 13.6, in which increasing amounts of sediment detrital material, relative to the biogenic silica content, appears to control silica accumulation in pore waters, both by consuming dissolved silica in authigenic mineral formation and by inhibiting biogenic silica dissolution by coating these particles with new Al-silicate phases.

In contrast, the formation of authigenic clay minerals in sediment on the Amazon continental shelf is likely silica limited. This appears to occur because the intense chemical weathering of crustal materials in such humid tropical settings produces riverine particles, and ultimately delta shelf sediments, that are relatively rich in weathered Fe and Al oxides (see discussions in Rude and Aller, 1989). Furthermore, the intense iron redox cycling that occurs in these sediments (Aller et al., 1986; Aller, 1998; also see section 7.5.2) leads to a situation in which both iron and perhaps Al are readily solubilized and thus made available for the formation of new clay minerals by rxn. (18.2).

In early studies of major ion pore-water profiles in deep-sea sediments, Sayles (1979, 1981) observed the uptake of K^+ and Mg^{2+} from sediment pore waters, and suggested that their uptake could be interpreted as evidence for the occurrence of reverse weathering reactions such as rxn. (18.1). In the context of the present discussion, Sayles's results illustrated two interesting points. The first is that the uptake of these elements showed geographic variability, with much higher uptake in sediments containing "a higher proportion of continental detritus" (Sayles, 1979). Secondly, he observed that siliceous oozes showed very little evidence of K^+ and Mg^{2+} uptake as compared to other deep-sea sediments (Sayles, 1981). These results therefore appear to be consistent with reverse weathering reactions such as eqn. (18.2), as well as with the suggestion that sources of Fe and Al limit the occurrence of authigenic clay mineral reactions in deep-sea sediments.

Although these results provide compelling evidence for the occurrence of authigenic mineral formation in marine sediments, of equal importance in this discussion is the quantitative significance of these processes in terms of balancing major element budgets in seawater. Martin and Sayles (1994) re-examined the pore-water data in Sayles (1979, 1981) and estimated that the uptake of Mg^{2+} and K^+ from sediment pore waters roughly balances the river input of these cations. Based on more recent studies of Amazon sediments (Rude and Aller, 1994; Michalopoulos and Aller, 2004), it is estimated that ~7–10% of the riverine fluxes of K^+ and F^- are taken up in Amazon sediments alone (here F^- either replaces OH^- in existing clays or simply is taken up in the formation of authigenic clays). Although not quantified in the same fashion, authigenic clay mineral formation in Amazon sediments also appears to involve Mg^{2+} uptake (Aller et al., 1986; Rude and Aller, 1989).

Processes occurring in muddy deltaic sediments of other tropical rivers are similar to those that occur in Amazon margin sediments (e.g., see sections 5.5 and 7.5.2). Tropical rivers also contribute a major fraction (~60%) of the terrigenous particulate flux to the oceans (see Schlünz and Scneider, 2000; McKee et al., 2004; and references therein). As a result, the observations in the previous paragraph suggest that authigenic clay mineral formation may represent quantitatively important uptake processes in the oceanic budgets of K and F, and perhaps Mg (also see discussions in section 18.2).

18.1 Sediment Silica Budgets

In their studies of Amazon sediments Michalopoulos and Aller (2004) concluded that the conventional procedure used to determine biogenic silica, i.e., a timed, alkaline (Na_2CO_3) leach (e.g., DeMaster et al., 1983), underestimates biogenic silica concentrations. This appears to occur because the presence of surface coatings such as metal oxides or authigenic clays on silica particles may inhibit their chemical solubilization by base. There is also uncertainty as to the extent to which this alkaline leach releases silica that has been incorporated into authigenic clays (see discussions in DeMaster and Aller, 2001).

To overcome these apparent difficulties, Michalopoulos and Aller (2004) used a two-step leaching procedure, i.e., a mild acid leach followed by an alkaline leach, to recover what they suggest is biogenic silica plus silica that has been incorporated into authigenic clays. They claim that this new procedure yields results that are "regular, repeatable, and consistent with a wide range of other qualitative and quantitative observations" (Michalopoulos and Aller, 2004). These results will be discussed here in terms of their implications for sedimentary silica cycling. However, the application of these techniques to the study of silica cycling in sediments other than in the Amazon will clearly be required to further validate the conclusions based on these new results.

Based on results obtained with this new leaching procedure, Michalopoulos and Aller (2004) suggest that ~90% of the biogenic silica deposited in Amazon delta sediments is either completely converted to authigenic clays or altered in some fashion. This alteration may occur, for example, by the formation of authigenic clays (or metal oxides) as particle coatings that then inhibit dissolution of the remaining biogenic silica (as was also discussed in section 13.4). This transfer of biogenic silica into authigenic clay fractions is also analogous to the sink-switching concept used to describe phosphorus burial and preservation in sediments (see section 16.3).

Previous estimates of silica burial in Amazon sediments assumed that biogenic silica determined by the timed alkaline leach represented the primary form of silica buried in these sediments (DeMaster et al., 1983). A comparison of such previous estimates of silica burial with those presented by Michalopoulos and Aller (2004) suggests that silica burial in Amazon sediments could be ~5–6-fold greater than previously estimated. More specifically, because ~3% of Amazon delta sediments appear to be authigenic clays (Michalopoulos and Aller, 1995), these clays then account for ~40–60% of the silica burial in this new estimate (Michalopoulos and Aller, 2004). When extrapolated to tropical sediments globally these results roughly double the silica burial efficiency with respect to tropical river input, from an assumed average value of ~10% (Tréguer et al., 1995) to ~22% (Michalopoulos and Aller, 2004).

18.2 Final Thoughts

Given the somewhat limited data used in the calculations discussed in this chapter, some care must be taken in interpreting these results.

Nevertheless, these recent results provide compelling evidence for the in situ occurrence of low-temperature authigenic clay mineral formation, and for its role in the oceanic cycles of major ions in seawater such as Mg^{2+} or K^+. Furthermore, the omission of these processes in geochemical budgets for silica may have underestimated silica sedimentary sinks. Clearly more work will be needed to critically examine these problems.

As a final comment, I note that discussions in chapter 15 (section 15.5) of carbon preservation in sediments suggest that strong linkages might exist among weathering processes on land, carbon burial in sediments, and the controls on global redox balance, i.e., atmospheric O_2 levels. At the same time, discussions in this chapter also suggest that reverse weathering reactions, which may play a role in balancing these weathering reactions, may be strongly linked to iron redox cycling and organic matter remineralization. Furthermore, studies of both reverse weathering and carbon preservation suggest that continental margin sediments are major sites for the occurrence of both processes. Thus, it is interesting to speculate that a connection might exist between weathering on land and carbon burial and authigenic clay mineral formation in continental margin sediments. Although highly speculative, this may prove to be an interesting direction for future research efforts.

≈ *Appendix* ≈

SOME OF THE FIELD SITES DISCUSSED IN THE TEXT

Since many of the field sites discussed in the text may not be known to all readers, brief descriptions of these sites are listed below. Also included is at least one general reference describing work done at the site (additional references can be found in the main body of the text).

Buzzards Bay is southwest of Cape Cod, Massachusetts (USA), and has an average water depth of 13–16 m in the central basin. The sediments in Buzzards Bay are silty clays extensively colonized by benthic macrofauna (Henrichs and Farrington, 1987; Martin and Banta, 1992).

The **California Borderlands** is a continental margin environment off the coast of southern California (between roughly Santa Barbara, California, and the US-Mexico border). It consists of a network of submarine basins separated by ridges, islands, and submarine sills (Emery, 1960; see map in Berelson et al. 1996). The biogeochemical characteristics of each basin are strongly influenced both by water circulation patterns (i.e., horizontal flow and exchange with overlying waters), and by the depth of the basin sill relative to the depth of the water column oxygen minimum (Berelson, 1991). As a result, basin bottom waters have dissolved oxygen concentrations that range from <5 μM to ~50–60 μM. In the inshore low oxygen basins (**San Pedro, Santa Monica,** and **Santa Barbara Basins**) the sediments are laminated (varved) and benthic macrofauna are virtually absent (Emery, 1960; Savrda et al., 1984; Hagadorn et al., 1995). In contrast, macrofauna are present in the outer basins (e.g., **Santa Catalina, San Clemente, Tanner,** and **San Nicolas Basins**) where bottom-water oxygen is greater than ~10–20 μM (see Smith et al., 1993; Thompson et al., 1993; and references cited therein). Additional information on the geochemistry of these sediments can be found in Berelson et al. (1987b, 1996), Bender et al. (1989), Jahnke (1990), Shaw et al. (1990), and McManus et al. (1997).

Cape Lookout Bight is a coastal embayment along the North Carolina (USA) Outer Banks, 110 km SW of Cape Hatteras. Fine-grained, organic-rich sediments accumulate at this site at a rate of ~10 cm yr^{-1}.

APPENDIX

The average water depth is 8 m (±1 m). Sulfate reduction and methanogenesis dominate organic matter remineralization in these sediments (Martens and Klump, 1984; Klump and Martens, 1989).

Chesapeake Bay is the largest estuary in the United States and located in the mid-Atlantic region of the east coast of North America. The average water depth of this coastal plain estuary is ~8 m, although much of the Bay is significantly shallower. In addition, a deep trough that runs down the center of much of the Bay can exceed 40 m in depth. Sediments in the Bay range from fine-grained organic-rich sediments (TOC >3–4%) to more coarse-grained silty sands (Cornwell and Sampou, 1995; Marvin-DiPasquale and Capone, 1998; Burdige, 2001).

Long Island Sound is an estuarine trough along the northeast coast of North America between Long Island and New York/Connecticut (USA). The **FOAM** site (Friends of Anoxic Mud) is an organic-rich sediment in the Sound at a water depth of ~9 m (Goldhaber et al., 1977; Aller, 1980a).

The MANOP sites. MANOP (the Manganese Nodule Program) was a multi-investigator project studying sediment geochemistry (in general) and the formation of manganese nodules (in particular) in Pacific Ocean pelagic and hemipelagic sediments (Emerson et al., 1980, 1982, 1985; Jahnke et al., 1982; Lyle, 1983; Bender and Heggie, 1984; Heggie et al., 1986). **MANOP site C** is a carbonate ooze located in the central equatorial Pacific beneath the equatorial upwelling zone (water depth of 4,400–4,900 m). **MANOP site M** is at a water depth of ~3,100 m in the eastern Equatorial Pacific, 20–30 km east of the spreading axis of the East Pacific Rise. The sediments at site M have continental (terrigenous), biogenic (primarily calcite), and hydrothermal sources. **MANOP site S** is a siliceous ooze/clay sediment (water depth of 4,910 m) with extensive manganese nodule coverage. **MANOP site H** is a hemipelagic sediment (2,590 m water depth) in the Guatemala Basin ~900 km east of the East Pacific Rise. **MANOP site R** is a red clay ooze at a water depth of 5,800 m in the North Pacific central gyre, north of Hawaii.

The **Madeira Abyssal Plain** (MAP) is in the Atlantic Ocean off the northwest coast of Africa. It is roughly halfway between the northwestern African continental shelf and the mid-Atlantic ridge, approximately 600 km south of the Azore Islands. The average water depth is 5,400 m. Sediments here consist of thick (1–4 m) turbidites interbedded with thin (3–15 cm) pelagic sediments. Based on geochemical data, turbidites in this region are classified as organic rich, volcanic, or calcareous. The calcareous and volcanic turbidites are relatively homogeneous, while the organic-rich turbidites show differing amounts of postdepositional oxidation or remineralization. These latter turbidites are derived from the northwest African continental margin (de Lange et al., 1987; Thomson et al., 1998).

The **Patton Escarpment** is in the eastern Pacific on the lower continental slope off southern California. Geochemical studies (e.g., Shaw et al., 1990; Berelson et al., 1996) have been carried out at a site on the escarpment at a water depth of ~3,700 m. Pore-water oxygen is depleted here at a sediment depth of ~2.5 cm and nitrate is depleted by ~4–10 cm. There is some bioturbation and bioirrigation in these sediments.

Saanich Inlet is an intermittently anoxic fjord along the coast of Vancouver Island, British Columbia (Canada), with a maximum water depth of ~240 m (Hamilton and Hedges, 1988; Cowie et al., 1992).

The **Skagerrak** and **Kattegat** are shallow continental margin seas between Denmark, Sweden, and Norway that connect the North Sea and Baltic Sea (Iversen and Jørgensen, 1985; Canfield et al., 1993a). The Norwegian Trench cuts through the entire Skaggerak to a maximum water depth of ~700m. In contrast, the Kattegat is much shallower, having an average depth of ~25 m. Some sediments in the Skaggerak are extremely rich in manganese (up to 4 wt%; compare with values in table 13.2).

Skan Bay is an anoxic embayment in the Aleutian Islands off Alaska in the North Pacific. The water depth of the bay is ~65 m (Alperin et al., 1992b).

≈ *References* ≈

Adler, M., Hensen, C., Wenzhöfer, F., Pfeifer, K., and Schulz, H.D., 2001. Modeling of calcite dissolution by oxic respiration in supralysoclinal deep-sea sediments. Mar. Geol., 177: 167–190.
Aiken, G.R., McKnight, D.M., Wershaw, R.L., and MacCarthy, P., 1985. *Humic Substances in Soil, Sediment and Water.* Wiley.
Albert, D.B., and Martens, C.S., 1997. Determination of low-molecular weight organic acid concentrations in seawater and pore-water samples via HPLC. Mar. Chem., 56: 27–37.
Albert, D.B., Martens, C.S., and Alperin, M.J., 1998. Biogeochemical processes controlling methane in gassy coastal sediments—Part 2: Groundwater flow control of acoustic turbidity in Eckernförde Bay Sediments. Cont. Shelf Res., 18: 1771–1793.
Aller, R.C., 1978. Experimental studies of changes produced by deposit feeders on pore water, sediment and overlying water chemistry. Am. J. Sci., 278: 1185–1234.
Aller, R.C., 1980a. Diagenetic processes near the sediment-water interface of Long Island Sound, I. Decomposition and nutrient element geochemistry (S,N,P). Adv. Geophys., 22: 235–348.
Aller, R.C., 1980b. Diagenetic processes near the sediment-water interface of Long Island Sound, II. Fe and Mn. Adv. Geophys., 22: 351–415.
Aller, R.C., 1980c. Quantifying solute distributions in the bioturbated zone of marine sediments by defining an average microenvironment. Geochim. Cosmochim. Acta, 44: 1955–1965.
Aller, R.C., 1982a. Carbonate dissolution in shallow water marine sediments: Role of physical and biological reworking. J. Geol., 90: 79–95.
Aller, R.C., 1982b. The effects of macrobenthos on chemical properties of marine sediments and overlying water. *In*: P.L. McCall and M.J.S. Tevesz (eds.), *Animal-Sediment Relations, The Biogenic Alteration of Sediments.* Plenum Press, pp. 53–102.
Aller, R.C., 1983. The importance of the diffusive permeability of animal burrow linings in determining marine sediment chemistry. J. Mar. Res., 41: 299–322.
Aller, R.C., 1988. Benthic fauna and biogeochemical processes in marine sediments: the role of burrow structures. *In*: T.H. Blackburn and J. Sørensen (eds.), *Nitrogen Cycling in Coastal Marine Environments.* John Wiley & Sons, pp. 301–338.
Aller, R.C., 1990. Bioturbation and manganese cycling in hemipelagic sediments. Phil. Trans. R. Soc. London, A, 331: 51–68.
Aller, R.C., 1994a. Bioturbation and remineralization of sedimentary organic matter: effects of redox oscillation. Chem. Geol., 114: 331–345.

Aller, R.C., 1994b. The sedimentary Mn cycle in Long Island Sound: its role as intermediate oxidant and the influence of bioturbation, O_2, and C_{org} flux on diagenetic reaction balances. J. Mar. Res., 52: 259–295.

Aller, R.C., 1998. Mobile deltaic and continental shelf muds as suboxic, fluidized bed reactors. Mar. Chem., 61: 143–155.

Aller, R.C., 2001. Transport and reactions in the bioirrigated zone. In: B.P. Boudreau and B.B. Jorgensen (eds.), *The Benthic Boundary Layer*. Oxford University Press, pp. 269–301.

Aller, R.C., 2004. Conceptual models of early diagenetic processes: the muddy seafloor as an unsteady, batch reactor. J. Mar. Res., 62: 815–835.

Aller, R.C., and Aller, J.Y., 1992. Meiofauna and solute transport in marine muds. Limnol. Oceanogr., 37: 1018–1033.

Aller, R.C., and Aller, J.Y., 1998. The effect of biogenic irrigation intensity and solute exchange on diagenetic reaction rates in marine sediments. J. Mar. Res., 56: 905–936.

Aller, R.C., Aller, J.Y., and Kemp, P.F., 2001. Effects of particle and solute transport on rates and extent of remineralization in bioturbated sediments. In: J.Y. Aller, S.A. Woodin, and R.C. Aller (eds.), *Organism-Sediment Interaction*. University of South Carolina Press, pp. 315–333.

Aller, R.C., and Benninger, L.K., 1981. Spatial and temporal patterns of dissolved ammonium, manganese, and silica fluxes from the bottom sediments of Long Island Sound, U.S.A. J. Mar. Res., 39: 295–314.

Aller, R.C., Blair, N.C., Xia, Q., and Rude, P.D., 1996. Remineralization rates, recycling and storage of carbon in Amazon shelf sediments. Cont. Shelf Res., 16: 753–786.

Aller, R.C., and Blair, N.E., 1996. Sulfur diagenesis and burial on the Amazon shelf: Major control by physical sedimentation processes. Geo-Mar. Lett., 16: 3–10.

Aller, R.C., and Blair, N.E., 2004. Early diagenetic remineralization of sedimentary organic C in the Gulf of Papua deltaic complex (Papua New Guinea): Net loss of terrestrial C and diagenetic fractionation of C isotopes. Geochim. Cosmochim. Acta, 68: 1815–1825.

Aller, R.C., and Cochran, J.K., 1976. $^{234}Th/^{238}U$ disequilibrium in near-shore sediment: particle reworking and diagenetic time scales. Earth Planet. Sci. Lett., 29: 37–50.

Aller, R.C., Hall, P.O.J., Rude, P.D., and Aller, J.Y., 1998. Biogeochemical heterogeneity and suboxic diagenesis in hemipelagic sediments of the Panama Basin. Deep-Sea Res., 45: 133–165.

Aller, R.C., Hannides, A., Heilbrun, C., and Panzeca, C., 2004. Coupling of early diagenetic processes and sedimentary dynamics in tropical shelf environments: The Gulf of Papua deltaic complex. Cont. Shelf Res., 24: 2455–2486.

Aller, R.C., and Mackin, J.E., 1984. Preservation of reactive organic matter in marine sediments. Earth Planet. Sci. Lett., 70: 260–266.

REFERENCES

Aller, R.C., and Mackin, J.E., 1989. Open-incubation, diffusion method for measuring solute reaction rates in sediments. J. Mar. Res., 47: 411–440.

Aller, R.C., Mackin, J.E., and R.T. Knox, J., 1986. Diagenesis of Fe and S in Amazon inner shelf muds: apparent dominance of Fe reduction and implications for the genesis of ironstones. Cont. Shelf Res., 6: 263–289.

Aller, R.C., and Rude, P.D., 1988. Complete oxidation of solid phase sulfides by manganese and bacteria in anoxic marine sediments. Geochim. Cosmochim. Acta, 52: 751–765.

Aller, R.C., and Yingst, J.Y., 1980. Relationships between microbial distributions and the anaerobic decomposition of organic matter in surface sediments of Long Island Sound, USA. Mar. Biol., 56: 29–42.

Alley, R.B., 2000. Ice-core evidence of abrupt climate changes. Proc. Natl. Acad. Sci., 97: 1331–1334.

Alongi, D.M., 1995. Decomposition and recycling of organic matter in muds of the Gulf of Papua, northern Coral Sea. Cont. Shelf Res., 15: 1319–1337.

Alperin, M.J., 2003. Sediment-water column interactions in a shallow, eutrophic estuary: a multi-species reaction-transport model for sediment and the lower benthic boundary layer, Abstract presented at the 2003 ASLO Aquatic Science Meeting, Salt Lake City, VT, Feb. 9–14.

Alperin, M.J., Albert, D.B., and Martens, C.S., 1994. Seasonal variations in production and consumption rates of dissolved organic carbon in an organic-rich coastal sediment. Geochim. Cosmochim. Acta, 58: 4909–4929.

Alperin, M.J., Blair, N.E., Albert, D.B., Hoehler, T.M., and Martens, C.S., 1992a. Factors that control the stable carbon isotopic composition of methane produced in an anoxic marine sediment. Global Biogeochem. Cycles, 6: 271–291.

Alperin, M.J., Hee, C.A., Albert, D.B., and Martens, C.S., 2000. Stable- and radio-carbon isotopic composition of porewater DOC in an organic-rich marine sediment. EOS, Trans. AGU, Fall Meet. Suppl., 81: OS11B-03 (abstr.).

Alperin, M.J., Martens, C.S., Albert, D.B., Suayah, I.B., Benninger, L.K., Blair, N.E., and Jahnke, R.A., 1999. Benthic fluxes and porewater concentration profiles of dissolved organic carbon in sediments from the North Carolina continental slope. Geochim. Cosmochim. Acta, 63: 427–448.

Alperin, M.J., and Reeburgh, W.S., 1984. Geochemical observations supporting anaerobic methane oxidation. *In*: R.L. Crawford and R.S. Hanson (eds.), *Microbial Growth on C-1 Compounds*. ASM, pp. 282–289.

Alperin, M.J., and Reeburgh, W.S., 1985. Inhibition experiments on anaerobic methane oxidation. Appl. Environ. Microbiol., 50: 940–945.

Alperin, M.J., Reeburgh, W.S., and Devol, A.H., 1992b. Organic carbon remineralization and preservation in sediments of Skan Bay, Alaska. *In*: J.K. Whelan and J.W. Farrington (eds.), *Productivity, Accumulation, and Preservation of Organic Matter in Recent and Ancient Sediments*. Columbia University Press, pp. 99–122.

Altabet, M.A., and Francois, R., 1994. Sedimentary nitrogen isotopic ratio as a recorder for surface ocean nitrate utilization. Global Biogeochem. Cycles, 8: 103–116.

Amon, R.M.W., and Benner, R., 1994. Rapid cycling of high-molecular-weight dissolved organic matter in the ocean. Nature, 369: 549–552.

An, S., and Gardner, W.S., 2002. Dissimilatory nitrate reduction to ammonium (DNRA) as a nitrogen link, versus denitrification as a sink in a shallow estuary (Laguna Madre/Baffin Bay, Texas). Mar. Ecol. Prog. Ser., 237: 41–50.

Anderson, D.M., and Archer, D., 2002. Glacial-interglacial stability of ocean pH inferred from foraminifer dissolution rates. Nature, 416: 70–73.

Anderson, I.C., Poth, M.A., Homstead, J., and Burdige, D.J., 1993. A comparison of NO and N2O production by autotrophic and heterotrophic nitrifiers. Appl. Environ. Microbiol., 59: 3525–3533.

Anderson, L., and Delaney, M.L., 2000. Sequential extraction and analysis of phosphorus in marine sediments: Streamlining of the SEDEX procedure. Limnol. Oceanogr., 45: 509–515.

Anderson, L.A., and Sarmiento, J.L., 1994. Redfield ratios of remineralization determined by nutrient data analysis. Global Biogeochem. Cycles, 8: 65–80.

Andrews, D., and Bennett, A., 1981. Measurement of diffusivity near the sediment-water interface with a fine-scale resistivity probe. Geochim. Cosmochim. Acta, 45: 2169–2175.

Anschutz, P., Sundby, B., Lefrançois, L., Luther, G.W., III, and Mucci, A., 2000. Interactions between metal oxides and species of nitrogen and iodine in bioturbated marine sediments. Geochim. Cosmochim. Acta, 64: 2751–2763.

Aplin, A.C., 1993. The composition of authigenic clay minerals in recent sediments: links to the supply of unstable reactants. *In*: D.A. Manning, P.L. Hall, and C.R. Hughes (eds.), *Geochemistry of Clay-Pore Fluid Interactions*. Chapman & Hall, pp. 81–106.

Archer, D., Emerson, S., and Smith, C.R., 1989a. Direct measurement of the diffusive sublayer at the deep sea floor using oxygen microelectrodes. Nature, 340: 623–626.

Archer, D., Kheshgi, H., and Maier-Reimer, E., 1998. Dynamics of fossil fuel CO_2 neutralization by marine $CaCO_3$. Global Biogeochem. Cycles, 12: 259–276.

Archer, D., Winguth, A., Lea, D., and Mahowald, N., 2000. What caused the glacial/interglacial atmospheric pCO_2 cycles? Rev. Geophys., 38: 159–189.

Archer, D.E., 1991a. Equatorial Pacific calcite preservation cycles: production or dissolution? Paleoceanography, 6: 561–571.

Archer, D.E., 1991b. Modeling the calcite lysocline. J. Geophys. Res., 96: 17037–17050.

Archer, D.E., 1996. An atlas of the distribution of calcium carbonate in sediments of the deep sea. Global Biogeochem. Cycles, 10: 159–174.

REFERENCES

Archer, D.E., and Devol, A., 1992. Benthic oxygen fluxes on the Washington shelf and slope: A comparison of in situ microelectrode and chamber flux measurements. Limnol. Oceanogr., 37: 614–629.

Archer, D.E., Emerson, S., and Reimers, C., 1989b. Dissolution of calcite in deep-sea sediments: pH and O_2 microelectrode results. Geochim. Cosmochim. Acta, 53: 2831–2845.

Archer, D.E., Lyle, M., Rodgers, K., and Froelich, P., 1993. What controls opal preservation in tropical deep-sea sediments. Paleoceanography, 8: 7–21.

Archer, D.E., Morford, J.L., and Emerson, S.R., 2002. A model of suboxic sedimentary diagenesis suitable for automatic tuning and gridded global domains. Global Biogeochem. Cycles, 16: 10.1029/2000GB001288.

Arnarson, T.S., and Keil, R.G., 2001. Organic-mineral interactions in marine sediments studied using density fractionation and X-ray photoelectron spectroscopy. Org. Geochem., 32: 1401–1415.

Arnosti, C., 2000. Substrate specificity in polysaccharide hydrolysis: Contrasts between bottom water and sediments. Limnol. Oceanogr., 45: 1112–1119.

Arnosti, C., 2004. Speed bumps in the carbon cycle: Subrate structural effects on carbon cycling. Mar. Chem., 92: 263–273.

Arnosti, C., and Holmer, M., 1999. Carbohydrate dynamics and contributions to the carbon budget of an organic-rich coastal sediment. Geochim. Cosmochim. Acta, 63: 353–403.

Arnosti, C., Repeta, D.J., and Blough, N.V., 1994. Rapid bacterial degradation of polysaccharides in anoxic marine sediments. Geochim. Cosmochim. Acta, 58: 2639–2652.

Arrhenius, G., 1952. Sediment cores from the east Pacific. Rep. Swed. Deep-Sea Exped. 1947–1948, 5: 1–228.

Aycard, M., Derenne, S., Largeau, C., Mongenot, T., Tribovillard, N., and Baudin, F., 2003. Formation pathways of proto-kerogens in Holocene sediments of the upwelling influenced Cariaco Trench, Venezuela. Org. Geochem., 34: 701–718.

Bacon, M.P., Belastock, R.A., and Bothner, M.H., 1994. ^{210}Pb balance and implications for particle transport on the continental shelf, U.S. Middle Atlantic Bight. Deep-Sea Res., 41: 511–535.

Bacon, M.P., and Rosholt, J.N., 1982. Accumulation rates of Th-230, Pa-231, and some transition metals on the Bermuda Rise. Geochim. Cosmochim. Acta, 46: 651–666.

Baldock, J.A., Oades, J.M., Waters, A.G., Peng, X., Vassallo, A.M., and Wilson, M.A., 1992. Aspects of the chemical structure of soil organic materials as revealed by solid state ^{13}C NMR spectroscopy. Biogeochemistry 16: 1–42.

Baldock, J.A., Masiello, C.A., Gelinas, Y., and Hedges, J.I., 2004. Cycling and composition of organic matter in terrestrial and marine ecosystems. Mar. Chem., 92: 39–64.

REFERENCES

Balistrieri, L., Brewer, P.G., and Murray, J.W., 1981. Scavenging residence times of trace metals and surface chemistry of sinking particles in the deep ocean. Deep-Sea Res., 28A: 101–121.

Balistrieri, L.S., and Murray, J.W., 1986. The surface chemistry of sediments from the Panama Basin: the influence of Mn oxides on metal adsorption. Geochim. Cosmochim. Acta, 50: 235–2243.

Balzer, W., 1982. On the distribution of iron and manganese at the sediment/water interface: thermodynamic versus kinetic control. Geochim. Cosmochim. Acta, 46: 1153–1161.

Barcelona, M.J., 1980. Dissolved organic carbon and volatile fatty acids in marine sediment pore waters. Geochim. Cosmochim. Acta, 44: 1977–1984.

Barron, E.J., and Whitman, J.M., 1981. Oceanic sediments in space and time. In: C. Emiliani (ed.), *The Sea*. Vol. 7, *The Oceanic Lithosphere*. Wiley-Interscience Pub., pp. 689–731.

Bauer, J.E., and Druffel, E.R.M., 1998. Ocean margins as a significant source of organic matter to the deep open ocean. Nature, 392: 482–485.

Bauer, J.E., Reimers, C.E., Druffel, E.R.M., and Williams, P.M., 1995. Isotopic constraints on carbon exchange between deep ocean sediments and sea water. Nature, 373: 686–689.

Behl, R.J., and Kennett, J.P., 1996. Brief interstadial events in the Santa Barbara basin, NE Pacific, during the past 60 kyr. Nature, 379: 243–246.

Belicka, L.L., Macdonald, R.W., and Harvey, H.R., 2002. Sources and transport of organic carbon to shelf, slope, and basin surface sediments of the Arctic Ocean. Deep-Sea Res., 49: 1463–1483.

Ben Yaakov, S., 1973. pH buffering of pore water of recent anoxic marine sediment. Limnol. Oceanogr., 18: 86–94.

Bender, M.L., 1971. Does upward diffusion supply the excess manganese in pelagic sediments? J. Geophys. Res., 76: 4212–4215.

Bender, M., Jahnke, R., Weiss, R., Martin, W., Heggie, D.T., Orchardo, J., and Sowers, T., 1989. Organic carbon oxidation and benthic nitrogen and silica dynamics in San Clemente Basin, a continental borderland site. Geochim. Cosmochim. Acta, 53: 685–697.

Bender, M.L., and Heggie, D.T., 1984. Fate of organic carbon reaching the deep sea floor: a status report. Geochim. Cosmochim. Acta, 48: 977–986.

Bender, M.L., Hudson, A., Graham, D.W., Barnes, R.O., Leinen, M., and Kahn, D., 1985/86. Diagenesis and convection reflected in pore water chemistry on the western flank of the East Pacific Rise, 20 degrees south. Earth Planet. Sci. Lett., 76: 71–83.

Bender, M.L., Ku, T., and Broecker, W.S., 1970. Accumulation rates of manganese in pelagic sediments and nodules. Earth Planet. Sci. Lett., 8: 143–148.

Bender, M.L., Ku, T.-L., and Broecker, W.S., 1966. Manganese nodules: their evolution. Science, 151: 325–328.

REFERENCES

Benner, R., 2002. Chemical composition and reactivity. *In*: D.A. Hansell and C.D. Carlson (eds.), *Biogeochemistry of Marine Dissolved Organic Matter*. Elsevier Science, pp. 59–90.

Benner, R., Maccubbin, A.E., and Hodson, R.E., 1984. Anaerobic biodegradation of the lignin and polysaccharide components of lignocellulose and synthetic lignin by sediment microflora. Appl. Environ. Microbiol., 47: 998–1004.

Bennett, R.H., O'Brien, N.R., and Hulbert, M.H., 1991. Determinants of clay and shale microfabric signatures: process and mechanisms. *In*: R.H. Bennett, N.R. O'Brien, and M.H. Hulbert (eds.), *Microstructure of Fine-Grained Sediments*. Springer-Verlag, pp. 5–32.

Benning, L.G., Wilkin, R.T., and Barnes, H.L., 2000. Reaction pathways in the Fe-S system below 100°C. Chem. Geol., 167: 25–51.

Berelson, W., McManus, J., Coale, K., Johnson, K., Burdige, D., Kilgore, T., Colodner, D., Chavez, F., Kudela, R., and Boucher, J., 2003. A time series of benthic flux measurements from Monterey Bay, CA. Cont. Shelf Res., 23: 457–481.

Berelson, W.M., 1991. The flushing of two deep-sea basins, southern California borderland. Limnol. Oceanogr., 36: 1150–1166.

Berelson, W.M., Anderson, R.F., Dymond, J., DeMaster, D., Hammond, D.E., Collier, R., Honjo, S., Leinen, M., Pope, R., Smith, C., and Stephens, M., 1997. Biogenic budgets of particle rain, benthic remineralization and sediment accumulation in the Equatorial Pacific. Deep-Sea Res. II, 44: 2251–2282.

Berelson, W.M., Buckholtz, M.R., Hammond, D.E., and Santschi, P.H., 1987a. Radon fluxes measured with the MANOP lander. Deep-Sea Res., 34: 1209–1228.

Berelson, W.M., and Hammond, D.E., 1986. The calibration of a new free vehicle flux chamber for use in the deep-sea. Deep-Sea Res., 33: 1439–1454.

Berelson, W.M., Hammond, D.E., and Cutter, G.A., 1990a. *In situ* measurements of calcium carbonate dissolution rates in deep-sea sediments. Geochim. Cosmochim. Acta, 54: 3013–3020.

Berelson, W.M., Hammond, D.E., and Fuller, C., 1982. Radon-222 as a tracer for mixing in the water column and benthic exchange in the southern California borderland. Earth Planet. Sci. Lett., 61: 41–54.

Berelson, W.M., Hammond, D.E., and Johnson, K.S., 1987b. Benthic fluxes and cycling of biogenic silica and carbon in two southern California borderland basins. Geochim. Cosmochim. Acta, 51: 1345–1363.

Berelson, W.M., Hammond, D.E., McManus, J., and Kilgore, T.E., 1994. Dissolution kinetics of calcium carbonate in equatorial Pacific sediments. Global Biogeochem. Cycles, 8: 219–235.

Berelson, W.M., Hammond, D.E., O'Neill, D., Xu, X.-M., Chin, C., and Zukin, J., 1990b. Benthic fluxes and pore water studies from sediments from the central equatorial north Pacific: nutrient diagenesis. Geochim. Cosmochim. Acta, 54: 3001–3012.

Berelson, W.M., McManus, J., Kilgore, T., Coale, K., Johnson, K.S., Burdige, D., and Pilskaln, C., 1996. Biogenic matter diagenesis on the sea floor: A comparison between two continental margin transects. J. Mar. Res., 54: 731–762.

Berelson, W.M., Townsend, T., Heggie, D., Ford, P., Longmore, A., Skyring, G., Kilgore, T., and Nicholson, G., 1999. Modelling bio-irrigation rates in the sediments of Port Phillip Bay. Mar. Freshwater Res., 50: 573–579.

Berg, P., Røy, H., Janssen, F., Meyer, V., Jørgensen, B.B., Hüttel, M., and de Beer, D., 2004. Oxygen uptake by aquatic sediments measured with a novel noninvasive eddy correlation technique. Mar. Ecol. Prog. Ser., 261: 75–83.

Berg, P., Rysgaard, S., Funch, P., and Sejr, M.K., 2001. Effects of bioturbation on solutes and solids in marine sediments. Aquatic Microb. Ecol., 26: 81–94.

Berg, P., Rysgaard, S., and Thamdrup, B., 2003. Dynamic modeling of early diagenesis and nutrient cycling. A case study in an Arctic marine sediment. Am. J. Sci., 303: 905–955.

Bergamaschi, B.A., Walters, J.S., and Hedges, J.I., 1999. Distributions of uronic acids and O-methyl sugars in sinking and sedimentary particles in two coastal marine environments. Geochim. Cosmochim. Acta, 63: 413–425.

Berger, W.H., 1976. Biogenous deep sea sediments: Production, preservation and interpretation. *In*: J.P. Riley and R. Chester (eds.), *Chemical Oceanography*. Vol. 5, 2nd ed. Academic Press, pp. 265–388.

Berger, W.H., Finkel, R.C., Killingley, J.S., and Marchig, V., 1983. Glacial-Holocene transition in deep-sea sediments: manganese-spike in the east-equatorial Pacific. Nature, 303: 231–233.

Berner, E.K., and Berner, R.A., 1996. *Global Environment: Water, Air, and Geochemical Cycles*. Prentice-Hall.

Berner, E.K., Berner, R.A., and Moulton, K.L., 2003. Plants and mineral weathering: present and past. *In*: Drever, J.I. (ed.), *Treatise on Geochemistry*. Vol. 5, Elsevier, pp. 169–188.

Berner, R.A., 1964. An idealized model of dissolved sulfate distribution in recent sediments. Geochim. Cosmochim. Acta, 28: 1497–1503.

Berner, R.A., 1970. Sedimentary pyrite formation. Am. J. Sci., 268: 1–23.

Berner, R.A., 1971. *Principles of Chemical Sedimentology*. McGraw-Hill.

Berner, R.A., 1972. Chemical kinetic models of early diagenesis. J. Geol. Ed., 20: 267–272.

Berner, R.A., 1976. Inclusion of adsorption in the modelling of early diagenesis. Earth Planet. Sci. Lett., 29: 333–340.

Berner, R.A., 1977. Stoichiometric models for nutrient regeneration in anoxic sediments. Limnol. Oceanogr., 22: 781–786.

Berner, R.A., 1980. *Early Diagenesis, A Theoretical Approach*. Princeton University Press.

Berner, R.A., 1981a. A new geochemical classification of sedimentary environments. J. Sed. Petrol., 51: 359–365.

Berner, R.A., 1981b. A rate model for organic matter decomposition during bacterial sulfate reduction in marine sediments, *Biogéochemie de la matière organique à l'interface eau-sédiment marin*. Colloques Internationaux du C.N.R.S. no. 293.

Berner, R.A., 1982. Burial of organic carbon and pyrite sulfur in the modern ocean: its geochemical and environmental significance. Am. J. Sci., 282: 451–473.

Berner, R.A., 1984. Sedimentary pyrite formation: an update. Geochim. Cosmochim. Acta, 48: 605–615.

Berner, R.A., 1989. Biogeochemical cycles of carbon and sulfur and their effect on atmospheric oxygen over Phanerozoic time. Paleogeogr. Paleoclimatol. Paleoecol., 75: 97–122.

Berner, R.A., 1999. Atmospheric oxygen over Phanerozoic time. Proc. Natl. Acad. Sci., 96: 10955–10957.

Berner, R.A., and Canfield, D.E., 1989. A new model for atmospheric oxygen over Phanerozoic time. Am. J. Sci., 289: 333–361.

Berner, R.A., and Raiswell, R., 1983. Burial of organic carbon and pyrite sulfur in sediments over Phanerozoic time: a new theory. Geochim. Cosmochim. Acta, 47: 855–862.

Berner, R.A., Scott, M.R., and Thomlinson, C., 1970. Carbonate alkalinity in the pore waters of anoxic marine sediments. Limnol. Oceanoogr., 15: 544–549.

Berner, R.A., and Westrich, J.T., 1985. Bioturbation and the early diagenesis of carbon and sulfur. Am. J. Sci., 285: 193–206.

Betts, J.N., and Holland, H.D., 1991. The oxygen content of ocean bottom waters, the burial efficiency of organic carbon, and the regulation of atmospheric oxygen. Paleogeogr. Paleoclimatol. Paleoecol., 73: 97–122.

Bianchi, T.S., Mitra, S., and McKee, B.A., 2002. Sources of terrestrially-derived organic carbon in lower Mississippi River and Louisiana shelf sediments: implications for differential sedimentation and transport at the coastal margin. Mar. Chem., 77: 211–223.

Bickert, T., 2000. Influence of geochemical processes on stable isotope distribution in marine sediments. *In*: H.D. Schultz and M. Zabel (eds.), *Marine Geochemistry*. Springer-Verlag, pp. 309–334.

Bigeleisen, J., 1965. Chemistry of isotopes. Science, 147: 463–471.

Blackburn, T.H., 1981. Seasonal variations in the rate of organic-N remineralization in anoxic marine sediments. *In*: *Biogéochemie de la matière organique à l'interface eau-sédiment marin*. Colloques Internationaux du C.N.R.S. 293.

Blackburn, T.H., 1988. Benthic mineralization and bacterial production. *In*: T.H. Blackburn and J. Sørensen (eds.), *Nitrogen Cycling in Coastal Marine Environments*. John Wiley & Sons, pp. 175–190.

REFERENCES

Blackburn, T.H., Hall, P.O.J., Hulth, S., and Landén, A., 1996. Organic-N loss by efflux and burial associated with a low efflux of inorganic N and with nitrate assimilation in Arctic sediments (Svalbard, Norway). Mar. Ecol. Prog. Ser., 141: 283–293.

Blair, N.E., and Aller, R.C., 1995. Anaerobic methane oxidation on the Amazon shelf. Geochim. Cosmochim. Acta, 59: 3707–3715.

Blair, N.E., Leithold, E.L., and Aller, R.C., 2004. From bedrock to burial: the evolution of particulate organic carbon across coupled watershed-continental margin systems. Mar. Chem., 92: 141–156.

Blair, N.E., Leithold, E.L., Ford, S.T., Peeler, K.A., Holmes, J.C., and Perkey, D.W., 2003. The persistence of memory: the fate of ancient sedimentary organic carbon in a modern sedimentary system. Geochim. Cosmochim. Acta, 67: 63–73.

Blair, N.E., Plaia, G.R., Boehme, S.E., DeMaster, D.J., and Levin, L.A., 1994. The remineralization of organic carbon on the North Carolina continental slope. Deep-Sea Res., 41: 755–766.

Bock, M.J., and Mayer, L.M., 2000. Mesodensity organo-clay associations in a near-shore sediment. Mar. Geol., 163: 65–75.

Boehme, S.E., Blair, N.E., Chanton, J.P., and Martens, C.S., 1996. A mass balance of ^{13}C and ^{12}C in an organic-rich methane-producing marine sediment. Geochim. Cosmochim. Acta, 60: 3835–3848.

Boetius, A., Ravenschlag, K., Schubert, C.J., Rickert, D., Widdel, F., Gieseke, A., Amann, R., Jorgensen, B.B., Witte, U., and Pfannkuche, O., 2000. A marine microbial consortium apparently mediating anaerobic oxidation of methane. Nature, 407: 623–626.

Bonatti, E., 1981. Metal deposits in the oceanic lithosphere. *In*: C. Emiliani (ed.), *The Sea*. Vol. 7, *The Oceanic Lithosphere*. Wiley-Interscience Pub., pp. 639–686.

Bond, G., Showers, W., Cheseby, M., Lotti, R., Almasi, P., deMenocal, P., Priore, P. Cullen, H., Hajdaas, I., and Bonani, G., 1997. A pervasive millennial-scale cycle in North Atlantic Holocene and glacial climates. Science, 278: 1257–1266.

Borch, N.H., and Kirchman, D.L., 1999. Protection of protein from bacterial degradation by submicron particles. Aquat. Microb. Ecol., 16: 265–272.

Borowski, W.S., Paull, C.K., and Ussler III, W., 1997. Carbon cycling within the upper methanogenic zone of continental rise sediments: an example from the methane-rich sediments overlying the Blake Ridge gas hydrate deposits. Mar. Geol., 57: 299–311.

Borowski, W.S., Paull, C.K., and Ussler III, W., 1999. Global and local variations of interstitial sulfate gradients in deep-water, continental margin sediments: Sensitivity to underlying methane and gas hydrates. Mar. Geol., 159: 131–154.

Boschker, H.T.S., Bertilsson, S.A., Dekkers, E.M.J., and Cappenberg, T.E., 1995. An inhibitor-based method to measure initial decomposition of naturally occurring polysaccharides in sediments. Appl. Environ. Microbiol., 61: 2186–2192.

Boudreau, B.P., 1984. On the equivalence of nonlocal and radial-diffusion models for porewater irrigation. J. Mar. Res., 42: 731–735.

Boudreau, B.P., 1986a. Mathematics of tracer mixing in sediments: I. Spatially-dependent, diffusive mixing. Am. J. Sci, 286: 199–238.

Boudreau, B.P., 1986b. Mathematics of tracer mixing in sediments: II. Nonlocal mixing and biological conveyor-belt phenomena. Am. J. Sci, 286: 161–198.

Boudreau, B.P., 1987. A steady-state diagenetic model for dissolved carbonate species and pH in the porewaters of oxic and suboxic sediments. Geochim. Cosmochim. Acta, 51: 1985–1996.

Boudreau, B.P., 1992. A kinetic model for microbe organic-matter decomposition in marine sediments. FEMS Microb. Ecol., 102: 1–14.

Boudreau, B.P., 1994. Is burial velocity a master parameter for bioturbation? Geochim. Cosmochim. Acta, 58: 1243–1249.

Boudreau, B.P., 1996. A method-of-lines code for carbon and nutrient diagenesis in aquatic sediments. Comput. Geosci., 22: 479–496.

Boudreau, B.P., 1997a. *Diagenetic Models and Their Implementation.* Springer-Verlag.

Boudreau, B.P., 1997b. A one-dimensional model for bed-boundary layer particle exchange. J. Mar. Syst., 11: 279–303.

Boudreau, B.P., 1998. Mean mixed depth of sediments: the wherefore and the why. Limnol. Oceanogr., 43: 524–526.

Boudreau, B.P., 1999. A theoretical investigation of the organic carbon-microbial biomass relation in muddy sediments. Aquat. Microb. Ecol., 17: 181–189.

Boudreau, B.P., 2000. The mathematics of early diagenesis: From worms to waves. Rev. Geophys., 38: 389–416.

Boudreau, B.P., 2004. What controls the mixed-layer depth in deep-sea sediments? The importance of particulate organic carbon flux. Limnol. Oceanogr., 49: 620–622.

Boudreau, B.P., and Bennett, R.H., 1999. New rheological and porosity equations for steady-state compaction. Am. J. Sci., 299: 517–528.

Boudreau, B.P., and Canfield, D.E., 1988. A provisional diagenetic model for pH in anoxic porewaters: Application to the FOAM site. J. Mar. Res., 46: 429–455.

Boudreau, B.P., and Canfield, D.E., 1993. A comparison of closed- and open-system models for porewater pH and calcite-saturation state. Geochim. Cosmochim. Acta, 57: 317–334.

Boudreau, B.P., Gardiner, B.S., and Johnson, B.D., 2001a. Rate of growth of isolated bubbles in sediments with a diagenetic source of methane. Limnol. Oceanogr., 46: 616–622.

REFERENCES

Boudreau, B.P., Huettel, M., Forster, S., Jahnke, R.A., McLachlan, A., Middelburg, J.J., Nielsen, P., Sansone, F., Taghon, G., Van Raaphorst, W., Webster, I., Marcin, J., Wiberg, P., and Sundby, B., 2001b. Permeable marine sediments, overturning an old paradigm. EOS, 82: 133.

Boudreau, B.P., and Imboden, D.M., 1987. Mathematics of tracer mixing in sediments. III. The theory of nonlocal mixing within sediments. Am. J. Sci., 287: 693–719.

Boudreau, B.P., and Jørgensen, B.B. (eds.), 2001. *The Benthic Boundary Layer: Transport Processes and Biogeochemistry*. Oxford University Press.

Boudreau, B.P., and Marinelli, R.L., 1994. A modelling study of discontinuous biological irrigation. J. Mar. Res., 52: 947–968.

Boudreau, B.P., Mucci, A., Sundby, B., Luther, G.W., and Silverberg, N., 1998. Comparative diagenesis at three sites on the Canadian continental margin. J. Mar. Res., 56: 1259–1284.

Boudreau, B.P., and Ruddick, B.R., 1991. On a reactive continuum representation of organic matter diagenesis. Am. J. Sci., 291: 507–538.

Boudreau, B.P., and Scott, M.R., 1978. A model for the diffusion controlled growth of deep-sea manganese nodules. Am. J. Sci., 278: 903–929.

Boudreau, B.P., and Westrich, J.T., 1984. The dependence of bacterial sulfate reduction on sulfate concentrations in marine sediments. Geochim. Cosmochim. Acta, 48: 2503–2516.

Brandes, J.A., and Devol, A.H., 1995. Simultaneous nitrate and oxygen respiration in coastal sediments: evidence for discrete diagenesis. J. Mar. Res., 53: 771–797.

Brandes, J.A., and Devol, A.H., 1997. Isotopic fractionation of oxygen and nitrogen in coastal marine sediments. Geochim. Cosmochim. Acta, 61: 1793–1801.

Brandes, J.A., and Devol, A.H., 2002. A global marine–fixed nitrogen isotopic budget: implications for Holocene nitrogen cycling. Global Biogeochem. Cycles, 16: 1120.

Brassell, S.C., 1993. Applications of biomarkers for delineating marine paleoclimatic fluctuations during the Pleistocene. *In*: M.H. Engel and S.A. Macko (eds.), *Organic Geochemistry*. Plenum Press, pp. 699–738.

Breitzke, M., 2000. Physical properties of marine sediments. *In*: H.D. Schulz and M. Zabel (eds.), *Marine Geochemistry*. Springer-Verlag, pp. 29–72.

Broecker, W.S., 1974. *Chemical Oceanography*. Harcourt Brace Jovanovich.

Broecker, W.S., and Clark, E., 2003. Pseudo dissolution of marine calcite. Earth Planet. Sci. Lett., 208: 291–296.

Broecker, W.S., and Peng, T.H., 1982. *Tracers in the Sea*. Eldigio Press.

Brown, F.S., Baedecker, M.J., Nissenbaum, A., and Kaplan, I.R., 1972. Early diagenesis in a reducing fjord, Saanich Inlet, British Columbia. Geochim. Cosmochim. Acta, 36: 1185–1203.

Brüchert, V., and Arnosti, C., 2003. Anaerobic carbon transformation: experimental studies with flow-through cells. Mar. Chem., 80: 171–183.

REFERENCES

Brüchert, V., and Pratt, L.M., 1996. Contemporaneous early diagenetic formation of organic and inorganic sulfur in estuarine sediments from St. Andrew Bay, Florida, USA. Geochim. Cosmochim. Acta, 60: 2325–2332.

Buckley, D.E., and Cranston, R.E., 1988. Early diagenesis in deep sea turbidites: the imprint of paleo-oxidation zones. Geochim. Cosmochim. Acta, 52: 2925–2939.

Buffett, B., and Archer, D., 2004. Global inventory of methane clathrate: sensitivity to changes in the deep ocean. Earth Planet. Sci. Lett., 227: 185–199.

Burdige, D.J., 1991. The kinetics of organic matter mineralization in anoxic marine sediments. J. Mar. Res., 49: 727–761.

Burdige, D.J., 1993. The biogeochemistry of manganese and iron reduction in marine sediments. Earth-Sci. Rev., 35: 249–284.

Burdige, D.J., 2001. Dissolved organic matter in Chesapeake Bay sediment pore waters. In: E. Canuel and T. Bianchi (eds.), *Organic Geochemical Tracers in Estuaries*. Org. Geochem. 32: 487–505.

Burdige, D.J., 2002. Sediment pore waters. In: D.A. Hansell and C.D. Carlson (eds.), *Biogeochemistry of Marine Dissolved Organic Matter*. Academic Press, pp. 611–663.

Burdige, D.J., 2005. The burial of terrestrial organic matter in marine sediments: A re-assessment. Global Biogeochem. Cycles, 19: 10-1029/200468002368.

Burdige, D.J., Alperin, M.J., Homstead, J., and Martens, C.S., 1992a. The role of benthic fluxes of dissolved organic carbon in oceanic and sedimentary carbon cycling. Geophys. Res. Lett., 19: 1851–1854.

Burdige, D.J., Berelson, W.M., Coale, K.H., McManus, J., and Johnson, K.S., 1999. Fluxes of dissolved organic carbon from California continental margin sediments. Geochim. Cosmochim. Acta, 63: 1507–1515.

Burdige, D.J., Dhakar, S.P., and Nealson, K.H., 1992b. Effects of manganese oxide mineralogy on microbial and chemical manganese reduction. Geomicrobiol. J., 10: 27–48.

Burdige, D.J., and Gardner, K.G., 1998. Molecular weight distribution of dissolved organic carbon in marine sediment pore waters. Mar. Chem., 62: 45–64.

Burdige, D.J., Gardner, K.G., and Skoog, A., 2000. Dissolved and particulate carbohydrates in contrasting marine sediments. Geochim. Cosmochim. Acta, 64: 1029–1041.

Burdige, D.J., Gardner, K.G., and Zheng, S., 1996. The molecular weight distribution of dissolved organic matter (DOM) in marine sediment pore waters. EOS, 76(3): OS11K (abstr.).

Burdige, D.J., and Gieskes, J.M., 1983. A porewater/solid phase diagenetic model for manganese in marine sediments. Am. J. Sci., 283: 29–47.

Burdige, D.J., and Homstead, J., 1994. Fluxes of dissolved organic carbon from Chesapeake Bay sediments. Geochim. Cosmochim. Acta, 58: 3407–3424.

Burdige, D.J., Kline, S.W., and Chen, W., 2004. Fluorescent dissolved organic matter in marine sediment pore waters. Mar. Chem., 89: 289–311.

Burdige, D.J., and Martens, C.S., 1988. Biogeochemical cycling in an organic-rich marine basin—10. The role of amino acids in sedimentary carbon and nitrogen cycling. Geochim. Cosmochim. Acta, 52: 1571–1584.

Burdige, D.J., and Martens, C.S., 1990. Biogeochemical cycling in an organic-rich marine basin—11. The sedimentary cycling of dissolved free amino acids. Geochim. Cosmochim. Acta, 54: 3033–3052.

Burdige, D.J., and Zheng, S., 1998. The biogeochemical cycling of dissolved organic nitrogen in estuarine sediments. Limnol. Oceanogr., 43: 1796–1813.

Burdige, D.J., and Zimmerman, R.C., 2002. Impact of seagrass density on carbonate dissolution in Bahamian sediments. Limnol. Oceanogr., 47: 1751–1763.

Burke, I.T., and Kemp, A.E.S., 2002. Microfabric analysis of Mn-carbonate laminae deposition and Mn-sulfide formation in the Gotland Deep, Baltic Sea. Geochim. Cosmochim. Acta, 66: 1589–1600.

Burnett, B.A., and Nealson, K.H., 1981. Organic films and microorganisms associated with manganese nodules. Deep-Sea Res., 28A: 637–645.

Burnett, W.C., Beers, M.J., and Roe, K.K., 1982. Growth rates of phosphate nodules from the continental margin off Peru. Science, 215: 1616–1618.

Burns, R.G., and Burns, V.M., 1979. Manganese minerals. *In*: R.G. Burns (ed.), *Marine Minerals*. MSA Short Course Notes. Vol. 6, pp. 1–46.

Burns, R.G., and Burns, V.M., 1981. Authigenic oxides. *In*: C. Emiliani (ed.), *The Sea*. Vol. 7, *The Oceanic Lithosphere*. Wiley-Interscience Pub. pp. 875–914.

Burns, S.J., 1998. Carbon isotopic evidence for coupled sulfate reduction–methane oxidation in Amazon Fan sediments. Geochim. Cosmochim. Acta, 62: 797–804.

Cai, W., and Sayles, F.L., 1996. Oxygen penetration depths and fluxes in marine sediments. Mar. Chem., 52: 123–131.

Cai, W.-J., and Reimers, C.E., 1993. The development of pH and $p\mathrm{CO}_2$ microelectrodes for studying the carbonate chemistry of pore waters near the sediment-water interface. Limnol. Oceanogr., 38: 1762–1773.

Cai, W.-J., and Reimers, C.E., 1995. Benthic oxygen flux, bottom water oxygen concentration and core top organic carbon content in the deep northeast Pacific Ocean. Deep-Sea Res., 42: 1681–1699.

Cai, W.-J., Reimers, C.E., and Shaw, T.J., 1995. Microelectrode studies of organic carbon degradation and calcite dissolution at a California continental rise site. Geochim. Cosmochim. Acta, 59: 497–511.

Calvert, S.E., 1978. Geochemistry of oceanic ferromanganese deposits. Phil. Trans. R. Soc. London, A, 290: 43–73.

Calvert, S.E., Pederen, T.F., and Karlin, R.E., 2001. Geochemical and isotopic evidence for post-glacial paleoceanographic changes in Saanich Inlet, British Columbia. Mar. Geol., 174: 287–305.

REFERENCES

Calvert, S.E., and Pedersen, T.F., 1992. Organic carbon accumulation and preservation in marine sediments: How important is anoxia? *In*: J.K. Whelan and J.W. Farrington (eds.), *Productivity, Accumulation, and Preservation of Organic Matter in Recent and Ancient Sediments*. Columbia University Press, pp. 231–263.

Calvert, S.E., and Pedersen, T.F., 1993. Geochemistry of recent oxic and anoxic marine sediments: Implications for the geological record. Mar. Geol., 113: 67–88.

Calvert, S.E., and Pedersen, T.F., 1996. Sedimentary geochemistry of manganese: Implication for the environment of formation of Manganiferous Black Shales. Econ. Geol., 91: 36–47.

Camacho-Ibar, V.F., Aveytua-Alcázar, L., and Carrquiry, J.D., 2003. Fatty acid reactivities in sediment cores from the northern Gulf of California. Org. Geochem., 34: 425–439.

Cane, M., and Clement, A.C., 1999. A role for the tropical Pacific coupled ocean-atmosphere system on Milankovitch and Millennial timescales. Part II: Global impacts. *In*: P.U. Clark, R.S. Webb, and L.D. Keigwin (eds.), *Mechanisms of Global Climate Change at Millennial Time Scales*. American Geophysical Union, pp. 373–383.

Canfield, D., and Thamdrup, B., 1994. The production of ^{34}S-depleted sulfide during bacterial disproportionation of elemental sulfur. Science, 266: 1973–1975.

Canfield, D.E., 1989a. Reactive iron in marine sediments. Geochim. Cosmochim. Acta, 53: 619–632.

Canfield, D.E., 1989b. Sulfate reduction and oxic respiration in marine sediments: implications for organic carbon preservation in euxinic environments. Deep-Sea Res., 36: 121–138.

Canfield, D.E., 1994. Factors influencing organic matter preservation in marine sediments. Chem. Geol., 114: 315–329.

Canfield, D.E., Jorgensen, B.B., Fossing, H., Glud, R., Gundersen, J., Ramsing, N.B., Thamdrup, B., Hansen, J.W., Nielsen, L.P., and Hall, P.O.J., 1993a. Pathways of organic carbon oxidation in three continental margin sediments. Mar. Geol., 113: 27–40.

Canfield, D.E., Raiswell, R., Westrich, J.T., Reaves, C.M., and Berner, R.A., 1986. The use of chromium reduction in the analysis of reduced inorganic sulfur in sediments and shales. Chem. Geol., 54: 149–155.

Canfield, D.E., and Raiswell, R., 1991a. Carbonate dissolution and precipitation: its relevance to fossil preservation. *In*: P.A. Allison and D.E.G. Briggs (eds.), *Taphony: Releasing the Data Locked in the Fossil Record*. Plenum Press, pp. 411–453.

Canfield, D.E., and Raiswell, R., 1991b. Pyrite formation and fossil preservation. *In*: P.A. Allison and D.E.G. Briggs (eds.), *Taphonomy: Releasing the Data Locked in the Fossil Record*. Plenum Press, pp. 337–387.

REFERENCES

Canfield, D.E., Raiswell, R., and Bottrell, S., 1992. The reactivity of sedimentary iron minerals toward sulfide. Am. J. Sci., 292: 659–683.

Canfield, D.E., Thamdrup, B., and Hansen, J.W., 1993b. The anaerobic degradation of organic matter in Danish coastal sediments: Fe reduction, Mn reduction, and sulfate reduction. Geochim. Cosmochim. Acta, 57: 2563–2570.

Canuel, E.A., and Martens, C.S., 1993. Seasonal variations in the sources and alteration of organic matter associated with recently-deposited sediments. Org. Geochem., 20: 563–577.

Canuel, E.A., and Martens, C.S., 1996. Reactivity of recently deposited organic matter: degradation of lipid compounds near the sediment-water interface. Geochim. Cosmochim. Acta, 60: 1793–1806.

Capone, D.G., and Kiene, R.P., 1988. Comparison of microbial dynamics in marine and freshwater sediments: Contrasts in anaerobic carbon catabolisms. Limnol. Oceanogr., 33: 725–749.

Chanton, J.P., 1985. Sulfur mass balance and isotopic fractionation in an anoxic marine sediment. Ph.D. thesis, Univ. of North Carolina, Chapel Hill.

Chanton, J.P., 1989. Gas transport from methane-saturated, tidal freshwater and wetland sediments. Limnol. Oceanogr., 34: 807–819.

Chanton, J.P., Martens, C.S., and Goldhaber, M.B., 1987a. Biogeochemical cycling in an organic-rich marine basin—7. Sulfur mass balance, oxygen uptake and sulfide retention. Geochim. Cosmochim. Acta, 51: 1187–1199.

Chanton, J.P., Martens, C.S., and Goldhaber, M.B., 1987b. Biogeochemical cycling in an organic-rich marine basin—8. A sulfur isotopic budget balanced by differential diffusion across the sediment-water interface. Geochim. Cosmochim. Acta, 51: 1201–1208.

Chanton, J.P., Martens, C.S., and Kipphut, G.W., 1983. Lead-210 sediment geochronology in a changing coastal environment. Geochim. Cosmochim. Acta, 47: 1791–1804.

Chen, C.-T.A., 2002. Shelf- vs. dissolution-generated alkalinity above the chemical lysocline. Deep-Sea Res. II, 49: 5365–5375.

Chen, R.F., Bada, J.L., and Suzuki, Y., 1993. The relationship between dissolved organic carbon (DOC) and fluorescence in anoxic marine porewaters: Implications for estimating benthic DOC fluxes. Geochim. Cosmochim. Acta, 57: 2149–2153.

Chen, Z., and Mayer, L.M., 1998. Mechanisms of Cu solubilization during deposit feeding. Environ. Sci. Tech., 32: 770–775.

Chester, R., 2000. *Marine Geochemistry*, 2nd ed. Blackwell.

Christensen, J.P., 1989. Sulfate reduction and carbon oxidation rates in continental shelf sediments, an examination of offshelf carbon transport. Cont. Shelf Res., 9: 223–246.

Christensen, J.P., 2000. A relationship between deep-sea benthic oxygen demand and oceanic primary productivity. Oceanologica Acta, 23: 65–82.

Christensen, J.P., Devol, A.H., and Smethie, W.M.J., 1984. Biological enhancement of solute exchange between sediments and bottom water on the Washington continental shelf. Cont. Shelf Res., 3: 9–23.

Christensen, J.P., Smethie, W.M.J., and Devol, A.H., 1987. Benthic nutrient regeneration and denitrification on the Washington continental shelf. Deep-Sea Res., 34: 1027–1047.

Claypool, G.E., and Kaplan, I.R., 1974. The origin and distribution of methane in marine sediments. *In*: I.R. Kaplan (ed.), *Natural Gases in Marine Sediments*. Plenum Press, pp. 99–139.

Cochran, J.K., 1985. Particle mixing rates in sediments of the eastern equatorial Pacific: evidence from ^{210}Pb, ^{239,240}Pu, and ^{137}Cs distributions at MANOP sites. Geochim. Cosmochim. Acta, 49: 1195–1210.

Codispoti, L.A., Brandes, J.A., Christensen, J.P., Devol, A.H., Naqvi, S.W.A., Paerl, H.W., and Yoshinari, T., 2001. The oceanic fixed nitrogen and nitrous oxide budgets: moving targets as we enter the anthropocene? Scientia Marina, 65: 85–105.

Colberg, P.J., 1988. Anaerobic microbial degradation of cellulose, lignin, oligolignols, and monoaromatic derivatives. *In*: A.J.B. Zehnder (ed.), *Biology of Anaerobic Microrganisms*. Wiley-Interscience, pp. 333–372.

Colbert, D., Coale, K.H., Berelson, W.M., and Johnson, K.S., 2001. Cadmium flux in Los Angeles/Long Beach harbours and at sites along the California continental margin. Est. Coastal Shelf Sci., 53: 169–180.

Cole, T.G., 1985. Composition, oxygen isotope geochemistry, and origin of smectite in the metalliferous sediments of the Bauer Deep, southeast Pacific. Geochim. Cosmochim. Acta, 49: 221–235.

Colley, S., and Thomson, J., 1985. Recurrent uranium relocations in distal turbidites emplaced in pelagic conditions. Geochim. Cosmochim. Acta, 49: 2339–2348.

Colley, S., Thomson, J., Wilson, T.R.S., and Higgs, N.C., 1984. Post-depositional migration of elements during diagenesis in brown clay and turbite sequences in the northeast Atlantic. Geochim. Cosmochim. Acta, 48: 1223–1235.

Collier, R., and Edmond, J., 1984. The trace element geochemistry of marine biogenic particulate matter. Prog. Oceanog., 13: 113–199.

Collins, M.J., Bishop, A.N., and Farrimond, P., 1995. Sorption by mineral surfaces: Rebirth of the classical condensation pathway for kerogen formation? Geochim. Cosmochim. Acta, 59: 2387–2391.

Colman, A.S., and Holland, H.D., 2000. The global diagenetic flux of phosphorus from marine sediments to the oceans: Redox sensitivity and the control of atmospheric oxygen levels. *In*: *Marine Authigenesis: From Global to Microbial*. SEPM Spec. Publ. no. 66, pp. 53–75.

Colman, A.S., Mackenzie, F.T., and Holland, H.D., 1997. Redox stabilization of the atmosphere and oceans and marine productivity. Science, 275: 406–407.

Colombo, J.C., Silverberg, N., and Gearing, J.N., 1998. Amino acid biogeochemistry in the Laurentian Trough: vertical fluxes and individual reactivity during early diagenesis. Org. Geochem., 29: 933–945.

Cooper, D.C., and Morse, J.W., 1998a. Biogeochemical controls on trace metal cycling in anoxic marine sediments. Environ. Sci. Tech., 32: 327–330.

Cooper, D.C., and Morse, J.W., 1998b. Extractability of metal sulfide minerals in acidic solutions: application to environmental studies of trace metal contamination. Environ. Sci. Tech., 32: 1076–1078.

Coppedge, M.L., and Balsam, W.L., 1992. Organic carbon distribution in the North Atlantic Ocean during the last glacial maximun. Mar. Geol., 105: 37–50.

Cornwell, J.C., Conley, D.J., Owens, M., and Stevenson, J.C., 1996. A sediment chronology of the eutrophication of Chesapeake Bay. Estuaries, 19: 488–499.

Cornwell, J.C., Kemp, W.M., and Kana, T.M., 1999. Denitrification in coastal ecosystems: methods, environmental controls, and ecosystem level controls, a review. Aquat. Ecol., 33: 41–54.

Cornwell, J.C., and Morse, J.W., 1987. The characterization of iron sulfide minerals in anoxic marine sediments. Mar. Chem, 22: 193–206.

Cornwell, J.C., and Sampou, P.A., 1995. Environmental controls on iron sulfide mineral formation in a coastal plain estuary. *In*: M.A. Vairavamurthy and M.A.A. Schoonen (eds.), *Geochemical Transformations of Sedimentary Sulfur*. American Chemical Society, pp. 224–242.

Cowan, J.L.W., and Boynton, W.R., 1996. Sediment-water oxygen and nutrient exchanges across the longitudinal axis of Chesapeake Bay: seasonal patterns, controlling factors and ecological significance. Estuaries, 19: 562–580.

Cowie, G.L., and Hedges, J.I., 1984. Carbohydrate sources in a coastal marine environment. Geochim. Cosmochim. Acta, 48: 2075–2087.

Cowie, G.L., and Hedges, J.I., 1992a. An improved method for quantification of amino acids: charge-matched recovery standards and reduced time requirements. Mar. Chem., 37: 223–238.

Cowie, G.L., and Hedges, J.I., 1992b. The role of anoxia in organic matter preservation in coastal sediments: relative stabilities of the major biochemicals under oxic and anoxic depositional conditions. Org. Geochem., 19: 229–234.

Cowie, G.L., and Hedges, J.I., 1992c. Sources and reactivities of amino acids in a coastal marine environment. Limnol. Oceanogr., 37: 703–724.

Cowie, G.L., and Hedges, J.I., 1994. Biochemical indicators of diagenetic alteration in natural organic matter mixtures. Nature, 369: 304–307.

Cowie, G.L., Hedges, J.I., and Calvert, S.E., 1992. Sources and reactivity of amino acids, neutral sugars, and lignin in an intermittently anoxic marine environment. Geochim. Cosmochim. Acta, 56: 1963–1978.

Cowie, G.L., Hedges, J.I., Prahl, F.G., and deLange, G.J., 1995. Elemental and biochemical changes across an oxidation front in a relict turbidite: an oxygen effect. Geochim. Cosmochim. Acta, 59: 33–46.

Crank, J., 1975. *The Mathematics of Diffusion*, 2nd ed. Clarendon Press.

Cranwell, P.A., 1982. Lipids of aquatic sediments and sedimenting particulates. Prog. Lipid Res., 21: 271–308.

Crill, P.M., and Martens, C.S., 1983. Spatial and temporal fluctuations of methane production in anoxic coastal marine sediments. Limnol. Oceanogr., 28: 1117–1130.

Crill, P.M., and Martens, C.S., 1986. Methane production from bicarbonate and acetate in an anoxic marine sediment. Geochim. Cosmochim. Acta, 50: 2089–2097.

Crill, P.M., and Martens, C.S., 1987. Biogeochemical cycling in an organic-rich marine basin—6. Temporal and spatial variations in sulfate reduction rates. Geochim. Cosmochim. Acta, 51: 1175–1186.

Crusius, J., and Thomson, J., 2003. Mobility of authigenic rhenium, silver, and selenium during postdepositional oxidation in marine sediments. Geochim. Cosmochim. Acta, 67: 265–273.

Curry, W.B., Duplessy, J.C., Labeyrie, L.D., and Shackleton, N.J., 1988. Changes in the distribution of $\delta^{13}C$ of deep water ΣCO_2 between the last glaciation and the Holocene. Paleoceanography, 3: 317–341.

D'Hondt, S., Rutherford, S., and Spivack, A.J., 2002. Metabolic activity of subsurface life in deep-sea sediments. Science, 295: 2067–2070.

Dansgaard, W., 1964. Stable isotopes in precipitation. Tellus, 16: 436–468.

Dauwe, B., and Middelburg, J.J., 1998. Amino acids and hexosamines as indicators of organic matter degradation state in North Sea sediments. Limnol. Oceanogr., 43: 782–798.

Dauwe, B., Middelburg, J.J., and Herman, P.M.J., 2001. Effect of oxygen on the degradablity of organic matter in subtidal and intertidal sediments of the North Sea area. Mar. Ecol. Prog. Ser., 215: 13–22.

Dauwe, B., Middelburg, J.J., Herman, P.M.J., and Heip, C.H.R., 1999. Linking diagenetic alteration of amino acids and bulk organic matter reactivity. Limnol. Oceanogr., 44: 1809–1814.

de Lange, G.J., 1986. Early diagenetic reactions in interbedded pelagic and turbiditic sediments in the Nares Abyssal Plain (western North Atlantic): Consequences for the composition of sediment and interstitial waters. Geochim. Cosmochim. Acta, 50: 2543–2561.

de Lange, G.J., 1992. Distribution of exchangeable, fixed, organic and total nitrogen in interbedded turbiditic/pelagic sediments of the Madeira Abyssal Plain, eastern North Atlantic. Mar. Geol., 109: 95–114.

de Lange, G.J., Jarvis, I., and Kuijpers, A., 1987. Geochemical characteristics and provenance of late Quaternary sediments from the Madeira Abyssal

Plain, N Atlantic. *In*: P.P.E. Weaver and J. Thomson (eds.), *Geology and Geochemistry of Abyssal Plains*, Geol. Soc. Spec. Publ. no. 31, pp. 147–165.

de Lange, G.J., Middelburg, J.J., and Pruysers, P.A., 1989. Middle and late quaternary depositional sequences and cycles in the eastern Mediterranean. Sedimentology, 36: 151–158.

de Lange, G.J., van Os, B., Pryusers, P.A., Middelburg, J.J., Castradori, D., van Santvoort, P., Müller, P.J., Eggenkamp, H., and Prahl, F.G., 1994. Possible early diagenetic alteration of paleo proxies. *In*: R. Zahn, T.F. Pedersen, M.A. Kaminski, and L. Labeyrie (eds.), *Carbon Cycling in the Glacial Ocean: Constraint on the Ocean's Role in Global Change*. vol. 117, NATO ASI Series. Springer-Verlag, pp. 225–258.

de Leeuw, J.W., and Largeau, C., 1993. A review of macromolecular organic compounds that comprise living organisms and their role in kerogen, coal, and petroleum formation. *In*: M.H. Engel and S.A. Macko (eds.), *Organic Geochemistry*. Plenum Press, pp. 23–72.

Dean, W.E., Gardner, J.V., and Hemphill-Haley, E., 1989. Changes in redox conditions in deep-sea sediments of the subarctic north Pacific Ocean: Possible evidence for the presence of north Pacific deep water. Paleoceanography, 4: 639–653.

Dean, W.E., and Ghosh, S.K., 1978. Factors contributing to the formation of ferromanganese nodules in Oneida Lake, New York. U.S. Geol. Surv. J. Res., 6: 231–240.

Dean, W.E., Leinen, M., and Stow, D.A.V., 1985. Classification of deep-sea, fine-grained sediments. J. Sed. Petr., 55: 250–256.

Deflandre, B., Mucci, A., Gagne, J., Guignard, C., and Sundby, B., 2002. Early diagenetic processes in coastal marine sediments disturbed by a catastrophic sedimentation event. Geochim. Cosmochim. Acta, 66: 2547–2558.

Degens, E.T., 1969. Biogeochemistry of stable carbon isotopes. *In*: G. Eglinton and M.T.J. Murphy (eds.), *Organic Geochemistry*. Springer-Verlag, pp. 304–329.

Delaney, M.L., 1998. Phosphorus accumulation in marine sediments and the oceanic phosphorus cycle. Global Biogeochem. Cycles, 12: 563–572.

Demaison, G.J., and Moore, G.T., 1980. Anoxic environments and oil source bed genesis. AAPG Bull., 64: 1179–1209.

DeMaster, D.J., and Aller, R.C., 2001. Biogeochemical processes on the Amazon shelf: Changes in dissolved and particulate fluxes during river/ocean mixing. *In*: M.E. McClain, R.L. Victoria and J.E. Richey (eds.), *The Biogeochemistry of the Amazon Basin*. Oxford University Press, pp. 328–357.

DeMaster, D.J., Brewster, D.C., McKee, B.A., and Nittrouer, C.A., 1991. Rates of particle scavenging, sediment reworking, and longitudinal ripple formation at the HEBBLE site based on measurements of ^{234}Th and ^{210}Pb. Mar. Geol., 99: 423–444.

DeMaster, D.J., and Cochran, J.K., 1982. Particle mixing rates in deep-sea sediments determined from excess ^{210}Pb and ^{32}Si profiles. Earth Planet. Sci. Lett., 61: 257–271.

DeMaster, D.J., Knapp, G.B., and Nittrouer, C.A., 1983. Biological uptake and accumulation of silica on the Amazon continental shelf. Geochim. Cosmochim. Acta, 47: 1713–1723.

DeMaster, D.J., Pope, R.H., Levin, L.A., and Blair, N.E., 1994. Biological mixing intensity and rates of organic carbon accumulation in North Carolina slope sediments. Deep-Sea Res. II, 41: 735–753.

Deming, J.W., and Baross, J.A., 1993. The early diagenesis of organic matter: bacterial activity. *In*: M.H. Engel and S.A. Macko (eds.), *Organic Geochemistry*. Plenum Press, pp. 119–144.

Derenne, S., Knicker, H., Largeau, C., and Hatcher, P., 1998. Timing and mechanisms of changes in nitrogen functionality during biomass fossilization. *In*: B.A. Stankiewicz and P.F. van Bergen (eds.), *Nitrogen-containing Macromolecules in the Bio- and Geosphere*. American Chemical Society, pp. 243–253.

Devol, A.H., 1978. Bacterial oxygen uptake kinetics as related to biological processes in oxygen deficient zones of the ocean. Deep-Sea Res., 25: 137–146.

Devol, A.H., 1987. Verification of flux measurements made with in situ bentic chambers. Deep-Sea Res., 34: 1007–1026.

Devol, A.H., 1991. Direct measurement of nitrogen gas fluxes from continental margin sediments. Nature, 349: 319–321.

Devol, A.H., Anderson, J.J., Kuivala, K., and Murray, J.W., 1984. A model for coupled sulfate reduction and methane oxidation in the sediments of Saanich Inlet. Geochim. Cosmochim. Acta, 48: 933–1004.

Devol, A.H., and Christensen, J.P., 1993. Benthic fluxes and nitrogen cycling in sediments of the continental margin of the eastern North Pacific. J. Mar. Res., 51: 345–372.

Devol, A.H., Codispoti, L.A., and Christensen, J.P., 1997. Summer and winter denitrification rates in western Arctic shelf sediments. Cont. Shelf Res., 17: 1029–1050.

Dhakar, S.P., 1995. A time dependent diagenetic model for manganese redox cycling in deep sea sediments. Ph.D. dissertation, Old Dominion University, Norfolk, Va.

Dhakar, S.P., and Burdige, D.J., 1996. A coupled, non-linear, steady state model for early diagenetic processes in pelagic sediments. Am. J. Sci., 296: 296–330.

Dickens, A.F., Gélinas, Y., and Hedges, J.I., 2004a. Physical separation of combustion and rock sources of graphitic black carbon in sediments. Mar. Chem., 92: 215–233.

Dickens, A.F., Gélinas, Y., Masiello, C.A., Wakeham, S.G., and Hedges, J.I., 2004b. Reburial of fossil organic carbon in marine sediments. Nature, 427: 336–339.

REFERENCES

Dickens, G.R., 2001. Sulfate profiles and barium fronts in sediment on the Blake Ridge: Present and past methane fluxes through a large gas hydrate resevoir. Geochim. Cosmochim. Acta, 65: 529–543.

Dietrich, R.V., and Skinner, B.J., 1979. *Rocks and Rock Minerals.* John Wiley & Sons.

Ding, X., and Henrichs, S.M., 2002. Adsorption and desorption of proteins and polyamino acids by clay minerals and marine sediments. Mar. Chem., 77: 225–237.

DiToro, D.M., 2001. *Sediment Flux Modeling.* Wiley-Interscience.

Dixit, S., and Van Cappellen, P., 2002. Surface chemistry and reactivity of biogenic silica. Geochim. Cosmochim. Acta, 66: 2559–2568.

Dixit, S., Van Cappellen, P., and van Bennekom, A.J., 2001. Processes controlling solubility of biogenic silica and pore water build-up of silicic acid in marine sediments. Mar. Chem., 73: 333–352.

Dollar, S.J., Smith, S.V., Vink, S.M., Obrebski, S., and Hollibaugh, J.T., 1991. Annual cycle of benthic nutrient fluxes in Tomales Bay, California, and contribution of the benthos to total carbon metabolism. Mar. Ecol. Prog. Ser., 79: 115–125.

Donat, J.R., and Bruland, K.W., 1995. Trace elements in the oceans. *In*: E. Steinnes and B. Salbu (eds.), *Trace Elements in Natural Waters.* CRC Press, pp. 247–281.

Dong, H., Kostka, J.E., and Kim, J., 2003a. Microscopic evidence for microbial dissolution of smectite. Clays and Clay Mins., 51: 502–512.

Dong, H., Kukkadapu, R.K., Fredrickson, J.K., Zachara, J.M., Kennedy, D.W., and Kostandarithes, H.M., 2003b. Microbial reduction of structural Fe(III) in illite and geothite. Environ. Sci. Tech., 37: 1268–1276.

Dymond, J., 1981. Geochemistry of Nazca plate surface sediments: an evaluation of hydrothermal, biogenic, detrital and hydrogenous sources. *In: Nazca Plate: Crustal Formation and Andean Convergence.* Geol. Soc. Am. Mem. 154: 133–173.

Dymond, J., 1984. Sediment traps, particle fluxes, and benthic boundary layer processes. *In: Global Ocean Flux Study, Proceedings of a Workshop.* Natl. Acad. Press, pp. 260–284.

Dymond, J., Corliss, J.B., Heath, G.R., Field, C.W., Dasch, E.J., and Veeh, H.H., 1973. Origin of metalliferous sediments from the Pacific Ocean. Geol. Soc. Am. Bull., 84: 3355–3372.

Dymond, J., Lyle, M., Finney, B., Piper, D.Z., Murphy, K., Conrad, R., and Pisias, N., 1984. Ferromanganese nodules from MANOP sites H, S, and R—control of mineralogical and chemical composition by multiple accretionary processes. Geochim. Cosmochim. Acta, 48: 1913–1928.

Edmond, J.M., Corliss, J.B., and Gordon, L.I., 1979a. Ridge crest hydrothermal metamorphism at the Galapagos spreading center and reverse weathering. *In*: M. Talwani (ed.), *Ewing Symposium*, vol. 2. AGU, pp. 383–390.

REFERENCES

Edmond, J.M., Measures, C., Mangum, B., Grant, B., Sclater, F.R., Collier, R., Hudson, A., Gordon, L.I., and Corliss, J.B., 1979b. On the formation of metal-rich deposits at ridge crests. Earth Planet. Sci. Lett., 46: 19–30.

Egeberg, P.K., and Dickens, G.R., 1999. Thermodyanmic and pore water halogen constraints on gas hydrate accumulation at ODP site 997 (Blake Ridge). Chem. Geol., 153: 53–79.

Eglinton, T.I., Benitez-Nelson, B.C., Pearson, A., McNichol, A.P., Bauer, J.E., and Druffel, E.R.M., 1997. Variability in radiocarbon ages of individual organic compunds from marine sediments. Science, 277: 796–799.

Ehrlich, H.L., 1996. *Geomicrobiology*. Marcel Dekker.

Elderfield, H., 1981. Metal-organic associations in interstitial waters of Narragansett Bay sediments. Am. J. Sci., 281: 1184–1196.

Elderfield, H., Luedtke, N., McCafrey, R.J., and Bender, M., 1981a. Benthic flux studies in Narragansett Bay. Am. J. Sci., 281: 768–787.

Elderfield, H., McCafrey, R.J., Luedtke, N., Bender, M., and Truesdale, V.W., 1981b. Chemical diagenesis in Narragansett Bay sediments. Am. J. Sci., 281: 1021–1055.

Elderfield, H., and Schultz, A., 1996. Mid-ocean ridge hydrothermal fluxes and the chemical composition of the ocean. Ann. Rev. Earth Planet. Sci., 24: 191–224.

Emerson, S., 1985. Organic carbon preservation in marine sediments. *In*: E.T. Sundquist and W.S. Broecker (eds.), *The Carbon Cycle and Atmospheric CO_2: Natural Variations Archaen to Present*. AGU Geophys. Monogr, 32: 78–87.

Emerson, S., and Bender, M.L., 1981. Carbon fluxes at the sediment-water interface: calcium carbonate preservation. J. Mar. Res., 39: 139–162.

Emerson, S., Fischer, K., Reimers, C., and Heggie, D., 1985. Organic carbon dynamics and preservation in deep-sea sediments. Deep-Sea Res., 32: 1–21.

Emerson, S., Grundmanis, V., and Graham, D., 1982. Carbonate chemistry in marine pore waters: MANOP sites C and S. Earth Planet. Sci. Lett., 61: 220–232.

Emerson, S., and Hedges, J.I., 1988. Processes controlling the organic carbon content of open ocean sediments. Paleoceanography, 3: 621–634.

Emerson, S., and Hedges, J.I., 2003. Sediment diagenesis and benthic flux. *In*: H. Elderfield (ed.), *Treatise on Geochemistry*. Vol. 6. Elsevier, pp. 293–319.

Emerson, S., Jahnke, R., Bender, M., Froelich, P., Klinkhammer, G., Bowser, C., and Setlock, G., 1980. Early diagenesis in sediments from the eastern equatorial Pacific, I. Porewater nutrient and carbonate results. Earth Planet. Sci. Lett., 49: 57–80.

Emerson, S., Jahnke, R., and Heggie, D., 1984. Sediment-water exchange in shallow water estuarine sediments. J. Mar. Res., 42: 709–730.

Emerson, S., Stump, C., Grootes, P.M., Stuiver, M., Farwell, G.W., and Schmidt., F.H., 1987. Estimates of degradable organic carbon in deep sea sediments from ^{14}C concentrations. Nature, 329: 51–53.

REFERENCES

Emerson, S.R., and Archer, D., 1990. Calcium carbonate preservation in the ocean. Phil. Trans. R. Soc. London, A, 331: 29–40.

Emery, K.O., 1960. *The Sea Off Southern California*. J. Wiley & Sons.

Emery, K.O., 1968. Relict sediments on continental shelves of the world. AAPG Bull., 52: 445–464.

Emiliani, C., 1955. Pleistocene temperatures. J. Geol., 63: 538–578.

Engelhardt, W.V., 1977. *The Origin of Sediments and Sedimentary Rocks*, 2nd ed. John Wiley & Sons.

Epstein, S., Buchsbaum, R., Lowenstam, H., and Urey, H.C., 1951. Carbonate-water isotopic temperature scale. Bull. Geol. Soc. Am., 62: 417–426.

Ertel, J.R., and Hedges, J.I., 1985. Sources of sedimentary humic substances: vascular plant debris. Geochim. Cosmochim. Acta, 49: 2097–2107.

Farrington, J.W., Henrichs, S.M., and Anderson, R., 1977. Fatty acids and Pb-210 geochronology of a sediment core from Buzzards Bay, Massachusetts. Geochim. Cosmochim. Acta, 41: 289–296.

Farrington, J.W., and Tripp, J.W., 1977. Hydrocarbons in western North Atlantic surface sediments. Geochim. Cosmochim. Acta, 41: 1627–1641.

Fauque, G.D., 1995. Ecology of sulfate-reducing bacteria. *In*: L.L. Barton (ed.), *Sulfate-Reducing Bacteria*. Plenum Press, pp. 217–241.

Faure, G., 1986. *Principles of Isotope Geology*, 2nd ed. John Wiley & Sons.

Faure, G., 1998. *Principles and Applications of Geochemistry*, 2nd ed. Prentice Hall.

Fenchel, T., King, G.M., and Blackburn, T.H., 1998. *Bacterial Biogeochemistry: The Ecophysiology of Mineral Cycling*. Academic Press.

Ferdelman, T.G., Fossing, H., Neumann, K., and Schulz, H.D., 1999. Sulfate reduction in surface sediments of the southeast Atlantic continental margin between 15°38'S and 27°57'S (Angola and Namibia). Limnol. Oceanogr., 44: 650–661.

Filippelli, G.M., and Delaney, M.L., 1996. Phosphorus geochemistry of equatorial Pacific sediments. Geochim. Cosmochim. Acta, 60: 1479–1495.

Finney, B.P., Lyle, M.W., and Heath, G.R., 1988. Sedimentation at MANOP site H (eastern equatorial Pacific) over the past 400,000 years: climatically induced redox variations and their effects on transition metal cycling. Paleoceanography, 3: 169–189.

Fogel, M.L., and Cifuentes, L.A., 1993. Isotope fractionation during primary production. *In*: S.A. Macko and M.H. Engel (eds.), *Organic Geochemistry*. Plenum Press, pp. 73–98.

Fogel, M.L., and Tuross, N., 1999. Transformation of plant biochemicals to geological macromolecules during early diagenesis. Oecologia, 120: 336–346.

Folk, R.L., 1980. *Petrology of Sedimentary Rocks*. Hemphill.

Föllmi, K.B., 1996. The phosphorus cycle, phosphogenesis and marine phosphate-rich deposits. Earth-Sci. Rev., 40: 55–124.

REFERENCES

Fornes, W.L., DeMaster, D.J., Levin, L.A., and Blair, N.E., 1999. Bioturbation and particle transport in Carolina slope sediments: A radiochemical approach. J. Mar. Res., 57: 335–355.

Fornes, W.L., DeMaster, D.J., and Smith, C.R., 2001. A particle introduction experiment in Santa Catalina Basin sediments: Testing the age-dependent mixing hypothesis. J. Mar. Res., 59: 97–112.

Forster, S., Glud, R.N., Gundersen, J.K., and Huettel, M., 1999. *In situ* study of bromide tracer and oxygen flux in coastal sediments. Est. Coastal Shelf Sci., 49: 813–827.

Forster, S., Huettel, M., and Ziebis, W., 1996. Impact of boundary layer flow velocity on oxygen utilisation in coastal sediments. Mar. Ecol. Prog. Ser., 143: 173–185.

Fossing, H., Ferdelman, T.G., and Berg, P., 2000. Sulfate reduction and methane oxidation in continental margin sediments influenced by irrigation (south-east Atlantic off Namibia). Geochim. Cosmochim. Acta, 64: 897–910.

Fowler, S.W., and Knauer, G.A., 1986. Role of large particles in the transport of elements and organic compounds through the oceanic water column. Prog. Oceanog., 16: 147–194.

Freeman, C., Ostle, N., and Kang, H., 2001. An enzymic "latch" on a global carbon store. Nature, 409: 149–150.

Friedman, G.M., and Sanders, J.E., 1978. *Principles of Sedimentology*. John Wiley & Sons.

Froelich, P.N., Arthur, M.A., Burnett, W.C., Deakin, M., Hensley, V., Jahnke, R., Kaul, L., Kim, K.-H., Roe, K., Soutar, A., and Vathakanon, C., 1988. Early diagenesis of organic matter in Peru continental margin sediments: phosphorite precipitation. Mar. Geol., 80: 309–343.

Froelich, P.N., Bender, M.L., Luedtke, N.A., Heath, G.R., and DeVries, T.A., 1982. The marine phosphorus cycle. Am. J. Sci., 282: 474–511.

Froelich, P.N., Kim, K.H., Jahnke, R., Burnett, W.C., Soutar, A., and Deakin, M., 1983. Pore water fluoride in Peru continental margin sediments: Uptake from seawater. Geochim. Cosmochim. Acta, 47: 1605–1612.

Froelich, P.N., Klinkhammer, G.P., Bender, M.L., Luedtke, N.A., Heath, G.R., Cullen, D., Dauphin, P., Hammond, D., Hartman, B., and Maynard, V., 1979. Early oxidation of organic matter in pelagic sediments of the eastern equatorial Atlantic: suboxic diagenesis. Geochim. Cosmochim. Acta, 43: 1075–1090.

Furukawa, Y., 2001. Biogeochemical consequences of macrofauna burrow ventilation. Geochem. Trans., 2: 83–91.

Furukawa, Y., Bentley, S.J., and Lavoie, D.L., 2001. Bioirrigation modeling in experimental benthic mesocosms. J. Mar. Res., 59: 417–452.

Furukawa, Y., Bentley, S.J., Shiller, A.M., Lavoie, D.L., and Van Cappellen, P., 2000. The role of biologically-enhanced pore water transport in early diagenesis: An example from carbonate sediments in the vicinity of North Key Harbor, Dry Tortugas National Park, Florida. J. Mar. Res., 58: 493–522.

Gächer, R., Meyer, J.S., and Mares, A., 1988. Contribution of bacteria to release and fixation of phosphorus in lake sediments. Limnol. Oceanogr., 33: 1542–1558.

Gagosian, R.B., and Peltzer, E.T., 1987. The importance of atmospheric input of terrestrial organic material to deep sea sediments. Org. Geochem., 10: 661–669.

Gallinari, M., Ragueneau, O., Corrin, L., DeMaster, D.J., and Tréguer, P., 2002. The importance of water column processes on the dissolution properties of biogenic silica in deep-sea sediments. I. Solubility. Geochim. Cosmochim. Acta, 66: 2701–2717.

Ganeshram, R.S., Calvert, S.E., Pedersen, T.F., and Cowie, G.L., 1999. Factors controlling the burial of organic carbon in laminated and bioturbated sediments off NW Mexico: implications for hydrocarbon preservation. Geochim. Cosmochim. Acta, 63: 1723–1734.

Ganeshram, R.S., Pedersen, T.F., Calvert, S.E., and Murray, J.W., 1995. Large changes in oceanic nutrient inventories from glacial to interglacial periods. Nature, 376: 755–758.

Gardner, W.D., Sullivan, L.G., and Thorndike, E.M., 1984. Long-term photographic, current, and nephelometer observations of manganese nodule environments in the Pacific. Earth Planet. Sci. Lett., 70: 95–109.

Garrels, R.M., Lerman, A., and Mackenzie, F.T., 1976. Controls on atmospheric O_2 and CO_2: past, present and future. Am. J. Sci., 64: 306–315.

Gehlen, M., Mucci, A., and Boudreau, B., 1999. Modelling the distribution of stable carbon isotopes in porewaters of deep-sea sediments. Geochim. Cosmochim. Acta, 63: 2763–2773.

Gélin, F., Volkman, J.K., Largeau, C., Derenne, S., Damste, J.S.S., and De Leeuw, J.W., 1999. Distribution of aliphatic, nonhydrolyzable biopolymers in marine microalgae. Org. Geochem., 30: 147–159.

Gélinas, Y., Baldock, J.A., and Hedges, J.I., 2001. Organic carbon composition of marine sediments: effect of oxygen exposure on oil generation potential. Science, 294: 145–148.

Gerino, M., Aller, R.C., Lee, C., Cockran, J.K., Aller, J.Y., Green, M.A., and Hirschberg, D., 1998. Comparison of different tracers and methods used to quantify bioturbation during a spring bloom: 234-thorium, luminophores and chlorophyll *a*. Est. Coast. Shelf Sci., 46: 531–547.

Gianguzza, A., Pelizzetti, E., and Sammartano, S. (eds.), 2002. *Chemistry of Marine Water and Sediments*. Springer-Verlag.

Gieskes, J.M., 1981. Deep-Sea Drilling interstitial water studies: Implication for chemical alteration of the oceanic crust, layers I and II. SEPM Spec. Pub., 32: 149–167.

Gieskes, J.M., and Lawrence, J.R., 1981. Alteration of volcanic matter in deep sea sediments: evidence from the chemical composition of interstitial waters from deep sea drilling cores. Geochim. Cosmochim. Acta, 45: 1687–1703.

Gilbert, F., Aller, R.C., and Hulth, S., 2003. The influence of macrofaunal burrow spacing and diffusive scaling on sedimentary nitrification and denitrification: An experimental simulation and model approach. J. Mar. Res., 61: 101–126.

Gingele, F.X., and Kasten, S., 1994. Solid-phase manganese in Southeast Atlantic sediments: Implications for the paleoenvironment. Mar. Geol., 121: 317–332.

Glasby, G., 2000. Manganese: predominant role of nodules and crusts. In: H.D. Schulz and M. Zabel (eds.), *Marine Geochemistry*. Springer Verlag, pp. 335–372.

Glasby, G.P. (ed.), 1977. *Marine Manganese Deposits*. Elsevier.

Glasby, G.P., 1984. Manganese in the marine environment. Oceanogr. Mar. Biol. Ann. Rev., 22: 169–194.

Glenn, C.R., Arthur, M.A., Yeh, H.-W., and Burnett, W.C., 1988. Carbon isotopic composition and lattice-bound carbonate of Peru-Chile margin phosphorites. Mar. Geol., 80: 287–307.

Glud, R.N., Gundersen, J.K., Jørgensen, B.B., Revsbech, N.P., and Schulz, H.D., 1994. Diffusive and total oxygen uptake of deep-sea sediments in the eastern South Atlantic Ocean: *in situ* and laboratory measurements. Deep-Sea Res., 41: 1767–1788.

Glud, R.N., Holby, O., Hoffmann, F., and Canfield, D.E., 1998. Benthic mineralization and exchange in Arctic sediments (Svalbard, Norway). Mar. Ecol. Prog. Ser., 173: 237–251.

Gobeil, C., MacDonald, R.W., and Sundby, B., 1997. Diagenetic separation of cadmium and manganese in suboxic continental margin sediments. Geochim. Cosmochim. Acta, 61: 4647–4654.

Gobeil, C., Sundby, B., Macdonald, R.W., and Smith, J.N., 2001. Recent change in organic carbon flux to Arctic Ocean deep basins: Evidence from acid volatile sulfide, manganese and rhenium discord in sediments. Geophys. Res. Lett., 28: 1743–1746.

Goldberg, E.D., 1985. *Black Carbon in the Environment*. Wiley.

Goldberg, E.D., and Koide, M., 1962. Geochronological studies of deep-sea sediments by the thorium-ionium method. Geochim. Cosmochim. Acta, 26: 417–450.

Goldhaber, M.B., 2003. Sulfur-rich sediments. In: F.T. Mackenzie (ed.), *Treatise on Geochemistry*. Vol. 7. Elsevier, pp. 257–268.

Goldhaber, M.B., Aller, R.C., Cockran, J.K., Rosenfeld, J.K., Martens, C.S., and Berner, R.A., 1977. Sulfate reduction, diffusion, and bioturbation in Long

Island Sound sediments: report of the FOAM Group. Am. J. Sci., 277: 193–237.

Goloway, F., and Bender, M., 1982. Diagenetic models of interstitial nitrate profiles in deep sea suboxic sediments. Limnol. Oceanogr., 27: 624–638.

Gong, C., and Hollander, D.J., 1997. Differential contribution of bacteria to sedimentary organic matter in oxic and anoxic environments, Santa Monica Basin, California. Org. Geochem., 26: 545–563.

Goñi, M.A., Ruttenberg, K.C., and Eglinton, T.I., 1998. A reassessment of the sources and importance of land-derived organic matter in surface sediments from the Gulf of Mexico. Geochim. Cosmochim. Acta, 62: 3055–3075.

Goñi, M.A., Yunker, M.B., Macdonald, R.W., and Eglinton, T.I., 2000. Distribution and sources of organic biomarkers in arctic sediments from the Mackenzie River and Beaufort Shelf. Mar. Chem., 71: 23–51.

Gordon, A., and Millero, F.J., 1985. Adsorption mediated decrease in the biodegradation rate of organic compounds. Microb. Ecol., 11: 289–298.

Gordon, E.S., and Goñi, M.A., 2003. Sources and distribution of terrigenous organic matter delivered by the Atchafalaya River to sediments in the northern Gulf of Mexico. Geochim. Cosmochim. Acta, 67: 2359–2375.

Gordon, E.S., and Goñi, M.A., 2004. Controls on the distribution of and accumulation of terrigenous organic matter in sediments from the Mississippi and Atchafalya river margins. Mar. Chem., 92: 331–352.

Gough, M.A., Fauzi, R., Mantoura, C., and Preston, M., 1993. Terrestrial plant biopolymers in marine sediments. Geochim. Cosmochim. Acta, 57: 945–964.

Gratton, Y., Edenborn, H.E., Silverberg, N., and Sundby, B., 1990. A mathematical model for manganese diagenesis in bioturbated sediments. Am. J. Sci., 290: 246–262.

Graybeal, A.L., and Heath, G.R., 1984. Remobilization of transition metals in surficial pelagic sediments from the eastern Pacific. Geochim. Cosmochim. Acta, 48: 965–975.

Green, M.A., and Aller, R.C., 1998. Seasonal patterns of carbonate diagenesis in nearshore terrigenous muds: Relation to spring phytoplankton bloom and temperature. J. Mar. Res., 56: 1097–1123.

Green, M.A., and Aller, R.C., 2001. Early diagenesis of calcium carbonate in Long Island Sound sediments: Benthic fluxes of Ca^{2+} and minor elements during periods of net dissolution. J. Mar. Res., 59: 769–794.

Griffin, J.J., Windom, H., and Goldberg, E.D., 1968. The distribution of clay minerals in the world ocean. Deep-Sea Res., 15: 433–459.

Grigg, N., Boudreau, B.P., Webster, I.T., and Ford, P.W., 2005. The nonlocal model of porewater irrigation: Limits to its equivalence with a cylinder diffusion model. J. Mar. Res., 63: 437–455.

Grim, R.E., 1968. *Clay Mineralogy*. McGraw-Hill.

Grundmanis, V., and Murray, J.W., 1982. Aerobic respiration in pelagic marine sediments. Geochim. Cosmochim. Acta, 46: 1101–1120.

Grutters, M., Van Raaphorst, W., and Helder, W., 2001. Total hydrolysable amino acid mineralisation in sediments across the northeastern Atlantic continental slope (Goban Spur). Deep-Sea Res., 48: 811–832.

Grutters, M., van Raaphorst, W., Epping, E., Helder, W., de Leeuw, J.W., Glavin, D.P., and Bada, J., 2002. Preservation of amino acids from in situ-produced bacterial cell wall peptidoglycans in northeastern Atlantic continental margin sediments. Limnol. Oceanogr., 47: 1521–1524.

Guinasso, N.L., Jr., and Schink, D.R., 1975. Quantitative esimates of biological mixing rates in abyssal sediments. J. Geophys. Res., 80: 3032–3043.

Gujer, W., and Zehnder, A.J.B., 1983. Conversion processes in anaerobic digestion. Water Sci. Technol., 15: 127–167.

Guo, L., and Santschi, P.H., 2000. Sedimentary sources of old high molecular weight dissolved organic carbon from the ocean margin nepheloid layer. Geochim. Cosmochim. Acta, 64: 651–600.

Gustafsson, O., and Gschwend, P.M., 1998. The flux of black carbon to surface sediments on the New England continental shelf. Geochim. Cosmochim. Acta, 62: 465–472.

Habicht, K.S., and Canfield, D.E., 1997. Sulfur isotope fractionation during bacterial sulfate reduction in organic-rich sediments. Geochim. Cosmochim. Acta, 61: 5351–5361.

Habicht, K.S., and Canfield, D.E., 2001. Isotope fractionation by sulfate-reducing natural populations and the isotopic composition of sulfide in marine sediments. Geology, 29: 555–558.

Habicht, K.S., Canfield, D.E., and Rethmeier, J., 1998. Sulfur isotope fractionation during bacterial reduction and disproportionation of thiosulfate and sulfite. Geochim. Cosmochim. Acta, 62: 2585–2595.

Haddad, R.I., 1989. Sources and reactivity of organic matter accumulating in an anoxic coastal marine sediment. Ph.D thesis, Univ. of North Carolina, Chapel Hill.

Haddad, R.I., and Martens, C.S., 1987. Biogeochemical cycling in an organic-rich marine basin—9. Sources and fluxes of vascular plant derived organic material. Geochim. Cosmochim. Acta, 51: 2991–3001.

Haddad, R.I., Martens, C.S., and Farrington, J.W., 1992. Quantifying early diagenesis of fatty acids in a rapidly accumulating coastal marine sediment. Org. Geochem., 19: 205–216.

Haese, R.R., 2000. The reactivity of iron. In: H.D. Schulz and M. Zabel (eds.), Marine Geochemistry. Springer-Verlag, pp. 233–261.

Haese, R.R., Wallimann, K., Dahmke, A., Kretzmann, U., Müller, P.J., and Schulz, H.D., 1997. Iron species determination to investigate early diagenetic reactivity in marine sediments. Geochim. Cosmochim. Acta, 61: 63–72.

Hagadorn, J.W., Stott, L.D., Sinha, A., and Rincon, M., 1995. Geochemical and sedimentologic variations in inter-annually laminated sediments from Santa Monica Basin. Mar. Geol., 125: 111–131.

Hales, B., and Emerson, S., 1996. Calcite dissolution in sediments of the Ontong-Java Plateau: In situ measurements of pore water O_2 and pH. Global Biogeochem. Cycles, 10: 527–541.

Hales, B., and Emerson, S., 1997a. Calcite dissolution in sediments of the Ceara Rise: In situ measurements of pore water O_2, pH and $CO_{2(aq)}$. Geochim. Cosmochim. Acta, 61: 501–514.

Hales, B., and Emerson, S., 1997b. Evidence supporting first-order calcite dissolution kinetics in seawater. Earth Planet. Sci. Lett., 148: 317–328.

Hales, B., Emerson, S., and Archer, D., 1994. Respiration and dissolution in the sediments of the western North Atlantic: estimates from models of *in situ* microelectrode measurements of porewater oxygen and pH. Deep-Sea Res., 41: 695–719.

Hall, P.O.J., Anderson, L.G., Rutgers van der Loeff, M.M., Sundby, B., and Westerlund, S.F.G., 1989. Oxygen uptake kinetics in the benthic boundary layer. Limnol. Oceanogr., 34: 734–746.

Hamilton, E.L., 1976. Variations of density and porosity with depth in deep-sea sediments. J. Sed. Petrol., 46: 280–300.

Hamilton, S.E., and Hedges, J.I., 1988. The comparative geochemistries of lignins and carbohydrates in an anoxic fjord. Geochim. Cosmochim. Acta, 52: 129–142.

Hammond, D.E., and Fuller, C., 1979. The use of radon-222 to estimate benthic exchange and atmospheric exchange rates in San Francisco Bay. *In*: *San Francisco Bay: The Urbanized Estuary*. California Academy of Sciences, pp. 213–230.

Hammond, D.E., Giodani, P., Berelson, W.M., and Poletti, R., 1999. Diagenesis of carbon and nutrients and benthic exchange in sediments of the northern Adriatic Sea. Mar. Chem., 66: 53–97.

Hammond, D.E., McManus, J., Berelson, W.M., Kilgore, T.E., and Pope, R.H., 1996. Early diagenesis of organic material in equatorial Pacific sediments: stoichiometry and kinetics. Deep-Sea Res., 43: 1365–1412.

Han, X., Jin, X., Yang, S., Fietzke, J., and Eisenhauer, A., 2003. Rhythmic growth of Pacific ferromanganese nodules and their Milankovitch climatic origin. Earth Planet. Sci. Lett., 211: 143–157.

Hannides, A.K., Dunn, S.M., and Aller, R.C., 2005. Diffusion of organic and inorganic solutes through macrofaunal mucus secretions and tube linings in marine sediments. J. Mar. Res., 63: 957–981.

Hannington, M.D., Jonasson, I.R., Herzig, P.M., and Petersen, S., 1995. Physical and chemical processes of seafloor mineralization at mid-ocean ridges. *In*: S.E. Humphris, R.A. Zierenberg, L.S. Mullineaux, and R.E. Thomson (eds.), *Seafloor Hydrothermal Systems*. Am. Geophys. Union, pp. 115–156.

Hansen, L.S., and Blackburn, T.H., 1991. Aerobic and anaerobic mineralization of organic material in marine sediment microcosms. Mar. Ecol. Progr. Ser., 75: 283–291.

Hartnett, H.E., and Devol, A.H., 2003. Role of a strong oxygen-deficient zone in the preservation and degradation of organic matter: A carbon budget for the continental margins of northwest Mexico and Washington State. Geochim. Cosmochim. Acta, 67: 247–264.

Hartnett, H.E., Keil, R.G., Hedges, J.I., and Devol, A.H., 1998. Influence of oxygen exposure time on organic carbon preservation in continental margin sediments. Nature, 391: 572–574.

Harvey, H.R., Fallon, R.D., and Patton, J.S., 1986. The effect of organic matter and oxygen on the degradation of bacterial membrane lipids in marine sediments. Geochim. Cosmochim. Acta, 50: 795–804.

Harvey, H.R., and Macko, S.A., 1997a. Catalysts or contributors? Tracking bacterial mediation of early diagenesis in the marine water column. Org. Geochem., 26: 531–544.

Harvey, H.R., and Macko, S.A., 1997b. Kinetics of phytoplankton decay during simulated sedimentation: changes in lipids under oxic and anoxic conditions. Org. Geochem., 27: 129–140.

Harvey, H.R., Tuttle, J.H., and Bell, J.T., 1995. Kinetics of phytoplankton decay during simulated sedimentation: Changes in biochemical composition and microbial activity under oxic versus anoxic conditions. Geochim. Cosmochim. Acta, 59: 3367–3377.

Hatcher, P.G., and Spiker, E.C., 1988. Selective degradation of plant biomolecules. *In*: F.C. Frimmel and R.C. Christman (eds.), *Humic Substances and Their Role in the Environment*. J. Wiley & Sons, pp. 59–74.

Hatcher, P.G., Spiker, E.C., Szeverenyi, N.M., and Maciel, G.E., 1983. Selective preservation and origin of petroleum-forming aquatic kerogen. Nature, 305: 498–501.

Hathaway, 1979. Clay minerals. *In*: R.G. Burns (ed.), *Marine Minerals*. Vol. 6. Mineralogical Society of America, pp. 123–174.

Haugen, J., and Lichtentaler, R., 1991. Amino acid diagenesis, organic carbon and nitrogen mineralization in surface sediments from the inner Oslofjord, Norway. Geochim. Cosmochim. Acta, 55: 1649–1661.

Hayes, J.M., 1993. Factors controlling ^{13}C contents of sedimentary organic compounds: principles and evidence. Mar. Geol., 113: 111–125.

Hayes, J.M., 2001. Fractionation of the isotopes of carbon and hydrogen in biosynthetic processes. *In*: J.W. Valley and D.R. Cole (eds.), *Stable Isotope Geochemistry*. Rev. Mineral. Geochem. Vol. 43. Mineral. Soc. Am., pp. 225–277.

Hayes, J.M., Freeman, K.H., Popp, B.N., and Hoham, C.H., 1990. Compound-specific isotopic analysis: a novel tool for reconstruction of ancient biogeochemical processes. Org. Geochem., 16: 1115–1128.

Heath, G.R., 1978. Burial rates, growth rates, and size distributions of deep-sea manganese nodules. Science, 205: 903–904.

Heath, G.R., and Dymond, J., 1977. Genesis and transformation of metalliferous sediments from the East Pacific Rise, Bauer Deep, and Central Basin, northwest Nazca plate. Geol. Soc. Am. Bull., 88: 723–733.

Hecky, R.E., Mopper, K., Kilham, P., and Degens, E.T., 1973. The amino acid and sugar composition of diatom cell-walls. Mar. Biol., 19: 323–331.

Hedges, J.I., 1988. Polymerization of humic substances in natural environments. *In*: F.C. Frimmel and R.C. Christman (eds.), *Humic Substances and Their Role in the Environment*. J. Wiley & Sons, pp. 45–58.

Hedges, J.I., 1992. Global biogeochemical cycles: Progress and problems. Mar. Chem., 39: 67–93.

Hedges, J.I., 2002. Sedimentary organic matter preservation and atmospheric O_2 regulation. *In*: A. Gianguzza, E. Pelizzetti, and S. Sammartano (eds.), *Chemistry of Marine Water and Sediments*. Springer-Verlag, pp. 105–123.

Hedges, J.I., Baldock, J.A., Gélinas, Y., Lee, C., Peterson, M., and Wakeham, S.G., 2001. Evidence for non-selective preservation of organic matter in sinking marine particles. Nature, 409: 801–803.

Hedges, J.I., Baldock, J.A., Gelinas, Y., Lee, C., Peterson, M.L., and Wakeham, S.G., 2002. The biochemical and elemental compositions of marine plankton: a NMR perspective. Mar. Chem., 78: 47–63.

Hedges, J.I., Clark, W.A., Quay, P.D., Richey, J.E., Devol, A.H., and Santos, U.d.M., 1986. Composition and fluxes of particulate organic material in the Amazon River. Limnol. Oceanogr., 31: 717–738.

Hedges, J.I., Eglinton, G., Hatcher, P.G., Kirchman, D.L., Arnosti, C., Derenne, S., Evershed, R.P., Kogel-Knabner, I., de Leeuw, J.W., Littke, R., Michaelis, W., and Rülkotter, J., 2000. The molecularly-uncharacterized component of nonliving organic matter in natural environments. Org. Geochem., 31: 945–958.

Hedges, J.I., Hu, F.S., Devol, A.H., Hartnett, H.E., Tsamakis, E. and Keil, R.G., 1999a. Sedimentary organic matter preservation: A test for selective degradation under oxic conditions. Am. J. Sci., 299: 529–555.

Hedges, J.I., and Keil, R.G., 1995. Sedimentary organic matter preservation: an assessment and speculative synthesis. Mar. Chem., 49: 81–115.

Hedges, J.I., and Keil, R.G., 1999. Organic geochemical perspectives on estuarine processes: sorption reactions and consequences. Mar. Chem., 65: 55–65.

Hedges, J.I., Keil, R.G., and Benner, R., 1997. What happens to terrestrial organic matter in the ocean? Org. Geochem., 27: 195–212.

Hedges, J.I., Keil, R.G., Lee, C., and Wakeham, S.G., 1999b. Invited lecture: Atmospheric O_2 control by a 'mineral conveyor belt' linking the continents and ocean. *In*: H. Ármannsson (ed.), *Geochemistry of the Earth's Surface*. Proc. 5th Int. Symp. Geochem. Earth's Surface. Balkema, pp. 241–244.

Hedges, J.I., and Mann, D.C., 1979. The characterization of plant tissues by their lignin oxidation products. Geochim. Cosmochim. Acta, 43: 1803–1807.

Hedges, J.I., and Oades, J.M., 1997. Comparative organic geochemistries of soils and marine sediments. Org. Geochem., 27: 319–361.

Hedges, J.I., and Prahl, F.G., 1993. Early diagenesis: consequences for applications of molecular biomarkers. *In*: M.H. Engel and S.A. Macko (eds.), *Organic Geochemistry*. Plenum Press, pp. 237–253.

Hedges, J.I., and Weliky, K., 1989. Diagenesis of conifer needles in a coastal marine environment. Geochim. Cosmochim. Acta, 53: 2659–2673.

Heggie, D., Kahn, D., and Fischer, K., 1986. Trace metals in metalliferous sediments, MANOP Site M: interfacial pore water profiles. Earth Planet. Sci. Lett., 80: 106–116.

Heggie, D., Maris, C., Hudson, A., Dymond, J., Beach, R., and Cullen, J., 1987. Organic carbon oxidation and preservation in NW Atlantic continental margin sediments. *In*: P.P.E. Weaver and J. Thomson (eds.), *Geology and Geochemistry of Abyssal Plains*. Blackwell Scientific, pp. 215–236.

Hein, J.R., Yeh, H., and Alexander, E., 1979. Origin of iron-rich montmorillontite from the manganese nodule belt of the north equatorial Pacific. Clays and Clay Mins., 27: 185–194.

Heller-Kallai, L., 1997. Reduction and reoxidation of nontronite: The data reassessed. Clays and Clay Minerals, 45: 476–479.

Henrichs, S.M., 1992. Early diagenesis of organic matter in marine sediments: progress and perplexity. Mar. Chem., 39: 119–149.

Henrichs, S.M., 1993. Early diagenesis of organic matter: the dynamics (rates) of cycling of organic compounds. *In*: M. Engel and S. Macko (eds.), *Organic Geochemistry*. Plenum Press, pp. 101–117.

Henrichs, S.M., 1995. Sedimentary organic matter preservation: an assessment and speculative synthesis—a comment. Mar. Chem., 49: 127–136.

Henrichs, S.M., and Doyle, A.P., 1986. Decomposition of ^{14}C-labelled organic substrates in marine sediments. Limnol. Oceanogr., 31: 765–778.

Henrichs, S.M., and Farrington, J.W., 1979. Amino acids in interstitial waters of marine sediments. Nature, 279: 319–322.

Henrichs, S.M., and Farrington, J.W., 1984. Peru upwelling region sediments near 15°S. 1. Remineralization and accumulation of organic matter. Limnol. Oceanogr., 29: 1–19.

Henrichs, S.M., and Farrington, J.W., 1987. Early diagenesis of amino acids and organic matter in two coastal marine sediments. Geochim. Cosmochim. Acta, 51: 1–15.

Henrichs, S.M., Farrington, J.W., and Lee, C., 1984. Peru upwelling region sediments near 15°S. 2. Dissolved free and total hydrolyzable amino acids. Limnol. Oceanogr., 29: 20–34.

Henrichs, S.M., and Reeburgh, W.S., 1987. Anaerobic mineralization of marine sediment organic matter: rates and the role of anaerobic processes in the oceanic carbon economy. Geomicrobiol. J., 5: 191–237.

Henrichs, S.M., and Sugai, S.F., 1993. Adsorption of amino acids and glucose by sediments of Resurrection Bay, Alaska, USA: Functional group effects. Geochim. Cosmochim. Acta, 57: 823–835.

Hensen, C., Landenberger, H., Zabel, M., and Schulz, H.D., 1998. Quantification of diffusive benthic fluxes of nitrate, phosphate, and silicate in the southern Atlantic Ocean. Global Biogeochem. Cycles, 12: 193–210.

Hensen, C., and Zabel, M., 2000. Early diagenesis at the benthic boundary layer: oxygen and nitrate in marine sediments. *In*: H.D. Schulz and M. Zabel (eds.), *Marine Geochemistry*. Springer-Verlag, pp. 209–231.

Hensen, C., Zabel, M., Pfeifer, K., Schwenk, T., Kasten, S., Riedinger, N., Schulz, H.D., and Boetius, A., 2003. Control of sulfate pore-water profiles by sedimentary events and the significance of anaerobic oxidation of methane for the burial of sulfur in marine sediments. Geochim. Cosmochim. Acta, 67: 2631–2647.

Hernes, P.J., Hedges, J.I., Peterson, M.L., Wakeham, S.G., and Lee, C., 1996. Neutral carbohydrate geochemistry of particulate material in the central equatorial Pacific. Deep-Sea Res., 43: 1181–1204.

Hines, M.E., Lyons, W.B., Armstrong, P.A., Orem, W.H., Spencer, M.J., and Gaudette, H.E., 1984. Seasonal metal remobilization in the sediments of Great Bay, New Hampshire. Mar. Chem., 15: 173–187.

Hoefs, J., 1997. *Stable Isotope Geochemistry*, 4th ed. Springer-Verlag.

Hoefs, M.J.L., Rijpstra, W.I.C., and Sinninghe Damsté, J.S., 2002. The influence of oxic degradation on the sedimentary biomarker record. I. Evidence from Madeira Abyssal Plain turbidites. Geochim. Cosmochim. Acta, 66: 2719–2735.

Hoefs, M.J.L., Sinninghe Damsté, J.S., De Lange, G.J., and de Leeuw, J.W., 1998. Changes in kerogen composition across an oxidation front in Madeira Abyssal Plain turbidites as revealed by pyrolysis GC-MS. Proc. ODP, Sci. Res., 157: 591–607.

Hoehler, T.M., Alperin, M.J., Albert, D.B., and Martens, C.S., 1994. Field and laboratory studies of methane oxidation in an anoxic marine sediment: evidence for a methanogen-sulfate reducer consortium. Global Biogeochem. Cycles, 8: 451–463.

Hoehler, T.M., Alperin, M.J., Albert, D.B., and Martens, C.S., 1998. Thermodynamic control on hydrogen concentrations in anoxic sediments. Geochim. Cosmochim. Acta, 62: 1745–1756.

Hoehler, T.M., Borowski, W.S., Alperin, M.J., Rodriquez, N.M., and Paull, C.K., 2000. Model, stable isotope, and radiotracer characterization of anaerobic methane oxidation in gas hydrate-bearing sediments of the Blake Ridge. Proc. ODP, Sci. Res., 164: 79–85.

REFERENCES

Holdren, G.R., Jr., Bricker, O.P., III, and Matisoff, G., 1975. A model for the control of dissolved manganese in the interstitial water of Chesapeake Bay. *In*: T.M. Church (ed.), *Marine Chemistry in the Coastal Environment*. ACS Symp. Series no. 18, pp. 364–381.

Honjo, S., 1980. Material fluxes and modes of sedimentation in the mesopelagic and bathypelagic zones. J. Mar. Res., 38: 53–96.

Huerta-Diaz, M.A., and Morse, J.W., 1992. Pyritization of trace metals in anoxic marine sediments. Geochim. Cosmochim. Acta, 56: 2681–2702.

Huettel, M., and Gust, G., 1992a. Impact of bioroughness on interfacial solute exchange in permeable sediments. Mar. Ecol. Prog. Ser., 89: 253–267.

Huettel, M., and Gust, G., 1992b. Solute release mechanisms from confined sediment cores in stirred benthic chambers and flume flows. Mar. Ecol. Prog. Ser., 82: 187–197.

Huettel, M., and Rusch, A., 2000. Transport and degradation of phytoplankton in permeable sediment. Limnol. Oceanogr., 45: 534–549.

Huettel, M., and Webster, I.T., 2001. Porewater flow in permeable sediments. *In*: B.P. Boudreau and B.B. Jørgensen (eds.), *The Benthic Boundary Layer*. Oxford University Press, pp. 144–179.

Huettel, M., Ziebis, W., and Forster, S., 1996. Flow-induced uptake of particulate matter in permeable sediments. Limnol. Oceanogr., 41: 309–322.

Huettel, M., Ziebis, W., Forster, S., and Luther, G.W., III, 1998. Advective transport affecting metal and nutrient distributions and interfacial fluxes in permeable sediments. Geochim. Cosmochim. Acta, 62: 613–631.

Hulth, S., Aller, R.C., and Gilbert, F., 1999. Coupled anoxic nitrification/manganese reduction in marine sediments. Geochim. Cosmochim. Acta, 63: 49–66.

Hulth, S., Tengberg, A., Landen, A., and Hall, P.O.J., 1997. Mineralization and burial of organic carbon in sediments of the southern Wedell Sea (Antarctica). Deep-Sea Res., 44: 955–981.

Hulthe, G., Hulth, S., and Hall, P.O.J., 1998. Effect of oxygen on degradation rate of refractory and labile organic matter in continental margin sediments. Geochim. Cosmochim. Acta, 62: 1319–1328.

Hunt, C.D., 1983. Variability in the benthic Mn flux in coastal marine ecosystems resulting from temperature and primary production. Limnol. Oceanogr., 28: 913–923.

Hunt, J.M., 1996. *Petroleum Geochemistry and Geology*, 2nd ed. W.H. Freeman.

Hyacinthe, C., Anschutz, P., Carbonel, P., Jouanneau, J.-M., and Jorissen, F.J., 2001. Early diagenetic processes in the muddy sediments of the Bay of Biscay. Mar. Geol., 177: 111–128.

Imbrie, J., and Imbrie, K.P., 1979. *Ice Ages, Solving the Mystery*. Harvard University Press.

Indermühle, A., Stocker, T.F., Joos, F., Fischer, H., Smith, H.J., Wahlen, M., Deck, B., Mastroianni, D., Tschumi, J., Blunier, T., Meyer, R., and Stauffer, B., 1999. Holocene carbon-cycle dynamics based on CO_2 trapped in ice at Taylor dome, Antarctica. Nature, 398: 121–126.

Ingall, E., and Jahnke, R., 1997. Influence of water-column anoxia on the elemental fractionation of carbon and phosphorus during sediment diagenesis. Mar. Geol., 139: 219–229.

Ingall, E.D., and Van Capellen, P., 1990. Relation between sedimentation rate and burial organic phosphorus and organic carbon in marine sediments. Geochim. Cosmochim. Acta, 54: 373–386.

Ingalls, A.E., Aller, R.C., Lee, C., and Sun, M., 2000. The influence of deposit-feeding on chlorophyll-a degradation in coastal marine sediments. J. Mar. Res., 58: 631–651.

Ingalls, A.E., Lee, C., Wakeham, S.G., and Hedges, J.I., 2003. The role of biominerals in the sinking flux and preservation of amino acids in the Southern Ocean along 170°W. Deep-Sea Res. II, 50: 713–738.

Ingalls, A.E., Aller, R.C., Lee, C., and Wakeham, S.G., 2004. Organic matter diagenesis in shallow water carbonate sediments. Geochim. Cosmochim. Acta, 68: 4363–4379.

Iversen, N., and Jørgensen, B.B., 1985. Anaerobic methane oxidation at the sulfate-methane transition in marine sediments from the Kattegat and Skagerrak (Denmark). Limnol. Oceanogr., 30: 944–955.

Jahnke, R., Heggie, D., Emerson, S., and Grundmanis, V., 1982. Pore waters of the central Pacific Ocean: nutrient results. Earth Planet. Sci. Lett., 61: 233–256.

Jahnke, R.A., 1984. The synthesis and solubility of carbonate fluoroapatite. Am. J. Sci., 284: 58–78.

Jahnke, R.A., 1990. Early diagenesis and recycling of biogenic debris at the seafloor, Santa Monica Basin, California. J. Mar. Res., 48: 413–436.

Jahnke, R.A., 1996. The global ocean flux of particulate organic carbon: Areal distribution and magnitude. Global Biogeochem. Cycles, 10: 71–88.

Jahnke, R.A., 2001. Constraining organic matter cycling with benthic fluxes. In: B.P. Boudreau and B.B. Jørgensen (eds.), The Benthic Boundary Layer: Transport Processes and Biogeochemistry. Oxford University Press, pp. 302–319.

Jahnke, R.A., Craven, D.B., and Galliard, J.-F., 1994. The influence of organic matter diagenesis on $CaCO_3$ dissolution at the deep-sea floor. Geochim. Cosmochim. Acta, 58: 2799–2809.

Jahnke, R.A., Craven, D.B., McCorkle, D.C., and Reimers, C.E., 1997. $CaCO_3$ dissolution in California continental margin sediments: the influence of organic matter remineralization. Geochim. Cosmochim. Acta, 61: 3587–3604.

Jahnke, R.A., Emerson, S.R., Cochran, J.K., and Hirschberg, D.J., 1986. Fine scale distributions of porosity and excess ^{210}Pb, organic carbon and $CaCO_3$

in surface sediments of the deep equatorial Pacific. Earth Planet. Sci. Lett., 77: 59–69.

Jahnke, R.A., Emerson, S.R., Reimers, C.E., Schuffert, J., Ruttenberg, K., and Archer, D., 1989. Benthic recycling of biogenic debris in the eastern tropical Atlantic Ocean. Geochim. Cosmochim. Acta, 53: 2947–2960.

Jahnke, R.A., and Jahnke, D.B., 2000. Rates of C, N, P, and Si recycling and denitrification at the US mid-Atlantic continental slope depocenter. Deep-Sea Res., 47: 1405–1428.

Jahnke, R.A., and Jahnke, D.B., 2004. Calcium carbonate dissolution in deep sea sediments: Reconciling microelectrode, pore water and benthic flux chamber results. Geochim. Cosmochim. Acta, 68: 47–59.

Jahnke, R.A., Nelson, J.R., Marinelli, R.L., and Eckman, J.E., 2000. Benthic fluxes of biogenic elements on the southeastern US continental shelf: influence of pore water advective transport and benthic microalgae. Cont. Shelf Res., 20: 109–127.

Jahnke, R.A., Reimers, C.E., and Craven, D.B., 1990. Intensification of recycling of organic matter at the sea floor near ocean margins. Nature, 348: 50–54.

Jarvis, I., and Higgs, N., 1987. Trace-element mobility during early diagenesis in distal turbidites: late Quaternary of the Madeira Abyssal Plain, N Atlantic. In: P.P.E. Weaver and J. Thomson (eds.), *Geology and Geochemistry of Abyssal Plains*. Geol. Soc. Spec. Publ. no. 31, pp. 179–213.

Jasper, J.P., and Gagosian, R.B., 1993. The relationship between sedimentary organic carbon isotopic composition and organic biomarker compound concentration. Geochim. Cosmochim. Acta, 57: 167–186.

Jenkins, M.C., and Kemp, W.M., 1984. The coupling of nitrification and denitrification in two estuarine sediments. Limnol. Oceanogr., 29: 609–619.

Jensen, H.S., McGlathery, K.J., Marino, R., and Howarth, R.W., 1998. Forms and availability of sediment phosphorus in carbonate sands of Bermuda seagrass beds. Limnol. Oceanogr., 43: 799–810.

Jensen, H.S., Mortensen, P.B., Anderson, F.Ø., Rasmussen, E., and Jensen, A., 1995. Phosphorus cycling in a coastal marine sediment, Aarhus Bay, Denmark. Limnol. Oceanogr., 40: 908–917.

Johnson, K.S., Berelson, W.M., Coale, K.H., Coley, T.L., Elrod, V.A., Farley, W.R., Iams, H.D., Kilgore, T.E., and Nowicki, J.L., 1992. Manganese flux from continental margin sediments in a transect through the oxygen minimum. Science, 257: 1242–1245.

Jørgensen, B.B., 1978. A comparison of methods for the quantification of bacterial sulfate reduction in coastal marine sediments. 2. Calculations from mathematical models. Geomicrobiol. J., 1: 29–51.

Jørgensen, B.B., 1979. A theoretical model of the stable sulfur isotope distribution in Maine sediments. Geochim. Cosmochim. Acta, 43.

Jørgensen, B.B., 1982. Mineralization of organic matter in the sea bed—the role of sulphate reduction. Nature, 296: 643–645.

REFERENCES

Jørgensen, B.B., 2001. Life in the diffusive boundary layer. *In*: B.P. Boudreau and B.B. Jørgensen (eds.), *The Benthic Boundary Layer: Transport Processes and Biogeochemistry*. Oxford University Press, pp. 348–373.

Jørgensen, B.B., and Bak, F., 1991. Pathways and microbiology of thiosulfate transformations and sulfate reduction in a marine sediment (Kattegat, Denmark). Appl. Environ. Microbiol., 57: 847–856.

Jørgensen, B.B., and Boudreau, B.P., 2001. Diagenesis and sediment-water exchange. in: B.P. Boudreau and B.B. Jørgensen (eds.), *The Benthic Boundary Layer*. Oxford University Press.

Jørgensen, B.B., and Revsbech, N.P., 1983. Colorless sulfur bacteria., *Beggiatoa* spp. and *Thiovulum* spp., in O_2 and H_2S microgradients. Appl. Environ. Microbiol., 45: 1261–1270.

Jørgensen, B.B., and Revsbech, N.P., 1985. Diffusive boundary layers and the oxygen uptake of sediments and detritus. Limnol. Oceanogr., 30: 111–122.

Joye, S.B., and Hollibaugh, J.T., 1995. Influence of sulfide inhibition of nitrification on nitrogen regeneration in sediments. Science, 270: 623–625.

Jumars, P.A., and Wheatcroft, R.A., 1989. Responses of benthos to changing food quality and quantity, with a focus on deposit feeding and bioturbation. *In*: W.H. Berger, V.S. Smetacek, and G. Wefer (eds.), *Productivity of the Ocean: Present and Past*. John Wiley & Sons, pp. 235–253.

Jung, M., Ilmberger, J., Mangini, A., and Emeis, K.-C., 1997. Why some Mediterranean sapropels survived burn-down (and others did not). Mar. Geol., 141: 51–60.

Kalhorn, S., and Emerson, S., 1984. The oxidation state of manganese in surface sediments of the deep sea. Geochim. Cosmochim. Acta, 48: 897–902.

Kasten, S., and Jørgensen, B.B., 2000. Sulfate reduction in marine sediments. *In*: H.D. Schulz and M. Zabel (eds.), *Marine Geochemistry*. Springer-Verlag, pp. 263–282.

Kastner, M., 1981. Authigenic silicates in deep-sea sediments: formation and diagenesis. *In*: C. Emiliani (ed.), *The Sea*. Vol. 7, *The Oceanic Lithosphere*. John Wiley & Sons, pp. 915–980.

Kastner, M., 1999. Oceanic minerals: their origin, nature of their environment, and significance. Proc. Nat. Acad. Sci., 96: 3380–3387.

Keil, R.G., Dickens, A.F., Arnarson, T., Nunn, B.L., and Devol, A.H., 2004. What is the oxygen exposure time of laterally transported organic matter along the Washington margin? Mar. Chem., 92: 157–165.

Keil, R.G., and Fogel, M.L., 2001. Reworking of amino acid in marine sediments: stable carbon isotopic composition of amino acids in sediments along the Washington coast. Limnol. Oceanogr., 46: 14–23.

Keil, R.G., Hu, F.S., Tsamakis, E.C., and Hedges, J.I., 1994a. Pollen grains deposited in marine sediments are degraded only under oxic conditions. Nature, 369: 639–641.

REFERENCES

Keil, R.G., Mayer, L.M., Quay, P.D., Richey, J.E., and Hedges, J.I., 1997. Loss of organic matter from riverine particles in deltas. Geochim. Cosmochim. Acta, 61: 1507–1511.

Keil, R.G., Tsamakis, C., and Hedges, J.I., 2000. Early diagenesis of particulate amino acids in marine systems. *In*: G. Goodfriend and M. Fogel (eds.), *Amino Acids in Geological Systems: A Tribute to Ed Hare*. Plenum, pp. 69–82.

Keil, R.G., Tsamakis, E., Fuh, C.B., Giddings, C., and Hedges, J.I., 1994b. Mineralogical and textural controls on organic composition of coastal marine sediments: hydrodynamic separation using SPLITT fractionation. Geochim. Cosmochim. Acta, 58: 879–893.

Keil, R.G., Tsamakis, E., Giddings, J.C., and Hedges, J.I., 1998. Biochemical distributions (amino acids, neutral sugars, and lignin phenols) among size-classes of modern sediments from the Washington coast. Geochim. Cosmochim. Acta, 62: 1347–1364.

Keir, R.S., 1980. The dissolution kinetics of biogenic calcium carbonates in seawater. Geochim. Cosmochim. Acta, 44: 241–252.

Kemp, W.M., and Boynton, W.R., 1984. Spatial and temporal coupling of nutrient inputs to estuarine primary production: The role of particulate transport and decomposition. Bull. Mar. Sci., 35: 522–535.

Kemp, W.M., and Boynton, W.R., 1992. Benthic-pelagic interactions: nutrient and oxygen dynamics. *In*: Smith, D.E., M. Leffler and G. Mackiernan (eds.), *Oxygen Dynamics in the Chesapeake Bay: A Synthesis of Recent Research*. Maryland Sea Grant Book, pp. 149–221.

Kemp, W.M., Sampou, P., Mayer, M., Henricksen, K., and Boyton, W.R., 1990. Ammonium recycling versus denitrification in Chesapeake Bay sediments. Limnol. Oceanogr., 35: 1545–1563.

Kennett, J.P., 1982. *Marine Geology*. Prentice-Hall.

Kennett, J.P., Cannariato, K.G., Hendy, I.L., and Behl, R.J., 2000. Carbon isotopic evidence for methane hydrate instability during Quaternary interstadials. Science, 288: 128–133.

Killops, S.D., and Killops, V.J., 2005. *Introduction to Organic Geochemistry*. 2nd ed. Blackwell.

Kineke, G.C., Sternberg, R.W., and Trowbridge, J.H., 1996. Fluid-mud processes on the Amazon continental shelf. Cont. Shelf Res., 16: 667–696.

King, G.M., Klug, M.J., and Lovely, D.R., 1983. Metabolism of acetate, methanol, and methylated amines in intertidal sediments of Lowes Cove, Maine. Appl. Environ. Microbiol., 45: 1848–1853.

King, S.L., Froelich, P.N., and Jahnke, R.A., 2000. Early diagenesis of germanium in sediments of the Antarctic South Atlantic: In search of the missing Ge sink. Geochim. Cosmochim. Acta, 64: 1375–1390.

Kipput, G.W., and Martens, C.S., 1982. Biogeochemical cycling in an organic-rich marine basin—3. Dissolved gas transport in methane-saturated sediments. Geochim. Cosmochim. Acta, 46: 2049–2060.

REFERENCES

Klinkhammer, G., Heggie, D.T., and Graham, D.W., 1982. Metal diagenesis in oxic marine sediments. Earth Planet. Sci. Lett., 61: 211–219.

Klinkhammer, G.P., 1980. Early diagenesis in sediments from the eastern equatorial Pacific. II. Pore water metal results. Earth Planet. Sci. Lett., 49: 81–101.

Klok, J., Bass, M., Cox, H.C., de Leeuw, J.W., Rijpstra, W.I.C., and Schenck, P.A., 1984. Qualitative and quantitative characterization of the total organic matter in a recent marine sediment (Part II). Org. Geochem., 6: 265–278.

Klump, J.V., and Martens, C.S., 1981. Biogeochemical cycling in an organic-rich marine basin—2. Nutrient sediment-water exchange processes. Geochim. Cosmochim. Acta, 46: 1575–1589.

Klump, J.V., and Martens, C.S., 1983. Benthic nitrogen regeneration. *In*: E.J. Carpenter and D.G. Capone (eds.), *Nitrogen in the Marine Environment*. Academic Press, pp. 411–457.

Klump, J.V., and Martens, C.S., 1987. Biogeochemical cycling in an organic-rich marine basin—5. Sedimentary nitrogen and phosphorus budgets based upon kinetic models, mass balances, and the stoichiometry of nutrient regeneration. Geochim. Cosmochim. Acta, 51: 1161–1173.

Klump, J.V., and Martens, C.S., 1989. The seasonality of nutrient regeneration in an organic rich coastal sediment: Kinetic modelling of changing pore water–nutrient and sulfate distributions. Limnol. Oceanogr., 34: 559–577.

Knicker, H., 2000. Solid-state 2-D double cross polarization magic angle spinning 2D ^{15}N ^{13}C NMR spectroscopy on degraded algal residues. Org. Geochem., 31: 337–340.

Knicker, H., del Río, J.C., Hatcher, P., and Minard, R.D., 2001. Identification of protein remnants in insoluble geopolymers using TMAH thermochemolysis/GC-MS. Org. Geochem., 32: 397–409.

Knicker, H., Frund, R., and Ludemann, H.-D., 1993. The chemical nature of nitrogen in native soil organic matter. Naturwissenschaften, 80: 219–221.

Knicker, H., and Hatcher, P.G., 1997. Survival of protein in an organic-rich sediment: possible protection by encapsulation in organic matter. Naturwissenschaften, 81: 231–234.

Knicker, H., Scaroni, A.W., and Hatcher, P.G., 1996. ^{13}C and ^{15}N NMR spectroscopic investigation on the formation of fossil algal residues. Org. Geochem., 24: 661–669.

Kohnen, M.E.L., Schouten, S., Damste, J.S.S., de Leeuw, J.W., Merrit, D., and Hayes, J.M., 1992. The combined application of organic sulphur and isotope geochemistry to assess multiple sources of palaeobiochemicals with identical carbon skeletons. Org. Geochem., 19: 403–419.

Koike, I., and Sørensen, J., 1988. Nitrate reduction and denitrification in marine sediments. *In*: T.H. Blackburn and J. Sørensen (eds.), *Nitrogen Cycling in Coastal Marine Sediments*. John Wiley & Sons, pp. 251–274.

Kok, M.D., Schouten, S., and Sinninghe Damste, J.S. 2000. Formation of insoluble, nonhydrolyzable, sulfur-rich macromolecules via incorporation of inorganic sulfur species into algal carbohydrates. Geochim. Cosmochim. Acta, 64: 2689–2699.

Kolowith, L.C., Ingall, E.D., and Benner, R., 2001. Composition and cycling of marine organic phosphorus. Limnol. Oceanogr., 46: 309–320.

Komada, T., Reimers, C.E., and Boehme, S.E., 1998. Dissolved inorganic carbon profiles and fluxes determined using pH and P_{CO2} microelectrodes. Limnol. Oceanogr., 43: 769–781.

Komada, T., Reimers, C.E., Luther, G.W., III, and Burdige, D.J., 2004. Factors affecting dissolved organic matter dynamics in mixed-redox to anoxic coastal sediments. Geochim. Cosmochim. Acta, 68: 4099–4111.

König, I., Drodt, M., and Trautwein, A.X., 1997. Iron reduction through the tan-green color transition in deep-sea sediments. Geochim. Cosmochim. Acta, 61: 1679–1683.

König, I., Haechel, M., Drodt, M., Suess, E., and Trautwein, A.X., 1999. Reactive Fe(II) layers in deep-sea sediments. Geochim. Cosmochim. Acta, 63: 1517–1526.

Koning, E., Brummer, G.-J., van Raaphorst, W., van Bennekom, J., Helder, W., and van Iperen, J., 1997. Settling, dissolution and burial of biogenic silica in the sediments off Somalia (northwestern Indian Ocean). Deep-Sea Res. II, 44: 1341–1360.

Koretsky, C.M., Meile, C., and Van Cappellen, P., 2002. Quantifying bioirrigation using ecological parameters: a stochastic approach. Geochem. Trans., 3: 17–30.

Kostka, J.E., Dalton, D.D., Skelton, H., Dollhopt, S., and Stucki, J.W., 2002. Growth of iron(III)-reducing bacteria on clay minerals as the sole electron acceptor and comparison of growth yields on a variety of oxidized iron forms. Appl. Environ. Microbiol., 68: 6256–6262.

Kostka, J.E., Haefele, E., Viehweger, R., and Stucki, J.W., 1999a. Respiration and dissolution of Iron(III)-containing minerals by bacteria. Environ. Sci. Tech., 33: 3127–3133.

Kostka, J.E., and Luther III, G.W., 1994. Partitioning and speciation of solid phase iron in saltmarsh sediments. Geochim. Cosmochim. Acta, 58: 1701–1710.

Kostka, J.E., Stucki, J.W., Nealson, K.H., and Wu, J., 1996. Reduction of structural Fe(III) in smectite by a pure culture of *Shewanella putrefaciens* strain MR-1. Clays and Clay Mins., 44: 522–529.

Kostka, J.E., Thamdrup, B., Glud, R.N., and Canfield, D.E., 1999b. Rates and pathways of carbon oxidation in permanently cold Arctic sediments. Mar. Ecol. Prog. Ser., 180: 7–21.

Kristensen, E., 1988. Benthic fauna and biogeochemical processes in marine sediments: microbial activities and fluxes. *In*: T.H. Blackburn and J. Sørensen (eds.), *Nitrogen Cycling in Coastal Marine Environments*. John Wiley & Sons, pp. 275–300.

REFERENCES

Kristensen, E., 2001. Impact of polychaetes (*Nereis* spp. and *Arenicola marina*) on carbon biogeochemistry in coastal marine sediments. Geochem. Trans., 2: 92–103.

Kristensen, E., Ahmed, S.I., and Devol, A.H., 1995. Aerobic and anaerobic decomposition of organic matter in marine sediments: Which is fastest? Limnol. Oceanogr., 40: 1430–1437.

Kristensen, E., Devol, A.H., and Hartnett, H.E., 1999. Organic matter diagenesis in sediments on the continental shelf and slope of the eastern tropical and temperate North Pacific. Cont. Shelf Res., 19: 1331–1351.

Kristensen, E., and Holmer, M.E., 2001. Decomposition of plant material in marine sediments exposed to different electron acceptors (O_2, NO_3^-, and SO_4^{2-}) with emphasis on substrate origin, degradation kinetics, and the role of bioturbation. Geochim. Cosmochim. Acta, 65: 419–433.

Kristensen, E., and Mikkelsen, O.L., 2003. Impact of the burrow-dwelling polychaete *Nereis diversicolor* on degradation of fresh and aged macroalgal detritus in coastal marine sediment. Mar. Ecol. Prog. Ser., 265: 141–153.

Kröger, N., Deutzmann, R., and Sumper, M., 2001. Silica-precipitating peptides from diatoms. J. Biol. Chem., 276: 26055–26070.

Krom, M.D., and Berner, R.A., 1981. The diagenesis of phosphorus in a nearshore marine sediment. Geochim. Cosmochim. Acta, 45: 207–216.

Krom, M.D., and Sholkovitz, E.R., 1977. Nature and reactions of dissolved organic matter in the interstitial waters of marine sediments. Geochim. Cosmochim. Acta, 41: 1565–1573.

Krom, M.D., and Westrich, J.T., 1981. Dissolved organic matter in the pore waters of recent marine sediments; a review, *Biogéochemie de la matière organique à l'interface eau-sédiment marin*. Colloques Internationaux du C.N.R.S. no. 293, pp. 103–111.

Kühl, M., and Revsbeck, N.-P., 2001. Biogeochemical microsensors for boundary layer studies. *In*: B.P. Boudreau and B.B. Jørgensen (eds.), *The Benthic Boundary Layer: Transport Processes and Biogeochemistry*. Oxford University Press, pp. 180–210.

Kump, L.R., Kasting, J.F., and Crane, R.G., 2003. *The Earth System*, 2nd ed. Prentice-Hall.

Kvenvolden, K.A., 1993. Gas hydrates—geological perspecive and global change. Rev. Geophys., 31: 173–187.

Kvenvolden, K.A., 1999. Potential effects of gas hydrate on human welfare. Proc. Natl. Acad. Sci., 96: 3420–3426.

Laanbroek, H.J., and Veldkamp, H., 1982. Microbial interactions in sediment communities. Phil. Trans. Royal Soc. London, B, 297: 533–550.

Landén-Hillmeyr, A., 1998. Nitrogen cycling in continental margin sediments with emphasis on dissolved organic nitrogen and amino acids. PhD dissertation, Göteborg University, Göteborg, Sweden.

REFERENCES

Larter, S.R., and Horsfield, B., 1993. Determination of structural components of kerogens by the use of analytical pyrolysis methods. *In*: M.H. Engal and S.A. Macko (eds.), *Organic Geochemistry*. Plenum Press, pp. 271–374.

Lasaga, A.C., and Holland, H.D., 1974. Mathematical aspects of non-steady-state diagenesis. Geochim. Cosmochim. Acta, 40: 257–266.

Lee, C., 1988. Amino acid and amine biogeochemistry in marine particulate material and sediments. *In*: T.H. Blackburn and J. Sørensen (eds.), *Nitrogen Cycling in Coastal Marine Environments*. John Wiley & Sons, pp. 125–141.

Lee, C., 1992. Controls on organic carbon preservation: The use of stratified water bodies to compare intrinsic rates of decompositin in oxic and anoxic systems. Geochim. Cosmochim. Acta, 56: 3233–3335.

Lee, C., Gagosian, R.B., and Farrington, J.W., 1977. Sterol diagenesis in recent sediments from Buzzards Bay, Massachusetts. Geochim. Cosmochim. Acta, 41: 985–992.

Lee, C., and Wakeham, S.G., 1988. Organic matter in seawater: biogeochemical processes. *In*: J.P. Riley and R. Chester (eds.), *Chemical Oceanography*. Vol. 9. Academic Press, pp. 1–51.

Lee, C., Wakeham, S.G., and Hedges, J.I., 2000. Composition and flux of particulate amino acids and chloropigments in equatorial Pacific seawater and sediments. Deep-Sea Res., 47: 1535–1568.

Lehmann, M.F., Sigman, D.M., and Berelson, W.M., 2004. Coupling the $^{15}N/^{14}N$ and $^{18}O/^{16}O$ of nitrate as a constraint on benthic nitrogen cycling. Mar. Chem., 88: 1–20.

Lerman, A., 1979. *Geochemical Processes, Water and Sediment Environments*. John Wiley & Sons.

Levin, I., and Hesshaimer, V., 2000. Radiocarbon—a unique tracer of global carbon cycle dynamics. Radiocarbon, 42: 69–80.

Lewis, E., and Wallace, D.W.R., 1998. Program developed for CO_2 system calculations. ORNL/CDIAC-105, Oak Ridge National Laboratory.

Li, Y.-H., 2000. *A Compendium of Geochemistry: From Solar Nebula to the Human Brain*. Princeton University Press.

Li, Y.-H., and Gregory, S., 1974. Diffusion of ions in seawater and in deep-sea sediments. Geochim. Cosmochim. Acta, 38: 703–714.

Li, Y.H., Bischoff, J., and Mathieu, G., 1969. The migration of manganese in Artic Basin sediment. Earth Planet. Sci. Lett., 7: 265.

Libes, S.M., 1992. *An Introduction to Marine Biogeochemistry*. John Wiley & Sons.

Liu, K.-K., Iseki, K., and Chao, S.-Y., 2000. Continental margin carbon fluxes. *In*: R.B. Hanson, H.W. Ducklow, and J.G. Field (eds.), *The Changing Ocean Carbon Cycle*. Cambridge University Press, pp. 187–240.

Lohse, L., Epping, E.H., Helder, W., and van Raaphorst, W., 1996. Oxygen pore water profiles in continental shelf sediments of the North Sea: Turbulent vs. molecular diffusion. Mar. Ecol. Prog. Ser., 145: 63–75.

REFERENCES

Lohse, L., Helder, W., Eping, E.H.G., and Balzer, W., 1998. Recycling of organic matter along a shelf-slope transect across the N.W. European continental margin (Goban Spur). Prog. Oceanogr., 42: 77–110.

Lomstein, B.A., Jensen, A.-G.U., Hansen, J.W., Andreasen, J.B., Hansen, L.S., Berntsen, J., and Kuzendorf, H., 1998. Budgets of sediment nitrogen and carbon cycling in the shallow water of Knebel Vig, Denmark. Aquat. Microbiol. Ecol., 14: 69–80.

Lovley, D.R., and Goodwin, S., 1988. Hydrogen concentrations as an indicator of the predominant terminal electron-accepting reactions in aquatic sediments. Geochim. Cosmochim. Acta, 52: 2993–3003.

Lovley, D.R., and Phillips, E.J.P., 1987. Competitive mechanisms for inhibition of sulfate reduction and methane production in the zone of ferric iron reduction in sediments. Appl. Environ. Microbiol., 53: 2636–2641.

Lowenstam, H.A., and Weiner, S., 1989. *On Biomineralization*. Oxford University Press.

Luff, R., Haeckel, M., and Wallmann, K., 2001. Robust and fast FORTRAN and MATLAB libraries to calculate pH distributions in marine systems. Comp. Geosci., 27: 157–169.

Luther, G.W., III, and Church, T.M., 1992. An overview of the environmental chemistry of sulphur in wetland systems. *In*: R.W. Howarth, J.W.B. Stewart, and M.V. Ivanov (eds.), *Sulphur Cycling on the Continents*. John Wiley & Sons, pp. 125–142.

Luther, G.W., III, Ferdelmen, T.G., Kostka, J.E., Tsamakis, E.J., and Church, T.M., 1991. Temporal and spatial variability of reduced sulfur species and porewater parameters in salt marsh sediments. Biogeochem., 14: 57–88.

Luther, G.W., III, Kostka, J.E., Church, T.M., Sulzberger, B., and Stumm, W., 1992. Seasonal iron cycling in the salt marsh sedimentary environment: the importance of ligand complexes with Fe(II) and Fe(III) in the dissolution of Fe(III) minerals and pyrite, respectively. Mar. Chem., 40: 81–103.

Luther, G.W., III, Sundby, B., Lewis, B.L., Brendel, P.J., and Silverberg, N., 1997. Interactions of manganese with the nitrogen cycle: alternative pathways to dinitrogen. Geochim. Cosmochim. Acta, 61: 4043–4052.

Lyle, M., 1983. The brown-green color transition in marine sediments: A marker of the Fe(III)-Fe(II) redox boundary. Limnol. Oceanogr., 28: 1026–1033.

Lyle, M., Heath, G.R., and Robbins, J.M., 1984. Transport and release of transition elements during early diagenesis: Sequential leaching of sediments from MANOP Sites M and H. Part 1. pH 5 acetic acid leach. Geochim. Cosmochim. Acta, 48: 1705–1715.

Lyle, M., Murray, D.W., Finney, B.P., Dymond, J., Robbins, J.M., and Brooksforce, K., 1988. The record of late Pleistocene biogenic sedimentation in the eastern tropical Pacific Ocean. Paleoceanography, 3: 39–59.

Lynn, D.C., and Bonatti, E., 1965. Mobility of manganese in diagenesis of deep sea sediments. Mar. Geol., 3: 457.

Mackenzie, F.T., and Garrel, R.M., 1966. Chemical mass balance between rivers and oceans. Am. J. Sci., 264: 507–525.

Mackenzie, F.T., and Kump, L.R., 1995. Reverse weathering, clay mineral formation, and oceanic element cycles. Science, 270: 586–7.

Mackenzie, F.T., Ristvet, B.L., Thorstenson, D.C., Lerman, A., and Leeper, R.H., 1981. Reverse weathering and chemical mass balance in coastal environments. *In*: J.M. Marten, J.D. Burton, and D. Eisma (eds.), *River Inputs to Ocean Systems*. UNEP and UNESCO, pp. 152–187.

Mackin, J.E., 1987. Boron and silica in salt-marsh sediments: implications for paleo-boron distributions and the early diagenesis of silica. Am. J. Sci., 287: 197–241.

Mackin, J.E., and Aller, R.C., 1984a. Ammonium adsorption in marine sediments. Limnol. Oceanogr., 29: 250–257.

Mackin, J.E., and Aller, R.C., 1984b. Dissolved Al in sediments and waters of the East China Sea: implications for authigenic mineral formation. Geochim. Cosmochim. Acta, 48: 281–297.

Mackin, J.E., and Aller, R.C., 1989. The nearshore and estuarine geochemistry of dissolved aluminium and rapid authigenic mineral precipitation. Rev. Aquat. Sci., 1: 537–554.

Mackin, J.E., Aller, R.C., and Ullman, W.J., 1988. The effects of iron reduction and nonsteady-state diagenesis on iodine, ammonium, and boron distributions in sediments from the Amazon continental shelf. Cont. Shelf Res., 8: 363–386.

Mackin, J.E., and Swider, K.T., 1989. Organic matter decomposition pathways and oxygen consumption in coastal marine sediments. J. Mar. Res., 47: 681–716.

Macko, S.A., Engel, M.H., and Parker, P.L., 1993. Early diagenesis of organic matter in sediments: assessment of mechanisms and preservation by the use of isotopic molecular approaches. *In*: M.H. Engel and S.A. Macko (eds.), *Organic Geochemistry*. Plenum Press, pp. 211–224.

Madigan, M.T., Martinko, J.M., and Parker, J., 1997. *Brock Biology of Microorganisms*, 8th ed. Prentice Hall.

Maita, Y., Montani, S., and Ishii, J., 1982. Early diagenesis of amino acids in Okhotsk Sea sediments. Deep-Sea Res., 29: 485–498.

Mangini, A., Jung, M., and Laukenmann, S., 2001. What do we learn from peaks of uranium and of manganese in deep sea sediments? Mar. Geol., 177: 63–78.

Mangini, A., Segl, M., Glasby, G.P., Stoffers, P., and Plüger, W.L., 1990. Element accumulation rates in and growth histories of manganese nodules from the southwestern Pacific Basin. Mar. Geol., 94: 97–107.

Manheim, F.T., and Waterman, L.S., 1974. Diffusimitry (difuse contant estimation) on sediment cores by resistivity probe. Initial Reports of the Deep Sea Drilling Project, 22: 663–670.

Marinelli, R.L., and Boudreau, B.P., 1996. An experimental and modeling study of pH and relaed solutes in an irrigated anoxic coastal sediment. J. Mar. Res., 54: 939–966.

Marinelli, R.L., Jahnke, R.A., Craven, D.B., Nelson, J.R., and Eckman, J.E., 1998. Sediment nutrient dynamics on the South Atlantic Bight continental shelf. Limnol. Oceanogr., 43: 1305–1320.

Martens, C.S., Albert, D.B., and Alperin, M.J., 1998. Biogeochemical processes controlling methane in gassy coastal sediments Part 1. A model coupling organic matter flux to gas production, oxidation and transport. Cont. Shelf Res., 18: 1741–1770.

Martens, C.S., and Berner, R.A., 1977. Interstitial water chemistry of anoxic Long Island Sound sediments. 1. Dissolved gases. Limnol. Oceanogr., 22: 10–25.

Martens, C.S., Berner, R.A., and Rosenfeld, J.K., 1978. Interstitial water chemistry of anoxic Long Island Sound sediments. 2. Nutrient regeneration and phosphate removal. Limnol. Oceanogr., 23: 605–617.

Martens, C.S., Haddad, R.I. and Chanton, J.P., 1992. Organic matter accumulation, remineralization and burial in an anoxic marine sediment. *In*: J.K. Whelan and J.W. Farrington (eds.), *Productivity, Accumulation, and Preservation of Organic Matter in Recent and Ancient Sediments*. Columbia University Press, pp. 82–98.

Martens, C.S., Kipphut, G.W., and Klump, J.V., 1980. Sediment-water chemical exchange in the coastal zone traced by in situ radon-222 flux measurements. Science, 208: 285–288.

Martens, C.S., and Klump, J.V., 1980. Biogeochemical cycling in an organic-rich marine basin—1. Methane sediment-water exchange processes. Geochim. Cosmochim. Acta, 44: 471–490.

Martens, C.S., and Klump, J.V., 1984. Biogeochemical cycling in an organic-rich marine basin—4. An organic carbon budget for sediments dominated by sulfate reduction and methanogenesis. Geochim. Cosmochim. Acta, 48: 1987–2004.

Martin, J.H., Knauer, G.A., Karl, D.M., and Broenkow, W.W., 1987. VERTEX: carbon cycling in the northeast Pacific. Deep-Sea Res., 34: 267–285.

Martin, W.R., and Banta, G.T., 1992. The measurement of sediment irrigation rates: A comparison of the Br$^-$ tracer and ^{222}Rn/Ra^{226} disequilibrium techniques. J. Mar. Res., 50: 125–154.

Martin, W.R., Bender, M., Leinen, M., and Orchardo, J., 1991. Benthic organic carbon degradation and biogenic silica dissolution in the central equatorial Pacific. Deep-Sea Res., 38: 1481–1516.

Martin, W.R., and Bender, M.L., 1988. The variability of benthic fluxes and sedimentary remineralization rates in response to seasonally variable organic carbon rain rates in the deep sea: a modeling study. Am. J. Sci., 288: 561–574.

Martin, W.R., McNichol, A.P., and McCorkle, D.C., 2000. The radiocarbon age of calcite dissolving at the sea floor: Estimates from pore water data. Geochim. Cosmochim. Acta, 64: 1391–1404.

Martin, W.R., and Sayles, F.L., 1987. Seasonal cycles of particle and solute transport processes in nearshore sediments: ^{222}Rn/^{226}Ra and ^{234}Th/^{238}U disequilibrium at a site in Buzzards Bay, MA. Geochim. Cosmochim. Acta, 51: 927–943.

Martin, W.R., and Sayles, F.L., 1994. Seafloor diagenetic fluxes. *In*: *Material Fluxes on the Surface of the Earth*. Natl. Acad. Press, pp. 143–163.

Martin, W.R., and Sayles, F.L., 1996. CaCO$_3$ dissolution in sediments of the Ceara Rise, western equatorial Atlantic. Geochim. Cosmochim. Acta, 60: 243–264.

Martin, W.R., and Sayles, F.L., 2003. The recycling of biogenic material at the sea floor. *In*: F.T. Machenzie (ed.), *Treatise on Geochemistry*. Vol. 7. Elsevier, pp. 37–65.

Marvin-DiPasquale, M.C., and Capone, D.G., 1998. Benthic sulfate reduction along the Chesapeake Bay central channel. I. Spatial trends and controls. Mar. Ecol. Prog. Ser., 168: 213–228.

Masiello, C.A., 2004. New directions in black carbon organic geochemistry. Mar. Chem., 92: 201–213.

Masiello, C.A., and Druffel, E.R.M., 1998. Black carbon in deep-sea sediments. Science, 280: 1911–1913.

Masiello, C.A., and Druffel, E.R.M., 2001. Carbon isotope geochemistry of the Santa Clara River. Global Biogeochem. Cycles, 15: 407–416.

Masiello, D.A., Druffel, E.R.M., and Currie, L.A., 2002. Radiocarbon measurements of black carbon in aerosols and ocean sediments. Geochim. Cosmochim. Acta, 66: 1025–1036.

Mayer, L.M., 1989. Extracellular proteolytic enzyme activity in sediments of an intertidal mudflat. Limnol. Oceanogr., 34: 973–981.

Mayer, L.M., 1994a. Relationships between mineral surfaces and organic carbon concentrations in soils and sediments. Chem. Geol., 114: 347–363.

Mayer, L.M., 1994b. Surface area control of organic carbon accumulation in continental margin sediments. Geochim. Cosmochim. Acta, 58: 1271–1284.

Mayer, L.M., 1995. Sedimentary organic matter preservation: an assessment and speculative synthesis—a comment. Mar. Chem., 49: 123–126.

Mayer, L.M., 1999. Extent of coverage of mineral surfaces by organic matter in marine sediments. Geochim. Cosmochim. Acta, 63: 207–215.

Mayer, L.M., 2004. The inertness of being organic. Mar. Chem., 92: 135–140.

Mayer, L.M., Chen, Z., Findlay, R.H., Fang, J., Sampson, S., Self, R.F.L., Jumars, P.A., Quetel, C., and Donard, O.F.X., 1996. Bioavailability of sedimentary contaminants subject to deposit-feeder digestion. Environ. Sci. Tech., 30: 2641–2645.

Mayer, L.M., Keil, R.G., Macko, S.A., Joye, S.B., Ruttenberg, K.C., and Aller, R.C., 1998. Importance of suspended particulates in riverine delivery of

REFERENCES

bioavailable nitrogen to coastal zones. Global Biogeochem. Cycles, 12: 573–579.

Mayer, L.M., and Rice, D.L., 1992. Early diagenesis of proteins: A seasonal study. Limnol. Oceanogr., 37: 280–295.

Mayer, L.M., Schick, L.L., Self, R.F.L., and Jumars, P.A., 1997. Digestive environments of benthic macroinvertebrate guts: enzymes, surfactants and dissolved organic matter. J. Mar. Res., 55: 785–812.

Mayer, L.M., Schick, L.L., and Setchell, F.W., 1986. Measurement of protein in nearshore marine sediments. Mar. Ecol. Prog. Ser., 30: 159–165.

McCaffrey, M.A., 1990. Sedimentary lipids as indicators of depositional conditions in the coastal Peruvian upwelling regine. Ph.D. dissertation, Massachusetts Institute of Technology/Woods Hole Oceanographic Institution.

McCarthy, M., Hedges, J., and Benner, R., 1998. Major bacterial contribution to marine dissolved organic nitrogen. Science, 281: 231–234.

McCorkle, D.C., Emerson, S.R., and Quay, P.D., 1985. Stable carbon isotopes in marine pore waters. Earth Planet. Sci. Lett., 74: 13–26.

McDuff, R.E., 1981. Major cation gradients in DSDP interstitial waters: the role of diffusive exchange between seawater and upper oceanic crust. Geochim. Cosmochim. Acta, 45: 1705–1713.

McDuff, R.E., 1985. The chemistry of interstitial waters, Deep Sea Drilling Project Leg 86. In: G.R. Heath and L.H. Burckle (eds.), *Init. Rep. Deep Sea Drilling Proj.* Vol. 86. U.S. Government Printing Office, pp. 675–680.

McDuff, R.E., and Ellis, R.A., 1979. Determining diffusion coefficients in marine sediments: a laboratory study of the validity of the resistivity techniques. Am. J. Sci., 279: 666–675.

McDuff, R.E., and Gieskes, J.M., 1976. Calcium and magnesium profiles in DSDP interstitial waters: Diffusion or reaction? Earth Planet. Sci. Lett., 33: 1–10.

McInerney, M.J., 1988. Anaerobic hydrolysis and fermentation of fats and proteins. In: A.J.B. Zehnder (ed.), *Biology of Anaerobic Microorganisms*. Wiley-Interscience, pp. 373–415.

McKee, B.A., Aller, R.C., Allison, M.A., Bianchi, T.S., and Kineke, G.C., 2004. Transport and transformation of dissolved and particulate materials on continental margins influenced by major rivers: benthic boundary layer and seabed processes. Cont. Shelf Res., 24: 899–926.

McManus, J., Berelson, W.M., Coale, K., Johnson, K., and Kilgore, T.E., 1997. Phosphorus regeneration in continental margin sediments. Geochim. Cosmochim. Acta, 61: 2891–2907.

McManus, J., Hammond, D.E., Berelson, W.M., Kilgore, T.E., DeMaster, D.J., Ragueneau, O.G., and Collier, R.W., 1995. Early diagenesis of biogenic opal: Dissolution rates, kinetics and paleoceanographic implications. Deep-Sea Res. II, 43: 871–903.

McNichol, A.P., Druffel, E.R.M., and Lee, C., 1991. Carbon cycling in coastal sediments: 2. An investigation of the sources of ΣCO_2 to pore waters using carbon isotopes. *In*: R.A. Baker (ed.), *Organic Substances in Sediment and Water.* Vol. 2, *Processes and Analytical.* Lewis Pub., pp. 249–272.

Meade, R.H., 1966. Factors influencing the early stages of the compaction of clays and sands—review. J. Sed. Petrol., 36: 1085–1101.

Means, R.E., and Parcher, J.V., 1964. *Physical Proportion of Soils.* Constable.

Meile, C., Berg, P., Van Capppellen, P., and Tuncay, K., 2005. Solute-specific pore water irrigation: Implications for chemical cycling in early diagenesis. J. Mar. Res., 63: 601–621.

Meile, C., Koretsky, C., and Van Cappellen, P., 2001. Quantifying bioirrigation in aquatic sediments: An inverse modeling approach. Limnol. Oceanogr., 46: 164–177.

Meile, C., and Van Cappellen, P., 2003. Global estimates of enhanced solute transport in marine sediments. Limnol. Oceanogr., 48: 777–786.

Mekik, F.A., Loubere, P.W., and Archer, D.E., 2002. Organic carbon flux and organic carbon to calcite flux ratio recorded in deep-sea sediments: Demonstration and a new proxy. Global Biogeochem. Cycles, 16: 10.1029/2001 GB001634.

Meybeck, M., 1982. Carbon, nitrogen and phosphorus transport by world's rivers. Am. J. Sci., 282: 401–450.

Meyers, P., Silliman, J.E., and Shaw, T.J., 1996. Effects of turbidity flows on organic matter accumulation, sulfate reduction, and methane generation in deep-sea sediments on the Iberia Abyssal Plain. Org. Geochem., 25: 69–78.

Meyers, P.A., 1997. Organic geochemical proxies of paleoceanographic, paleolimnologic, and paleoclimatic processes. Org. Geochem., 27: 213–250.

Meysman, F.J.R., Boudreau, B.P., and Middelburg, J.J., 2003a. Relations between local, non-local, discrete and continuous models of bioturbation. J. Mar. Res., 61: 391–410.

Meysman, F.J.R., Middelburg, J.J., Herman, M.J., and Heip, H.R., 2003b. Reactive transport in surface sediments. II. Media: an object-oriented problem-solving environment for early diagenesis. Comp. Geosc., 29: 301–318.

Meysman, F.J.R., Boudreau, B.P., and Middelburg, J.J., 2005. Modelling reactive transport in sediments subject to bioturbation and compaction. Geochim. Cosmochim. Acta, 69: 3601–3617.

Michalopoulos, P., and Aller, R.C., 1995. Rapid clay mineral formation in Amazon delta sediments: reverse weathering and elemental cycles. Science, 270: 614–617.

Michalopoulos, P., and Aller, R.C., 2004. Early diagenesis of biogenic silica in the Amazon delta: alteration, authigenic clay formation, and storage. Geochim. Cosmochim. Acta, 68: 1061–1085.

Michalopoulos, P., Aller, R.C., and Reeder, R.J., 2000. Conversion of diatoms to clays during early diagenesis in tropical continental shelf muds. Geology, 28: 1095–1098.

Middelburg, J.J., 1989. A simple rate model for organic matter decomposition in marine sediments. Geochim. Cosmochim. Acta, 53: 1577–1581.

Middelburg, J.J., 1993. Turbidites provide a unique oppportunity to study diagenetic processes. Geol. en Mijnb., 72: 15–21.

Middelburg, J.J., de Lange, G.J., and van der Weijden, C.H., 1987. Manganese solubility control in marine pore waters. Geochim. Cosmochim. Acta, 51: 759.

Middelburg, J.J., Nieuwenhuize, J., and van Breugel, P., 1999. Black carbon in marine sediments. Mar. Chem., 65: 245–252.

Middelburg, J.J., Soetaert, K., and Herman, P.M.J., 1997. Empirical relationships for use in global diagenetic models. Deep-Sea Res., 44: 327–344.

Middelburg, J.J., Soetaert, K., Herman, P.M.J., and Heip, C.H.R., 1996. Denitrification in marine sediments: a model study. Global Biogeochem. Cycles, 10: 661–673.

Middelburg, J.J., Vlug, T., and van der Nat, F.J.W.A., 1993. Organic matter mineralization in marine systems. Global Planet. Change, 8: 47–58.

Millero, F.J., 1996. *Chemical Oceanography*, 2nd ed. CRC Press.

Milliman, J.D., 1993. Production and accumulation of calcium carbonate in the ocean: budget of a nonsteady state. Global Biogeochem. Cycles, 7: 927–957.

Milliman, J.D., Trou, P.J., Balch, W.M., Adams, A.K., Li, Y.-H., and Mackenzie, F.T., 1999. Biologically mediated dissolution of calcium carbonate above the chemical lysocline. Deep-Sea Res., 46: 1653–1669.

Mix, A.C., 1989. Pleistocene paleoproductivity: Evidence from organic carbon and foraminiferal species. *In*: W.H. Berger, V.S. Smatacek, and G. Wefer (eds.), *Productivity of the Ocean: Present and Past*. John Wiley & Sons, pp. 313–340.

Moeslund, L., Thamdrup, B., and Jorgensen, B.B., 1994. Sulfur and iron cycling in a coastal sediment: radiotracer studies and seasonal dynamics. Biogeochemistry, 27: 129–152.

Montoya, J., 1994. Nitrogen isotope fractionation in the modern ocean: implications for the sedimentary record. *In*: R. Zahn, T.F. Pedersen, M.A. Kaminski, and L. Labeyrie (eds.), *Carbon Cycling in the Glacial Ocean: Constraints on the Ocean's Role in Global Change*. Springer-Verlag, pp. 259–279.

Moore, W.S., 1981. Iron-manganese banding in Oneida Lake ferromanganese nodules. Nature, 292: 233–235.

Moore, W.S., Dean, W.E., Krishnaswami, S., and Borole, D.V., 1980. Growth rates of manganese nodules in Oneida Lake, New York. Earth Planet. Sci. Lett., 46: 191–200.

Moore, W.S., Ku, T.-L., MacDougall, J.D., Burns, V.M., Burns, R., Dymond, J., Lyle, M.W., and Piper, D.Z., 1981. Fluxes of metals to a manganese nodule: radiochemical, chemical, structural, and mineralogical studies. Earth Planet. Sci. Lett., 52: 151–171.

Mopper, K., Schultz, C.A., Chevolot, L., Germain, C., Revuelta, R., and Dawson, R., 1992. Determination of sugars in unconcentrated seawater and other natural waters by liquid chromatography and pulsed amperometric detection. Environ. Sci. Tech., 26: 133–138.

Morel, F.M.M., and Hudson, R.J.M., 1985. The geobiological cycle of trace elements in aquatic systems: Redfield revisited. *In*: W. Stumm (ed.), *Chemical Processes in Lakes*. Wiley-Interscience, pp. 251–281.

Morford, J.L., and Emerson, S., 1999. The geochemistry of redox sensitive trace metals in sediments. Geochim. Cosmochim. Acta, 63: 1735–1750.

Morse, J.W., 1991. Oxidation kinetics of sedimentary pyrite in seawater. Geochim. Cosmochim. Acta, 55: 3665–3667.

Morse, J.W., 1994. Release of toxic metals via oxidation of authigenic pyrite in resuspended sediments. *In*: C.N. Alpers and D.W. Blowes (eds.), *Environmental Geochemistry of Sulfide Oxidation*, ACS Symposium Series no. 550. American Chemical Society, pp. 289–297.

Morse, J.W., 2003. Formation and diagenesis of carbonate sediments. *In*: F.T. Mackenzie (ed.), *Treatise on Geochemistry*. Vol. 7. Elsevier, pp. 67–85.

Morse, J.W., and Arvidson, R.S., 2002. The dissolution of kinetics of major sedimentary carbonate minerals. Earth-Sci. Rev., 58: 51–84.

Morse, J.W., and Berner, R.A., 1995. What determines sedimentary C/S ratios? Geochim. Cosmochim. Acta, 59: 1073–1077.

Morse, J.W., and Luther III, G.W., 1999. Chemical influences on trace metal–sulfide interaction in anoxic sediments. Geochim. Cosmochim. Acta, 63: 3373–3378.

Morse, J.W., and Mackenzie, F.T., 1990. *Geochemistry of Sedimentary Carbonates*. Elsevier.

Morse, J.W., Millero, F.J., Cornwell, J.C., and Rickard, D., 1987. The chemistry of the hydrogen sulfide and iron sulfide systems in natural waters. Earth-Sci. Rev., 24: 1–42.

Morse, J.W., and Wang, Q., 1997. Pyrite formation under conditions approximating those in anoxic sediments: II. Influence of precursor iron minerals and organic matter. Mar. Chem., 57: 187–193.

Mortimer, R.J.G., Krom, M.D., Harris, S.J., Hayes, P.J., Davies, I.M., Davison, W., and Zhang, H., 2002. Evidence for suboxic nitrification in recent marine sediments. Mar. Ecol. Prog. Ser., 236: 31–35.

Mucci, A., Boudreau, B., and Guignard, C., 2003. Diagenetic mobility of trace elements in sediments covered by a flash flood deposit: Mn, Fe and As. Appl. Geochem., 18: 1011–1026.

Müller, P.J., 1977. C/N ratios in Pacific deep-sea sediments: effect of inorganic ammonium and organic nitrogen compounds sorbed by clays. Geochim. Cosmochim. Acta, 41: 765–776.

Müller, P.J., Kirst, G., Ruhland, G., von Storch, I., and Rosell-Mele, A., 1998. Calibration of the alkenone paleotemperature index $U^{K'}_{37}$ based on coretops from the eastern South Atlantic and the global ocean (60°N–60°S). Geochim. Cosmochim. Acta, 62: 1757–1772.

Müller, P.J., and Suess, E., 1979. Productivity, sedimentation rate and sedimentary organic matter in the oceans—I. Organic carbon preservation. Deep-Sea Res., 26A: 1347–1362.

Mulsow, S., Boudreau, B.P., and Smith, J.N., 1998. Bioturbation and porosity gradients. Limnol. Oceanogr., 43: 1–9.

Murray, J., and Irvine, R., 1895. On the manganese oxide and manganese nodules in marine deposits. Trans. R. Soc. Edinburgh 37: 721–742.

Murray, J.W., 1979. Iron oxides. In: R.G. Burns (ed.), *Marine Minerals*. MSA Short Course Notes. Vol. 6, pp. 47–98.

Murray, J.W., and Kuivala, K.M., 1990. Organic matter diagenesis in the northeast Pacific: transition from aerobic red clay to suboxic hemipelagic sediments. Deep-Sea Res., 37: 59–80.

Murray, R.W., and Leinen, M., 1996. Scavenged excess aluminum and its relationship to bulk titanium in biogenic sediment from the central equatorial Pacific Ocean. Geochim. Cosmochim. Acta, 60: 3869–3878.

Myers, C.R., and Nealson, K.H., 1988. Bacterial manganese reduction and growth with manganese oxide as the sole electron acceptor. Science, 240: 1319–1321.

Nagata, T., Fukuda, R., Koike, I., Kogure, K., and Kirchman, D.L., 1998. Degradation by bacteria of membrane and soluble protein in seawater. Aquat. Microb. Ecol., 14: 29–37.

Nagata, T., and Kirchman, D.L., 1996. Bacterial degradation of protein adsorbed to model submicron particles in seawater. Mar. Ecol. Prog. Ser., 132: 241–248.

Nealson, K., and Berelson, W., 2003. Layered microbial communities and the search for life in the universe. Geomicrobiol. J., 20: 451–462.

Nealson, K.H., 1997. Sediment bacteria: Who's there, what are they doing, and what's new? Ann. Rev. Earth Planet. Sci., 25: 403–434.

Nealson, K.H., and Saffarini, D., 1994. Iron and manganese in anaerobic respiration: environmental significance, physiology, and regulation. Ann. Rev. Microbiol., 48: 311–343.

Nelson, D.M., Tréguer, P., Brzezinski, M.A., Leynaert, A., and Quéguiner, B., 1995. Production and dissolution of biogenic silica in the ocean: Revised global estimates, comparison with regional data and relationship to biogenic sedimentation. Global Biogeochem. Cycles, 9: 359–372.

Nevin, K.P., and Lovley, D.R., 2002. Mechanisms for Fe(III) oxide reduction in sedimentary environments. Geomicrobiol. J., 19: 141–159.

Nguyen, R.T., and Harvey, H.R., 1998. Protein preservation during early diagenesis in marine waters and sediments. In: B.A. Stankiewicz and P.F. van Bergen (eds.), *Nitrogen-Containing Macromolecules in the Bio- and Geosphere*. American Chemical Society, pp. 88–112.

Nguyen, R.T., and Harvey, H.R., 2001. Preservation of protein in marine systems: hydrophobic and other noncovalent associations as major stabilizing forces. Geochim. Cosmochim. Acta, 65: 1467–1480.

Nie, Y., Suayah, I.B., Benninger, L.K., and Alperin, M.J., 2001. Modeling detailed sedimentary ^{210}Pb and fallout $^{239,\,240}$Pu profiles to allow episodic events: An application in Chesapeake Bay. Limnol. Oceanogr., 46: 1425–1437.

Niewöhner, C., Hensen, C., Kasten, S., Zabel, M., and Schulz, H.D., 1998. Deep sulfate reduction completely mediated by anaerobic methane oxidation in sediments of the upwelling area off Namibia. Geochim. Cosmochim. Acta, 62: 455–464.

Nissenbaum, A., and Kaplan, I.R., 1972. Chemical and isotopic evidence for the *in situ* origin of marine humic substances. Limnol. Oceanogr., 17: 570.

Nixon, S.W., 1981. Remineralization and nutrient cycling in coastal marine ecosystems. *In*: B.J. Neilson and L.E. Cronin (eds.), *Estuaries and Nutrients*. Humana Press, pp. 111–138.

Novelli, P.C., Michelson, A.R., Scranton, M.I., Banta, G.T., Hobbie, J.E., and Howarth, R.W., 1988. Hydrogen and acetate cycling in two sulfate-reducing sediments: Buzzards Bay and Town Cove, Mass. Geochim. Cosmochim. Acta, 52: 2477–2486.

Odin, G.S. (ed.), 1988. *Green Marine Clays*. Elsevier.

Onstad, G.D., Canfield, D.E., Quay, P.D., and Hedges, J.I., 2000. Sources of particulate organic matter in rivers from the continental USA: Lignin phenol and stable carbon isotope compositions. Geochim. Cosmochim. Acta, 64: 3539–3546.

Open University, 1989. *The Ocean Basins: Their Structure and Evolution*. Pergamon Press.

Orphan, V.J., House, C.H., Hinrichs, K.-U., McKeegan, K.D., and DeLong, E.F., 2001. Methane-consuming archaea revealed by directly-coupled isotopic and phylogenetic analysis. Science, 293: 484–487.

Pancost, R.D., Hopmans, E.C., Damste, J.S.S., and Party, T.M.S.S., 2001. Archaeal lipids in Mediterranean cold seeps: molecular proxies for anaerobic methane oxidation. Geochim. Cosmochim. Acta, 65: 1611–1627.

Pantoja, S., and Lee, C., 1999. Peptide decomposition by extracellular hydrolysis in coastal seawater and salt marsh sediment. Mar. Chem., 63: 273–291.

Parkes, R.J., Cragg, B.A., Getliff, J.M., Harvey, S.M., Fry, J.C., Lewis, C.A., and Rowland, S.J., 1993. A quantitative study of microbial decomposition of biopolymers in recent sediments from the Peru Margin. Mar. Geol., 113: 55–66.

Passier, H.F., Middelburg, J.J., van Os, R.J.H., and deLange, G.J., 1996. Diaenetic pyritisation under eastern Mediterranean sapropels caused by downward sulphide diffusion. Geochim. Cosmochim. Acta, 60: 751–763.

Patience, R.L., Baxby, M., Bartle, K.D., Perry, D.L., Reiss, G.W., and Rowland, S.J., 1992. The functionality of organic nitrogen in some recent sediments from the Peru upwelling region. Org. Geochem., 18: 161–169.

Paytan, A., and Arrigo, K.R., 2000. The sulfur-isotopic composition of Cenozoic seawater sulfate: Implications for pyrite burial and atmospheric oxygen. Int. Geol. Rev., 42: 491–498.

Paytan, A., Cade-Menum, B.J., McLaughlin, K., and Faul, K.L., 2003. Selective phosphorus regeneration of sinking marine particles: evidence from ^{31}P-NMR. Mar. Chem., 82: 55–70.

Pearson, A., and Eglinton, T.I., 2000. The origin of n-alkanes in Santa Monica Basin surface sediment: a model based on compound-specific Δ^{14}C and δ^{13}C data. Org. Geochem., 31: 1103–1116.

Pearson, A., McNichol, A.P., Benitez-Nelson, B.C., Hayes, J.M., and Eglinton, T.I., 2001. Origins of lipid biomarkers in Santa Monica Basin surface sediment: a case study using compound-specific Δ^{14}C analysis. Geochim. Cosmochim. Acta, 65: 3123–3137.

Pedersen, A.U., Thomsen, T.R., Lomstein, B.A., and Jorgensen, N.O.G., 2001. Bacterial influence on amino acid enantiomerization in a coastal marine sediment. Limnol. Oceanogr., 46: 1358–1369.

Pedersen, T.F., 1983. Increased productivity in the eastern equatorial Pacific during the last glacial maximum (19,000 to 14,000 yr B.P.). Geology, 11: 16–19.

Pedersen, T.F., and Calvert, S.E., 1990. Anoxia vs. productivity: What controls the formation of organic-carbon-rich sediments and sedimentary rocks? AAPG Bull., 74: 454–466.

Pedersen, T.F., Pickering, M., Vogel, J.S., Southon, J.N., and Nelson, D.E., 1988. The response of benthic foraminifera to productivity cycles in the eastern equatorial Pacific: Faunal and geochemical constraints on glacial bottom water oxygen levels. Paleocean., 3: 157–168.

Pedersen, T.F., and Price, N.B., 1982. The geochemistry of manganese carbonate in Panama Basin sediments. Geochim. Cosmochim. Acta, 46: 49.

Pedersen, T.F., Vogel, J.S., and J.R.Southon, 1986. Copper and manganese in hemipelagic sediments at 21°N, East Pacific Rise: Diagenetic contrasts. Geochim. Cosmochim. Acta, 50: 2019–2031.

Peterson, L.C., and Prell, W.L., 1985. Carbonate dissolution in recent sediments of the eastern equatorial Indian Ocean: Preservation patterns and carbonate loss above the lysocline. Mar. Geol., 64: 259–290.

Petsch, S.T., and Berner, R.A., 1998. Coupling of the geochemical cycles of C, P, Fe, and S: The effect on atmospheric O_2 and the isotopic records of carbon and sulfur. Am. J. Sci., 298: 246–262.

Petsch, S.T., Berner, R.A., and Eglinton, T.I., 2000. A field study of the chemical weathering of ancient sedimentary organic matter. Org. Geochem., 31: 475–487.

Pilson, M.E.Q., 1998. *An Introduction to the Chemistry of the Sea*. Prentice Hall.

Piper, D.Z., and Williamson, M.E., 1981. Mineralogy and composition of concentric layers within a manganese nodule from the north Pacific Ocean. Mar. Geol., 40: 255–268.

Pope, R.H., DeMaster, D.J., Smith, C.R., and Seltmann Jr., H., 1996. Rapid bioturbation in equatorial Pacific sediments: evidence from excess ^{234}Th measurements. Deep-Sea Res., 43: 1339–1364.

Popp, B.N., Parekh, P., Tilbrook, B., Bidigare, R.R., and Laws, E.A., 1997. Organic carbon $\delta^{13}C$ variations in sedimentary rocks as chemostratigraphic and paleoenvironmental tools. Paleogeogr. Paleoclimatol. Paleoecol., 132: 119–132.

Post, J.E., 1999. Manganese oxide minerals: Crystal structures and economic and environmental significance. Proc. Natl. Acad. Sci., 96: 3447–3454.

Postma, D., and Jakobsen, R., 1996. Redox zonation: Equilibrium constraints on the Fe(III)/SO_4^{2-} reduction interface. Geochim. Cosmochim. Acta, 60: 3169–3175.

Poulton, S.W., Krom, M.D., and Raiswell, R., 2004. A revised scheme for the reactivity of iron (oxyhydr)oxide minerals towards dissolved sulfide. Geochim. Cosmochim. Acta, 68:3703–3715.

Prahl, F.G., 1985. Chemical evidence of differential particle dispersal in the southern Washington coastal environment. Geochim. Cosmochim. Acta, 49: 2533–2539.

Prahl, F.G., Cowie, G.L., De Lange, G.J., and Sparrow, M.A., 2003. Selective organic matter preservation in "burn-down" turbidites on the Madeira Abyssal Plain. Paleoceanography, 18: 10.1029/2002PA000853.

Prahl, F.G., de Lange, G.H., Scholten, S., and Cowie, G.L., 1997. A case of postdepositional aerobic degradation of terrestrial organic matter in turbidite deposits from the Madeira Abyssal Plain. Org. Geochem., 27: 141–152.

Prahl, F.G., Ertel, J.R., Goni, M.A., Sparrow, M.A., and Eversmeyer, B., 1994. Terrestrial organic carbon contributions to sediments on the Washington margin. Geochim. Cosmochim. Acta, 58: 3035–3048.

Prahl, F.G., and Muehlhausen, L.A., 1989. Lipid biomarkers as geochemical tools for paleoceanographic study. In: W.H. Berger, V.S. Smetacek, and G. Wefer (eds.), *Productivity of the Ocean: Present and Past.* John Wiley & Sons, pp. 271–289.

Premuzic, E.T., Benkovitz, C.M., Gaffney, J.S., and Walsh, J.J., 1982. The nature and distribution of organic matter in the surface sediments of world oceans and seas. Org. Geochem., 4: 63–77.

Presley, B.J., and Kaplan, I.R., 1968. Changes in dissolved sulfate calcium and carbonate from interstitial water of near-shore sediments. Geochim. Cosmochim. Acta, 32: 1037–1048.

Price, B.A., 1988. Equatorial Pacific sediments: A chemical approach to ocean history. Ph.D. dissertation, Scripps Inst. of Oceanogr., University of California, San Diego.

Rabouille, C., Gaillard, J., Tréguer, P., and Vincendeau, M., 1997. Biogenic silica recycling in surficial sediments across the Polar Front of the Southern Ocean (Indian Sector). Deep-Sea Res. II, 44: 1151–1176.

Rabouille, C., and Gaillard, J.-F., 1991a. A coupled model representing the deep-sea organic carbon mineralization and oxygen consumption in surficial sediments. J. Geophys. Res., 96: 2761–2776.

Rabouille, C., and Gaillard, J.-F., 1991b. Towards the EDGE: Early diagenetic global explanation. A model depicting the early diagenesis of organic matter, O_2, NO_3, Mn and PO_4. Geochim. Cosmochim. Acta, 55: 2511–2525.

Rabouille, C., Witbaard, R., and Duineveld, G.C.A., 2001. Annual and interannual variability of sedimentary recycling studied with a non-steady-state model: application to the North Atlantic Ocean (BENGAL site). Prog. Oceanogr., 50: 147–170.

Ragueneau, O., Tréguer, P., Leynaert, A., Anderson, R.F., Brzezinski, M.A., DeMaster, D.J., Dugdale, R.C., Dymond, J., Fischer, G., François, R., Heinze, C., Maier-Reimer, E., Marin-Jézéquel, V., Nelson, D.M., and Quéguiner, B., 2000. A review of the Si cycle in the modern ocean: recent progress and missing gaps in the application of biogenic opal as a paleoproductivity proxy. Global Planet. Change, 26: 317–365.

Raiswell, R., and Canfield, D.E., 1998. Sources of iron for pyrite formation in marine sediments. Am. J. Sci., 298: 219–245.

Raiswell, R., Canfield, D.E., and Berner, R.A., 1994. A comparison of iron extraction methods for the determination of degree of pyritisation and the recognition of iron-limited pyrite formation. Chem. Geol., 111: 101–110.

Ransom, B., Bennett, R.H., Baerwald, R., and Shea, K., 1997. TEM study of in situ organic matter on continental margins: occurrence and the "monolayer" hypothesis. Mar. Geol., 138: 1–9.

Rasheed, M., Wild, C., Franke, U., and Huettel, M., 2004. Benthic photosynthesis and oxygen consumption in permeable carbonate sediments at Heron Island, Great Barrier Reef, Australia. Est. Coast. Shelf Sci., 59: 139–150.

Rashid, M.A., 1985. *Geochemistry of Marine Humic Compounds*. Springer-Verlag.

Rasmussen, H., and Jørgensen, B.B., 1992. Microelectrode studies of seasonal oxygen uptake in a coastal sediment: role of molecular diffusion. Mar. Ecol. Prog. Ser., 81: 289–303.

Raymond, P.A., and Bauer, J.E., 2001. Use of ^{14}C and ^{13}C natural abundances for evaluating riverine, estuarine, and coastal DOC and POC sources and cycling: a review and synthesis. Org. Geochem., 32: 469–485.

Reeburgh, W.S., 1982. A major sink and flux control for methane in marine sediments: anaerobic consumption. *In*: K.A. Fanning and F.T. Manheim (eds.), *The Dynamics of the Ocean Floor*. D.C. Heath and Co., pp. 203–218.

Reeburgh, W.S., 1983. Rates of biogeochemical processes in anoxic sediments. Ann. Rev. Earth Planet. Sci., 11: 269–298.

Reimers, C., Stecher, H.A., III, Taghon, G.L., Fuller, C.M., Huettel, M., Rusch, A., Ryckelynck, N., and Wild, C., 2004. In situ measurments of advective solute transport in permeable shelf sands. Cont. Shelf Res., 24: 183–201.

Reimers, C.E., Jahnke, R.A., and McCorkle, D.C., 1992. Carbon fluxes and burial rates over the continental slope and rise off central California with implications for the global carbon cycle. Global Biogeochem. Cycles, 6: 199–224.

Reimers, C.E., Jahnke, R.A., and Thomsen, L., 2001. In situ sampling in the benthic boundary layer. In: B.P. Boudreau and B.B. Jørgensen (eds.), *The Benthic Boundary Layer*. Oxford University Press, pp. 245–268.

Reimers, C.E., Kalhorn, S., Emerson, S., and Nealson, K.H., 1984. Oxygen consumption rates in pelagic sediments from the central Pacific: first estimates from microelectrode profiles. Geochim. Cosmochim. Acta, 48: 903–910.

Reimers, C.E., Ruttenberg, K.C., Canfield, D.E., Christiansen, M.B., and Martin, J.B., 1996. Porewater pH and authigenic phases formed in the uppermost sediments of the Santa Barbara Basin. Geochim. Cosmochim. Acta, 60: 4037–4057.

Rempel, A.W., and Buffett, B.A., 1997. Formation and accumulation of gas hydrate in porous media. J. Geophys. Res., 102(B5): 10151–10164.

Rhoads, D.C., 1974. Organism-sediment relations on the muddy sea floor. Oceanogr. Mar. Biol. Ann. Rev., 12: 263–300.

Rice, D.L., and Hanson, R.B., 1984. A kinetic model for detritus nitrogen: role of the associated bacteria in nitrogen accumulation. Bull. Mar. Sci., 35: 326–340.

Rickard, D., 1997. Kinetics of pyrite formation by the H_2S oxidation of iron (II) monosulfide in aqueous solutions between 25 and 125°C: The rate equation. Geochim. Cosmochim. Acta, 61: 115–134.

Rickard, D., Schoonen, M.A.A., and Luther, G.W., 1995. Chemistry of iron sulfides in sedimentary environments. In: M.A. Vairavamurthy and M.A.A. Schoonen (eds.), *Geochemical Transformation of Sedimentary Sulfur*. American Chemical Society.

Riedel, G.F., Sanders, J.G., and Osman, R.W., 1997. Biogeochemical controls on the flux of trace elements from estuarine sediments: water column oxygen concentrations and benthic infauna. Est. Coastal Shelf Sci., 44: 23–38.

Robbins, J.A., 1978. Geochemical and geophysical application of radioactive lead. In: J.O. Nriagu (ed.), *The Biogeochemistry of Lead in the Environment*. Elsevier/North-Holland Biomedical Press, pp. 285–353.

Robbins, J.A., 1986. A model for particle-selective transport of tracers in sediments with conveyor belt deposit feeders. J. Geophys. Res., 91: 8542–8558.

Robbins, J.A., and Callender, E., 1975. Diagenesis of manganese in Lake Michigan sediments. Am. J. Sci., 275: 512–533.

Roden, E.E., and Tuttle, J.H., 1996. Carbon cycling in mesohaline Chesapeake Bay sediments 2: Kinetics of particulate and dissolved organic carbon turnover. J. Mar. Res., 54: 343–383.

Roden, E.E., Tuttle, J.H., Boynton, W.R., and Kemp, W.M., 1995. Carbon cycling in mesohaline Chesapeake Bay sediments. 1. POC deposition rates and mineralization pathways. J. Mar. Res., 53: 799–819.

REFERENCES

Romankevich, E.A., 1984. *Geochemistry of Organic Matter in the Ocean.* Springer-Verlag.

Rosenfeld, J.K., 1979a. Amino acid diagenesis and adsorption in nearshore anoxic marine sediments. Limnol. Oceanogr., 24: 1014–1021.

Rosenfeld, J.K., 1979b. Ammonium adsorption in nearshore anoxic sediments. Limnol. Oceanogr., 24: 356–364.

Rubinsztain, Y., Ioselis, P., Ikan, R., and Aizenshtat, Z., 1984. Investigations on the structural units of melanoidins. Org. Geochem., 6: 791–804.

Ruddiman, W.F., and Glover, L.K., 1972. Vertical mixing of ice-rafted vocanic ash in North Atlantic sediments. Bull. Geol. Soc. Am., 83: 2817–2836.

Rude, P.D., and Aller, R.C., 1989. Early diagenetic alteration of lateritic particle coatings in Amazon continental shelf sediments. J. Sed. Petr., 59: 704–716.

Rude, P.D., and Aller, R.C., 1991. Fluorine mobility during early diagenesis of carbonate sediments: An indicator of mineral transformations. Geochim. Cosmochim. Acta, 55: 2491–2509.

Rude, P.D., and Aller, R.C., 1994. Fluorine uptake by continental shelf sediment and its impact on the global fluorine cycle. Cont. Shelf Res., 14: 883–907.

Rullkötter, J., 2000. Organic matter: the driving force for early diagenesis. *In*: H.D. Schulz and M. Zabel (eds.), *Marine Geochemistry.* Springer-Verlag, pp. 129–172.

Rutgers van der Loeff, M.M., 1990. Oxygen in pore waters of deep-sea sediments. Phil. Trans. R. Soc. Lond., 331: 69–84.

Ruttenberg, K.C., 1992. Development of a sequential extraction method for different forms of phosphorus in marine sediments. Limnol. Oceanogr., 37: 1460–1482.

Ruttenberg, K.C., 1993. Reassessment of the oceanic residence time of phosphorus. Chem. Geol., 107: 405–409.

Ruttenberg, K.C., 2003. The global phosphorus cycle. *In*: W.C. Schlesinger (ed.), *Treatise on Geochemistry.* Vol. 8. Elsevier, pp. 585–643.

Ruttenberg, K.C. and Berner, R.A., 1993. Authigenic apatite formation and burial in sediments from non-upwelling, continental margin environments. Geochim. Cosmochim. Acta, 57: 991–1007.

Ruttenburg, K.C. and Goñi, M.A., 1997. Phosphorus distribution, C:N:P ratios, and $\delta^{13}C_{oc}$ in arctic, temperate, and tropical coastal sediments: tools for characterizing bulk sedimentary organic matter. Mar. Geol., 139: 123–145.

Rysgaard, S., Risgaard-Petersen, N., Sloth, N.P., Jensen, K., and Nielsen, L.P., 1994. Oxygen regulation of nitrification and denitrification in sediments. Limnol. Oceanogr., 39: 1643–1652.

Salwan, J.J., and Murray, J.W., 1983. Trace metal remobilization in the interstitial waters of red clay and hemi-pelagic marine sediments. Earth Planet. Sci. Lett., 64: 213–230.

Sampou, P., and Oviatt, C.A., 1991. A carbon budget for a eutrophic marine ecosystem and the role of sulfur metabolism in sedimentary carbon, oxygen and energy dynamics. J. Mar. Res., 49: 825–849.

Sanderson, B., 1985. How bioturbation supports manganese nodules at the sediment-water interface. Deep-Sea Res., 32: 1281–1285.

Sansone, F.J., Andrews, C.A., and Okamoto, M.Y., 1987. Adsorption of short-chain organic acids onto nearshore sediments. Geochim. Cosmochim. Acta, 51: 1889–1896.

Santschi, P.H., Anderson, R.F., Fleisher, M.Q., and Bowles, W., 1991. Mesurements of diffusive sublayer thicknesses in the ocean by alabaster dissolution, and their implications for the measurements of benthic fluxes. J. Geophys. Res., 96: 10641–10657.

Santschi, P.H., Bower, P., Nyffeler, U.P., Azevedo, A., and Broecker, W., 1983. Estimates of the resistance to chemical transport posed by the deep-sea boundary layer. Limnol. Oceanogr., 28: 899–912.

Santschi, P.H., Guo, L., Baskaran, M., Trumbore, S., Southon, J., Bianchi, T.S., Honeyman, B., and Cifuentes, L., 1995. Isotopic evidence for the contemporary origin of high-molecular-weight organic matter in oceanic environments. Geochim. Cosmochim. Acta, 59: 625–631.

Savin, S.M., and Yeh, H., 1981. Stable isotopes in ocean sediments. *In*: C. Emiliani (ed.), *The Sea*. Vol. 7, *The Oceanic Lithosphere*. John Wiley & Sons, pp. 1521–1554.

Savrda, C.E., Bottjer, D.J., and Gorsline, D.S., 1984. Development of a comprehensive oxygen-deficient biofacies model: Evidence from Santa Monica, San Pedro and Santa Barbara Basins, California Continental Borderland. AAPG Bull., 68: 1179–1192.

Sayles, F.L., 1979. The composition and diagenesis of interstitial solutions. I. Fluxes across the seawater-sediment interface in the Atlantic Ocean. Geochim. Cosmochim. Acta, 43: 527–545.

Sayles, F.L., 1981. The composition and diagenesis of interstitial solutions. II. Fluxes and diagenesis at the water-sediment interface in the high latitude North and South Atlantic. Geochim. Cosmochim. Acta 45: 1061.

Sayles, F.L., and Curry, W.B., 1988. $\delta^{13}C$, TCO_2 and the metabolism of organic carbon in deep sea sediments. Geochim. Cosmochim. Acta, 52: 2963–2978.

Sayles, F.L., and Mangelsdorf, P.C., Jr., 1977. The equilibration of clay minerals with seawater: exchange reactions. Geochim. Cosmochim. Acta, 41: 951–960.

Sayles, F.L., and Mangelsdorf, P.C., Jr., 1979. Cation-exchange characteristics of Amazon River suspended sediment and its reaction with seawater. Geochim. Cosmochim. Acta, 43: 767–779.

Sayles, F.L., and Martin, W.R., 1995. *In situ* tracer studies of solute transport across the sediment-water interface at the Bermuda time series site. Deep-Sea Res., 42: 31–52.

Sayles, F.L., Martin, W.R., Chase, Z., and Anderson, R.F., 2001. Benthic remineralization and burial of biogenic SiO_2, $CaCO_3$, organic carbon, and detrital material in the Southern Ocean along a transect at 170° West. Deep-Sea Res., 48: 4323–4383.

Sayles, F.L., Martin, W.R., and Deuser, W.G., 1994. Response of benthic oxygen demand to particulate organic carbon supply in the deep sea near Bermuda. Nature, 371: 686–689.

Schenau, S.J., and De Lange, G.J., 2000. A novel chemical method to quantify fish debris in marine sediments. Limnol. Oceanogr., 45: 963–971.

Schenau, S.J., Slomp, C.P., and de Lange, G.J., 2000. Phosphogenesis and active phosphorite formation in sediments from the Arabian Sea oxygen minimum zone. Mar. Geol., 169: 1–20.

Schink, B., 1988. Principles and limits of anaerobic degradation: environmental and technological aspects. *In*: A.J.B. Zehnder (ed.), *Biology of Anaerobic Microorganisms*. John Wiley & Sons, pp. 771–846.

Schink, D.R., Guinasso Jr., N.L., and Fanning, K.A., 1975. Processes affecting the concentration of silica at the sediment-water interface of the Atlantic Ocean. J. Geophys. Res., 80: 3013.

Schippers, A., and Jørgensen, B.B., 2002. Biogeochemistry of pyrite and iron sulfide oxidation in marine sediments. Geochim. Cosmochim. Acta, 66: 85–92.

Schlünz, B., and Scneider, R.R., 2000. Transport of terrestrial organic carbon to the oceans by rivers: re-estimating flux- and burial rates. Int. Jour. Earth Sci., 88: 599–606.

Schlüter, M., Sauter, E.J., Schäfer, A., and Ritzrau, W., 2000. Spatial budget of organic carbon flux to the seafloor of the northern North Atlantic (60°N–80°N). Global Biogeochem. Cycles, 14: 329–340.

Schmidt, J.L., Deming, J.W., Jumars, P.A., and Keil, R.G., 1998. Constancy of bacterial abundance in surficial marine sediments. Limnol. Oceanogr., 43: 976–982.

Schoonen, M.A.A., and Barnes, H.L., 1991. Reactions forming pyrite and marcasite from solution: II. Via FeS precursors below 100°C. Geochim. Cosmochim. Acta, 55: 1505–1514.

Schouten, S., Hoefs, M.J.L., and Sinninghe Damste, J.S., 2000. A molecular and stable carbon isotopic study of lipids in late Quaternary sediments from the Arabian Sea. Org. Geochem., 31: 509–521.

Schrag, D.P., Adkins, J.F., McIntyre, K., Alexander, J.L., Hodell, D.A., Charles, C.D., and McManus, J.F., 2002. The oxygen isotopic composition of seawater during the Last Glacial Maximum. Quat. Sci. Rev., 21: 331–342.

Schrag, D.P., and DePaolo, D.J., 1993. Determination of $\delta^{18}O$ of seawater in the deep ocean during the Last Glacial Maximum. Paleoceanography, 8: 1–6.

Schubert, C.J., and Calvert, S.E., 2001. Nitrogen and carbon isotopic composition of marine and terrestrial organic matter in Arctic Ocean sediments:

implications for nutrient utilization and organic matter composition. Deep-Sea Res., 48: 789–810.

Schuffert, J.D., Jahnke, R.A., Kastner, M., Leather, J., Sturz, A., and Wing, M.R., 1994. Rates of formation of modern phosphorite off wesern Mexico. Geochim. Cosmochim. Acta, 58: 5001–5010.

Schuffert, J.D., Kastner, M., and Jahnke, R.A., 1998. Carbon and phosphorus burial associated with modern phosphorite formation. Mar. Geol., 146: 21–31.

Schulz, H.D., and Zabel, M. (eds.), 2000. *Marine Geochemistry*. Springer-Verlag.

Schulz, H.N., and Jørgensen, B.B., 2001. Big bacteria. Ann. Rev. Microbiol., 55: 105–137.

Seibold, E., and Berger, W.H., 1996. *The Sea Floor, An Introduction to Marine Geology*, 3rd ed. Springer-Verlag.

Seitzinger, S.P., 1988. Denitrification in freshwater and marine ecosystems: ecological and geochemical significance. Limnol. Oceanogr., 33: 702–724.

Shackleton, N.J., 1977. Carbon-13 in *Uvigerina*: tropical rainforest history and the equatorial Pacific carbonate dissolution cycles. *In*: N.R. Andersen and A. Malahoff (eds.), *The Fate of Fossil Fuel CO_2 in the Oceans*. Plenum, pp. 401–427.

Shackleton, N.J., 2000. The 100,000-year ice-age cycle identified and found to lag temperature, carbon dioxide, and orbital eccentricity. Science, 289: 1897–1902.

Shackleton, N.J., Hall, M.A., Line, J., and Shuxi, C., 1983. Carbon isotope data in core V19-30 confirm reduced carbon dioxide concentration in the ice age atmosphere. Nature, 306: 319–322.

Shackleton, N.J., Le, J., Mix, A., and Hall, M.A., 1992. Carbon isotope records from Pacific surface waters and atmospheric carbon dioxide. Quat. Sci. Rev., 11: 387–400.

Shaw, T.J., Gieskes, J.M., and Jahnke, R.A., 1990. Early diagenesis in differing depositional environments: The response of transition metals in pore waters. Geochim. Cosmochim. Acta, 54: 1233–1246.

Sherwood, B.A., Sager, S.L., and Holland, H.D., 1987. Phosphorus in foraminiferal sediments from North Atlantic Ridge cores and in pure limestones. Geochim. Cosmochim. Acta, 51: 1861–1866.

Shimmield, G.B., and Pedersen, T.F., 1990. The geochemistry of reactive trace metals and halogens in hemipelagic continental margin sediments. Rev. Aquat. Sci., 3: 255–279.

Shimmield, G.B., and Price, N.B., 1986. The behaviour of molybdenum and manganese during early sediment diagenesis—offshore Baha California, Mexico. Mar. Chem., 19: 261–280.

Shine, J.P., Ika, R., and Ford, T.E., 1998. Relationship between oxygen consumption and sediment-water fluxes of heavy metals in coastal marine sediments. Environ. Toxicol. Chem., 17: 2325–2337.

Sholkovitz, E., 1973. Interstitial water chemistry of the Santa Barbara Basin. Geochim. Cosmochim. Acta, 37: 2043–2073.

Showers, W.J., and Angle, D.G., 1986. Stable isotopic characterization of organic carbon accumulation on the Amazon continental shelf. Cont. Shelf Res., 6: 227–244.

Shum, K.T., and Sundby, B., 1996. Organic matter processing in continental shelf sediments—the subtidal pump revisited. Mar. Chem., 53: 81–87.

Sillén, L.G., 1967. The ocean as a chemical system. Science, 156: 1189–1197.

Sinninghe Damsté, J.S., and de Leeuw, J.W., 1989. Analysis, structure and geochemical significance of organically-bound sulphur in the geosphere: state of the art and future research. Org. Geochem., 16: 1077–1101.

Skoog, A., and Benner, R., 1998. Aldoses in various size fractions of marine organic matter: Implications for carbon cycling. Limnol. Oceanogr., 42: 1803–1813.

Skrabal, S.A., Donat, J.R., and Burdige, D.J., 1997. Fluxes of copper-complexing ligands from estuarine sediments. Limnol. Oceanogr., 42: 992–996.

Skrabal, S.A., Donat, J.R., and Burdige, D.J., 2000. Pore water distributions of dissolved copper and copper-complexing ligands in estuarine and coastal marine sediments. Geochim. Cosmochim. Acta, 64: 1843–1857.

Slomp, C.P., Epping, E.H.G., Helder, W., and van Raaphorst, W., 1996. A key role for iron-bound phosphorus in authigenic apatite formation in North Atlantic continental platform sediments. J. Mar. Res., 54: 1179–1205.

Slomp, C.P., Malschaert, J.F.P., Lohse, L., and Van Raaphorst, W., 1997. Iron and manganese cycling in different sedimentary environments on the North Sea continental margin. Cont. Shelf Res., 17: 1083–1117.

Smethie, W.M.J., Nittrouer, C.A., and Self, R.F.L., 1981. The use of radon-222 as a tracer of sediment irrigation and mixing on the Washington continental shelf. Mar. Geol., 42: 173–200.

Smith, C.R., DeMaster, D.J., and Fornes, W.L., 2001. Mechanisms of age-dependent bioturbation on the bathyl California margin: The young and the restless. *In*: J.Y. Aller, S.A. Woodin, and R.C. Aller (eds.), *Organism-Sediment Interactions*. University of South Carolina Press, pp. 263–277.

Smith, C.R., Hoover, D.J., Doan, S.E., Pope, R.H., DeMaster, D.J., Dobbs, F.C., and Altabet, M.A., 1996. Phytodetritus at the abyssal seafloor across 10° of latitude in the central equatorial Pacific. Deep-Sea Res., 43: 1309–1338.

Smith, C.R., Pope, R.H., DeMaster, D.J., and Magaard, L., 1993. Age-dependent mixing in deep-sea sediments. Geochim. Cosmochim. Acta, 57: 1473–1488.

Smith, C.R., and Rabouille, C., 2002. What controls the mixed-layer depth in deep-sea sediments? The importance of POC flux. Limnol. Oceanogr., 47: 418–426.

Smith, C.R., and Rabouille, C., 2004. Reply to comment by Boudreau on: What controls the mixed-layer depth in deep-sea sediments? The importance of POC flux. Limnol. Oceanogr., 49: 623–624.

Smith, D.J., and Eglinton, G., 1983. The lipid chemistry of an interfacial sediment from the Peru Continental Shelf: fatty acids, alcohols, aliphatic ketones and hydrocarbons. Geochim. Cosmochim. Acta, 47: 2225–2232.

Smith, J.N., Boudreau, B.P., and Noshkin, V., 1986/87. Plutonium and ^{210}Pb distributions in northeast Atlantic sediments: subsurface anomalies caused by non-local mixing. Earth Planet. Sci. Lett., 81: 15–28.

Smith, K.L., Baldwin, R.J., and Williams, P.M., 1992. Reconciling particulate organic carbon flux and sediment community oxygen consumption in the deep North Pacific. Nature, 359: 313–316.

Smith, K.L., Jr., White, G.A., and Laver, M.B., 1979. Oxygen uptake and nutrient exchange of sediments measured in situ using a free vehicle grab respirometer. Deep-Sea Res., 26A: 337–346.

Soetaert, K., Herman, M.J., Middelburg, J.J., Heip, C., deStigter, H.S., van Weering, T.C.E., Epping, E., and Helder, W., 1996a. Modeling ^{210}Pb-derived mixing activity in ocean margin sediments: Diffusive versus nonlocal mixing. J. Mar. Res., 54: 1207–1227.

Soetaert, K., Herman, P.M.J., and Middelburg, J.J., 1996b. Dynamic response of deep-sea sediments to seasonal variations: A model. Limnol. Oceanogr., 41: 1651–1668.

Soetaert, K., Herman, P.M.J., and Middelburg, J.J., 1996c. A model for early diagenetic processes from the shelf to abyssal depths. Geochim. Cosmochim. Acta, 60: 1019–1040.

Somasundaran, P., and Agar, G.E., 1967. The zero point of charge of calcite. J. Coll. Interf. Sci., 24: 433–440.

Sorem, R.K., and Fewkes, R.H., 1977. Internal characteristics. In: G.P. Glasby (ed.), *Marine Manganese Deposits*. Elsevier, pp. 147–184.

Sørensen, J., 1987. Nitrate reduction in marine sediment: pathways and interactions with iron and sulfur cycling. Geomicrobiol. J., 5: 401–421.

Sørensen, J., Hydes, D.J., and Wilson, T.R.S., 1984. Denitrification in a deep-sea sediment core from the eastern equatorial Atlantic. Limnol. Oceanogr., 29: 653–657.

Srodon, J., 1999. Nature of mixed-layer clays and mechanisms of their formation and alteration. Ann. Rev. Earth Planet. Sci., 27: 19–53.

Stankiewicz, B.A., and van Bergen, P.F. (eds.), 1998. *Nitrogen-Containing Macromolecules in the Bio- and Geosphere*. ACS Symposium Series, 707. American Chemical Society.

Starikova, N.D., 1970. Vertical distribution patterns of dissolved organic carbon in sea water and interstitial solutions. Oceanology, 10: 796–807.

Steinberg, S.M., Venkatesan, M.I., and Kaplan, I.R., 1987. Organic geochemistry of sediments from the continental margin off southern New England,

U.S.A. Part I. Amino acids, carbohydrates and lignin. Mar. Chem., 21: 249–265.

Stephens, M.P., and Kadko, D.C., 1997. Glacial/Holocene calcium carbonate dissolution at the central equatorial Pacific seafloor. Paleoceanography, 12: 797–804.

Stevenson, F.J., 1994. *Humus Chemistry: Genesis, Composition, Reactions*, 2nd ed. John Wiley & Sons.

Stone, A.T., Godtfredsen, K.L., and Deng, B., 1994. Sources and reactivity of reductants encountered in aquatic environments. *In*: G. Bidoglio and W. Stumm (eds.), *Chemistry of Aquatic Systems: Local and Global Perspectives*. Kluwer Publishers, pp. 337–374.

Stott, L.D., Berelson, W., Douglas, R., and Gorsline, D., 2000. Increased dissolved oxygen in Pacific intermediate waters due to lower rates of carbon oxidation in sediments. Nature, 407: 367–370.

Stuiver, M., and Polach, H.A., 1977. Reporting of ^{14}C data. Radiocarbon, 19: 355–363.

Stumm, W., and Morgan, J.J., 1996. *Aquatic Chemistry: Chemical Equilibria and Rates in Natural Waters*, 3rd ed. Wiley-Interscience.

Suess, E., 1973. Interactions of organic compounds with calcium carbonate. II. Organo-carbonate interactions in recent sediments. Geochim. Cosmochim. Acta, 37: 2435–2447.

Suess, E., 1979. Mineral phases formed in anoxic sediments by microbial decomposition of organic matter. Geochim. Cosmochim. Acta, 43: 339–341.

Suess, E., 1981. Phosphate regeneration from sediments of the Peru continental margin by dissolution of fish debris. Geochim. Cosmochim. Acta, 45: 577–588.

Suess, E., and Müller, P.J., 1981. Productivity, sedimentation rate and sedimentary organic matter in the oceans: II. Elemental fractionation. *In*: *Biogéochemie de la matière organique à l'interface eau-sédiment marin*. Colloques Internationaux du C.N.R.S. no. 293, pp. 17–26.

Sugai, S.F., and Henrichs, S.M., 1992. Rates of amino acid uptake and remineralization in Resurrection Bay (Alaska) sediments. Mar. Ecol. Prog. Ser., 88: 129–141.

Sun, M., Aller, R.C., and Lee, C., 1991. Early diagenesis of chlorophyll-*a* in Long Island Sound sediments: A measure of carbon flux and particle reworking. J. Mar. Res., 49: 379–401.

Sun, M., Aller, R.C., Lee, C., and Wakeham, S.G., 2002. Effects of oxygen and redox oscillation on degradation of cell-associated lipids in surficial marine sediments. Geochim. Cosmochim. Acta, 66: 2003–2012.

Sun, M., and Wakeham, S.G., 1994. Molecular evidence for degradation and preservation of organic matter in the anoxic Black Sea Basin. Geochim. Cosmochim. Acta, 58: 3395–3406.

Sun, M., Wakeham, S.G., and Lee, C., 1997. Rates and mechanisms of fatty acid degradation in oxic and anoxic coastal marine sediments of Long Island Sound, New York, USA. Geochim. Cosmochim. Acta, 61: 341–355.

Sunda, W.G., and Kieber, D.J., 1994. Oxidation of humic substances by manganese oxides yields low-molecular-weight organic substrates. Nature, 367: 62–64.

Sundby, B., Anderson, L.G., Hall, P.O.J., Iverfeldt, A., van der Loeff, M.M.R., and Westerlund, S.F.G., 1986. The effect of oxygen on the release of cobalt, manganese, iron and phosphate at the sediment-water interface. Geochim. Cosmochim. Acta, 50: 1281–1288.

Sundby, B., Gobeil, C., Silverberg, N., and Mucci, A., 1992. The phosphorus cycle in coastal marine sediments. Limnol. Oceanogr., 37: 1129–1145.

Sundquist, E.T., 1990. Influence of deep-sea benthic processes on atmospheric CO_2. Phil. Trans. Royal Soc. London, A, 331: 155–165.

Sweeney, R.E., and Kaplan, I.R., 1973. Pyrite framboid formation: Laboratory synthesis and marine sediments. Econ. Geol., 68: 618–634.

Swider, K.T., and Mackin, J.E., 1989. Transformations of sulfur compounds in marsh-flat sediments. Geochim. Cosmochim. Acta, 53: 2311–2323.

Swinbanks, D.D., and Shirayama, Y., 1984. Burrow stratigraphy in relation to manganese diagenesis in modern deep-sea sediments. Deep-Sea Res., 31: 1197–1223.

Tebo, B.M., Ghiorse, W.C., van Waasbergen, L.G., Siering, P.L., and Caspi, R., 1997. Bacterially mediated mineral formation: insights into manganese(II) oxidation from molecular genetic and biochemical studies. *In*: J.F. Banfield and K.H. Nealson (eds.), *Geomicrobiology: Interactions between Microbes and Minerals*. Rev. Mineral, 35. Mineral. Soc. Am., pp. 225–266.

Tegelaar, E.W., de Leeuw, J.W., Derenne, S., and Largeau, C., 1989. A reappraisal of kerogen formation. Geochim. Cosmochim. Acta, 53: 3103–3106.

Tengberg, A., et al., 1995. Benthic chamber and profiling landers in oceanography: A review of design, technical solutions and functioning. Prog. Oceanog., 35: 253–294.

Thamdrup, B., 2000. Bacterial manganese and iron reduction in aquatic sediments. Adv. Microbiol. Ecol., 16: 41–83.

Thamdrup, B., and Canfield, D.E., 1996. Pathways of carbon oxidation in continental margin sediments off central Chile. Limnol. Oceanogr., 41: 1629–1650.

Thamdrup, B., and Dalsgaard, T., 2002. Production of N_2 through anaerobic ammonium oxidation coupled to nitrate reduction in marine sediments. Appl. Environ. Microbiol., 68: 1312–1318.

Thamdrup, B., Glud, R.N., and Hansen, J.W., 1994. Manganese oxidation and in situ manganese fluxes from a coastal sediment. Geochim. Cosmochim. Acta, 58: 2563–2570.

REFERENCES

Thiel, V., Peckmann, J., Richnow, H.H., Luth, U., Reitner, J., and Michaelis, W., 2001. Molecular signals for anaerobic methane oxidation in Black Sea seep carbonates and a microbial mat. Mar. Chem., 73: 97–112.

Thimsen, C.A., and Keil, R.G., 1998. Potential interactions between sedimentary dissolved organic matter and mineral surfaces. Mar. Chem., 62: 65–76.

Thistle, D., 2003. The deep-sea floor: An overview. In: P.A. Tyler (ed.), *Ecosystems of the Deep Oceans*. Elsevier, pp. 5–37.

Thompson, B., Dixon, J., Schroeter, S., and Reish, D.J., 1993. Benthic invertebrates. In: M.D. Dailey, D.J. Reish, and J.W. Anderson (eds.), *Ecology of the Southern California Bight: A Synthesis and Interpretation*. University of California Press, pp. 369–458.

Thomson, J., Carpenter, M.S.N., Colley, S., Wilson, T.R.S., Kennedy, H., and Elderfield, H., 1984a. Metal accumulation rates in northwest Atlantic pelagic sediments. Geochim. Cosmochim. Acta, 48: 1935–1948.

Thomson, J., Colley, S., Higgs, N.C., Hydes, D.J., Wilson, T.R.S., and Sørensen, J., 1987. Geochemical oxidation fronts in NE Atlantic distal turbidites and their effects in the sedimentary record. In: P.P. Weaver and J. Thomson (eds.), *Geology and Geochemistry of Abyssal Plains*. Geol. Soc. Spec. Publ. no. 31, pp. 167–177.

Thomson, J., Higgs, N.C., and Colley, S., 1989. A geochemical investigation of reduction haloes developed under turbidites in brown clay. Mar. Geol., 89: 315–330.

Thomson, J., Higgs, N.C., and Colley, S., 1996. Diagenetic redistribution of redox-sensitive elements in NE Atlantic glacial/interglacial transition sediments. Earth Planet. Sci. Lett., 139: 365–377.

Thomson, J., Higgs, N.C., Croudace, I.W., Colley, S., and Hydes, D.J., 1993. Redox zonation of elements at an oxic/post-oxic boundary in deep-sea sediments. Geochim. Cosmochim. Acta, 57: 579–595.

Thomson, J., Higgs, N.C., Jarvis, I., Hydes, D.J., Colley, S., and Wilson, T.R.S., 1986. The behaviour of manganese in Atlantic carbonate sediments. Geochim. Cosmochim. Acta, 50: 1807–1818.

Thomson, J., Jarvis, I., Green, D.R.H., Green, D.A., and Clayton, T., 1998. Mobility and immobility of redox-sensitive elements in deep-sea turbidites during shallow burial. Geochim. Cosmochim. Acta, 62: 643–656.

Thomson, J., Wilson, T.R.S., Culkin, F., and Hydes, D.J., 1984b. Non-steady state diagenetic record in eastern equatorial Atlantic sediments. Earth Planet. Sci. Lett., 71: 23–30.

Thorstenson, D.C., and Mackenzie, F.T., 1974. Time variability of pore water chemistry in recent carbonate sediments, Devil's Hole, Harrington Sound, Bermuda. Geochim. Cosmochim. Acta, 38: 1–19.

Thurman, E.M., 1985. *Organic Geochemistry of Natural Waters*. Martinus Nijhoff/Dr. W. Junk Pub.

REFERENCES

Tissot, B., and Welte, D.H., 1978. *Petroleum Occurence and Formation.* Springer-Verlag.

Tobias, C.R., Anderson, I.C., Canuel, E.A., and Macko, S.A., 2001. Nitrogen cycling through a fringing marsh-aquifer ecotone. Mar. Ecol. Prog. Ser., 210: 25–39.

Tréguer, P., Nelson, D.M., Van Bennekom, A.J., DeMaster, D.J., Leynaert, A., and Quéguiner, B., 1995. The silica balance in the world ocean: a reestimate. Science, 268: 375–379.

Tribble, G.W., Sansone, F.J., and Smith, S.V., 1990. Stoichiometric modeling of carbon diagenesis within a coral reef framework. Geochim. Cosmochim. Acta, 54: 2439–2449.

Tromp, T.K., van Cappellen, P., and Key, R.M., 1995. A global model for the early diagenesis of organic carbon and organic phosphorus in marine sediments. Geochim. Cosmochim. Acta, 59: 1259–1284.

Trumbore, S.E., 1993. Comparison of carbon dynamics in tropical and temperate soils using radiocarbon measurements. Global Biogeochem. Cycles, 7: 275–290.

Trumbore, S.E., and Druffel, E.R.M., 1995. Carbon isotopes for characterizing sources and turnover of nonliving organic matter. *In*: R.G. Zepp and C. Sonntag (eds.), *Role of Nonliving Organic Matter in the Earth's Carbon Cycle.* John Wiley & Sons, pp. 7–22.

Turchyn, A.V., and Schrag, D.P., 2004. Oxygen isotope constraints on the sulfur cycle over the past 10 million years. Science, 303: 2004–2007.

Turekian, K.K., Benoit, G.J., and Benninger, L.K., 1980. The mean residence time of plankton-derived carbon in a Long Island Sound sediment core: a correction. Est. Coastal Shelf Sci., 11: 583.

Turekian, K.K., and Wedepohl, K.H., 1961. Distribution of elements in some major units of the Earth's crust. Geol. Soc. Am. Bull., 72: 175–192.

Ullman, W.J., and Aller, R.C., 1982. Diffusion coefficients in nearshore marine sediments. Limnol. Oceanogr., 27: 552–556.

Vairavamurthy, A., and Wang, S., 2002. Organic nitrogen in geomacromolecules: Insights on speciation and transformation with K-edge XANES spectroscopy. Environ. Sci. Tech., 36: 3050–3056.

Vairavamurthy, M.A., Orr, W.L., and Manowitz, B., 1995. Geochemical transformations of sedimentary sulfur: An introduction. *In*: M.A. Vairavamurthy and M.A.A. Schoonen (eds.), *Geochemical Transformations of Sedimentary Sulfur.* American Chemical Society, pp. 1–14.

Valentine, D.L., and Reeburgh, W.S., 2000. New perspectives on anaerobic methane oxidation. Environ. Microbiol., 2: 477–484.

Van Cappellen, P., and Berner, R.A., 1988. A mathematical model for the early diagenesis of phosphorus and fluorine in marine sediments: Apatite precipitation. Am. J. Sci., 288: 289–333.

REFERENCES

Van Cappellen, P., Dixit, S., and Gallinari, M., 2004. Biogenic silica dissolution and the marine Si cycle: Kinetics, surface chemistry and preservation. Océanis, 28: 66–106.

Van Cappellen, P., and Ingall, E.D., 1996. Redox stabilization of the atmosphere and oceans by phosphorus-limited marine productivity. Science, 271: 493–496.

Van Cappellen, P., and Qiu, L., 1997a. Biogenic silica dissolution in sediments of the Southern Ocean. II. Kinetics. Deep-Sea Res. II, 44: 1129–1149.

Van Cappellen, P., and Qiu, L., 1997b. Biogenic silica dissolution in sediments of the Southern Ocean. I. Solubility. Deep-Sea Res. II, 44: 1109–1128.

Van Cappellen, P., and Wang, Y., 1996. Cycling of iron and manganese in surface sediments: a general theory for the coupled transport and reaction of carbon, oxygen, nitrogen, sulfur, iron, and manganese. Am. J. Sci., 296: 197–243.

Van der Zee, C., Roberts, D.R., Rancourt, D.G., and Slomp, C.P., 2003. Nanogoethite is the dominant reactive oxyhydroxide phase in lake and marine sediments. Geology, 31: 993–996.

Van Liew, H.D., 1962. Semilogarithmic plots of data which reflect a continuum of exponential processes. Science, 138: 682–683.

Van Os, R., Visser, H.-J., Middelburg, J.J., and de Lange, G.J., 1993. Occurrence of thin, metal-rich layers in deep-sea sediments: A geochemical characterization of copper remobilization. Deep-Sea Res., 40: 1713–1730.

Van Vleet, E.S., and Quinn, J.G., 1979. Early diagenesis of fatty acids and isoprenoid alcohols in estuarine and coastal sediments. Geochim. Cosmochim. Acta, 43: 289–303.

Vanderborght, J.-P., Wollast, R., and Billen, G., 1977. Kinetic models of diagenesis in disturbed sediments. Part 1. Mass transfer properties and silica diagenesis. Limnol. Oceanogr., 22: 787–793.

Verardo, D.J., and McIntyre, A., 1994. Production and destruction: Controls of biogenous sedimentation in the tropical Atlantic 0–300,000 years B.P. Paleoceanography, 9: 63–86.

Volkman, J.K., Barrett, S.M., Blackburn, S.I., Mansour, M.P., Sikes, E.L., and Gelin, F., 1998. Microalgal biomarkers: a review of recent research developments. Org. Geochem., 29: 1163–1179.

Volkman, J.K., Farrington, J.W., and Gagosian, R.B., 1987. Marine and terrigenous lipids in coastal sediments from the Peru upwelling region at 15°S: Sterols and triterpene alchohols. Org. Geochem., 11: 463–477.

Wakeham, S.G., Damste, J.S.S., Kohnen, M.E.L., and de Leeuw, J.W., 1995. Organic sulfur compounds formed during early diagenesis in Black Sea sediments. Geochim. Cosmochim. Acta, 59: 521–533.

Wakeham, S.G., Hedges, J.I., Lee, C., Peterson, M.L., and Hernes, P.J., 1997a. Compositions and transport of lipid biomarkers through the water column and surficial sediments of the equatorial Pacific Ocean. Deep-Sea Res., 44: 2131–2162.

Wakeham, S.G., Lee, C., Hedges, J.I., Hernes, P.J., and Peterson, M.L., 1997b. Molecular indicators of diagenetic status in marine organic matter. Geochim. Cosmochim. Acta, 61: 5363–5369.

Waksman, S.A., 1938. *Humus, Origin, Chemical Compositions and Importance in Nature.* Williams & Wilkins Co.

Walker, J.C.G., and Kasting, J.F., 1992. Effects of fuel and forest conservation on future level of atmospheric carbon dioxide. Paleogeogr. Paleoclimatol. Paleoecol., 97: 151–189.

Wallace, H.E., Thomson, J., Wilson, T.R.S., Weaver, P.P.E., Higgs, N.C., and Hydes, D.J., 1988. Active diagenetic formation of metal-rich layers in N.E. Atlantic sediments. Geochim. Cosmochim. Acta, 52: 1557–1569.

Walsh, I., Fischer, K., Murray, D., and Dymond, J., 1988. Evidence for resuspension of rebound particles from near-bottom sediment traps. Deep-Sea Res., 35: 59–70.

Walter, L.M., Bishop, S.A., Patterson, W.P., and Lyons, T.W., 1993. Dissolution and recrystallization in modern carbonates: evidence from pore water and solid phase chemistry. Phil. Trans. R. Soc. Lond., A, 344: 27–36.

Walter, L.M., and Burton, E.A., 1990. Dissolution of recent platform carbonate sediments in marine pore fluids. Am. J. Sci., 290: 601–643.

Wang, Q., and Morse, J.W., 1996. Pyrite formation under conditions approximating those in anoxic sediments I. Pathway and morphology. Mar. Chem., 52: 99–121.

Wang, X., Druffel, E.R.M., Griffin, S., Lee, C., and Kashgarian, M., 1998. Radiocarbon studies of organic compound classes in plankton and sediment of the northeastern Pacific Ocean. Geochim. Cosmochim. Acta, 62: 1365–1378.

Wang, X.-C., Druffel, E.R.M., and Lee, C., 1996. Radiocarbon in organic compound classes in particulate organic matter and sediment in the deep northeast Pacific Ocean. Geophys. Res. Lett., 23: 3583–3586.

Wang, X.-C., and Lee, C., 1993. Adsorption and desorption of aliphatic amines, amino acids and acetate by clay minerals in marine sediments. Mar. Chem., 44: 1–23.

Wang, Y., and Van Capellen, P., 1996. A multicomponent reactive transport model of early diagenesis: Application to redox cycling in coastal sediments. Geochim. Cosmochim. Acta, 60: 2993–3014.

Wangersky, P.J., 1962. Sedimentation in three carbonate cores. J. Geol., 70: 364.

Warnken, K.W., Gill, G.A., Griffin, L.L., and Santschi, P.H., 2001. Sediment-water exchange of Mn, Fe, Ni, and Zn in Galveston Bay, Texas. Mar. Chem., 73: 215–231.

Wedepohl, K.H., 1960. Spurenanalytische untersuchungen an tiefseetonen aus dem Atlantik: Ein beitrag zur deutung der geochemischen sonderstellung von pelagischen tonen. Geochim. Cosmochim. Acta, 18: 200–231.

Wehmiller, J.F., 1993. Applications of organic geochemistry for Quaternary research. *In*: M.H. Engel and S.A. Macko (eds.), *Organic Geochemistry*. Plenum Press, pp. 755–783.

Weiss, M.S., Abele, U., Weckesser, J., Welte, W., Schultz, E., and Schultz, G.E., 1991. Molecular architecture and electrostatic properties of a bacterial porin. Science, 254: 1627–2690.

Wellsbury, P., Goodman, K., Barth, T., Cragg, B.A., Barnes, S.P., and Parkes, R.J., 1997. Deep marine biosphere fuelled by increasing organic matter availability during burial and heating. Nature, 388: 573–576.

Wenzhöfer, F., and Glud, R.N., 2002. Benthic carbon mineralization in the Atlantic: a synthesis based on in situ data from the last decade. Deep-Sea Res., 49: 1255–1279.

Werne, J.P., Lyons, T.W., Hollander, D.J., Formolo, M.J., and Sininghe-Damsté, J.S., 2003. Reduced sulfur in euxinic sediments of the Cariaco Basin: sulfur istope constraints on organic sulfur formation. Chem. Geol., 195: 159–179.

Westerlund, S.F.G., Anderson, L.G., Hall, P.O.J., Iverfeldt, A., Loeff, M.M.R.v.d., and Sundby, B., 1986. Benthic fluxes of cadmium, copper, nickel, zinc, and lead in the coastal environment. Geochim. Cosmochim. Acta, 50: 1289–1296.

Westrich, J.T., and Berner, R.A., 1984. The role of sedimentary organic matter in bacterial sulfate reduction: The G model tested. Limnol. Oceanogr., 29: 236–249.

Westrich, J.T., and Berner, R.A., 1988. The effect of temperature on rates of sulfate reduction in marine sediments. Geomicrobiol. J., 6: 99–117.

Wheatcroft, R.A., 1992. Experimental tests for particle size–dependent bioturbation in the deep ocean. Limnol. Oceanogr., 37: 90–104.

Wheatcroft, R.A., Jumars, P.A., Smith, C.R., and Nowell, A.R.M., 1990. A mechanistic view of the particulate biodiffusion coefficient: Step lengths, rest periods and transport directions. J. Mar. Res., 48: 177–207.

Whelan, J.K., 1977. Amino acids in a surface sediment core of the Atlantic abyssal plain. Geochim. Cosmochim. Acta, 41: 803–810.

Whelan, J.K. and Emeis, K.-C., 1992. Sedimentation and preservation of amino compounds and carbohydrates in marine sediments. *In*: J.K. Whelan and J.W. Farrington (eds.), *Productivity, Accumulation, and Preservation of Organic Matter in Recent and Ancient Sediments*. Columbia University Press, pp. 176–200.

Whelan, J.K., and Thompson-Rizer, C.L., 1993. Chemical methods for assessing kerogen and protokerogen types and maturity. *In*: M.H. Engel and S.A. Macko (eds.), *Organic Geochemistry*. Plenum Press, pp. 289–353.

Widdel, F., 1988. Microbiology and ecology of sulfate- and sulfur-reducing bacteria. *In*: A.J.B. Zehnder (ed.), *Biology of Anaerobic Microorganisms*. John Wiley & Sons.

REFERENCES

Wilkin, R.T., and Barnes, H.L., 1996. Pyrite formation by reactions of iron monosulfides with dissolved inorganic and organic sulfur species. Geochim. Cosmochim. Acta, 60: 4167–4179.

Wilson, T.R.S., Thomson, J., Colley, S., Hydes, D.J., Higgs, N.C., and Sørensen, J., 1985. Early organic diagenesis: The significance of progressive subsurface oxidation fronts in pelagic sediments. Geochim. Cosmochim. Acta, 49: 811–822.

Wilson, T.R.S., Thomson, J., Hydes, D.J., Colley, S., Culkin, F., and Sørensen, J., 1986. Oxidation fronts in pelagic sediments: Diagenetic formation of metal-rich layers. Science, 232: 972–975.

Winfrey, M.R., and Zeikus, J.G., 1977. Effect of sulfate on carbon and electron flow during microbial methanogenesis in freshwater sediments. Appl. Environ. Microbiol., 33: 275–281.

Woese, C.R., Kandler, O., and Wheelis, 1990. Towards a natural system of organisms: proposals for the domains Archaea, Bacteria and Eucarya. Proc. Natl. Acad. Sci., 87: 4576–4579.

Wolin, M.J., 1979. The rumen fermentation: a model for microbial interactions in anaerobic ecosystems. Adv. Microbial Ecol., 3: 49–78.

Xu, W., and Ruppel, C., 1999. Predicting the occurrence, distribution and evolution of methane gas hydrate in porous marine sediments. J. Geophys. Res., 104(B3): 5081–5095.

Young, L.Y., and Frazer, A.C., 1987. The fate of lignin and lignin-derived compounds in anaerobic environments. Geomicrobiol. J., 5: 261–293.

Yu, E.-F., Francois, R., and Bacon, M.P., 1996. Similar rates of modern and last-glacial ocean thermohaline circulation inferred from radiochemical data. Nature, 379: 689.

Zabel, M., Dahhmke, A., and Schulz, H.D., 1998. Regional distribution of diffusive phosphate and silicate fluxes through the sediment-water interface: the eastern South Atlantic. Deep-Sea Res., 45: 277–300.

Zachos, J., Pagani, M., Sloan, L., Thomas, E., and Billups, K., 2001. Trends, rhythms, and aberration in global climate 65 Ma to Present. Science, 292: 686–693.

Zang, X., Nguyen, R.T., Harvey, H.R., Knicker, H., and Hatcher, P.G., 2001. Preservation of proteinaceous material during the degradation of the green alga *Botryococcus braunii*: A solid-state 2D ^{15}N ^{13}C NMR spectroscopy study. Geochim. Cosmochim. Acta, 65: 3299–3305.

Zegouagh, Y., Derenne, S., Largeau, C., P.Bertrand, Sicre, M., Saliot, A., and Rousseau, B., 1999. Refractory organic matter in sediments from the northwest African upwelling system: abundance, chemical structure and origin. Org. Geochem., 30: 83–99.

Zhu, A., Aller, R.C., and Mak, J., 2002. Stable carbon isotope cycling in mobile coastal muds of Amapá, Brazil. Cont. Shelf Res., 22: 2065–2079.

REFERENCES

Ziebis, W., Huettel, M., and Forster, S., 1996. Impact of biogenic sediment topography on oxygen fluxes in permeable seabeds. Mar. Ecol. Prog. Ser., 140: 227–237.

Zimmerman, A.R., and Canuel, E.A., 2001. Bulk organic matter and lipid biomarker composition of Chesapeake Bay surficial sediments as indicators of environmental processes. Est. Coastal Shelf Sci., 53: 319–341.

≈ Index ≈

abiotic condensation, 439–41
accelerator mass spectrometry (AMS), 45
accumulation rates, 95–96
acid volatile sulfides (AVSs), 479, 507–8
adenosine tripohsphate (ATP), 98, 106
adsorption, 20–24, 415–16; diagenesis and, 83–84; dissolved organic matter (DOM) and, 234–36; equilibrium coefficients and, 235; phosphorus and, 465
advection, 152; causes of, 78; compaction and, 78–83; diagenesis and, 78–83; sediment-water interface and, 285
aerobic processes: biogeochemical processes and, 99–104, 114–15; decomposition dynamics and, 134–41; pelagic sediments and, 328–32; respiration and, 105; sediment-water interface and, 278 (*see also* sediment-water interface); suboxic/anoxic conditions and, 442–51
age-dependent mixing, 298
alcohol, 99
aldehyde group, 189
Aleutian Islands, 519
algaenans, 212
algal mats, 13
aluminum: cation-rich, 6; clays and, 15–24; opaline silica and, 355–57; reverse weathering and, 511–14
Amazon region, 71, 119; authigenic clays and, 512–16; continental shelf, 373–74, 425; Guiana mudbelt, 70; River of, 23–24, 244
amino acids, 9, 171–72, 174; analytical procedures for, 181, 183, 187–88; carbon and, 180–81, 183; classification of, 179; concentrations of, 183; degradation index and, 178, 185–88; D-enantiomers and, 185; D-form, 180; dissolved organic matter (DOM) and, 231–32; fermentation and, 99; functions of, 179–80; L-form, 180; lipids and, 204; mole percentages and, 182;

N-acetylmuramic acid and, 185; nonprotein, 181, 187; organic carbon preservation and, 418; peptide linkages and, 179; protein, 179–86; remineralization and, 183–84, 188–89; sources of, 185
ammonium, 106; anoxic sediments and, 92–94; DNRA and, 106–7; nitrogen cycling and, 452–64; oxidation and, 115; sediment-water interface and, 278
anabolism, 98–99
anaerobic methane oxidation (AMO), 113, 491, 493–94, 499
anaerobic processes, 97, 101–4, 116; foodchains of, 135–39; methane oxidation and, 491, 493–94, 499; obligate, 101; zonation and, 99–104
anoxic sediments, 129–30, 395, 404; ammonium production and, 92–94; carbonate chemistry and, 442–51; pore water pH and, 442–51; trace metal cycling and, 504–5; zonation and, 102–3
apatite: carbonate fluoroapatite (CFA) and, 14, 466, 468, 471–77; detrital fluoroapatite and, 466
apparent diffusion coefficient, 323
aragonite, 9–12
Archaea, 106
Archie's Law, 77
Arctic Ocean, 246–48
arsenic, 507–8
assimilatory metabolism, 98–99
Atchafalaya River, 246
Atlantic Ocean, 25, 246, 331–32, 378–80, 402, 406
atoms: Brownian motion and, 59; isotopes and, 27–45; potential energy and, 30; radioactive decay and, 27–28, 40–45; vibrational energy and, 29–30
authigenic minerals, 12; carbonate fluoroapatite (CFA) and, 14, 466, 468, 471–77; clays and, 6, 15–24, 34–35, 509–16; cobalt, 14; copper, 14; iron,

INDEX

authigenic minerals (*continued*)
13–14; isotopes and, 34–35; manganese, 13–14; nickel, 14; nonbiogenic carbonates, 13; phosphorites, 14; redox boundary and, 14; reverse weathering and, 509–16; sulfides, 15; trace metal diagenesis and, 332–43
autotrophic metabolism, 98–99
Azore Islands, 519

bacteria: aerobic processes and, 105; ATP and, 98, 106; bacterial biomass in sediments, 250–53; decomposition models and, 143–50; denitrification and, 105–7; iron reduction and, 107–10; manganese nodules and, 345; manganese reduction and, 107–10; metabolism and, 98–99; methanogenesis and, 111–14; organic matter repackaging in sediments and, 237; phosphorus cycling and, 472–73, 476–77; respiration and, 99–114; sulfate reduction and, 15, 110–11; sulfur cycling and, 478–90; zonation and, 99–104
Baltic Sea, 519
basalt, 19
Beaufort Shelf, 246
benthic macrofauna, 2, 9; bioturbation and, 65–67; organic carbon preservation and, 419–21
biodiffusion. *See* bioturbation
biogenic material, 8–12
biogenic methane, 111
biogeochemical processes: aerobic, 99–105, 114–15, 134–41; anaerobic, 97, 101–4, 116, 135–39; ATP and, 98, 106; bacterial metabolism and, 98–99; bacterial respiration and, 99–114; carbon and, 99–103, 442–51; chemolithotrophic reactions, 114–20; continental margin sediments and, 442–405 (*see also* continental margin sediments); decomposition dynamics and, 134–41; denitrification, 100, 105–7; depth scales and, 120–24; energy in, 98–104; half cell potential and, 101–3; iron and, 102–3, 107–10, 132;

manganese and, 102–3, 107–10, 132–33; methanogenesis, 111–14; multi-G model and, 143–50; nonsteady-state processes and, 373 (*see also* nonsteady-state processes); organic matter and, 116–41; oxidation and, 98–104; redox conditions and, 98, 139–41; remineralization and, 116–33; respiration and, 99; in situ conditions and, 104–6, 113; suboxic, 102; sulfate reduction, 110–11; turbidites and, 382–95; water column depth and, 124–33; zonation and, 99–104, 111–14, 120–24
bioirrigation, 65–67, 73, 118–20; Aller's radial-diffusion tube model and, 304–13; apparent diffusion coefficient and, 323; benthic flux enhancement factor and, 316–26; burrowing organisms and, 302–13; diffusion coefficient and, 323; diffusive openness of sediments, 313–16; flux enrichment factors and, 323–26; mixed redox conditions and, 139–41; nitrogen cycling and, 452–64; nonsteady-state processes and, 374; organic carbon preservation and, 421; phosphorus cycling and, 464–77; pyrite and, 484–88; quantification methods for, 316–19; rates of, 319–23; sediment-water interface and, 302–26; silica dissolution and, 309–10; sulfur and, 484–88; trace metal cycling and, 504–6; zonation and, 120–24
biological monomers, 210
biomineralization, 8–9
bioturbation, 43, 65–66, 152; advection and, 78–83; age-dependent mixing and, 300; biodiffusion paradox and, 288–89; boundary conditions and, 86–87; chlorophyll distribution and, 293; coefficient determination and, 67–68, 291–92; compaction and, 78–83; definition of, 286–89; deposit-feeding organisms and, 286–87; depth profile effects and, 271; diagenesis and, 78–90, 293, 296, 301–2; egestion coefficient and, 301; flux and, 72–73; frequency

594

criterion and, 287; gradient destruction and, 293; interphase mixing and, 81, 292–93; intraphase mixing and, 81–82; models of, 289–99; nitrogen cycling and, 452–64; nonlocal, 288–89, 299–302; nonsteady-state processes and, 374, 376–77; organic carbon preservation and, 421; as particle reworking, 287; particulate organic matter (POC) and, 289, 291, 300; phosphorus cycling and, 464–77; quantification of, 291–99; radiogeochemical approach and, 292–99; random particle movement and, 287–88; remineralization and, 88–90; sediment-water interface and, 271, 286–302 (*see also* sediment-water interface); sulfur and, 486–88; trace metal cycling and, 504–6; trace metal diagenesis and, 340–41; zonation and, 120–24

Black Sea, 70, 201–2

Blake Ridge, 496

bomb radiocarbon, 44

bottom simulating reflector (BSR), 498

boundary conditions, 86–87; Aller's radial-diffusion tube model and, 304–13; ammonium production and, 92–94; benthic fluxes and, 274–83; diffusive, 275–76, 278; remineralization and, 88–92; sediment-water interface and, 271–327 (*see also* sediment-water interface); trace metal diagenesis and, 332–43; turbidites and, 382–95

Brownian motion, 59

brucite, 17

Buzzards Bay, 188–89, 202, 319, 517

bypass zones, 70

cadmium, 332, 391, 406, 502, 506

calcareous oozes, 24

calcite, 2–3; biogenic carbonates and, 9–10; fractionation and, 35–36; isotopes and, 35–36; sediment-water interface and, 278

calcite compensation depth (CCD), 12, 18, 364, 366

calcite saturation horizon, 361

calcium carbonate: biogenic, 9–10; burial efficiency and, 370–71; calcite compensation depth (CCD) and, 12, 18, 364, 366; carbonate burial efficiency and, 370–71; diagenesis and, 359–72; dissolution rate and, 361–72; flux and, 371–72; lysocline and, 364, 366–67; pelagic sediments and, 359–72; pH profiles and, 363; remineralization and, 11–12; in situ conditions and, 367–68; suboxic/anoxic conditions and, 451; undersaturation and, 11–12, 369–70

California Borderlands, 517

Calvin cycle, 37

Canada, 14

CANDI model, 169–70

Cape Lookout Bight, 78, 266; amino acids and, 188–89; carbohydrates and, 192; description of, 517–18; dissolved organic matter (DOM) and, 223; fulvic acids and, 211–12; gas ebullition and, 327; lipids and, 198–204; location of, 517; methanogenesis and, 491; nitrogen cycling and, 461; nonsteady-state processes and, 382; reactive component budgets and, 153–61

carbohydrates, 171–72, 175; analysis techniques and, 191; cellulose and, 190; chitin and, 190–91; diagenesis and, 192; dissolved (DHCOs), 231; dissolved organic matter (DOM) and, 231; organic carbon preservation and, 418; particulate, 191–92; pore waters and, 191; remineralization and, 191–92; sugar structures and, 189–90; uses of, 190

carbon, 71; amino acids and, 180–81, 183, 188; ammonium production and, 92–94; anoxic conditions and, 442–51; biogenic carbonates and, 9–10; biogeochemical processes and, 99–103, 119; bioturbation and, 65–67; black, 206–7, 237, 249–50; burial efficiency (CBE), 156, 161–62, 171, 370–71, 408–11, 423, 426, 428–32, 435–39; carbonate fluoroapatite (CFA) and, 466, 468, 471–77; classification of, 204–6; cycling and, 1,

carbon (*continued*)
   45, 220; dissolved inorganic (DIC), 35–36; dissolved organic (DOC), 218–28, 233, 266–67, 503; isotopes and, 35–39, 43–45, 238–41; lipids and, 198–99; methanogenesis and, 491; mixed redox conditions and, 139–41; molecularly uncharacterized organic matter (MU-OM) and, 172, 174; multi-G model and, 143–50; nitrogen ratio and, 253, 255–56, 261–70, 330, 446–47, 477; nitrogen ratios and, 149, 240–48; nonbiogenic carbonates, 13; organic carbon compensation depth (OCCD) and, 432; organic carbon preservation and, 408–41 (*see also* organic carbon preservation); phosphorus ratio and, 468, 470, 477; photosynthesis and, 37; plant fixation and, 239; pore water pH and, 442–51; PWSR model and, 220–22; radioactive, 43–45, 239–40; rain and, 454–55, 472; Redfield ratio and, 100; redox conditions and, 139–41; sediment-water interface and, 285; suboxic conditions and, 442–51; sulfur ratio and, 488; total organic (TOC), 65, 88 (*see also* total organic carbon (TOC)); turbidites and, 383, 385, 387, 392, 394
carbonate fluoroapatite (CFA), 14, 466, 468, 471–77
carbon dioxide, 6, 12; benthic flux stoichiometric models and, 260–61; biogeochemical processes and, 99; calcium carbonate diagenesis and, 359–70; isotopic composition and, 261–65; margin sediments and, 442–77 (*see also* margin sediments); organic carbon preservation and, 408; suboxic/anoxic conditions and, 442–51; sulfur and, 489–90
carbon exchange capacity (CEC), 21
carbonic acid, 6
Cariaco Trench, 212, 214
carrier phase, 332–33
Cascadia Basin, 423
catabolism, 98–99
catagenic methane, 111

cations, 6, 17, 20–24
Ceara Rise, 363–64
cellulose, 190
chemical weathering, 5–6, 509–16
chemoautotrophs, 114
chemolithotrophic reactions: aerobic processes and, 114–16; anaerobic processes and, 116; organic matter remineralization and, 116–20; redox conditions and, 98, 139–41 (*see also* redox conditions)
Chesapeake Bay, 263, 464, 472, 518
chitin, 190–91
chlorite, 7, 18
clays, 6; adsorption and, 20–24; aluminum and, 355–57, 511–14; components of, 15–18; deep-sea red, 7–8; definitions of, 15; distribution of, 18–20; ion exchange and, 20–24; iron hydroxide and, 8; isotope fractionation and, 34–35; metal reduction and, 110; mineral neoformation and, 512; mud and, 47; oxidation and, 513, 516; porosity and, 50; reverse weathering and, 509–16; silica and, 512–16
cobalt, 333, 336, 341, 345, 390–91
Columbia River, 246
cometabolism, 140
compaction: diagenesis and, 78–83, 95–96; sediment accumulation rates, 95–96
continental margin sediments, 3–4; anaerobic methane oxidation and, 490–500; anoxic conditions and, 442–51; carbonate chemistry and, 442–51; continental, 493–500; methanogenesis and, 490–500; nitrogen cycling and, 452–64; pelagic sediments and, 442; phosphorus cycling and, 464–77; pore water pH and, 442–51; shallow, 490–93; suboxic conditions and, 442–51; sulfur cycling and, 478–90; trace metal cycling and, 500–508
conveyor-belt mixing, 67, 299–300, 432
co-oxidation, 140
copper, 332–33, 336, 341, 345, 391, 506–8

INDEX

cross-coupling, 75
Curie point pyrolysis, 213
cycling: anaerobic methane oxidation and, 490–500; carbon, 1, 45, 220; chemical weathering and, 5–6; diagenesis models and, 162–70; dissolved organic matter (DOM) and, 218–36; layered/coupled models and, 162–70; lipids and, 198; manganese nodules and, 345; methanogenesis and, 490–500; nitrogen, 254, 452–64; organic carbon preservation and, 441 (*see also* organic carbon preservation); oxygen exposure time and, 421–32; phosphorus, 464–77; Pleistocene climate and, 402–4; reverse weathering and, 509–19; squalene, 198; sulfur, 478–90; tectonic, 429; trace metal, 332–43, 500–8; undersaturation and, 11–12; weather, 429

Damkohler number, 69–70, 152
Darcian flow, 58
Deep Sea Drilling Project (DSDP), 77, 493
deep sea sediments. *See* pelagic sediments
degree of pyritization (DOP), 487–88, 500–502
degree of trace metal pyritization (DTMP), 500–502, 507–8
denitrification, 1, 442; biogeochemical processes and, 100, 105–7; coupled models and, 167–69; nitrogen cycling and, 452–64
density: depth and, 51–52; description of, 47–55; diagenesis and, 73–74
depolymerization, 441
depositional signal, 1
depth-integrated rates (DIRs), 157
depth scales, 120–24
Desulfobacter, 110–11
Desulfobrio, 110–11
detrital material, 5–8
diagenesis, 2–3, 132–33; adsorption and, 83–84; advection and, 78–83; amino acids and, 231–32; ammonium production and, 92–94; biogeochemical processes and, 97 (*see also* biogeochemical processes); bioturbation and, 78–90, 293, 296, 301–2; boundary conditions and, 86–94; calcium carbonate and, 359–72; carbohydrates and, 192; carrier phase and, 332–33; compaction and, 78–83; coupled models and, 162–70; deep-sea turbidites and, 382–95; diffusion and, 74–78; dissolved organic matter (DOM) and, 218–36; equation solutions and, 84–96; fjord, 398n3; flux and, 72–73, 81–82, 272–73; formation factor and, 76–77; functional classification and, 70–71; general equations of, 72–84; Glacial-Holocene transition and, 400–402, 406; layered models and, 162–70; molecularly uncharacterized organic matter (MU-OM) and, 172, 174; multi-G model and, 143–50; multiple manganese peaks and, 395–407; nitrogen and, 215–17; nonsteady-state processes and, 395–407; opaline silica and, 352–59; oxic, 347, 349; paleoceanographic change and, 395–407; Peclet number and, 85; pelagic sediments and, 328–43, 352–72; pore waters and, 73–74, 254–60; reactive component budget and, 150–61; remineralization and, 88–92; remobilization and, 341–42; suboxic, 347; sulfate reduction and, 91–92; tortuosity and, 75–77; trace metal, 332–43, 504
diagenetic maturity, 172–74
diatoms, 24, 183, 185
diffusion, 59–61, 152; Aller's radial-diffusion tube model and, 304–13; benthic flux enhancement factor and, 316–26 (*see also* flux); bioirrigation and, 304–16 (*see also* bioirrigation); bioturbation and, 287 (*see also* bioturbation); boundary layer and, 275–83; conductivity and, 75–77; cross-coupling and, 75; diagenesis and, 74–78; dissolved organic matter (DOM) and, 218–36; Fick's First Law of, 61, 82; Fick's Second Law of, 72–73; formation factor and, 76–77; iron, 449; irrigation coefficient and, 323; manganese, 449; nitrogen, 452–64;

597

INDEX

diffusion (*continued*)
  nonsteady-state processes and, 375–76; phosphorus, 464–77; porosity and, 74–78; remineralization and, 88–90; sediment-water interface and, 271–27 (*see also* sediment-water interface); sublayer and, 275; tortuosity and, 75–76, 75–77; trace metal cycling and, 500–508 (*see also* cycling); turbidites and, 382–95
diffusive boundary layer, 87, 275–83
dissimilatory metabolism, 98–99
dissimilatory nitrate reduction to ammonium (DNRA), 106–7
dissolved carbohydrates (DHCOs), 231
dissolved combined amino acids (DCAAs), 231–32
dissolved free amino acids (DFAAs), 231–32
dissolved inorganic carbon (DIC), 35–36
dissolved inorganic nitrogen (DIN), 463–64
dissolved organic carbon (DOC), 266–67; benthic fluxes and, 233; cycling and, 218–28; diagenesis and, 218, 220; PWSR model and, 220–22; trace metal cycling and, 503
dissolved organic matter (DOM), 135–37, 147, 154–55; adsorption and, 234–36; asymptotic concentrations of, 223–26; carbohydrates and, 231; compositional data for, 228–32; concentration of, 222–23; diagenetic models of, 227–28; fermentation and, 222; fluxes and, 232–34; humiclike fluorescence and, 222–23; nitrogen and, 218–20, 231–34, 452–64; organic carbon preservation and, 433, 436–37; phosphorus cycling and, 464–77; PWSR model and, 220–22, 230; remineralization and, 218, 226; short-chain organic acids and, 230. *See also* high molecular weight (HMW) DOM; low molecular weight (LWM) DOM
dissolved organic nitrogen (DON), 218–20, 231–34, 463–64
dolomite, 13

East Pacific Rise, 406
ebullition, 326–27
ectoenzymes, 135
egestion coefficient, 301
Einstein-Smoluchowski equation, 67
elements: isotopes and, 27–45; radioactive, 27–28, 40–45. *See also specific kind*
eolian transport, 6
equations, mathematical: accumulation rates, 95–96; adsorption, 83–84; Aller's radial-diffusion tube model, 305–7, 311–12; Archie's Law, 77; bacterial respiration, 104, 112–13; benthic DOC flux, 233; benthic flux enhancement factor, 316; calcium carbonate diagenesis, 360–61; Carman-Kozeny, 58; Damkohler number, 69; decomposition models, 142–43, 145–46, 148–49; degree of pyritization, 487; depth profile, 145; Dhakar-Burdige, 167; diagenesis, 72–85, 88–96, 225, 360–61; diffusion, 61, 75–77, 276; dissolved organic matter (DOM) cycling, 225, 227–28, 233; Einstein-Smoluchowski, 67; Fick's First Law of Diffusion, 61; Fick's Second Law of Diffusion, 72–73; first-order decay, 143, 146; formation factor, 76; half cell potential, 103; hydraulic conductivity, 55; ion exchange, 22–23; isotopes, 30–33, 41–42; mass flux sediment, 80; Michaelis-Menten, 112–13, 142; nonsteady-state processes, 375, 377–78, 380; Peclet number, 68; permeability, 55, 58; porosity, 47, 49–50, 54; pore water model, 255–56, 259–60; radioactive decay, 41–42; reactive component budgets, 150–53, 156–57, 159; sediment accumulation rates, 95–96; sediment-water interface, 274, 276, 288, 293–98, 301, 305–7, 311–12, 316, 318, 322–23; Shaw-Hanratty, 276; silica dissolution, 354; steady-state compaction, 79; steady-state diagenesis, 225, 330; time/space scales, 67–69; tortuosity, 75; total benthic flux, 272; total organic matter burial rate, 436
equations, stoichiometric, 166; amino

INDEX

acid fermentation, 99; anaerobic methane oxidation, 491; anoxic, 443–45, 448–50; bacterial respiration, 99–100, 106, 112–13; bacterial sulfate reduction, 15; benthic flux models, 261; calcium carbonate diagenesis, 360, 367; carbonate system equilibrium, 444; carbonic ion exchange, 20–21; chemolithotrophic reactions, 115; denitrification, 442; diagenesis models, 166–67; dissolution of silicates, 6; equilibrium exchange reactions, 31; geopolymerization, 211; glucose fermentation, 99; iron diffusion, 449; iron reduction, 444; manganese diffusion, 449; manganese reduction, 444; negative alkalinity, 444; phosphoric acid dissociation, 445; pore water models, 254, 259; propionate fermentation, 99; radioactive decay, 41, 43; Redfield-Ketchum-Richards (RKR), 100; reverse weathering, 509, 511; reversibility and, 161; silica budgets, 514–15; suboxic, 443–45, 448–50; sulfate reduction, 91; sulfide precipitation, 448; sulfur cycling, 481, 484–85, 489–90; thiosulfate disporportionation, 481; trace metal cycling, 501
equilibrium exchange reactions, 31
euhedra, 479

facultative anaerobes, 101
Faraday's constant, 103
fatty acids, 171. *See also* lipids
feldspar, 6
Fenton's reagent, 141
fermentation, 99, 110–11; decomposition dynamics and, 134–41; dissolved organic matter (DOM) and, 222
ferromanganese concretions, 344
Fick's First Law of Diffusion, 61, 82
Fick's Second Law of Diffusion, 72–73
fission, 42
fjord sediments, 398n3, 519
fluorine, 71
flux, 59–61, 80, 150; Aller's radial-diffusion tube model and, 304–13; analysis techniques for, 272; benthic boundary layer and, 274–83; benthic enrichment factors and, 323–26; benthic flux enhancement factor and, 316–26; benthic flux stoichiometric models and, 260–61; bioirrigation and, 272, 323–26; bioturbation and, 272 (*see also* bioturbation); boundary conditions and, 89–90; calcium carbonate diagenesis and, 371–72; carbon burial efficiency and, 156; carbon profiles and, 155–56; depth-integrated rates (DIRs) and, 157; determination of benthic, 272–73; diagenesis and, 72–73, 81–82, 89–90; diffusive boundary layer and, 275–83; dissolved organic matter (DOM) and, 232–34; dissolved organic nitrogen (DON), 463–64; kerogen and, 249–50; methanogenesis and, 490–500; molecular, 272; nitrogen cycling and, 452–64; organic composition and, 253–70; phosphorus cycling and, 464–77; pressure gradient and, 283–84; reactive component budgets and, 150–61; redeposition and, 272; sediment oxidation efficiency and, 156; sediment-water interface and, 272–83; sulfur cycling and, 478–90; trace metals and, 332–43, 500–508; water column depth and, 124–33
foraminifera, 9, 24, 36
formation factor, 76–77
fossil fuels, 12
fractionation (isotope): calcite and, 35–36; clays and, 34–35; equilibrium exchange reactions and, 31–32; hydrosphere and, 32–34; ice cores and, 32–34; inorganic matter and, 32–36; interatomic distance and, 29–30; potential energy and, 30; principles of, 28–32
fractionation (organic matter): decomposition models and, 143–50; molecularly uncharacterized organic matter (MU-OM) and, 172, 174; multi-G model and, 143–50
framboids, 479

599

francolite, 466
friction, 276, 278
fulvic acids, 210–12

gas hydrates, 496, 498–500
Geobacter, 109
geochemistry, 3; amino acids and, 9, 172, 174, 179–89; anoxic conditions and, 442–51; biochemical classes and, 171–72, 174; bioturbation measurement and, 292–99; bulk chemical composition of sediments, 175–78; carbohydrates and, 189–92; carbonate chemistry and, 442–51; carbon/nitrogen ratio and, 240–48, 253, 256, 261–70; carbon preservation and, 171–72; fractionation and, 172 (*see also* fractionation); humic substances and, 204–15; isotopes and, 27–45; lignins and, 193–94; lipids and, 194–204; methanogenesis and, 490–500 (*see also* methanogenesis); molecularly uncharacterized organic matter (MU-OM) and, 172, 174; nitrogen diagenesis and, 215–17; organic composition and, 253–70; pelagic sediments and, 328–32 (*see also* pelagic sediments); radioactive decay and, 27–28, 43–45; Redfield-Ketchum-Richards equation and, 175–77; reverse weathering and, 509–16; suboxic conditions and, 442–51; trace metal diagenesis and, 332–43; turbidites and, 382–95
geopolymerization, 207, 209–12
gibbsite, 17
glacial erosion, 7
Glacial-Holocene boundary, 395, 400–402, 406
glucose, 99
glutamic acid, 231
glycerol, 194–95, 197
glycine, 183, 185
goethite, 108
grain size, 46–47
gravity waves, 284
Green River Shale, 216
greigite, 479

Gulf of California, 199
Gulf of Mexico, 246

half-cell potential, 101–3
half-saturation constant, 112–13, 142–43
hematite, 108
hemipelagic sediments, 26, 347, 402–7
heterotrophic metabolism, 98–99
high molecular weight (HMW) DOM, 135, 220, 222, 227–28, 231, 465
HMS *Challenger*, 7–8, 344
Holocene sediments, 404–7
humic substances, 175; black carbon and, 206–7; dissolved organic matter (DOM) and, 218–36; geopolymerization and, 209–12; molecularly uncharacterized organic matter (MU-OM) and, 206–9; physical protection of, 213–15; refractory biomacromolecule preservation and, 212–13; uncharacterized organic matter and, 204–15
hydothermal vents, 14
hydraulic conductivity, 55–56
hydrochloric acid, 479, 502
hydrogen, 138; aerobic processes and, 114–16; consumption and, 444; gas hydrates and, 496, 498–99; half-saturation constant and, 112–13; interspecies transfer and, 137; methanogenesis and, 111–14; negative alkalinity and, 444; pH and, 442–51 (*see also* pH); pore water stoichiometric models and, 254–60; suboxic/anoxic conditions and, 442–51; syntrophy and, 137
hydrogenous precipitation, 347
hydrogen sulfide, 15
hydrologic cycle, 32–34
hydrosphere, 32–34
hydrothermal phases, 333
hydrous ferric oxide, 108
hydroxyl, 189–92

ice cores, 32–34
illite, 18–19
Indian Ocean, 344

inorganic matter, 3; biogenic material and, 8–12; radioactive decay and, 27–28, 40–45
interphase mixing, 81
interspecies hydrogen transfer, 137
intraphase mixing, 81–82
ion exchange, 20–24
iron, 13; biogeochemical processes and, 107–10, 119, 132; clays and, 8, 15–24, 509–16; coupled models and, 168–69; diffusion of, 449; Fenton's reagant and, 141; Glacial-Holocene boundary and, 400–402; hematite and, 108; hydrous ferrous oxide and, 108; lepidochrosite and, 108; manganese nodules and, 344–52; mixed redox conditions and, 140–41; Oneida Lake and, 350–51; organic acids and, 109–10; organic carbon preservation and, 441; oxidation and, 13–14, 127; phosphorus adsorption and, 465; phosphorus-bound, 468, 471–77; pyrite and, 111, 478–90, 500–501; reactive, 487; reduction of, 71, 102–3, 107–10, 115–16, 444, 452; suboxic/anoxic conditions and, 448–51; sulfur cycling and, 478–90; trace metals and, 5–6, 332–43, 500–502; turbidites and, 383, 389–90; zonation and, 120–24
irrigation diffusion coefficient, 323
isotopes, 448; advection and, 78; benthic flux enhancement factor and, 316–26; carbon, 35–39, 43–45; carbon tracers and, 238–41; clays and, 34–35; description of, 27–28; distribution of, 28; elemental ratios and, 261–65; equilibrium exchange reactions, 31; fractionation and, 28–38; heavy, 30; hydrosphere and, 32–34; ice cores and, 32–34; interatomic distance and, 29–30; kinetic energy and, 29–30; light, 30; nitrogen, 39–40, 238–41; organic matter and, 36–38; oxygen, 35–36, 38–39; photosynthesis and, 37; pore water and, 38–39, 261–65; potential energy and, 30; radioactive, 27–28, 40–45; stable, 27–28; sulfur, 40, 478, 489–90

kaolinite, 18
Kattegat Sea, 519
kerogen, 162, 209, 249–50
ketone group, 189
kinetics, 29–30; anaerobic processes and, 116; bacteria and, 143; biogeochemical processes and, 97 (*see also* biogeochemical processes); fission and, 42; flux and, 150–61; layered/coupled models and, 162–70; trace metal cycling and, 500–508. *See also* remineralization

lead, 96, 502, 506
lepidochrosite, 108
lignins, 171, 193–94
limestone, 5
lipids, 171, 175; amino acids and, 204; as biomarkers, 203–4; carbon and, 198–99; classification of, 198; compound varieties of, 194; concentration of, 198–201; cycling and, 198; degradation of, 202–3; description of, 194; reactivity and, 198–201; saponification and, 194–95; solvent extraction and, 195, 197; steroids and, 198; terpenoids and, 197–98; uses of, 194–95, 197, 203–4; wax and, 197
lithic fragments, 7
lithogenic phases, 332–33
Long Island Sound, 138, 160, 450–51, 518
low molecular weight (LMW) DOM, 135, 137–38, 222–24, 227–28, 230
lysocline, 364, 366–67

Mackenzie River, 246
mackinawite, 479
Madeira Abyssal Plain (MAP), 38, 197, 390–92, 518–19
magnesium, 10; clays and, 15–24; reverse weathering and, 513–16
Maillard-type condensation reactions, 217
manganese, 5, 13–14, 71; biogeochemical processes and, 107–10, 119, 132–33; crusts and, 344–52; diffusion of, 449; ferromanganese concretions and, 344;

manganese (*continued*)
  Glacial-Holocene transition and, 400–402, 406; Holocene sediments and, 404–7; MANOP sites and, 247, 347, 402–4, 518; mixed redox conditions and, 140–41; nodules and, 344–52; nonsteady-state processes and, 395–407; Oneida Lake and, 350–51; organic acids and, 109–10; organic carbon preservation and, 441; oxidation and, 127; paleoceanographic change and, 395–407; pelagic sediments and, 344–52; Pleistocene climate cycles and, 402–4; redox downshift and, 397–98; reduction and, 102–3, 107–10, 115–16, 444, 452–53; sediment-water interface and, 278; suboxic/anoxic conditions and, 442–51; trace metals and, 332–43, 502; turbidites and, 383–91
MANOP (Manganese Nodule Program), 247, 347, 402–4, 518
margin sediments. *See* continental margin sediments
matrices, 8–9
mechanical weathering, 7
mercury, 211, 213, 215, 507–8
mesopore, 415–16
metabolic dissolution efficiency (MDE), 368–70
methane clathrates, 496, 498
methane gas ebullition, 326–27
Methanobacterium, 112
Methanococcus, 112
methanogenesis, 39, 111–14, 157; bottom simulating reflector (BSR) and, 498; carbon remineralization and, 491; continental margin sediments and, 493–500; gas hydrates and, 496, 498–500; normal, 491; Pleistocene age and, 499–500; shallow sediments and, 490–93; sulfur and, 490–91, 493, 498
Methylococcus, 115
Methylomonas, 115
methylotrophy, 39
Mexican continental slope, 423
mica, 6
Michaelis-Menten equation, 112–13, 142

Miocene age, 394
Mississippi River, 35, 246
moles, 91
molybdenum, 332, 400, 507–8
Monetery Bay, California, 456
Monod equation, 112–13, 142, 166
monolayer equivalent (ME) concentration, 413–14
monomers, 210
montmorillonite, 18–19
Mössbauer spectroscopy, 108
muddy sediments, 47, 70, 373–74
multi-G model, 143–50

Namibian shelf, 14
nannofossils, 24
negative alkalinity, 444
nickel, 332–33, 341, 345, 390–91, 506
nitrate fermentation, 106
Nitrobacter, 115
Nitrococcus, 115
Nitrocystis, 115
nitrogen: aerobic processes and, 115; amino acids and, 183; ammonium production and, 92–94; biogeochemical processes and, 105–7, 119; bioturbation and, 65–67; carbon ratio and, 149, 240–48, 253, 255–56, 261–70, 330, 446–47, 477; coupled models and, 167–69; cycling and, 1, 254, 452–64; denitrification and, 1, 100, 105–7, 167–69, 442, 452–64; diagenesis and, 215–17; dissolved organic (DON), 218–20, 231–34, 463–64; DNRA and, 106–7; fulvic acids and, 210–12; isotopes and, 39–40, 238–41; pelagic sediments and, 328–32; pore water nitrates and, 167–68; Redfield ratio and, 100; regeneration of, 444; suboxic/anoxic conditions and, 442–51; turbidites and, 383–84, 390–91
Nitrosomonas, 115
Nitrospirina, 115
NMR studies, 214–16
nonlocal transport, 67
nonsteady-state processes, 1–4; bioturbation and, 376–77; diagenesis and,

395–407; diffusion and, 375–76; Glacial-Holocene transition and, 400–402, 406; Holocene sediments and, 404–7; multiple manganese peaks and, 395–407; paleoceanographic change and, 395–407; periodic input, 374–77; phosphorus cycling and, 472–73; Pleistocene climate cycles and, 402–4; redox conditions and, 395–407; remineralization and, 375–82; scaling for, 374, 376; seasonality and, 378–82; time scales and, 373–82, 387, 394–95, 406–7; turbidites and, 382–95

North Sea, 519

Norwegian Trench, 519

nutrients: Cape Lookout Bight example, 153–61; flux and, 150–61, 260–61; pore water stoichiometric models for, 254–60; RKR equation and, 254

obligate anaerobes, 101

Ocean Drilling Program, 493

oceans: calcium carbonate diagenesis and, 359–72; continental shelf and, 25; gravity waves and, 284; Holocene sediments and, 404–7; hydrologic cycle and, 32–34; hydrothermal vents and, 14; isotopes and, 35–36, 39; manganese nodules and, 344–52; margin sediments and, 493–500 (see also margin sediments); nitrogen cycling and, 452–64; oxygen exposure time and, 421–32; paleoceanographic change and, 395–407; pelagic sediments and, 328–32 (see also pelagic sediments); phosphorus cycling and, 464–77; Pleistocene climate cycles and, 402–4; reverse weathering and, 509–16; spreading centers and, 12; subtidal pump and, 284; trace metal diagenesis and, 332–43; turbidites and, 382–95; undersaturation and, 11–12

OC:SA ratios, 423, 426–29

oligosaccharides. See carbohydrates

Oneida Lake, New York, 350–51

Ontong-Java Plateau, 363–64

opal, 11; burial efficiency (OBE), 358–59; opaline silica, 352–59

organic carbon compensation depth (OCCD), 432

organic carbon preservation: abiotic condensation and, 439–41; benthic macrofaunal processes and, 419–21; burial efficiency and, 408–11, 423, 426, 428–32, 435–39; environmental attributes and, 411–12; metric for, 417–19; monolayer (ME) concentration and, 413–14; organic carbon compensation depth (OCCD) and, 432; organic matter-mineral interactions and, 412–17; oxygen exposure time and, 421–32; physical protection and, 439–41; remineralization and, 408–9, 417–19; sediment composition and, 432–39; surface adsorption/mesopore protection hypothesis and, 415–16; tectonic cycle and, 429; terrestrial organic matter (TOM) and, 432–39; turbidites and, 424; weather cycle and, 429

organic matrix-mediated precipitation, 8–9

organic matter: amino acids and, 9, 172, 174, 179–89; bacterial biomass production and, 250–53; bacterial sulfate reduction and, 15; benthic flux stoichiometric models and, 260–61; benthic macrofauna and, 2; biogeochemical processes and, 116–41; biomineralization and, 8; bioturbation and, 65–67; carbohydrates and, 189–92; carbon and, 36–38, 238–41, 412–17 (see also carbon); as carrier phase, 332; composition of, 175–78, 253–70, 432–39; concentrations of, 174–75; decomposition dynamics of, 134–50; deep-sea turbidites and, 382–95; deposit-feeding organisms and, 286–87; detrital fluorapatite and, 466; diagenesis and, 2 (see also diagenesis); dissolved (DOM), 135–37, 147 (see also dissolved organic matter (DOM)); distribution of, 5; ectoenzymes and, 135; elemental ratios and, 241–44, 261–65; exposure time

INDEX

organic matter (*continued*) and, 178, 421–32; fractionation and, 36–38, 172; geochemistry and, 171–217 (*see also* geochemsitry); hotspots and, 123; humic substances and, 175, 204–15; intragranular pores and, 416; layered/coupled models and, 162–70; lignins and, 193; manganese nodules and, 344–52; mineral interactions and, 412–17; mixed redox conditions and, 139–41; molecularly uncharacterized organic matter (MU-OM) and, 172, 174, 204–15, 252, 439–41; molecular weight and, 135–38; multi-G model and, 143–50; organic acids and, 109–10; packaging and, 172; particulate, 177, 187, 207, 210, 212; pelagic sediments and, 328–32; photosynthesis and, 37; physical protection of, 213–15; pore water stoichiometric models and, 254–60; reactivity degrees and, 143–44; recycled kerogen and, 249–50; Redfield-Ketchum-Richards equation and, 175–77; remineralization and, 2–3, 88–92, 116–33, 328–32; respiration and, 38; sources of, 174–75, 237–53; spatial trends in, 244–48; sulfate reduction and, 91–92; surface adsorption/mesopore protection hypothesis and, 415–16; terrestrial (TOM), 244–48, 432–39

oxic diagenesis, 347, 349

oxidation: abiotic condensation and, 439–41; ammonium, 115; anaerobic methane, 113–14, 491, 493–94, 499; atmospheric oxygen and, 428–32; authigenic clays and, 513, 516; benthic flux enhancement factor and, 316–26; biogeochemical processes and, 98–104; carbon burial and, 156, 161–62; chemical weathering and, 5–6; co-oxidation and, 140; decomposition dynamics and, 134–50; denitrification and, 100, 105–7; detrital components and, 5–6; diffusive boundary layer and, 275–83; dissolved organic nitrogen and, 218–20; Fenton's reagant and, 141; fermentation and, 99; hydrochloric acid and, 479; iron and, 13–14, 107–9, 115–16, 127; isotopes and, 35–39; kerogen and, 162, 249–50; layered/coupled models and, 162–70; manganese and, 13–14, 107–9, 115–16, 127, 345, 400; mixed redox conditions and, 139–41; multi-G model and, 143–50; nitrogen cycling and, 452–64; organic acids and, 109–10; organic carbon preservation and, 408, 417–32, 439–41; oxygen exposure time and, 421–32; pelagic sediments and, 328–32; phosphorus cycling and, 464–77; pore water stoichiometric models and, 254–60; postoxic sediments and, 102–3; sediment oxidation efficiency and, 156; sediment-water interface and, 275–83, 285 (*see also* sediment-water interface); suboxic/anoxic conditions and, 102–3, 442–51; sulfur and, 478–90; trace metals and, 332–43, 506–8; turbidites and, 383–87, 390–95

oxygen exposure time (OET): atmospheric, 428–32; carbon burial and, 428–32; meaning of, 425–27; organic carbon compensation depth (OCCD) and, 432; organic carbon preservation and, 421–32; tectonic cycle and, 429

oxyhydroxides, 345

Pacific Ocean, 25, 344, 423; calcium carbonate and, 361, 371–72; manganese nodules and, 351; nitrogen cycling and, 457–58; nonsteady-state processes and, 378–80, 406

paleoceanographic records, 1–2; Holocene sediments and, 400–7; multiple manganese peaks and, 395–407; nonsteady-state processes and, 395–407; Pleistocene climate cycles and, 402–4; redox conditions and, 395–407. *See also* pore waters

paleothermometry, 32

Panama Basin, 133

Papua New Guinea, 70, 133, 264

particulate organic matter (POM), 177, 187, 207, 210, 212; bacterial biomass production and, 250–53; bioturbation and, 289, 291, 300; carbon/nitrogen ratio and, 240–48, 261–70; isotopic tracers and, 238–41; pore water stoichiometric models and, 254–60; sources of, 237–53

Patton Escarpment, 519

Peclet number, 68, 85

pelagic sediments, 3–4, 9; bioturbation and, 340–41; calcium carbonate and, 359–72; carbon/nitrogen ratio and, 330; diagenesis and, 328–43, 352–72; hemipelagic sediments and, 26, 347, 402–7; manganese and, 344–52; margin sediments and, 442; opaline silica and, 352–59; organic matter remineralization and, 328–32; oxygen exposure time and, 425–27; turbidites and, 382–95

peptidoglycan, 179–80

percolation zone, 70

permeability: Darcian flow and, 58; description of, 55–58; hydraulic conductivity and, 55–56; sediment-water interface, 283–86 (*see also* sediment-water interface)

Peru margin, 14

pH, 9, 115–16, 209; anoxic conditions and, 442–51; calcium carbonate and, 363; carbonate chemistry and, 442–51; carbon/nitrogen ratio and, 446–47; ion exchange and, 20–24; negative alkalinity and, 444; suboxic conditions and, 442–51; sulfur and, 479

Phanerozoic age, 429, 432

phosphogenesis, 466, 475

phospholipids, 197

phosphorites, 14–15, 466–67

phosphorus, 1, 5, 13; acid dissociation and, 445; adsorption and, 465; bacteria and, 472–73, 476–77; biogenic hydroxyapatite and, 466; burial and, 473–77; carbonate fluoroapatite (CFA) and, 466, 468, 471–77; carbon ratio and, 468, 470, 477; conceptual model for, 467–68;

cycling and, 464–77; detrital fluorapatite and, 466; iron-bound, 468, 471–77; mineral-bound, 465; nonsteady-state processes and, 472–73; pelagic sediments and, 331; Redfield ratio and, 100, 470–72; sources of, 464–65; turbidites and, 390

photoautotrophs, 98–99

photosynthesis, 37

phyllosilicates. *See* clays

physical protection, 439–41

phytoplankton, 175, 179, 237–39, 348

Plank's constant, 30

plant litter, 237

Pleistocene age, 394, 402–4, 499–500

polysaccharides. *See* carbohydrates

pore waters: advection and, 78–83, 285; Aller's radial-diffusion tube model and, 304–13; authigenic clays and, 19, 513–14; bacterial biomass production and, 250–53; benthic flux enhancement factor and, 316–26; calcium and, 2–3; carbohydrates and, 191; cross-coupling and, 75; denitrification and, 100, 105–7; Dhakar-Burdige diffusion and, 74–78; diagenesis and, 73–74, 359–72; dissolved organic matter (DOM) and, 135–37, 147, 154–55, 218–36; fluxes and, 150–61; grain size and, 46–47; isotopes and, 38–39; manganese nodules and, 344–52; multi-G model and, 143–50; nitrogen and, 167–68, 452–64; opaline silica and, 352–59; pelagic sediments and, 328–32; pH and, 442–51; phosphorus cycling and, 464–77; RKR equation and, 254; sediment-water interface and, 283–86 (*see also* sediment-water interface); stoichiometric nutrient models and, 254–60; suboxic/anoxic conditions and, 442–51; tortuosity and, 75–76; trace metals and, 332–43; turbidites and, 382–95; zonation and, 99–104

pore water size-reactivity (PWSR) model, 220–22, 230

porosity: compaction and, 78–83; cross-coupling and, 75; depth and, 51–52, 79;

INDEX

porosity (*continued*)
   description of, 47–55; diagenesis and, 73–74; diffusion and, 74–78; grain orientation and, 50–51; sands and, 51, 54
positron, 41
postoxic sediments, 102–3
potassium, 71, 513–14, 516
potential energy, 30, 98
property-property plots, 255–56
propionate fermentation, 99
protein: amino acids and, 9, 171–72, 174, 179–89; carbohydrates and, 189–92; lignins and, 193–94; organic carbon preservation and, 418
pyrite, 5, 111, 332; bioirrigation and, 484–88; burial efficiency and, 486–88; degree of pyritization and, 487–88, 500–502; degree of trace metal pyritization (DTMP) and, 500–502, 507–8; formation mechanisms for, 484–85; hydrochloric acid and, 479; iron availability and, 487; oxidation state and, 482–90; reactive metal fraction and, 500–501; sulfur cycling and, 478–90; trace metal cycling and, 507–8
pyroxene, 6

quartz, 7, 513
Quaternary age, 35, 180, 395

radial-diffusion tube model, 304–13
radioactive decay: carbon and, 43–45; fission and, 42; half-life and, 42–43; isotopes and, 27–28, 40–45; principles of, 40–43
radiolarian oozes, 24
rain rate, 87
random walk, 59–60
Redfield C/N soichiometry, 149
Redfield-Ketchum-Richards (RKR) equation, 100, 175–77, 254
Redfield ratio, 100, 470–72
redox conditions, 14, 98, 139–41; downshift and, 397–98; iron and, 107–9; layered/coupled models and, 162–70; lipids and, 202–3; manganese and, 107–9, 395–407; methanogenesis and, 490–500; nitrogen cycling and, 452–64; organic carbon preservation and, 421, 441 (*see also* organic carbon preservation); paleoceanographic diagenesis and, 395–407; phosphorus cycling and, 464–77; reverse weathering and, 509–16; sulfur cycling and, 478–90; trace metal diagenesis and, 332–43; turbidites and, 382–95
remineralization, 1, 3, 174; amino acids and, 183–84, 188–89; bacteria and, 145; benthic flux stoichiometric models and, 260–61; benthic macrofaunal processes and, 419–21; biogeochemical processes and, 97, 116–33 (*see also* biogeochemical processes); bioturbation and, 88–90; boundary conditions and, 88–92; carbohydrates and, 191–92; carbon/nitrogen ratio and, 240–44, 253, 256, 261–70; decomposition dynamics and, 134–41; diagenesis and, 2, 88–92; dissolved organic matter (DOM) and, 218, 226; elemental ratios and, 261–65; flux and, 150–61 (*see also* flux); layered/coupled models and, 162–70; methanogenesis and, 491, 494; multi-G model and, 143–50; nitrogen cycling and, 452–64; nonsteady-state processes and, 375–82; organic carbon preservation and, 408–9, 417–19; organic composition and, 253–70; oxygen exposure time and, 421–32; pelagic sediments and, 328–32; phosphorus cycling and, 464–77; pore waters stoichiometric models and, 254–60; reactive component budget and, 150–61; reverse weathering and, 509–16; RKR equation and, 254; stoichiometric nutrient models and, 254–60; suboxic/anoxic conditions and, 442–51; sulfate reduction and, 91–92; terrestrial organic matter (TOM) and, 435–36; turbidites and, 387; water column depth and, 124–33
respiration, 38, 99
reverse conveyor-belt feeders, 67, 299–300
reverse weathering, 509–16

INDEX

rhenium, 343
RuBP carboxylase, 37
Russia, 14

Saanich Inlet, 519
salt, 5, 50
saltmarsh grasses, 37
sands, 7, 24, 51, 54
Santa Monica Basin, 223
saponifcation, 194–95
Sargasso Sea, 18
Schmidt number, 278
seasonal behavior, 378–82
sediments: accumulation and, 61–65, 95–96, 148–50; advection and, 78–83; aliquots, 50; anoxic, 13, 92–94, 102–3, 129–30, 395, 404; authigenic minerals and, 12–15; bacterial biomass production and, 250–53; biogenic components and, 8–12; biogeochemical processes and, 97–141 (*see also* biogeochemical processes); bioirrigation and, 65–67; bioturbation and, 65–68; chemical weathering and, 5–6; classification of, 24–26; clays and, 15–24; compaction and, 78–83; cycling and, 1, 5 (*see also* cycling); Damkohler number and, 69–70; definition of, 6; density and, 47–55; detrital components and, 5–8; diagenesis and, 72–96 (*see also* diagenesis); diffusion and, 59–61; geochemistry and, 171–217 (*see also* geochemistry); glacial, 7, 395, 400–402; grain size and, 46–47; Holocene, 404–7; isotopes and, 27–45; margin, 442–77 (*see also* continental margin sediments); mud and, 47, 70, 373–74; nonlocal mixing and, 299–302; nonsteady-state processes and, 1–2, 373–407; oxidation and, 156 (*see also* oxidation); paleoceanographic records and, 1–2; pelagic, 328–32 (*see also* pelagic sediments); permeability and, 55–58, 283–86; porosity and, 47–55; postoxic, 102–3; radioactive decay and, 27–28, 40–45; reactive component budgets and, 150–61; redox boundary and, 14; reference frames and, 61–65; remineralization and, 142–70 (*see also* remineralization); suboxic, 102–3, 129–30, 442–51; terrigenous, 6–7; thickness of, 26; time/space scales for, 67–70; transport processes and, 59–71; volcanic ash layers and, 374; zonation and, 99–104
sediment-water interface: advection and, 285; Aller's radial-diffusion tube model and, 304–13; benthic flux enhancement factor and, 272–83, 316–26; bioirrigation and, 302–26; bioturbation and, 286–92; diffusive boundary layer and, 275–83; friction and, 276–77; gravity waves and, 284; manganese and, 278; methane gas ebullition and, 326–27; oxidation and, 275–85; permeable sediments and, 283–86; pressure gradient and, 283–84; in situ conditions and, 276; silica and, 278–79, 282; subtidal pump and, 284
serine, 183, 185
Shaw-Hanratty equation, 276, 278
shear velocity, 276, 278
Shewanella, 109
short-chain organic acids (SCOAs), 230
silica, 6; biogenic, 10; clays and, 15–24, 512–16; diagenesis and, 352–59; distribution of, 11; opaline, 352–59; pelagic sediments and, 352–72; reverse weathering and, 514–15; sediment-water interface and, 278–79, 282
silicon, 331
silts, 7, 24
Skagerrak Sea, 133, 519
Skan Bay, 263, 493, 519
smectites, 18–19, 35, 510
soil formation, 7
Southern Ocean, 240, 358
spreading centers, 12
steady-state processes: rarity of, 373; reactive component budget and, 150–61; sediments accumulation, 61–65
steroids, 198
sterols, 197
suboxic diagenesis, 347

subtidal pump, 284
sugars, 171. *See also* carbohydrates
sulfur, 5, 9, 13, 15, 71, 109, 391; acid volatile sulfides (AVSs) and, 479, 507–8; biogeochemical processes and, 110–11, 119; bioirrigation and, 484–88; bioturbation and, 65–67, 486–88; burial efficiency and, 486–88; Cape Lookout Bight and, 153–61; carbon ratio and, 488; conceptual model for, 480–82; cycling and, 478–90; greigite and, 479; isotopes and, 40, 478, 489–90; long-term cycle changes in, 489–90; mackinawite and, 479; methanogenesis and, 490–91, 493, 498; multi-G model and, 146; oxidation state and, 479–90; pH and, 479; precipitation and, 448; pyrite and, 479–90; reduction and, 91–92, 110–11, 146, 157; sediment color and, 479; sources of, 478; thiosulfate and, 479–81; trace metal cycling and, 502; zero-valent, 479
surface adsorption/mesopore protection hypothesis, 415–16
syntrophy, 137

tectonic cycle, 429
terpenoids, 197–98
terrestrial organic matter (TOM): burial efficiency and, 435–39; diagenesis and, 244–48; organic carbon preservation and, 432–39; rate of burial and, 436; remineralization and, 435
terrigenous sediments, 6–7
thermodynamics, 2–3, 32; biogeochemical processes and, 97, 107–9 (*see also* biogeochemical processes); decomposition dynamics and, 134–41; dolomite and, 13; Pleistocene climate cycles and, 402–4; seasonality and, 378–82; trace metal cycling and, 501–2
thermogenic methane, 111
Thiobacillus ferooxidans, 115–16
thiosulfate, 479–81
Thioploca, 106
threonine, 183, 185
tortuosity, 75–77

total dissolved free amino acids (TD-FAAs), 230–31
total organic carbon (TOC), 65, 88, 127; amino acids and, 183, 188; Arctic Ocean and, 247–48; benthic flux stoichiometric models and, 260–61; concentration of, 174–75; isotopic tracers and, 238–41; lipids and, 198–99; manganese nodules and, 348; organic carbon preservation and, 409–19, 429, 432, 438; remineralization and, 143, 145–46, 150, 155; sulfur and, 482, 488; turbidites and, 383, 385, 392, 394
toxic metals, 1
trace metals: benthic macrofauna and, 506–7; bioirrigation and, 504–6; bioturbation and, 504–6; concentration of, 504; conceptual model for, 502–3; cycling and, 500–507; degree of trace metal pyritization (DTMP) and, 500–502; diagenesis and, 332–43; dissolution/reprecipitation processes and, 503; reactive metal fraction and, 500–501
transport processes: accumulation, 61–65; bioirrigation, 65–67 (*see also* bioirrigation); bioturbation, 65–67 (*see also* bioturbation); conveyor-belt, 67, 299–300, 432; Damkohler number and, 69–70; diffusion, 59–61; scaling of, 67–70
turbidites, 65; anoxic sediments and, 395; burn-down and, 383, 385, 387, 389–91; carbon and, 383, 385, 387, 392, 394; conceptual model for, 383–87; diagenesis and, 391–95; gravity-driven flows and, 382–83; iron and, 383, 389–90; Madeira Abyssal Plain and, 390–92; manganese and, 383–87, 389–91; MAP f-turbidite and, 391–92, 424; Miocene age and, 394; nitrogen and, 383–84, 390; organic carbon preservation and, 424; oxidation and, 383–87, 390–95; pelagic sediments and, 382–87; Pleistocene-age f-turbidite and, 394; remineralization and, 387;

time scales and, 394–95; uranium and, 389–90

undersaturation, 11–12, 369–70
uranium, 71, 343, 389–91, 400
uronic acids, 191

vanadium, 343, 391, 400
viscosity, 75
volcanic ash layer, 374
volcanic glass, 19
vulcanization, 198

Washington State, 246, 319, 464
water: biogeochemical processes and, 124–33; bioirrigation and, 65–67 (*see also* bioirrigation); hydraulic conductivity and, 55–56; hydrologic cycle and, 32–34; ice cores and, 32–34; isotopes and, 32–34; salinity and, 50; sediment accumulation and, 61–65. *See also* pore waters
wax, 197
weathering cycle, 429

XANES, 216
X-ray studies, 41, 107, 216, 511

zinc, 332–33, 336, 341, 390–91, 502, 506
zonation: bacterial respiration and, 99–104; biogeochemical processes and, 120–24; depth scales of, 120–24; layered/coupled models and, 161–70; methanogenesis and, 111–14; nitrogen cycling and, 452–64; pelagic sediments and, 328–32; phosphorus cycling and, 464–77
zooplankton, 179